COMPLEX VARIABLES

Principles and Problem Sessions

COMPLEX VARIABLES

Principles and Problem Sessions

A K KAPOOR

University of Hyderabad, India

NEW JERSEY • LONDON • SINGAPORE • BEIJING • SHANGHAI • HONG KONG • TAIPEI • CHENNAI

Published by

World Scientific Publishing Co. Pte. Ltd.
5 Toh Tuck Link, Singapore 596224
USA office: 27 Warren Street, Suite 401-402, Hackensack, NJ 07601
UK office: 57 Shelton Street, Covent Garden, London WC2H 9HE

Library of Congress Cataloging-in-Publication Data
Kapoor, A. K.
Complex variables : principles and problem sessions / A.K. Kapoor.
p. cm.
ISBN-13 978-981-4313-52-0 (hardcover) -- ISBN-10 981-4313-52-1 (hardcover)
ISBN-13 978-981-4313-53-7 (pbk.) -- ISBN-10 981-4313-53-X (pbk.)
1. Functions of complex variables--Textbooks. I. Title.
QA331.7.K37 2011
515'.9--dc22
2010054043

British Library Cataloguing-in-Publication Data
A catalogue record for this book is available from the British Library.

Typeset by Stallion Press
Email: enquiries@stallionpress.com

Printed in Singapore by World Scientific Printers.

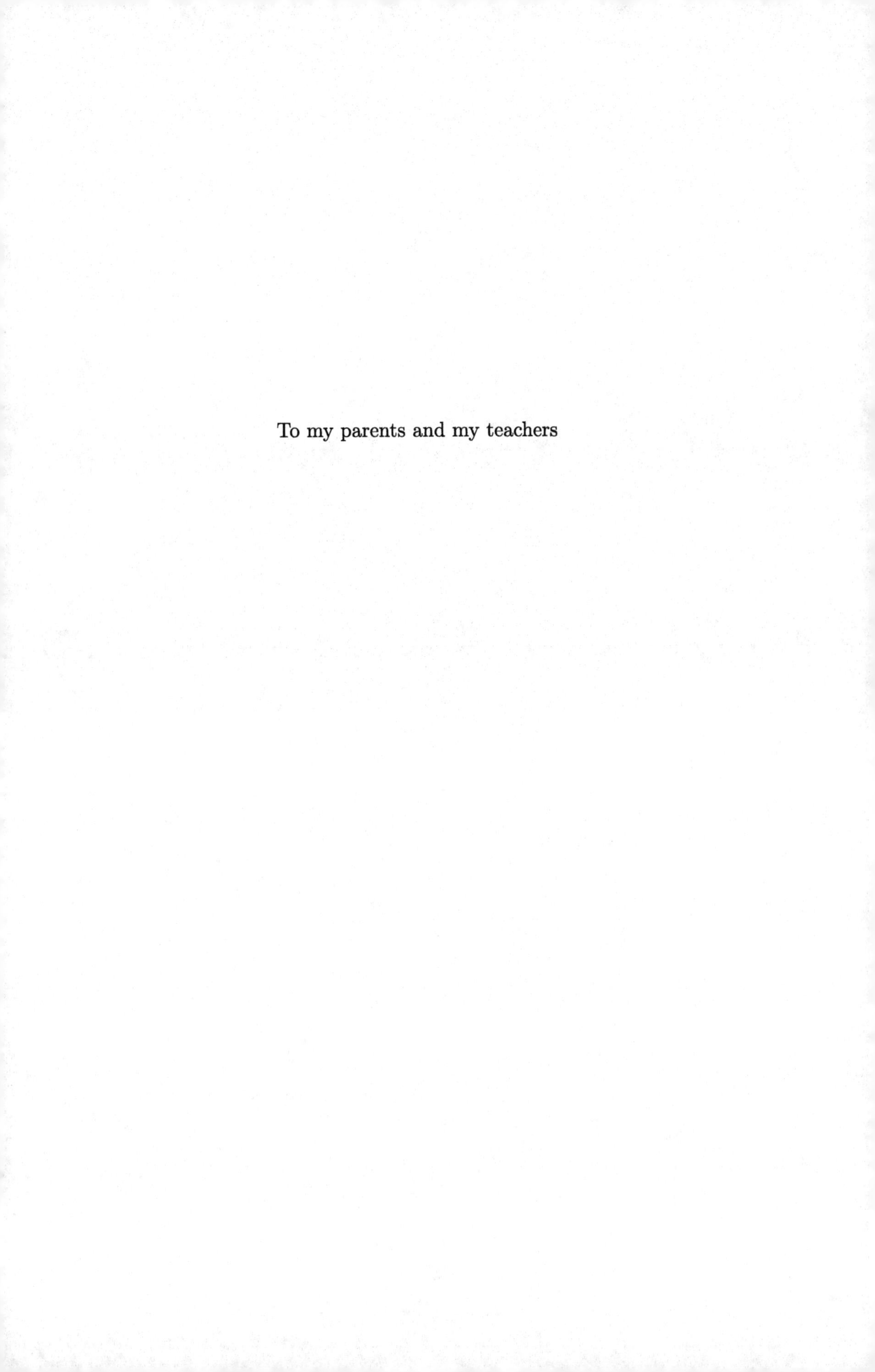

To my parents and my teachers

PREFACE

Mathematics is the language of physicists and is shared by scientists and engineers working in several branches; it provides a framework for making precise models of the real world and opens the road to making far-reaching predictions and applications of such models. It has always been a puzzle to me that several branches of mathematics (originally considered esoteric), created for reasons not even remotely connected with any application to the real world, have often turned out to be extremely useful. The theory of functions of a complex variable is one example.

In a subject like complex variables, the requirements and preferences of students and teachers of science and engineering are somewhat different from those of pure mathematics. While the long term objective of a future mathematician is to move toward general results and abstract concepts, students of science and engineering are expected to learn concrete results and apply them to specialized problems. Thus, the approach to teaching and learning of mathematics for beginners falls into two broad categories. One approach is to go through a rigorous cycle of learning definitions, theorems, proofs, and their analysis by means of examples and counterexamples. The other approach, advocated by many scientists and engineers, often reduces to learning only the recipes and using them to solve problems encountered in the real world. The path followed in the first part of this book attempts to take the best of these two approaches. While every effort is made to present the definitions and results precisely, the language used is not always a formal one. Inspired by the course charted in [Fraleigh (1973)], a rigorous presentation of heavy duty proofs is postponed to a later stage, or completely skipped, and references are provided. On the other hand, the seeds of mastering tools and techniques needed for specialized applications are sown at the earliest possible stage.

Problem-solving sessions are an integral part of teaching and learning mathematics. This is especially important for a course at the introductory level for the students who have not yet acquired the maturity to grasp essentials of principles and techniques needed for later applications. A majority of the students absorb the abstract theoretical concepts and results by going through regular sessions of solved examples, tutorials, practice, assessment, and evaluation. While there are excellent textbooks giving the theory, the organization of available textbooks and problem books does not fully meet this requirement. It is, therefore, not surprising that many professors prefer to "create problems" for the students, specially designed to meet certain well-defined objectives of teaching and evaluation. This book attempts to fill that gap by providing a richer variety of problems organized into focused sessions, each having a very specific objective. Rather than merely providing problems which can by copied and pasted onto assignment sheets in a course on complex variables, one of the objectives of the book is to arrange problems in a variety of templates for problem sessions which could be useful for students of other branches of science and engineering as well. It is hoped that this format will prove valuable to those who have to learn the subject without the assistance of a teacher or a subject expert.

This book is organized into two parts. The theory of functions of a complex variables is developed in the first part, which also has a good variety of short examples and fully solved problems. An idea of the topics included in the book can be had from the table of contents. All the topics usually covered in a first course on complex variables are included. The topic of multivalued functions, usually skipped in a first course, is treated in Ch. 3, where all essentials are covered in a manner not found in many textbooks. For some other topics in later chapters, asymptotic expansions being an important example, the book provides the reader with an initial breakthrough, and for proofs and many technical details the reader is redirected to other excellent books.

This book differs from other textbooks in its collection of problems in the second part of the book. The problems are organized into problem sessions, details of which are explained separately. Many problem sets are included to extend the usefulness of the book beyond the requirements of classroom instruction. An effort is made to encourage the reader to explore connections and alternative routes. Several problem sessions are included which may appear difficult in the first instance but are, in fact, within the reach of the reader who has received good training in the subject.

The Internet provides a rich source of information and technical articles not available in any textbook or journal. By providing references to the URLs of selected websites, it is hoped that the reader will be motivated to go beyond textbooks. A textbook need not be, and cannot be, the last word on all the topics covered. This book is no exception.

ACKNOWLEDGMENTS

A number of problems and examples in this book were specially designed for the class while I was teaching a course on mathematical physics; many others are modifications of problems taken from existing books. A work like this necessarily relies heavily and builds on existing literature. The books that were consulted frequently, many of them classics by themselves, are listed in the bibliography.

I thank Prasant Panigrahi, R. S. Bhalla, S. Sree Ranjani, and K. G. Geojo for the many fruitful discussion sessions during the course of research work carried out in their collaboration. These discussions have gone a long way towards improving my own understanding of the several topics in complex variables.

It is a pleasure to acknowledge the advice received from many friends, colleagues, and teachers, which played a vital role in shaping and planning the book. I am grateful to H. S. Mani, Ranjan Shrivastava, V. Srinivasan, Subhash Chaturvedi, P. P. Divakaran, and A. P. Balachandran for their encouragement and their helpful advice from time to time.

Numerous friends and colleagues have shown continous interest during the writing of the book and I thank A. K. Bhatnagar, G. Rajasekaran, V. Gupta, Ashok Das, U. P. Sukhatme, Gautam Desiraju, Bindu Bambah, Chandrasekhar Mukku, and R. Sandhya for their unceasing interest and support during the preparation of the book.

The comments and responses received from the students of mathematical physics courses I have taught were very useful. I acknowledge the active participation of all the students who helped me in assessing the usefulness of the problems to them.

I thank Subhash Chaturvedi, Sarita Vig, S. Sree Ranjani, and K. G. Geojo, who have painstakingly read large portions of the book. I am grateful for their time and efforts, which played an important role in eliminating a large number of mistakes and improving the presentation.

I am grateful to the staff of World Scientific Publishing Co, in particular to Chee-Hok Lim for providing valuable assistance during the production of the book.

The timely help and advice from Prasanta Panigrahi and Alok Patnaik has been crucial in ensuring that this book sees the light of day; I am immensely grateful to them. Last but not least, I thank my wife, Kiran, and my daughter, Sakhi, whose love, patience and support contributed in many ways.

ACKNOWLEDGMENTS

TO THE READER

This book consists of two parts. Part I covers the principles of complex variables. A good idea of the topics covered can be had from the table of contents. The book has been written with the objective of covering the standard results in a coherent fashion. At the same time, it provides ready reference material useful for solving problems. In addition, every chapter contains short examples and fully solved problems. An effort is made to provide either a short example or a solved problem on every aspect of theory and of techniques for solving problems.

The fully solved problems, easily identified by the symbol ☑, contain a statement of the problem and the solution. They are designed to explain usage of tools and methods for solving problems.

Each set of short examples, marked with ⚡, consists of closely related statements. While some short examples are meant to clarify important concepts, others illustrate the main tools and techniques for problem solving. Most of the statements in the short examples become obvious after a little bit of thinking, while some others do require careful thinking. In almost all cases no detailed computation is needed to follow the statements given.

Each chapter in Part II of the book is subdivided into problem sessions. Each problem session consists of a small number of problems, and hints and answers for the problems in the session. Frequently comments on problems are included, which will be fully appreciated by a student only after having solved a problem set completely. These are marked with bullets, ⊙. Full solutions have not been given, so as to keep the size of the book reasonable. The answers to the problem are provided only for selected problem sessions, mostly Tutorials, Exercise, Quiz and Short Question sessions. No answer is provided where a proof of statement is asked for. The figures are suppressed from almost all the answers.

The chapters in each part of the book are numbered from 1 to 10; a chapter in the Part II has problem sessions for the topics in the corresponding chapter in Part I of the book. The sections in the two parts are marked with symbols § and §§; so §5.4 and §§5.4 refers to Section 4 of Chapter 5 in Part I and Part II respectively. Figures are numbered consecutively in a chapter. A reference to a figure, such as Fig. 9.1, appearing in any part refers to the figure in the *same part* of the book. The numbering of Short Examples and Problems in Part I is reset in every section; so

☑6.5.1 is the 6th Short example in §5.1. Similary, ☑6.7.1 refers to the 6th Problem in §7.1.

The problem sessions are named Tutorial, Exercise, Question, Quiz, Open-Ended, Mixed Bag, etc. The Tutorial, Exercise, and Question sets are primarily intended for beginners, and an experienced reader may skip them after going through them. The sets titled Quiz, Questions, and Mined are for those who want to test their understanding of a topic. An experienced reader looking for something new may skip these sets and go directly to Open-Ended, Mixed-Bag, and Applied sessions. An important feature of these sessions is that usually each session focuses on a small number of theoretical concepts or problem-solving strategies. The following description will give a fair idea of main features of each problem session.

Tutorial sets. These are designed to provide the reader with training in basic techniques of solving problems on complex variables. Frequently, the student is asked to use a particular method, or a sequence of steps, to arrive at the solution. Effort is made to ensure that, while attempting to solve a problem, the reader concentrates on only the main sequence of steps in solving the problem and that there is no digression. *An important aspect of tutorial sessions is that each session is designed so that a well-prepared reader can complete all the problems in an approximately hour-long tutorial session.*

Exercise sets. The *Exercise Sessions* provide problems for further practice of the techniques learnt in the tutorial sessions. *Most of the problems have been carefully selected so as to keep computational complexity at a moderate level.* Problems which may require a long, tedious computation, not serving any useful purpose, are avoided.

Questions. The *Question Sessions* are designed to check the student's understanding of the basic concepts of the subject. For these sets long, detailed answers, or "essays," are not required and are discouraged. In most of the question sets, only short answers, quite often Yes/No or True/False, are expected from the students. However, the students should be able to defend their answers in a detailed oral cross-examination. Many of the questions from these sets can also be put to the students during the lectures.

Quiz. The *Quiz Sessions* contain questions requiring short answers only. *Each set is designed to be completed in 15–30 minutes and is focused on a very specific topic.* These include tests of the student's ability to identify the best way of attacking a problem.

Mined. A *Mined Session* contains a few sample solutions along with a statement of each problem. The reader is not expected to provide a full solution to the problems as an answer in the *mined problem sessions* and instead he or she is required to locate mines (errors), if any, and make comments on the sample solution. Sometimes several sample solutions are given and the reader is asked to rank them. These sets are aimed at clearing up common mistakes and misconceptions.

Problem. Part II of the book has a few problem sessions. These contain applications of complex variables to other areas of science and engineering. The topics touched upon are those which are within the reach of the reader and are meant to give illustrative applications of the theory covered in Part I.

Open-Ended. Several chapters contain problem sessions called *Open-Ended.* The problems in these sessions are designed to provoke the reader to think beyond what is available in most textbooks. In order that the reader is not robbed of the pleasure of arriving at a solution, answers or hints are not provided. It is hoped that the reader will find these problems challenging.

Mixed Bag. At the end of almost all the chapters, there is a collection — a *Mixed Bag* — of problems.

Suggestions for using this book appear in different parts of the book and are marked with the symbols ▷▷, ▷, etc., as explained in the table on pages xv and xvi.

NOTATION AND SYMBOLS

Symbol	Meaning
ϵ, δ, ρ	Positive, infinitesimally small real constants, and frequently the limit $\epsilon, \delta, \rho \to 0$ is to be taken.
$a, b, c, \ldots$	Real constants.
$p, q, r, s, \ldots$	Positive constants.
$i, j, k, l, n, r, \ldots$	All integers.
$M, N, \ldots$	Natural numbers $1, 2, 3, \ldots$
$\alpha, \beta, \gamma, \phi, \ldots$	Angles, in general, can assume positive or negative values.
$\alpha, \beta, \lambda, \mu, \nu, \xi, \eta, \zeta$	Complex constants.
z, x, y, r, θ	Complex variable, real part of z, imaginary part of z, and polar variables.
$\log z$	Logarithm of a complex number z without reference to any specific svb; also used for a set of all values w satisfying $z = \exp(w)$.
$\ln x$	Logarithm of a *positive* number x.
svb	Single-valued branch.
$\log z$	Principal value of the logarithm of a complex number.
$\mathrm{Sin}^{-1} z, \mathrm{Tan}^{-1} z$, etc.	Principal branches of the inverse functions.
$\mathrm{Re}\, z, \mathrm{Im}\, z$	Real and imaginary parts of z.
⊘	Instructions and comments given at the beginning of a problem set.
§, §§	Sections or subsections in Parts I and II of the book.
↯	A set of loosely related short examples.
☑	A fully solved problem.
⊙	Hints, comments, remarks, etc., given along with the answers.
⧐	Suggestion to try a problem set.
▌	End of proof or solution, etc.

↬	Left as an exercise.
◀	Reminder about an earlier text which is crucial to understanding of the subject matter here.
▼	Can be skipped without loss of continuity.
✍	Points to remember.
#, ⋆, ∗, ◇, ◊, □, ∘	Different bullets like these are used as markers for lists of statements or remarks.

CONTENTS

Part I

Principles

Chapter 1

COMPLEX NUMBERS

§1.1 Introduction

We begin with some elementary properties of complex numbers. After introducing geometric representation of complex numbers linear transformations of translation, scaling and rotation are described. These transformations together with inversion constitute building blocks of bilinear transformations which are important class of conformal mappings to be taken up in detail in the last two chapters. The point at infinity and related stereographic projection is outlined. All the important concepts, techniques, and results described here are illustrated by means of examples and solved problems.

§1.1.1 *Complex conjugate*

The *complex conjugate* of the complex number z will be denoted as $\bar{z}$. Thus, if $z = x + iy$, then $\bar{z} = x - iy$, where $x, y \in \mathbb{R}$. With $|z|$ denoting the absolute value, $\sqrt{x^2 + y^2}$, some simple properties involving the complex conjugate are as follows:

(*1) $\overline{(\bar{z})} = z$,
(*2) $\overline{(z_1 + z_2)} = \bar{z}_1 + \bar{z}_2$,
(*3) $\overline{(z_1 z_2)} = \bar{z}_1 \bar{z}_2$,
(*4) $\overline{\left(\frac{z_1}{z_2}\right)} = \dfrac{\bar{z}_1}{\bar{z}_2}$, $z_2 \neq 0$,
(*5) $(\bar{z})^N = \overline{(z^N)}$,
(*6) $\mathrm{Re}\,(z) = \frac{1}{2}(z + \bar{z})$,
(*7) $\mathrm{Im}\,(z) = \frac{1}{2i}(z - \bar{z})$,
(*8) $\frac{1}{z} = \frac{\bar{z}}{|z|^2}$, $z \neq 0$,
(*9) $|\bar{z}| = |z|$,
(*10) $|z|^2 = z\bar{z}$.

§1.1.2 *Polar form*

Geometrically the complex number $z = x + iy$ is represented by a point, with (x, y) as the coordinates, in a two-dimensional plane called the *complex plane* or the *Argand diagram.* Let r and θ denote the polar coordinates of a point, z; we have

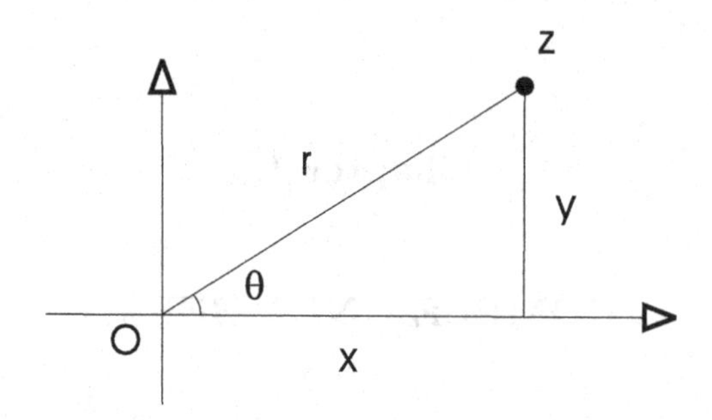

Fig. 1.1

(Fig. 1.1)

$$r = \sqrt{x^2 + y^2}, \qquad \theta = \tan^{-1}\left(\frac{y}{x}\right), \tag{1.1}$$

and the complex number z itself can be written as

$$z = r(\cos\theta + i\sin\theta). \tag{1.2}$$

The numbers r and θ are called the *modulus* and the *argument* of the complex number z and will be denoted as mod (z) and arg(z), respectively. Note that mod(z) is just $|z|$, introduced above. Given a nonzero complex number z, its argument does not have a unique value; there are infinite values differing by multiples of 2π. For many purposes, it is useful to restrict the allowed range for the polar angle θ so as to get a unique value of θ for a given z. If the value of $\theta = \arg(z)$ is restricted to lie in the range $-\pi < \theta \le \pi$, we obtain the *principal value* of arg(z), which will be denoted as Arg(z), i.e.

$$\text{Arg}(z) = \theta, \quad -\pi < \theta \le \pi.$$

It must be noted that arg (z) is not defined for $z = 0$. In the following we summarize a few properties of the modulus and the argument. Using r_1, r_2, and r_3 to denote the moduli and θ_1, θ_2, and θ_3 to denote the arguments of the complex numbers z_1, z_2, and z_3, respectively, we have:

(∘1) If $z_3 = z_1 z_2$, $r_3 = r_1 r_2$ and $\theta_3 = \theta_1 + \theta_2$;
(∘2) If $z_3 = \dfrac{z_1}{z_2}$, $r_3 = \dfrac{r_1}{r_2}$ and $\theta_3 = \theta_1 - \theta_2$, $r_2 \neq 0$;
(∘3) If $z_2 = \dfrac{1}{z_1}$, $r_2 = \dfrac{1}{r_1}$ and $\theta_2 = -\theta_1$, $r_1 \neq 0$;
(∘4) If $z_2 = z_1^N$, $r_2 = r_1^N$ and $\theta_2 = N\theta_1$.

The relations between the arguments given above hold modulo 2π. Here it is useful to recall De Moivre's theorem,

$$(\cos\theta + i\sin\theta)^N = \cos N\theta + i\sin N\theta, \tag{1.3}$$

which is needed to prove the result (∘4). We define the exponential of a *complex number* $z = x + iy$, exp (z), in terms of the exponential, the sine and the cosine functions of a *real variable* as follows:

$$\exp(z) = \exp(x)(\cos y + i \sin y). \tag{1.4}$$

Then the complex number z itself can be written as

$$z = r \exp(i\theta), \tag{1.5}$$

in terms of its argument θ and modulus r. This form of complex numbers is called the *polar form.*

§1.1.3 *Shifted polar form*

It turns out to be useful to introduce *shifted polar variables* defined w.r.t. a point z_0 instead of the origin. Treating z_0 as the "new origin," for a given complex number z, we introduce shifted polar variables ρ and ϕ as the distance and the angle marked in Fig. 1.2. Then

$$\rho = |z - z_0|, \quad \phi = \arg(z - z_0) \tag{1.6}$$

and a little exercise in geometry shows that

$$z = z_0 + \rho\big(\cos\phi + i \sin\phi\big) \tag{1.7}$$

and the number z has a "shifted polar form":

$$z - z_0 = \rho \exp(i\phi). \tag{1.8}$$

§1.1.4 *Representation of angles*

Having represented complex numbers as points in the complex plane, we can turn around and get geometrical quantities such as angles and distances in terms of polar or shifted polar variables.

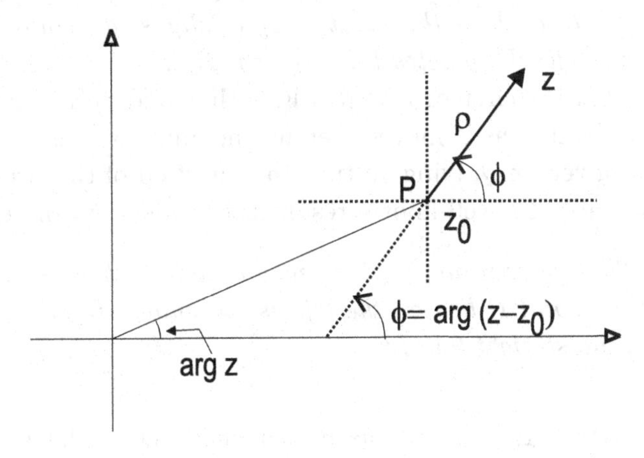

Fig. 1.2

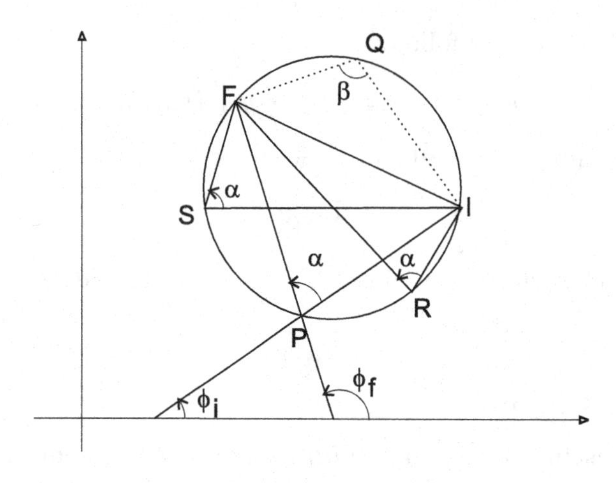

Fig. 1.3

Consider two directed line segments PI and PF drawn from a point P, see Fig. 1.3, at z_0 to two points I and F. We wish to write an expression for the angle α between the two rays in terms of the corresponding shifted polar angles ϕ_i and ϕ_f,

$$\phi_i = \arg(z_i - z_0), \qquad \phi_f = \arg(z_f - z_0), \tag{1.9}$$

of the two numbers z_i and z_f. The angle α is given by the difference

$$\phi_f - \phi_i = \arg(z_f - z_0) - \arg(z_i - z_0) = \arg\frac{z_f - z_0}{z_i - z_0}.$$

Therefore, we define

$$\angle \mathrm{IPF} = \mathrm{Arg}\left(\frac{z_f - z_0}{z_i - z_0}\right), \tag{1.10}$$

where z_i and z_f correspond to the points I and F, respectively. *The above definition includes the sign of the angle, the angle being positive when a positive — i.e. anticlockwise — rotation about the point P by α aligns the "initial" directed line segment PI with the "final" directed line segment PF.*

It is apparent that the circular arc IRPSF, with z_i and z_f as the two end points, is the locus of the point P as it moves keeping the angle α constant (see Fig. 1.3). If β is the angle at a vertex Q, lying on the other arc FQI of the circle, $|\alpha| + |\beta| = \pi$ holds as a consequence of a well-known result in Euclidean geometry.

✍ *To remember the sign convention, it is useful to note that the expression in* (1.10) *is positive, if the point P lies on the left as one moves from the point z_i to the point z_f along the straight line IF.*

⊳⊳ Exercises §§1.1 and §§1.2, on polar form, argument, and modulus.

§1.2 Examples

1.2.1 Short examples. We shall now give some simple examples utilizing properties of the modulus and argument of complex numbers:

(a) The sets $|z| = 1, 2, 3, 4$ are all circles with the center at $z = 0$ and have the radii $1, 2, 3$, and 4. They are shown in Fig. 1.4(a).

(b) $|z - 1 - i| = 1, 2, 3, 4$ are circles, all with the center at $1 + i$, and have the radii $1, 2, 3$, and 4, respectively. All these circles are shown in Fig. 1.4(b).

(c) The points satisfying $\arg(z) = \frac{1}{6}\pi$ lie on a half line, also called a *ray*, drawn from the origin and making an angle of $\pi/6$ with the real axis. Similarly, other rays corresponding to $\arg(z) = \frac{1}{3}\pi, \frac{3}{4}\pi, \frac{5}{4}\pi, \frac{5}{3}\pi, \frac{11}{6}\pi$ can be drawn. The answers are shown in Fig. 1.5(a). The ray $\arg = -\frac{1}{6}\pi$ coincides with the ray $\arg(z) = \frac{11}{6}\pi$, as the values of the arg function differ by multiples of 2π.

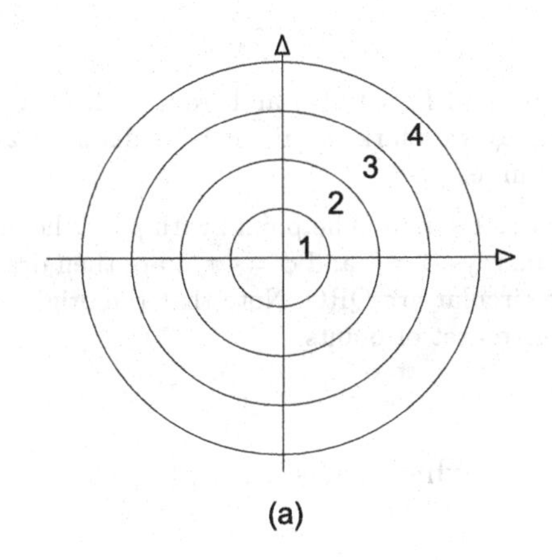

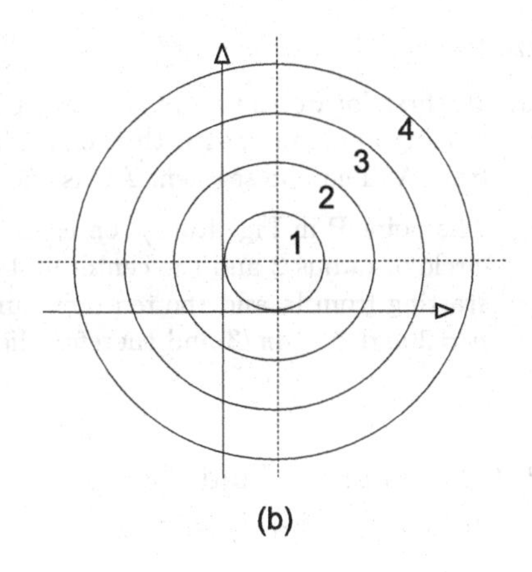

Fig. 1.4

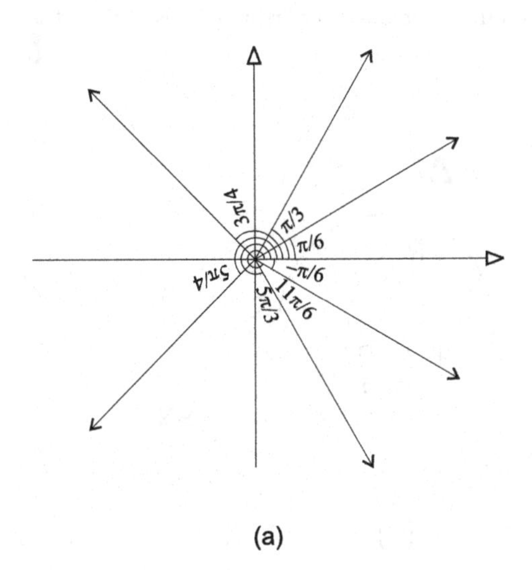

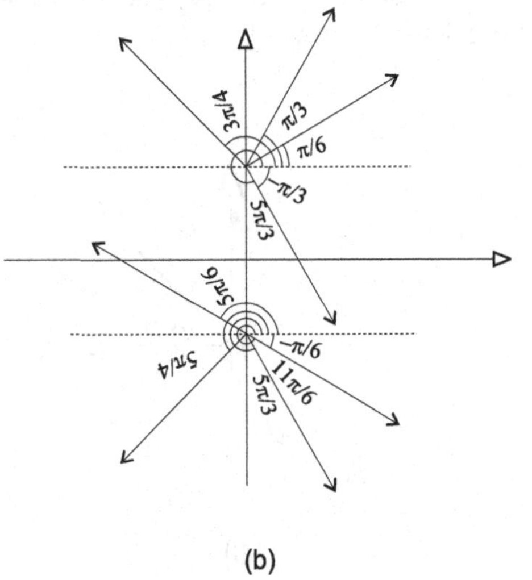

Fig. 1.5

(d) A ray drawn from the point $z = i$ at an angle α gives the the sketch for the set $\arg(z-i) = \alpha$. The rays $\arg(z-i) = -\frac{1}{3}\pi, \frac{1}{6}\pi, \frac{1}{3}\pi, \frac{3}{4}\pi, \frac{5}{3}\pi$ are shown in Fig. 1.5(b). Note that the ray $\arg(z-i) = -\frac{1}{3}\pi$ is the same as the ray $\arg(z-i) = \frac{5}{3}\pi$.

(e) In order to represent the sets $\arg(z+i) = -\frac{1}{6}\pi, -\frac{1}{3}\pi, \frac{5}{6}\pi, \frac{5}{4}\pi, \frac{11}{6}\pi$, one must draw rays at the specified angles with $z = -i$ as the starting point of the rays. Again, it should be noted that the rays $\arg(z+i) = -\frac{1}{6}\pi, \frac{11}{6}\pi$ are the same rays, and also that the rays $\arg(z+i) = \frac{5}{3}\pi$ and $-\frac{1}{3}\pi$ coincide.

☑**1.2.2 Problem.** Let ρ and ϕ be the shifted polar variables for a complex number z, w.r.t. ξ, defined by $z - \xi = \rho\exp(i\phi)$.

(a) Assume now that $\xi = -1 - i$ and draw the curve traced by z in the complex plane, as ϕ is held fixed equal to $\pi/4$ and ρ is varied from 0 to $2\sqrt{2}$.

(b) For $\xi = 1 + i\sqrt{3}$ draw the curve traced by z in the complex plane, as ρ is held fixed equal to 2 and ϕ is varied from 0 to $4\pi/3$.

Solution.

(a) We first locate the point $-1 - i$. This point is A in Fig. 1.6(a) and from A draw a ray making an angle $\pi/4$ with the x axis. On the ray we mark the point B at distance $2\sqrt{2}$ from A. The line segment AB is the required curve.

(b) The point P in Fig. 1.6(b) represents the number $1 + i\sqrt{3}$. The points with $\rho = 2$ lie on a circle of radius 2 and the center at $1 + i\sqrt{3}$. The rays $\phi = 0$ and $\phi = 4\pi/3$ are then drawn starting from P, and the required curve is the circular arc QRO. Note that the origin has $\rho = 2$ and $\phi = 4\pi/3$ and therefore lies in the given set of points.

∎

☑**1.2.3 Problem.** Find the complex number ξ which satisfies $\arg(\xi - 1 - i) = -\pi/6$ and $\arg(\xi - 1 + i) = \pi/6$.

Solution. To locate the point ξ geometrically, draw a ray from $1 + i$ at an inclination $-\pi/6$ and another ray from the point $1 - i$ at an angle $\pi/6$. These two rays will intersect at the point ξ (see Fig. 1.7). By the reflection symmetry of the figure about the x axis, the value of ξ is real and is easily found to be $\xi = 1 + \sqrt{3}$. ∎

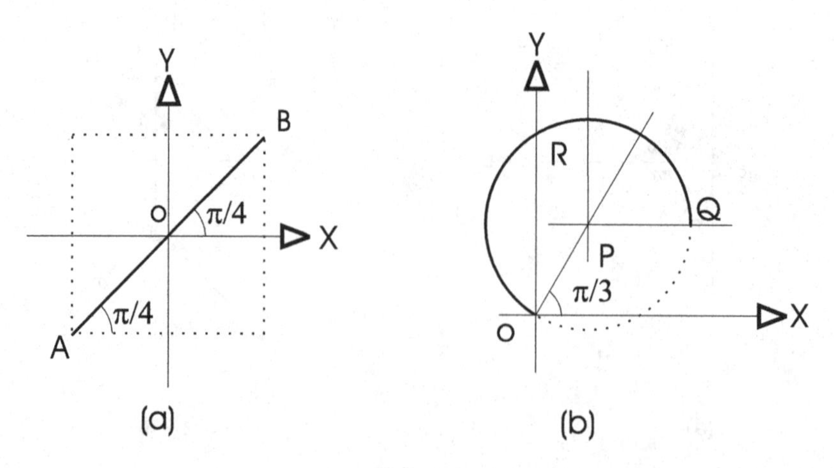

Fig. 1.6

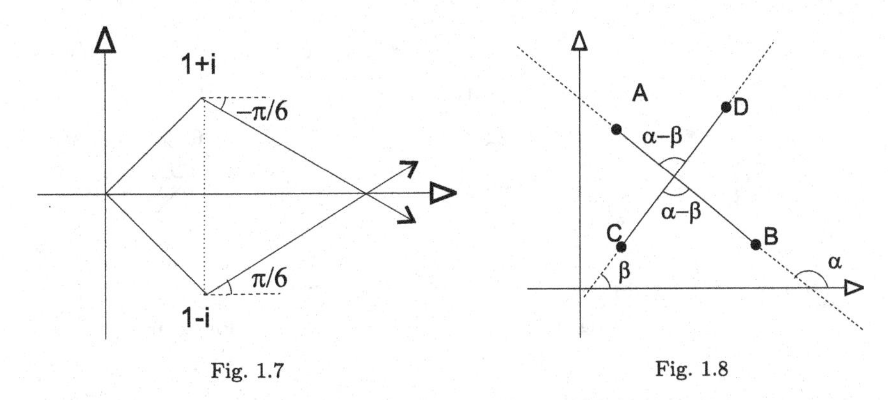

Fig. 1.7 Fig. 1.8

☑**1.2.4 Problem.** Show that two lines with end points z_1, z_2 and z_3, z_4 are perpendicular if and only if

$$\text{Arg}\left(\frac{z_1 - z_2}{z_3 - z_4}\right) = \pm\frac{\pi}{2}.$$

Solution. Let AB and CD be the two lines as shown in Fig. 1.8, and the numbers z_1, z_2 correspond to the end points of AB and the numbers z_3, z_4 correspond to the end points of CD. The angle made by the line AB with the real axis is $\alpha = \arg(z_1 - z_2)$, and the line CD, joining z_3 and z_4, makes an angle $\beta = \arg(z_4 - z_3)$ with the real axis. Therefore, the two lines are perpendicular if and only if

$$\arg(z_1 - z_2) - \arg(z_4 - z_3) = \pm\frac{\pi}{2},$$

and therefore we must have

$$\text{Arg}\left(\frac{z_1 - z_2}{z_3 - z_4}\right) = \pm\frac{\pi}{2}$$

as the necessary and sufficient condition for the two lines being perpendicular. ∎

✍ *It should be remembered that the "function"* $\arg(z)$ *does not have a unique value and the transition to Arg, in an equation involving* $\arg(z)$*, must be handled with care.*

1.2.5 Short examples. The angles between two lines, in Euclidean geometry, have only magnitudes. However, they carry a sign as well when defined in terms of the argument by means of Eq. (1.10), which has the range $(-\pi, \pi]$. Therefore, some care is needed when one is translating expressions written in terms of the Arg function into relations between angles. The following examples show this relationship in some simple situations:

(a) When three points O, I, and F are concurrent, we have the two cases shown in Fig. 1.9. Let ξ_i and ξ_f represent vectors **OI** and **OF**. Then in case (a)

$$\text{Arg}(\xi_f) = \text{Arg}(\xi_i) \quad \text{i.e.} \quad \text{Arg}(z_f - z_0) = \text{Arg}(z_i - z_0), \tag{1.11}$$

and in case (b)

$$\text{Arg}(\xi_f) = \pm\pi + \text{Arg}(\xi_i) \quad \text{i.e.} \quad \text{Arg}(z_f - z_0) = \pm\pi + \text{Arg}(z_i - z_0). \tag{1.12}$$

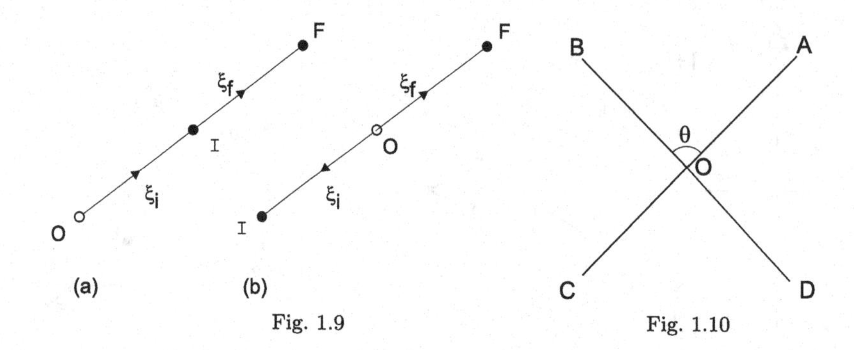

Fig. 1.9 Fig. 1.10

Of the two signs in the last term $\pm\pi$ in the above equation, the correct sign will depend on the numerical values of the arguments of ξ_i and ξ_f.

(b) Let z_A and z_B be complex numbers corresponding to the two end points of a straight line AB, and z be any other point not on AB. As z tends to a point P on the line AB and between A and B, we have the limiting value

$$\lim \operatorname{Arg}\left(\frac{z_{\mathrm{B}}-z}{z_{\mathrm{A}}-z}\right)=\begin{cases}+\pi & \text{if } z \text{ remains on the left of AB,}\\ -\pi & \text{if } z \text{ remains on the right of AB.}\end{cases} \tag{1.13}$$

(c) For four complex numbers $z_{\mathrm{A}}, z_{\mathrm{B}}, z_{\mathrm{C}}$, and z_{D} representing the points A, B, C, and D in Fig. 1.10. Then we have

$$\operatorname{Arg}(z_{\mathrm{A}}-z_{\mathrm{O}})=\operatorname{Arg}(z_{\mathrm{O}}-z_{\mathrm{C}}), \tag{1.14}$$

$$\operatorname{Arg}(z_{\mathrm{B}}-z_{\mathrm{O}})=\operatorname{Arg}(z_{\mathrm{O}}-z_{\mathrm{D}}). \tag{1.15}$$

We can therefore write

$$\operatorname{Arg}\left(\frac{z_{\mathrm{B}}-z_{\mathrm{O}}}{z_{\mathrm{A}}-z_{\mathrm{O}}}\right)=\operatorname{Arg}\left(\frac{z_{\mathrm{O}}-z_{\mathrm{D}}}{z_{\mathrm{A}}-z_{\mathrm{O}}}\right)=\operatorname{Arg}\left(\frac{z_{\mathrm{O}}-z_{\mathrm{D}}}{z_{\mathrm{O}}-z_{\mathrm{C}}}\right)=\theta, \tag{1.16}$$

$$\operatorname{Arg}\left(\frac{z_{\mathrm{O}}-z_{\mathrm{B}}}{z_{\mathrm{A}}-z_{\mathrm{O}}}\right)=\operatorname{Arg}\left((-1)\frac{z_{\mathrm{B}}-z_{\mathrm{O}}}{z_{\mathrm{A}}-z_{\mathrm{O}}}\right)=-\pi+\operatorname{Arg}\left(\frac{z_{\mathrm{B}}-z_{\mathrm{O}}}{z_{\mathrm{A}}-z_{\mathrm{O}}}\right)=-\pi+\theta, \tag{1.17}$$

where θ denotes the *magnitude* of the angle indicated. That $-\pi$, not $+\pi$, is the correct choice in Eq. (1.17) can be seen by recalling that $\operatorname{Arg} z$ has a range from $-\pi$ to π.

☑1.2.6 Problem.

(a) Show that the sum of two interior angles of a triangle is equal to the exterior angle at the third vertex.

(b) Verify that the sum of all the three interior angles of a triangle equals π.

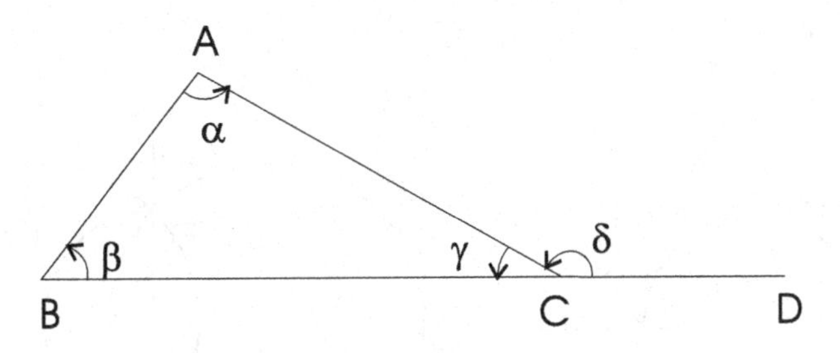

Fig. 1.11

Solution. Labeling the three vertices of the triangle as A, B, and C in anticlockwise order, the three interior angles of the triangle ABC and the exterior angle at C are given by

$$\alpha = \mathrm{Arg}\left(\frac{z_3 - z_1}{z_2 - z_1}\right), \qquad \beta = \mathrm{Arg}\left(\frac{z_1 - z_2}{z_3 - z_2}\right), \qquad \gamma = \mathrm{Arg}\left(\frac{z_2 - z_3}{z_1 - z_3}\right),$$

and the exterior angle is

$$\delta = \mathrm{Arg}\left(\frac{z_1 - z_3}{z_4 - z_3}\right) = \mathrm{Arg}\left(\frac{z_1 - z_3}{z_3 - z_2}\right) \qquad [\because \arg(z_4 - z_3) = \arg(z_3 - z_2)],$$

where z_1, z_2, z_3, and z_4 are the complex numbers corresponding to the four points A, B, C, and D, respectively. It may be noted that the above expressions for the four angles are positive when the vertices A, B, and C and the point D are taken as in Fig. 1.11.

(a) The sum of the angles α and β is

$$\alpha + \beta = \mathrm{Arg}\left(\frac{z_3 - z_1}{z_2 - z_1}\right) + \mathrm{Arg}\left(\frac{z_1 - z_2}{z_3 - z_2}\right) \tag{1.18}$$

$$= \mathrm{Arg}\left(\frac{z_3 - z_1}{z_2 - z_1}\frac{z_1 - z_2}{z_3 - z_2}\right) \tag{1.19}$$

$$= \mathrm{Arg}\left(\frac{z_3 - z_1}{z_2 - z_3}\right) = \delta. \tag{1.20}$$

(b) The sum of the three angles of the triangle is

$$\alpha + \beta + \gamma = \mathrm{Arg}\left(\frac{z_3 - z_1}{z_2 - z_1}\right) + \mathrm{Arg}\left(\frac{z_1 - z_2}{z_3 - z_2}\right) + \mathrm{Arg}\left(\frac{z_2 - z_3}{z_1 - z_3}\right) \tag{1.21}$$

$$= \arg\left(\frac{z_3 - z_1}{z_2 - z_1}.\frac{z_1 - z_2}{z_3 - z_2}.\frac{z_2 - z_3}{z_1 - z_3}\right) \tag{1.22}$$

$$= \arg(-1) = \pi. \tag{1.23}$$

▮

⧐ Exercise §§1.3, on proving a sample of standard theorems in geometry using complex variables. Some are nontrivial!

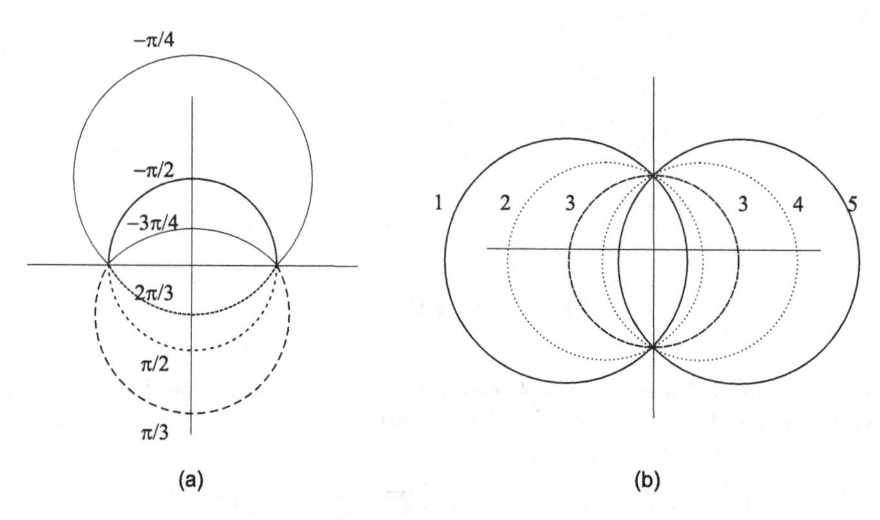

Fig. 1.12

▼ ✍**1.2.7 Short examples.** The following examples bring out representation of angles in terms of the Arg function and give a way of describing circles. Given two complex numbers, $z_1 \neq z_2$, we use S_α to denote the set

$$S_\alpha = \left\{ z \middle| \operatorname{Arg}\left(\frac{z - z_2}{z - z_1}\right) = \alpha \right\}, \qquad |\alpha| \leq \pi.$$

(a) The set S_α, for $\alpha = \pi$, is the line segment joining z_1 and z_2.

(b) The set S_α, for $0 < |\alpha| < \pi$, is an arc of a circle passing through the points z_1 and z_2 excluding the points z_1 and z_2. Figure 1.12(a) shows a few such arcs for $z_1 = 1, z_2 = -1$. For the positive values $\alpha = \frac{\pi}{3}, \frac{\pi}{2}, \frac{2\pi}{3}$, the arcs are on the left of the line from z_1 to z_2. Also shown in the figure are the arcs for the negative values $\alpha = -\frac{3\pi}{4}, -\frac{\pi}{2}, -\frac{\pi}{4}$, which lie on the right of the line from z_1 to z_2.

(c) As α tends to zero, the circular arcs become bigger and bigger, and in the limit $\alpha \to 0$ the set S_α coincides with the union of the two nonoverlapping line segments running to infinity in opposite directions, while one of these two line segments starts from z_1 and the other from z_2.

(d) The arcs S_α and $S_{-(\pi-\alpha)} (\alpha > 0)$ together form a circle passing through the two points z_1 and z_2. Thus C_α, given by the union

$$C_\alpha = S_\alpha \cup S_{-(\pi-\alpha)}, \quad \alpha > 0,$$

is a circle, excluding the points z_1 and z_2.

(e) A few members of the family of circles C_α are shown in Fig. 1.12(b) for $z_1 = -i$, $z_2 = i$. Each circle C_α *passes* through the two points z_1 and z_2 and has its center on the perpendicular bisector of the line joining the two points. While the circles marked 1 and 5 correspond to $\alpha = \pm\frac{\pi}{4}$, the circles for $\alpha = \pm\frac{\pi}{3}$ are labeled 2 and 4. The two semicircular arcs of the circle marked 3 correspond to $\alpha = \pm\frac{\pi}{2}$.

▼ ☑**1.2.8 Problem.** Show that the equation

$$|z - z_1| = k|z - z_2| \tag{1.24}$$

represents a circle with center z_0 and radius R, given by

$$z_0 = \frac{z_1 - k^2 z_2}{1-k^2}, \qquad R = \frac{k|z_1 - z_2|}{|1-k^2|}. \tag{1.25}$$

Solution. Substituting $z = x+iy$ and $z_1 = x_1 + iy_1$ and $z_2 = x_2 + iy_2$ in Eq. (1.24) and squaring gives

$$(x - x_1)^2 + (y - y_1)^2 = k^2\{(x - x_2)^2 + (y - y_2)^2\}. \tag{1.26}$$

Rearranging Eq. (1.26) we get

$$(1 - k^2)(x^2 + y^2) - 2x(x_1 - k^2 x_2) - 2y(y_1 - k^2 y_2) + x_1^2 + y_1^2 - k^2(x_2^2 + y_2^2) = 0. \tag{1.27}$$

This equation, when written in the standard form

$$(x - x_0)^2 + (y - y_0)^2 = R^2$$

for a circle, gives the desired results for the radius and the center of the circle,

$$R = \frac{k|z_1 - z_2|}{|1-k^2|}, \tag{1.28}$$

and

$$x_0 = \frac{x_1 - k^2 x_2}{1-k^2}, \qquad y_0 = \frac{y_1 - k^2 y_2}{1-k^2} \Rightarrow z_0 = \frac{z_1 - k^2 z_2}{1-k^2}. \tag{1.29}$$

∎

§1.2.9 Short examples. A few members of the family of circles

$$C_k \equiv \{z \big| |z - z_1| = k|z - z_2|\},$$

for $z_1 = -1$, $z_2 = 1$ and different values of k, are shown in Fig. 1.13. The following points should be noted:

(a) C_k, for $k = 1$, is a straight line which bisects the line joining the two points z_1 and z_2. It divides the complex plane into two half planes.
(b) When $k \approx 0$, C_k is a small circle enclosing the point z_1; as k increases, the circle becomes bigger and bigger and tends to the bisector as $k \to 1$.
(c) For $k > 1$, the circle C_k lies in the half plane containing the point z_2. As k increases, the circle becomes smaller and shrinks to the point z_2 when $k \to \infty$.
(d) The symmetry of Fig. 1.13, under reflection in the bisector, corresponds to the fact that the equation $|z - z_1| = k|z - z_2|$ remains invariant under simultaneous replacements $z_1 \leftrightarrow z_2, k \to 1/k$.

↬ The family of circles C_α, C_k, in the above examples, are known as Steiner circles. The two families of circles, C_α and C_k, are orthogonal; see Q [16] in Mixed Bag §§1.9.

§1.3 Linear Transformations

In this section we introduce *geometrical representation* of complex numbers as vectors and some important transformations on complex numbers.

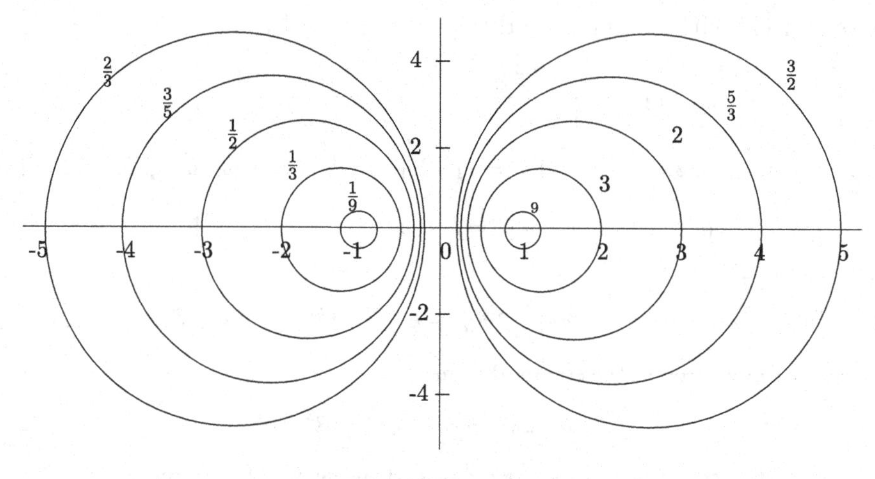

Fig. 1.13

§1.3.1 *Complex numbers as vectors in a plane*

It has been noted that complex numbers have a representation as *points* in a plane. Since the points in a plane are in a one-to-one correspondence with *directed line segments*, or vectors in two dimensions, the *directed line segments* in a plane give an alternate geometric representation of complex numbers. Thus, in Fig. 1.14, in addition to the point P representing a complex number $z = x + iy$, one may also think of the directed line segment **OP** as representing the same complex number, z. While for a point P its coordinates (x, y) give the real and imaginary parts, in the case of a directed line segment the projections on the two coordinate axes have the values x and y. Not only **OP** represents z; every other directed line segment, such as **QR**, drawn parallel to **OP** and equal to it in length (to be called *parallel translation* of **OP**), will also represent the same complex number, z. Here, in Fig. 1.14, the *directed line segments* **OP** and **QR** furnish two alternate ways of representing the same complex number z corresponding to the point P. With **OQ** representing another complex number ξ, the parallelogram rule of vector addition of two vectors states that **OR** corresponds to the sum $z + \xi$ and **QP** gives a representation for the difference $z - \xi$.

Finally, it is useful to remember that every vector having length r and making an angle θ with the real axis represents complex number $z = r\exp(i\theta)$.

§1.3.2 *Translation, scaling, and rotation*

A *linear transformation* is a mapping by a general linear function:

$$f(z) = \lambda z + \mu. \tag{1.30}$$

If we write $\lambda = |\lambda|e^{i\alpha}$, we have

$$f(z) = |\lambda|e^{i\alpha}z + \mu \tag{1.31}$$

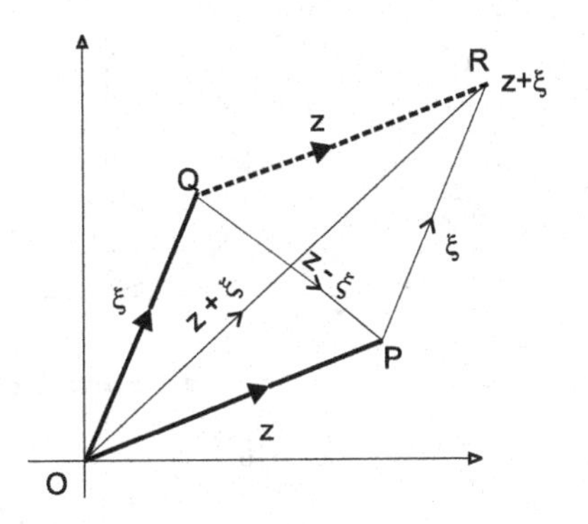

Fig. 1.14

and arbitrary linear mapping is seen to be a result of the composition of the simpler linear mappings scaling, rotation, and translation, defined below:

(*1) In Fig. 1.14 we have shown two vectors **OP** and **OQ**, as well as the parallel translation (**QR**) of the first vector **OP** drawn from the terminal point of the second vector **OQ**. This process of performing a parallel translation has taken the point P to R. If the points P and Q correspond to the complex numbers z and ξ, respectively, the point R will correspond to $z+\xi$ and hence the map $z \to z+\xi$ will be called *translation* by ξ.

(*2) Under the mapping $z \to w = sz$, $s > 0$, we have

$$|w| = s|z|, \qquad \arg w = \arg z.$$

Thus, the vector w is obtained by changing the length of the vector z by a factor of s. This mapping is called *scaling* or *stretching* w.r.t. the origin.

(*3) An anticlockwise rotation of the vector **OP**, about the origin O by an angle α, preserves the length and increases $\arg(z)$ by α. If the transformed vector **OQ** is the complex number w, we have (Fig. 1.15)

$$|w| = |z|, \qquad \arg(w) = \arg(z) + \alpha$$

and

$$w = (\cos\alpha + i\sin\alpha)z.$$

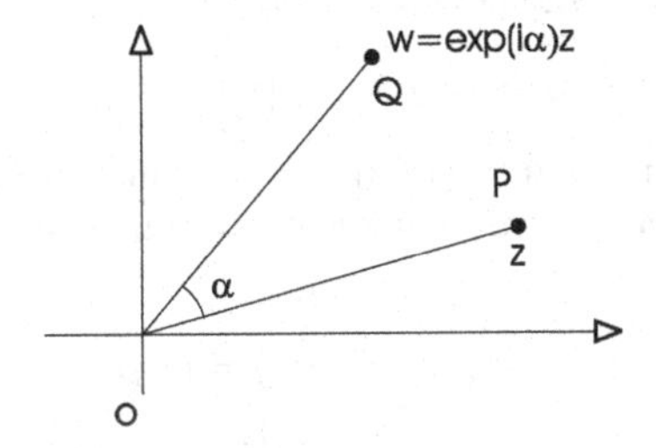

Fig. 1.15

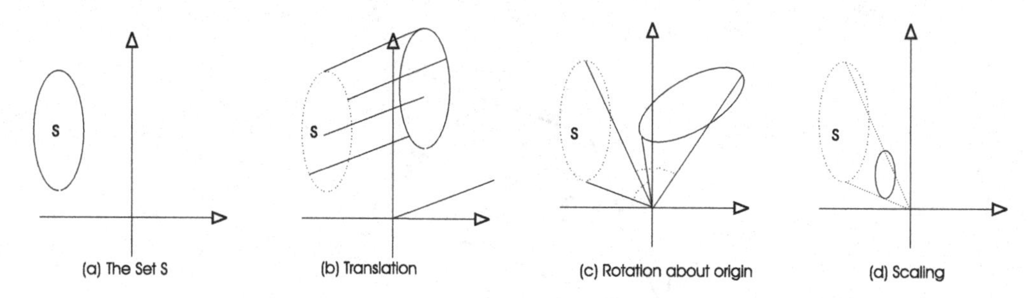

Fig. 1.16

Hence, the transformation $z \to w = (\cos\alpha + i\sin\alpha)z$ will be called a *rotation* by an angle α about the origin. Noting Euler's formula $\cos\alpha + i\sin\alpha = \exp(i\alpha)$, the rotation by an angle α can be written as $w = e^{i\alpha}z$.

Given a function $f(z)$ and a subset of complex numbers S, its image under the transformation $z \to w = f(z)$ is the set of complex numbers $w = f(z)$ for each $z \in S$. The image set so obtained will frequently be represented in a separate copy of the complex plane. Thus Figs. 1.16(a)–1.16(d) show the set S and its images under translation, a rotation, and a scaling, respectively.

♮1.3.1 Short examples. As a simple application of representation of complex numbers as vectors in two dimensions, we prove some inequalities involving complex numbers using geometric ideas:

(a) We note that $r = |z|$ in Fig. 1.1, and hence

$$|z| \geq \operatorname{Re} z, \qquad |z| \geq \operatorname{Im} z.$$

(b) If two consecutive sides of a triangle represent complex numbers z_A and z_B, the third side will represent $z_A + z_B$ and hence, using the well-known result from geometry that the length of a side of a triangle is less than the sum of the lengths of the other two sides, we get,

$$|z_A + z_B| \leq |z_A| + |z_B|. \tag{1.32}$$

(c) Similarly, the length of a side being greater than the difference of the lengths of the other two sides is translated into the inequality

$$|z_A - z_B| \leq ||z_A| - |z_B||.$$

The above inequalities are known as triangle inequalities.

(d) The inequality (1.32) is easily generalized to give

$$|z_1 + z_2 + \cdots + z_n| \leq |z_1| + |z_2| + \cdots + |z_n|. \tag{1.33}$$

(e) It is easily seen that the inequality (1.32) becomes an equality if and only if the *vectors* z_A and z_B are parallel. In other words, any two of the following three equations imply the third equation:

$$\arg z_1 = \arg z_2 = \arg z_3, \tag{1.34}$$

$$|z_1| + |z_2| = |z_3|, \tag{1.35}$$

$$z_1 + z_2 = z_3. \tag{1.36}$$

(f) The inequality

$$|z-1| \leq \big||z|-1\big| + |z||\mathrm{Arg}(z)|$$

can be easily proven from geometrical considerations.

↬ Proofs of the inequalities by algebraic method are left as an exercise for the reader.
⊳ Tutorial §§1.4 and Quiz §§1.5.

§1.4 Reflection and Inversion

Reflection Under complex conjugation mapping $z \to w = \bar{z}$, the point w is located at the mirror image of the point z in the real axis. The mirror image of a point in a straight line will be called the image under *reflection in the straight line.* Thus, the image of a point P under *reflection in a straight line* is a point Q such that the given line is the perpendicular bisector of PQ. It turns out to be useful to extend the above definition to reflection in a circle. The image of a point P under *reflection in a circle* is defined as a point Q on the ray drawn from the center of the circle to the point P such that the product of the distances OP × OQ equals the square of the radius of the circle. Figure 1.17 shows a geometrical construction to locate the reflection of a point P, when P lies outside the circle. First, the tangents PT and PS to the circle are drawn from P. The point Q is just the intersection of ST with the line joining the point P to the center O of the circle. From the definition, it is obvious that, in both the cases of a line and a circle, if Q is the image of P, P is the image of Q under reflection. We shall call two points z^* and z a pair of *symmetric points* w.r.t. a line, or a circle, if one of them is the image of the other point under reflection.

Inversion The map $w = 1/z, z \neq 0$, is called *inversion*. It is not difficult to see that $w = \overline{z^*}$, where z^* is the reflection of z in the unit circle $|z| = 1$. Inversion is thus equivalent to a reflection in the unit circle followed by complex conjugation. The inversion map satisfies the symmetry principle, which states that under inversion the symmetric points of a straight line (circle) go over to symmetric points of the image of the straight line (circle).

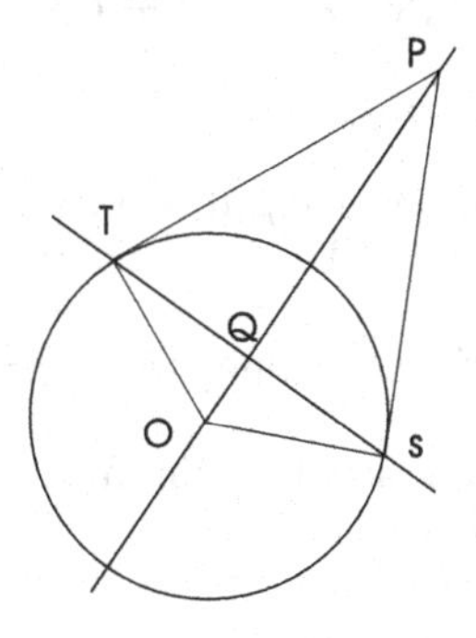

Fig. 1.17

↬ In what sense is the reflection in a circle a generalization of mirror reflection in a straight line? Do the two reflections share any property?

§1.5 Examples

⨍1.5.1 Short examples. We now give some examples of reflection.

(a) Points on a line remain unchanged under reflection in the line. Also, points on a circle remain invariant under reflection in the circle.
(b) The center of a circle and the point at infinity are symmetric w.r.t. reflection in the circle.
(c) If two points z_1 and z_2 are symmetric w.r.t. reflection in a line, every point on the line is equidistant from the two points z_1 and z_2; hence, the equation of the line is clearly given by $|z - z_1| = |z - z_2|$.
(d) If z_1 and z_2 are symmetric w.r.t. reflection in a circle, the circle is described by

$$|z - z_1| = k|z - z_2|$$

and d_1 and d_2 are the distances of the two points from the center of the circle, then

$$d_1 d_2 = R^2; \qquad d_1/d_2 = k^2$$

where R is the radius of the circle.
(e) The equation

$$a\, z\bar{z} + \bar{\lambda} z + \lambda \bar{z} + b = 0, \qquad a, b \in \mathbb{R}, \tag{1.37}$$

describes a circle if $a \neq 0$, and a straight line if $a = 0$. If ξ and ξ^* are a pair of symmetric points w.r.t. it, then they are related by the equation

$$a\, \xi\bar{\xi}^* + \bar{\lambda}\xi + \lambda\bar{\xi}^* + b = 0, \tag{1.38}$$

obtained by making the replacements $z \to \xi$ and $\bar{z} \to \overline{\xi^*}$ in Eq. (1.37). Note that, as expected, Eq. (1.38) is perfectly symmetric under the exchange $\xi \leftrightarrow \xi^*$.

↬ A proof of some of these and many other properties is left as an exercise for the reader.

⨍1.5.2 Short examples. A few simple properties of the inversion map are given below. Introducing polar variables

$$z = r\exp(i\theta) \qquad w = \rho\exp(i\phi), \tag{1.39}$$

the inversion map $w = \frac{1}{z}$ gives $\rho = \frac{1}{r}$, $\phi = -\theta$.

(a) A point in the upper half z plane is mapped to a point in the lower half w plane.
(b) A point inside the unit circle $|z| = 1$ has its image outside the unit circle $|w| = 1$ in the w plane.
(c) The image of a ray from the origin to infinity, $\text{Arg}(z) = \alpha$, is again a similar ray, $\text{Arg}(w) = -\alpha$, in the w plane.
(d) A straight line passing through $z = 0$ is again mapped to a straight line passing through the origin $w = 0$ in the w plane.
(e) The image of a circle with the center at $z = 0$ is again a circle with the center at the origin $w = 0$. The points inside a circle $z = R$ are mapped to points outside the image circle $w = \frac{1}{R}$.

(f) The circles, and also straight lines, can be described by an equation of the form

$$|z - z_1| = p|z - z_2|, \qquad p > 0.$$

Under the inversion map this equation takes the form

$$|w - w_1| = q|w - w_2|, \qquad q > 0,$$

and we see that the image of a circle, or a straight line, is again a circle or a straight line.

The above properties may be used to geometrically locate the image of straight lines and circles under the inversion map. It is straightforward to extend these properties to the transformation $w = \frac{1}{z-\xi}$, which may be regarded as inversion relative to the point ξ as the center.

☑**1.5.3 Problem.** Find the equation of a straight line in terms of its perpendicular distance p from the origin, and the angle α which the perpendicular line makes with the real axis. Use the symmetry principle for linear maps to find the reflection ξ^* of a point ξ in the straight line.

Solution. In Fig. 1.18 AB is a straight line and MN is perpendicular to AB and passes through the origin, so that OP $= p$ and the angle POX $= \alpha$. We shall find a linear map, so that MN and AB coincide with the real and imaginary axes. This map is easy to construct. A clockwise rotation about the origin by angle α, $z \to u = \exp(-i\alpha)z$, aligns the lines MN and AB parallel to the axes in the u plane. The translation $u \to w = u - p$ maps the point P to the origin in the w plane. The resultant map is

$$z \to w = e^{-i\alpha}z - p. \tag{1.40}$$

In the w plane, the equation of the imaginary axis is $\operatorname{Re} w = 0$, or $w + \bar{w} = 0$. Expressing w back in terms of z gives the required equation of the line:

$$e^{-i\alpha}z + e^{i\alpha}\bar{z} = 2p. \tag{1.41}$$

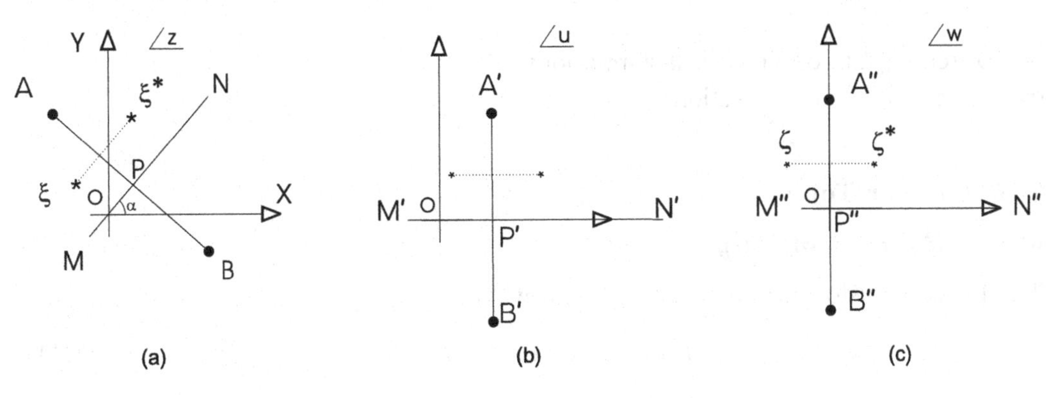

Fig. 1.18

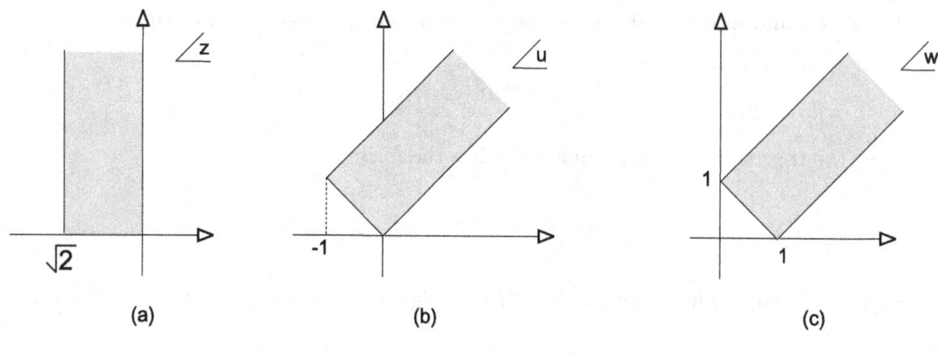

Fig. 1.19

Under the map (1.40) the points ξ and ξ^*, in the z plane, are mapped to ζ and ζ^* in the w plane, where

$$\zeta = e^{-i\alpha}\xi - p, \qquad \zeta^* = e^{-i\alpha}\xi^* - p. \tag{1.42}$$

The symmetry principle for linear maps implies that ζ^* is the reflection of ζ in the imaginary axis in the w plane and hence $\zeta^* = -\bar{\zeta}$. This gives

$$e^{-i\alpha}\xi^* + e^{i\alpha}\bar{\xi} = 2p. \tag{1.43}$$

This result verifies the statement in ⨍1.11(e). ∎

▶ A detailed discussion on properties of the inversion map and examples of properties will appear in Ch. 9.

⨍1.5.4 Short examples. We will find maps which will take the semi-infinite strip $\{z|-\sqrt{2} < x < 0, y > 0\}$ in Fig. 1.19(a) to semi-infinite strips drawn in Figs. 1.19(b) and 1.19(c).

(a) The result of applying clockwise rotation by $\pi/4$ on the region $\sqrt{2} < \operatorname{Re} z < 0$ is shown in Fig. 1.19(b). This map is given by $u = \frac{1-i}{\sqrt{2}}z$.
(b) The mapping $w = u + 1$ maps the region in Fig. 1.19(b) to the set shown in Fig. 1.19(c).
(c) Thus, the transformation obtained by applying the above two transformations in succession will take the semi-infinite strip in Fig. 1.19(a) to the strip in Fig. 1.19(c). This mapping is given by $w = \frac{1}{\sqrt{2}}(1-i)z + 1$.

⇛ Exercise §§1.6, on linear transformations.
⇛ Exercise §§1.7, on reflections.

§1.6 Further Topics

§1.6.1 *Point at infinity*

The linear transformation $z \to w = f(z)$, where

$$f(z) = \lambda z + \mu, \qquad \lambda \neq 0, \tag{1.44}$$

is a one-to-one mapping of the finite complex plane onto itself. This is not true of the inversion map $z \to w = 1/z$. Writing in polar forms, $z = re^{i\theta}, w = \rho e^{i\phi}$, we have

$\rho = 1/r$. Therefore, the points close to the origin in the z plane, $r \approx 0$, are mapped onto points far away from the origin in the w plane. All the points inside a disk of small radius ϵ, in the z plane, are mapped onto points outside a disk of large radius $1/\epsilon$, in the w plane. As $\epsilon \to 0$, the disk in the z plane shrinks to the origin and there is no image of $z = 0$ in the w plane. Similarly, as the point z moves farther and farther away from the origin, its image in the w plane moves closer and closer to the origin in w plane, but there is no point in the z plane which can be assigned $w = 0$ as the image under inversion.

It turns out to be useful to introduce the concept of a *point at infinity*, or $z = \infty$, as a formal image of $z = 0$ under the inversion map $w = 1/z$. The point $z = 0$ can then be regarded as the image of the point at infinity. The use of $z = \infty$ will always be understood in terms of a limiting process $w \to 0$, where $w = 1/z$. Every statement about the behavior of a function $f(z)$ at $z = \infty$ will mean the behavior of the function $f(1/w)$ as $w \to 0$. Thus, for example, we say that the function $f(z) = \frac{2z-1}{z-3}$ tends to 2 as $z \to \infty$, because $f(1/w) = \frac{2-w}{1-3w}$ tends to 2 as $w \to 0$.

✍ *Throughout this book, a complex number z will mean a number with finite real and imaginary parts unless stated otherwise. Similarly, a point in the complex plane will always mean a point with finite coordinates. If a statement applies to $z = \infty$, it will be stated so explicitly.*

§1.6.2 *Stereographic projection*

We know that complex numbers can be represented as points and as vectors in a two-dimensional plane. The stereographic projection described here gives another way of representing complex numbers geometrically.

Let (x, y, u) be a point in three-dimensional space with the X–Y plane identified with the complex plane. Consider the sphere

$$x^2 + y^2 + \left(\frac{u-1}{2}\right)^2 = \frac{1}{4}, \tag{1.45}$$

which touches the X–Y plane at the origin. The point $(0, 0, 1)$ on the sphere will be called the *north pole* and the origin the *south pole*. Given a complex number $z = x + iy$, we draw a line joining the point $(x, y, 0)$ to the north pole. The intersection of this line with the sphere is taken as the image of the complex number z and defines a one-to-one mapping of the complex plane onto the sphere with the north pole removed. If we take the image of the north pole as the point at infinity, we get a one-to-one mapping of the sphere to the extended complex plane. If (ξ, η, ζ) are the coordinates of the image of the point $(x, y, 0)$ under stereographic projection, one can prove that

$$\xi = \frac{x}{1+\rho^2}, \qquad \eta = \frac{y}{1+\rho^2}, \qquad \zeta = \frac{\rho^2}{1+\rho^2}, \tag{1.46}$$

where $\rho = |z| = \sqrt{x^2 + y^2}$. Conversely, one has

$$x = \frac{\xi}{1-\zeta}, \qquad y = \frac{\eta}{1-\zeta}, \qquad |z|^2 = \frac{\zeta}{1-\zeta}. \tag{1.47}$$

§1.6.3 *Bilinear transformations*

An arbitrary sequence of translation, rotation, scaling, and inversion is equivalent to a transformation by a function of the form

$$f(z) = \frac{\alpha z + \beta}{\gamma z + \delta}, \qquad \text{with} \qquad \alpha\delta - \beta\gamma \neq 0. \tag{1.48}$$

The function $f(z)$ gives a mapping $z \to w = f(z)$ which can also be described by a relation of the form

$$\lambda wz + \mu w + \rho z + \sigma = 0. \tag{1.49}$$

Such a transformation is called a *fractional linear transformation* or a *bilinear transformation.* Conversely, a general bilinear transformation can be decomposed into a product of translation, rotation, scaling, and inversion mappings. Three important properties, given below, are shared by linear transformations and inversion and hence are true for all bilinear transformations as well.

(*1) Linear transformations preserve the shape of a curve, but inversion does not have this property. However, the inversion mapping has the important property that the image of a straight line (or of a circle) is always a straight line or a circle. *It then follows that the property that the family of all circles and straight lines is mapped onto itself holds for every bilinear transformation.*

(*2) Let Γ be a straight line or a circle, and z_1 and z_2 be two points and Γ', z_1', and z_2' denote their images under a bilinear transformation. If z_1 and z_2 are symmetric points w.r.t. Γ, the image points z_1' and z_2' will be symmetric points w.r.t. the image curve Γ'. This property is called the *symmetry principle.*

(*3) As already noted, under a linear transformation the image of a set is geometrically similar to the original set, in particular, the angle between two curves at a point of intersection is preserved under a linear mapping. The property, that the angle between the images of two curves equals the angle between the two curves holds, also for inversion at all points except $z = 0$. In fact, both the magnitude and the sense of the angle are preserved. *Thus, the angle-preserving property, known as the conformal property, also holds for the inversion map and hence for a general bilinear transformation.*

⇸ We leave it as an exercise for the reader to think why the requirement $\alpha\delta - \beta\gamma \neq 0$ should be imposed in Eq. (1.48).

► We shall come back to the bilinear transformations along with a larger class of mappings and discuss their properties and applications in Chs. 9 and 10.

§1.7 Notes and References

(◇1) Many books give a good discussion on stereographic projection. See, for example, [Ablowitz (2003); Polya (1974)] .

(◇2) For a gallery of interesting geometric images created by multiple inversion of circles and straight lines, see http://xahlee.org/SpecialPlaneCurves_dir/InversionGallery_dir/inversion Gallery.htm

(◇3) A Web search for stereographic projection yields a large number of Internet sites with a variety of images and examples. See, for instance, www.uwgb.edu/dutchs/structge/sphproj.htm for some other types of projections on the sphere.

(◇4) Graphics packages in Mathematica can be used to generate a variety of stereographic maps. See, for example, Web pages mathworld.wolfram.com/StereographicProjection.html, and mathworld.wolfram.com/HopfMap.html on the Wolfram Mathematica site for stereographic projections and related topics.

Chapter 2

ELEMENTARY FUNCTIONS AND DIFFERENTIATION

In this chapter we begin with the definition and properties of elementary transcendental functions, *viz.* the exponential, trigonometric, and hyperbolic functions of a complex variable. These functions are many-to-one, and complex solutions to simple equations involving these functions will be discussed in §2.2. This leads to introduction of "multivalued functions", a topic to be treated in detail in the next chapter. We introduce differentiation w.r.t. a complex variable and prove that Cauchy Riemann equations are necessary for differentiability. The concept of analyticity is defined and harmonic functions are briefly discussed. The last section contains various graphical representations of functions of a complex variable.

We will assume that the reader is familiar with the limit, continuity, and differentiation of functions of one and two real variables.

§2.1 Exponential, Trigonometric, and Hyperbolic Functions

The exponential and trigonometric functions of a complex variable can be conveniently defined in terms of their power series. However, the reader should be more comfortable about handling a definition in terms of the corresponding functions of a real variable. Therefore, we define the exponential function $\exp(z)$ of a complex variable $z = x + iy$ by

$$\exp(z) = e^x(\cos y + i \sin y), \tag{2.1}$$

in terms of the sine and cosine functions of a real variable. We shall obtain the power series representation for these functions in Ch. 5. Some simple but useful properties of the exponential function, given below, can be derived using the properties of the exponential and trigonometric functions of a real variable.

(*1) For real values of θ we have Euler's formula

$$\exp(i\theta) = \cos\theta + i\sin\theta \tag{2.2}$$

and therefore

$$\sin\theta = \frac{1}{2i}(e^{i\theta} - e^{-i\theta}) = \text{Im}\,\exp(i\theta), \tag{2.3}$$

$$\cos\theta = \frac{1}{2}(e^{i\theta} + e^{-i\theta}) = \text{Re}\,\exp(i\theta). \tag{2.4}$$

(*2) The modulus of $\exp(i\alpha)$, for real α, is 1:

$$|\exp(i\alpha)| = 1, \quad \alpha \in \mathbb{R}. \tag{2.5}$$

For more general situations, we have

$$|\exp(z)| = \exp(x), \quad \arg(\exp(z)) = y. \tag{2.6}$$

(*3) The exponential function has an important property:

$$\exp(z_1 + z_2) = \exp(z_1)\exp(z_2). \tag{2.7}$$

(*4) The result in Eq. (2.7) immediately implies De Moivre's theorem:

$$(\cos\theta + i\sin\theta)^n = \cos n\theta + i\sin n\theta. \tag{2.8}$$

(*5) The exponential function is never zero, as is shown by its property:

$$\exp(z)\exp(-z) = 1. \tag{2.9}$$

The sine and cosine functions of a complex variable, also the hyperbolic sine and hyperbolic cosine functions, are defined in terms of $\exp(z)$:

$$\sin z = \frac{1}{2i}(e^{iz} - e^{-iz}), \quad \cos z = \frac{1}{2}(e^{iz} + e^{-iz}), \tag{2.10}$$

$$\sinh z = \frac{1}{2}(e^{z} - e^{-z}), \quad \cosh z = \frac{1}{2}(e^{z} + e^{-z}). \tag{2.11}$$

The above definitions can be combined to produce a complex version of De Moivre's theorem:

$$\exp(inz) = (\cos z + i\sin z)^n = \cos nz + i\sin nz. \tag{2.12}$$

Other trigonometric and hyperbolic functions of a complex variable are defined in terms of the above functions, as usual. Thus, for example,

$$\tan z = \frac{\sin z}{\cos z}, \quad \cot z = \frac{\cos z}{\sin z}, \tag{2.13}$$

$$\operatorname{cosec} z = \frac{1}{\sin z}, \quad \sec z = \frac{1}{\cos z}, \tag{2.14}$$

and the hyperbolic functions, tanh, coth, cosech, and sech, are defined similarly. Some of the important properties of the trigonometric and hyperbolic functions are summarized below and their proofs are left as simple exercises for the reader.

(*1) The following properties are generalizations of the corresponding properties for the real variable case:

(a) $\cos^2 z + \sin^2 z = 1$, (b) $\cosh^2 z - \sinh^2 z = 1$,

(c) $1 + \tan^2 z = \sec^2 z$, (d) $1 - \tanh^2 z = \text{sech}^2 z$,

(e) $\cot^2 z + 1 = \text{cosec}^2 z$, (f) $\coth^2 z - 1 = \text{cosech}^2 z$.

(*2) Some of the relations between the trigonometric and hyperbolic functions are given below; others can be derived using them:

(a) $\sin(iz) = i \sinh z$, (b) $\cos(iz) = \cosh z$,

(c) $\sinh(iz) = i \sin z$, (d) $\cosh(iz) = \cos z$.

(*3) The trigonometric functions are periodic with period 2π and the hyperbolic functions are periodic with period $2\pi i$. While sin and sinh are odd functions, cos and cosh are even ones.

(a) $\sin(2\pi + z) = \sin z$, (b) $\sinh(2\pi i + z) = \sinh z$,

(c) $\cos(2\pi + z) = \cos z$, (d) $\cosh(2\pi i + z) = \cosh z$,

(e) $\sin(-z) = -\sin z$, (f) $\sinh(-z) = -\sinh z$,

(g) $\cos(-z) = \cos z$, (h) $\cosh(-z) = \cosh z$.

(*4) The basic identities for functions of the sum of two variables are as follows:

(a) $\sin(z_1 + z_2) = \sin z_1 \cos z_2 + \cos z_1 \sin z_2$,

(b) $\cos(z_1 + z_2) = \cos z_1 \cos z_2 - \sin z_1 \sin z_2$,

(c) $\sinh(z_1 + z_2) = \sinh z_1 \cosh z_2 + \cosh z_1 \sinh z_2$,

(d) $\cosh(z_1 + z_2) = \cosh z_1 \cosh z_2 + \sinh z_1 \sinh z_2$.

(*5) With $z = x + iy$, the real and imaginary parts of the $\cos z, \sin z, \cosh z$, and $\sinh z$ functions are found to be

(a) $\cos z = \cos x \cosh y - i \sin x \sinh y$,

(b) $\sin z = \sin x \cosh y + i \cos x \sinh y$,

(c) $\cosh z = \cosh x \cos y + i \sinh x \sin y$,

(d) $\sinh z = \sinh x \cos y + i \cosh x \sin y$.

Almost all the above identities are easily proved by making use of the expressions of the trigonometric and hyperbolic functions in terms of the exponential function.

⊳ §§2.1 Exercise for problems on properties of trigonometric and hyperbolic functions.

We give a few quick examples to show that the results given above can be proven using De Moivre's theorem and Euler's formula.

2.1.1 Short examples. All the examples below make use of De Moivre's theorem.

(a) $\cos^2 z + \sin^2 z = \frac{1}{4}[(e^{iz}+e^{-iz})^2 - (e^{iz}-e^{-iz})^2] = 1$, because on expansion of the parentheses only the cross terms survive.

(b) To get a proof of

$$\sin(z_1 + z_2) = \sin z_1 \cos z_2 + \cos z_1 \sin z_2,$$

start from right hand side and use Euler's formula:

$$\begin{aligned}
\sin z_1 \cos z_2 + \cos z_1 \sin z_2 &= \frac{1}{4i}[(e^{iz_1} - e^{-iz_1})(e^{iz_2} + e^{-iz_2}) \\
&\quad + (e^{iz_1} + e^{-iz_1})(e^{iz_2} - e^{-iz_2})] \\
&= \frac{1}{2i}[e^{i(z_1+z_2)} - e^{-i(z_1+z_2)}] \\
&= \sin(z_1 + z_2).
\end{aligned} \tag{2.15}$$

(c) To compute the real and imaginary parts of $\sin z$, proceed as follows:

$$\begin{aligned}
\sin z &= \frac{1}{2i}\left(e^{i(x+iy)} - e^{-i(x+iy)}\right) = \frac{1}{2i}[(\cos x + i \sin x)e^{-y} - (\cos x - i \sin x)e^{y}] \\
&= \sin x \cosh y + i \cos x \sinh y.
\end{aligned}$$

(d) The real and imaginary parts of z^3 are given by $x^3 - 3xy^2 + i(3x^2y - y^3)$. Hence,

$$\sin z^3 = \sin[(x^3 - 3xy^2) + i(3x^2y - y^3)]$$

and using the result on the real and imaginary parts of $\sin z$, given above, the real and imaginary parts of $\sin z^3$ are seen to be

$$\begin{aligned}
\text{Re } \sin z^3 &= \sin(x^3 - 3xy^2)\cosh(3x^2y - y^3), \\
\text{Im } \sin z^3 &= \cos(x^3 - 3xy^2)\sinh(3x^2y - y^3).
\end{aligned}$$

2.1.2 Problem. Prove that

$$\cos^5 z = \frac{1}{16}(\cos 5z + 5\cos 3z + 10 \cos z)$$

and hence find the real and imaginary parts of $\cos^5 z$.

Solution. We use Euler's formula and first express $\cos z$ in terms of exponentials and expand the resulting expression using the binomial theorem which gives $\cos^5 z$ in terms of the cosine of multiples of z. Thus,

$$\begin{aligned}
\cos^5 z &= \frac{1}{32}[\exp(iz) + \exp(-iz)]^5 \\
&= \frac{1}{32}(e^{5iz} + 5e^{3iz} + 10e^{iz} + 10e^{-iz} + 5e^{-3iz} + e^{-5iz}) \\
&= \frac{1}{16}(\cos 5z + 5\cos 3z + 10 \cos z).
\end{aligned}$$

In the last step, Euler's formula has been used again to combine pairs of exponentials. Then the desired real and imaginary parts are obtained using the property (*5) and we get

$$\text{Re} \cos^5 z = \frac{1}{16}(\cos 5x \cosh 5y + 5 \cos 3x \cosh 3y + 10 \cos x \cosh y),$$

$$\text{Im} \cos^5 z = -\frac{1}{16}(\sin 5x \sinh 5y + 5 \sin 3x \sinh 3y + 10 \sin x \sinh y). \quad \blacksquare$$

⊳ §§2.2, for more problems on using the real and imaginary parts of expressions involving trigonometric and hyperbolic functions. See also Q[2]–Q[6] in §§2.13, Mixed Bag.

⊳ §§2.3, for some short questions to check for some common misconceptions.

§2.2 Solutions to Equations

Apart from the functions introduced in the previous section, many functions are defined implicitly through solutions to equations. Here we take a first look at solutions to some common equations in the complex plane. These solutions will be needed in later chapters on integration, where one will have to try to determine the singularities of expressions commonly encountered in many of the applications. Almost all equations have very different properties when their solutions are allowed to be complex. For example, a polynomial may not have any root in real numbers; the fundamental theorem of algebra asserts that every polynomial of degree N has exactly N roots in the complex plane when the roots are counted according to their multiplicities.

Solutions of $\exp(w) = z$. The equation $\exp(w) = p$, for $p > 0$, has a unique real solution for w and leads to introduction of the logarithm of positive real numbers through $w = \ln p$. However, the solution is no longer unique when w is allowed to become complex. Here we seek complex solutions w to the equation

$$\exp(w) = z, \tag{2.16}$$

when z is a given complex number. In general, one would hope to set the value of $\log z$ equal to w obtained by solving the above equation. As will be seen below, we are led to a set of an infinite number of complex values for w, which we shall call *the set of values of* $\log z$.

We substitute $w = a + ib$ and $z = r \exp(i\theta)$ in Eq. (2.16), obtaining

$$e^a e^{ib} = re^{i\theta}. \tag{2.17}$$

Comparing the moduli and the arguments of the two sides of this equation, we immediately get

$$e^a = r \Rightarrow a = \ln r, \tag{2.18}$$

$$b = \theta + 2\pi n, \tag{2.19}$$

where n is an integer. Hence, all solutions to Eq. (2.16) are given by

$$\boxed{w_n = \ln r + i(\theta + 2\pi n)} \tag{2.20}$$

where $n \in \mathbb{Z}$ is an integer. Thus, we find that, for a fixed z, there are infinite solutions w_n given by Eq. (2.20) and this set of all solutions will be called the *set of all values of* $\log z$.

✍ *Throughout the book we will use* $\ln x$ *to denote the value of the Napierian logarithm, when the variable* x *is a positive real number. The notation* $\log z$ *will be used for the logarithm of a complex number* z.

Roots of unity. We now introduce the roots of unity. The Nth roots of unity are the N solutions to the equation

$$w^N = 1. \tag{2.21}$$

Writing w in the polar form $\rho \exp(i\phi)$ and using $|w|^N = \rho^N$, we get

$$\rho^N = 1 \quad \exp(iN\phi) = 1, \tag{2.22}$$

which gives $\rho = 1, N\phi = 2\pi i m$, $m \in \mathbb{Z}$. Only N of these solutions are distinct, which we take as

$$\boxed{w_m = \exp\left(\frac{2\pi i m}{N}\right)} \tag{2.23}$$

with $m = 0, 1, 2, \ldots, N-1$.

Roots of $w^N = \xi$. To obtain the set of all solutions, to be called the set of all values of $\xi^{1/N}$, of the equation

$$w^N = \xi, \quad \xi \neq 0, \quad N = \text{integer}, \tag{2.24}$$

we first use the polar forms $w = \rho \exp(i\phi)$ and $\xi = |\xi| \exp(i\alpha)$ in Eq. (2.24) and rewrite it as

$$\rho^N \exp(iN\phi) = |\xi| \exp(i\alpha); \tag{2.25}$$

on taking the modulus of both sides, Eq. (2.25) gives $\rho = |\xi|^{1/N}$ and further we get

$$\exp(iN\phi) = \exp(i\alpha), \tag{2.26}$$

implying that $N\phi = \alpha + 2\pi i m, m \in \mathbb{Z}$. Of these solutions, only N solutions are distinct, which we take as

$$w_m = |\xi|^{1/N} e^{i\alpha/N} e^{2\pi i m/N}, \quad m = 0, 1, \ldots, (N-1). \tag{2.27}$$

The solutions can be described in terms of the root of unity as follows. We write ξ in the polar form $\xi = \rho \exp(i\alpha)$ and one of the desired solutions is $\eta_0 = \rho^{1/N} \exp(i\alpha/N)$.

The set of all solutions to Eq. (2.24) is then obtained by multiplying η_0 with the Nth roots of unity. A similar discussion leads to the set of values of rational power of a complex number ξ. However, when λ is an irrational number or a complex number, the set of all values of ξ^λ requires use of the set of values of the logarithm by means of the relation

$$\xi^\lambda = \exp(\lambda \log \xi). \tag{2.28}$$

This equation should be understood in the sense of giving all values of ξ^λ when the values of $\log \xi$ are inserted in the right hand side.

Simple trigonometric equations. One can arrive at the set of values for inverse trigonometric functions as the solutions to a corresponding equation involving the trigonometric function. So the solutions of

$$\sin w = \xi$$

give all possible values of $\sin^{-1} z$. All solutions to common equations involving the trigonometric and hyperbolic functions are most easily obtained by first writing the function $\sin w$ in terms of the exponential function and solving for e^w, and then solving for w. It turns out that the set of all values of the inverse trigonometric and hyperbolic functions can be written in terms of values of the logarithm.

✍ *While we have introduced the set of all values of* $\log z$*, for a complex number* z*, it must be realized that this process, as yet, does not define the logarithm of* z*, i.e.* $\log z$ *as a function of a complex variable. Similarly, arbitrary power* z^λ*, as a function of complex variable* z*, is not yet defined. This is due to the fact that in order to have any well-defined function of a complex variable one must have a rule which assigns a unique value. The above discussion does not lead to such rules for logarithm functions, fractional or complex powers, and inverse trigonometric and hyperbolic functions. We shall come back to these and related issues in the next chapter.*

§2.3 Examples

⚡2.3.1 Short examples. We shall give some simple examples of the set of all solutions in the complex plane for a few equations:

(a) The equation $\exp(z) = -p$ has no *real solution* for $p > 0$ but has an infinite number of solutions,

$$z = \ln p + (2n+1)\pi i, \quad n \in \mathbb{Z},$$

in the complex plane.

(b) The set of all values of the logarithm of a number ξ has been introduced as a solution to the equation $\exp(w) = \xi$, so $\log 1$ has a unique *real value* 0. However, when complex values are allowed, the set of all values of $\log 1$ becomes an *infinite set*, $\{\xi = 2n\pi i | n \in \mathbb{Z}\}$.

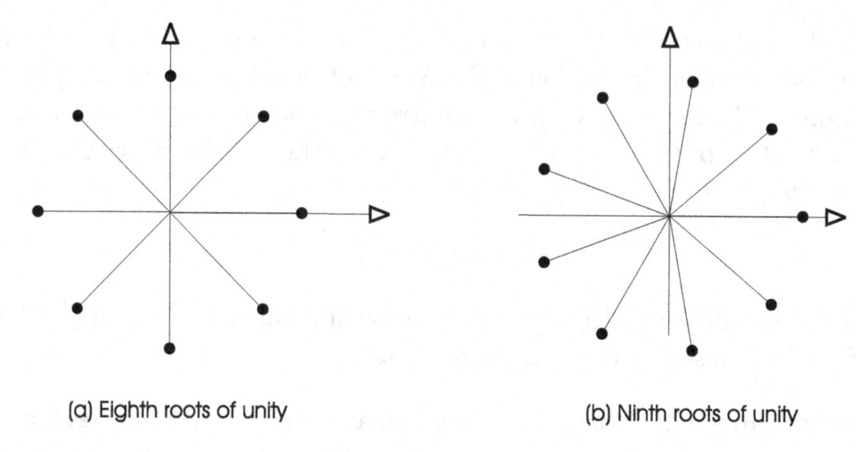

(a) Eighth roots of unity
(b) Ninth roots of unity

Fig. 2.1

(c) The equation $\sin z = \frac{1}{2}$ has *only real solutions,*

$$z = n\pi + (-1)^n \frac{\pi}{6}, \quad n \in \mathbb{Z},$$

and there are no other solutions in the complex plane.

(d) There is *no real* ξ satisfying the equation $\cosh \xi = 0$, but an *infinite number* of solutions $\xi = i(2n+1)\pi/2$ exist in the complex plane.

(e) There is only *one real number*, -1, satisfying $z^{2n+1} = -1$, but there are, in all, $(2n+1)$ *solutions* when the complex numbers are also counted.

(f) The *two real roots* of $z^{2N} = 1$ are ± 1, but we get $2N$ *solutions* when complex numbers are included.

(g) Equations such as $\sin z = 5$ and $\cos z = 12 + i13$ have no real solutions for z, but an infinite number of solutions exist in the complex plane.

⚡2.3.2 Short examples. We shall now give some properties of the roots of unity and of a complex number:

(a) The Nth roots of unity lie on the unit circle and have argument θ increasing in steps of $2\pi/N$.

(b) The Nth roots of unity lie on corners of a regular polygon with N sides having the center of the polygon at the origin and one of the corners at $z = 1$. For example, the eighth roots and the ninth roots of unity are shown in Fig. 2.1(a) and 2.1(b).

(c) The Nth roots of a complex number ξ lie on the circle radius $|\xi|$ and the center at the origin.

(d) The Nth roots of a complex number ξ can also be found by first finding one of the roots. The remaining, $N-1$ roots are located by applying $N-1$ successive rotations, each by angle $2\pi/N$.

(e) The Nth roots of ξ are located by applying a combination of scale transformation and rotation on the Nth roots of unity; the scale factor is seen to be $|\xi|^{1/N}$ and the angle of rotation is equal to $\arg \xi/N$.

☑2.3.3 Problem. Find all complex solutions to the following equations:

(a) $\exp(z) = -1 + i\sqrt{3}$, (b) $\sinh z = 2i$,

(c) $\tanh z = \sqrt{3}i$, (d) $\cot z = 2 + i$.

Solution.

(a) To solve $\exp(z) = -1 + i\sqrt{3}$, we write the right hand side as

$$\exp(z) = 2\exp\left(\frac{2i\pi}{3}\right)$$

$$z = \ln 2 + \frac{2\pi i}{3} + 2n\pi i = \ln 2 + \frac{2(3n+1)}{3}\pi i.$$

(b) To solve $\sinh z = 2i$, we express, as before, the equation in terms of e^z and solve for e^z. Thus, we get

$$\frac{e^z - e^{-z}}{2} = 2i \quad \Rightarrow \quad e^{2z} - 4ie^z - 1 = 0.$$

We therefore have

$$e^z = \frac{4i \pm \sqrt{-16+4}}{2},$$

$$\therefore \quad e^z = (2 \pm \sqrt{3})\exp\left(\frac{\pi i}{2}\right) \quad \Rightarrow \quad z = \ln(2 \pm \sqrt{3}) + \frac{\pi i}{2} + 2n\pi i.$$

(c) The equation $\tanh z = \sqrt{3}i$, when written in terms of $\exp(z)$, gives

$$\exp(2z) = \frac{1 + i\sqrt{3}}{1 - i\sqrt{3}} = \frac{-1 + i\sqrt{3}}{2}.$$

Thus, we get $\exp(2z) = \exp(2\pi i/3)$ and the solution for z is

$$2z = \frac{2\pi i}{3} + 2n\pi \Rightarrow z = \left(n + \frac{1}{3}\right)\pi i.$$

(d) Writing $\cot z = 2 + i$ in terms of exponentials, we get

$$i\left(\frac{e^{iz} + e^{-iz}}{e^{iz} - e^{-iz}}\right) = 2 + i \quad \Rightarrow \quad i\frac{e^{2iz} + 1}{e^{2iz} - 1} = 2 + i.$$

Solving for $\exp(2iz)$ gives

$$\exp(2iz) = 1 + i = \sqrt{2}\exp\left(\frac{i\pi}{4}\right)$$
$$= \exp\left(\frac{1}{2}\ln 2 + i\frac{\pi}{4}\right),$$

giving

$$2iz = \frac{1}{2}\ln 2 + i\frac{\pi}{4} + 2n\pi i \quad \Rightarrow \quad z = n\pi + \frac{\pi}{8} - \left(\frac{i}{4}\right)\ln 2.$$

∎

☑**2.3.4 Problem.** Relate the roots of the following equations to suitable roots of unity and plot your answers:

(a) $z^3 + 27 = 0$, (b) $z^4 + 16 = 0$, (c) $z^5 + 243(1+i) = 0$.

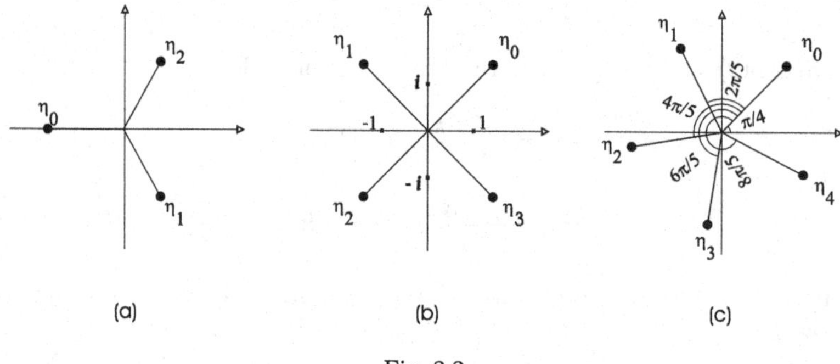

Fig. 2.2

Solution.

(a) We write the equation in the form $z^3 = -27$. Therefore, one root of this equation is -3, to be denoted as η_0. This root η_0 multiplied by the three cube roots of unity, 1, ω, ω^2, gives the final roots as $\eta_0 = -3, \eta_1 = -3\exp(2\pi i/3), \eta_2 = -3\exp(4\pi i/3)$. These values are shown in Fig. 2.2(a).

(b) To solve $z^4 + 16 = 0$, note that one root of this equation is $2\exp(i\pi/4) \equiv \eta_0$. The set of all the four roots is obtained by multiplying the fourth roots of unity $\{1, i, -1, -i\}$ by the root already obtained. The four roots of $z^4 + 16 = 0$ are then given by

$$\eta_0 = 2\exp(\pi i/4), \quad \eta_1 = 2\exp(3\pi i/4),$$
$$\eta_2 = 2\exp(5\pi i/4), \quad \eta_3 = 2\exp(7\pi i/4).$$

Geometrically, multiplication by $2\exp(i\pi/4)$ can be viewed as a scaling by a factor of 2 and anticlockwise rotation by $\pi/4$. We show the roots of unity and the required roots in Fig. 2.2(b).

(c) We have to find the fifth roots of $-243(1+i)$, which we write as $3^5\sqrt{2}\exp(5\pi i/4)$. One root is easily found to be $3\sqrt[10]{2}\exp(i\pi/4)$. We denote this root by η_0. All the required roots are obtained by starting with the answer η_0 already found and multiplying it by fifth roots of unity. Thus, we get the five roots as $\eta_0, \ldots, \eta_4$, where

$$\begin{aligned}
\eta_1 &= e^{2\pi i/5}\eta_0, \\
\eta_2 &= e^{2\pi i/5}\eta_1 = e^{4\pi i/5}\eta_0, \\
\eta_3 &= e^{2\pi i/5}\eta_2 = e^{6\pi i/5}\eta_0, \\
\eta_4 &= e^{2\pi i/5}\eta_3 = e^{8\pi i/5}\eta_0.
\end{aligned}$$

Multiplying η_0 with different fifth roots of unity is equivalent to applying rotation by $2\pi/5$ repeatedly. The complex numbers η_0 together with the four complex numbers η_1, η_2, η_3, and η_4, obtained by applying four successive rotations by $2\pi/5$, are shown in Fig. 2.2(c).

∎

✍ *Writing solutions to an equation such as* $z^3 + 27 = 0$ *in the form* $(-27)^{1/3}$ *is not giving a solution at all. Similarly,* $\log(-1 + i\sqrt{3})$ *as an answer for* ☑2.3.4(a), *or* $\cot^{-1}(2+i)$ *for* ☑2.3.3(d), *is merely a restatement of the problem.*

⊳ Exercise §§2.4, on solutions to equations involving trigonometric functions.

⊳ Mined set §§2.5, on solutions to equations involving trigonometric functions.

⇛ Tutorial §§2.6, for problems on roots of unity.
⇛ Quiz §§2.7, on roots of unity.

§2.4 Open Sets, Domains, and Regions

Let $\gamma(z_0, R)$ denote a circle, of radius R and with the center at z_0, given by $|z - z_0| = R$. The set of all points enclosed by $\gamma(z_0, R)$, excluding the points on the circle itself, is called a *neighborhood of the point* z_0, if $R > 0$. A subset S of complex numbers is called an *open set* if for every point ξ in the set it also contains a neighborhood of the point ξ. Thus, the set $\{z||z - z_0| < R\}$, called an open disk, is an open set. The upper half plane, $\{z|\text{Im}\, z > 0\}$, and the inside of a square, $\{z| - L < \text{Re}\, z < L, -L < \text{Im}\, z < L\}$, are other examples of open sets.

A subset of the complex plane is called *connected* if every two points in the subset can be joined by a continuous curve lying inside the subset. A connected open set is called a *domain.* A *region* is defined as a subset which is either a domain or a union of a domain D with a subset, or all, of the boundary of D. For example, the set all points enclosed by two overlapping circles is connected. The set of all points inside two nonoverlapping circles is not connected. An annular region between two concentric circles is also a connected set. An example of a region is the set consisting of an open disk and a part of its boundary.

A connected set S is called *simply connected* if for every closed curve in S, all points enclosed by the curve are also elements of S. A *multiply connected set* is a connected set which is not simply connected. An annular region, $\{z|R_1 < |z - z_0| < R_2\}$, lying between two concentric circles is an example of a multiply connected set. A set obtained by removing a point from a simply connected set is another example of a multiply connected set.

§2.5 Limit, Continuity, and Differentiation

We assume that the reader is familiar with the limit, continuity, and differentiation of functions of one and two real variables. A function of a complex variable $z = x + iy$ can also be regarded as a function of two real variables x and y. With this in mind, it may be remembered that results on the limit and continuity of functions of two real variables coincide with those for functions of a complex variable. We begin by writing the basic definitions and a few results for functions of a complex variable.

Definitions

Limit of a sequence. A sequence $\{z_n\}$ has a limit α if for every $\epsilon > 0$ there exists an integer N such that

$$|z_n - \alpha| < \epsilon, \quad \text{for all } n > N. \tag{2.29}$$

Limit of a function. For a function $f(z)$ of a complex variable, we say that the limit of the function $f(z)$ as $z \to w$ exists and has value w_0, if for arbitrarily small positive ϵ there exists a $\delta > 0$ such that

$$|f(z) - w_0| < \epsilon \tag{2.30}$$

holds for all z satisfying $|z - z_0| < \delta$.

Continuity of a function. A function $f(z)$ is continuous at a point if $\lim_{z \to z_0} f(z)$ exists and is equal to the value of the function at z_0. It then follows that a function is not continuous at a point z_0 if

(i) $\lim\limits_{z\to 0} f(z)$ does not exist, or
(ii) $\lim\limits_{z\to 0} f(z)$ exists but is not equal to the value of the function at the point z_0.

Properties of continuous functions

(*1) An important property of the limit is that if the limit of a sequence or of a function exists, the value of the limit is unique and does not depend on the way the limit is computed. Different ways of taking limits must give the same answer. If one gets different answers by taking the limit in different ways, the limit does not exist. This is a useful way of proving that a given function is not continuous at a specified point.
(*2) If two functions are continuous at a point, their linear combination $\alpha f + \beta g$ and the product fg are also continuous at that point.
(*3) The ratio of two functions f/g is continuous at any point, where f and g are separately continuous and where g does not vanish.
(*4) The function $f(g(z))$ is continuous at a point $z = \alpha$, and if g is continuous at α and f is continuous at $\beta = g(\alpha)$.
(*5) A function defined by the sum of a convergent power series, $\sum_0^\infty a_n (z - z_0)^n$, is continuous at all those points where the series converges.

Differentiation

Let $f(z)$ be a function of a complex variable z. If the limit

$$\lim_{\Delta z \to 0} \frac{f(z + \Delta z) - f(z)}{\Delta z} \tag{2.31}$$

exists, the value of the limit is the derivative of the function at z and is denoted by $\frac{df}{dz}$.

Properties of the derivative

(*1) If two functions $f(z)$ and $g(z)$ are differentiable at a point, the linear combination $\alpha f(z) + \beta g(z)$ is differentiable at that point.
(*2) If two functions $f(z)$ and $g(z)$ are differentiable at a point, the product $f(z)g(z)$ is differentiable at the same point.
(*3) If two functions $f(z)$ and $g(z)$ are differentiable at a point and if the function $g(z)$ does not vanish at that point, the function $f(z)/g(z)$ is differentiable.

(*4) If $g(z)$ is differentiable at the point z and if $f(w)$ is differentiable at $w = g(z)$, the composite function $f(g(z))$ is differentiable at z.

⊳ Exercise §§2.8, on continuity and differentiation.

Demanding the existence of the derivative of a function $f(z)$ w.r.t. z in an open set leads to many results that have no anolog in the theory of functions of one or two real variables and gives rise to many far-reaching consequences, which will unfold in different parts of this book. For now, we go over to the Cauchy–Riemann equations, a set of necessary conditions for the existence of the derivative of a function w.r.t. a complex variable.

§2.6 Cauchy–Riemann Equations

In the above section we have seen examples of a large class of differentiable functions of a complex variable. The Cauchy–Riemann equations give a simple necessary condition for differentiability and are obtained by taking the limit $\Delta z \to 0$ of the expression

$$\lim_{\Delta z \to 0} \frac{f(z + \Delta z) - f(z)}{\Delta z} \tag{2.32}$$

in two different ways and demanding that the two answers be the same. We will write $\Delta z = h + ik$ and consider the following two possible ways of taking the limit $\Delta z \to 0$.

(a) $h \to 0$ keeping $k = 0$,
(b) $k \to 0$ keeping $h = 0$,

and the two results must be equal if the limit in Eq. (2.32) exists. This will give us a necessary condition for the existence of the derivative of $f(z)$ w.r.t. z.

Case (a) : $h \to 0$, *keeping* $k = 0$ *fixed*

We write

$$f(z) = u(x, y) + iv(x, y) \tag{2.33}$$

and consider

$$\lim_{\Delta z \to 0} \frac{f(z + \Delta z) - f(z)}{\Delta z} = \lim_{h,k \to 0} \frac{f(x + h, y + k) - f(x, y)}{h + ik}$$

$$= \lim_{h \to 0} \frac{f(x + h, y) - f(x, y)}{h} \quad (\because k = 0) \tag{2.34}$$

$$= \frac{\partial f}{\partial x} = \frac{\partial u}{\partial x} + i\frac{\partial v}{\partial x}. \tag{2.35}$$

Case (b) : $k \to 0$, keeping $h = 0$ fixed

$$\lim_{\Delta z \to 0} \frac{f(z+\Delta z) - f(z)}{\Delta z} = \lim_{h,k\to 0} \frac{f(x+h, y+k) - f(x,y)}{h+ik}$$

$$= \lim_{k\to 0} \frac{f(x, y+k) - f(x,y)}{ik} \quad (\because h = 0)$$

$$= \frac{1}{i}\frac{\partial f}{\partial y} = \frac{\partial v}{\partial y} - i\frac{\partial u}{\partial y}. \tag{2.36}$$

Equating the two answers in the expressions (2.35) and (2.36), we get the required Cauchy–Riemann equations

$$\frac{\partial u(x,y)}{\partial x} = \frac{\partial v(x,y)}{\partial y}, \quad \frac{\partial u(x,y)}{\partial y} = -\frac{\partial v(x,y)}{\partial x}. \tag{2.37}$$

There are infinitely many ways of taking the limit $\Delta z \to 0$ and these equations cannot guarantee that all the methods will give the same answer. These are *only necessary conditions* for the existence of the derivative. In fact, it is easy to give examples of functions satisfying the Cauchy–Riemann equations at a point but for which the derivative does not exist.

§2.6.1 *Cauchy–Riemann equations in polar form*

Regarding a function $f(z)$ as a function of two variables $r = |z|$ and $\theta = \arg(z)$, one can easily prove that the following Cauchy–Riemann equations in polar form are necessary in order that the function $f(z)$ may be differentiable w.r.t. z:

$$\frac{\partial u(r,\theta)}{\partial r} = \frac{1}{r}\frac{\partial v(r,\theta)}{\partial \theta}, \quad \frac{1}{r}\frac{\partial u(r,\theta)}{\partial \theta} = -\frac{\partial v(r,\theta)}{\partial \theta} \quad (r \neq 0). \tag{2.38}$$

§2.6.2 *Sufficiency conditions*

As has already been mentioned, many examples can be given to show that the Cauchy–Riemann equations are not sufficient conditions for the existence of the derivative w.r.t. z. The following results give sufficient conditions. A function $f(z)$ is differentiable at a point $z = x + iy$ if

(a) The Cauchy–Riemann equations (2.37) hold;
(b) The four partial derivatives appearing in Eqs. (2.37) are continuous at the point (x, y).

Using the above sufficiency conditions it is, for example, very easy to show that $\exp(z)$ is differentiable everywhere.

In the polar form, the sufficiency requirement can be stated as follows. A function is differentiable at a point $z_0 = r_0(\cos\theta_0 + i\sin\theta_0)$, with $r_0 \neq 0$, if the following conditions hold in a neighborhood of r_0, θ_0.

(a) The real and imaginary parts of $f(z)$, *viz.* $u(r,\theta)$ and $v(r,\theta)$, are single-valued;
(b) Cauchy–Riemann equations in the polar form, Eqs. (2.38) are obeyed;
(c) The four partial derivatives appearing in Eqs. (2.38) are continuous functions of (r,θ).

The polar form of sufficiency conditions, very useful for many *multivalued functions*, require single-valuedness, a concept which will be explained in the next chapter. For a proof of the polar form of sufficiency conditions, the reader is referred to textbooks.

§2.7 Examples

The examples in this section concern continuity, differentiability of functions of complex variables, and the use of Cauchy–Riemann equations. Note that Cauchy–Riemann equations, being necessary but not sufficient, can be used effectively only to conclude that a function is not differentiable. In order to use Cauchy–Riemann equations to prove differentiability, the continuity of the four partial derivatives must also be verified.

⁄2.7.1 Short examples. We give examples of classes of functions which are easily seen to be differentiable.

(a) A function $f(z) = c$, which is constant everywhere, has a derivative equal to zero for all z.
(b) The function $f(z) = z$ has a derivative equal to 1 for all values of z. Using the properties (*1) and (*2) on page 36, we see that all polynomials are differentiable at all values of z.
(c) If $P(z)$ and $Q(z)$ are polynomials, the ratio $P(z)/Q(z)$ is differentiable at all those points where $Q(z)$ does not vanish.
(d) The real and imaginary parts of the functions $\exp(\alpha z)$, $\sin z$, $\cos z$, $\sinh z$, and $\cosh z$ satisfy Cauchy–Riemann equations and have continuous partial derivatives for all values of x and y. Hence, these functions are differentiable for all values of z.
(e) Other trigonometric and hyperbolic functions, such as $\tan z$ and $\coth z$, can be written as ratios of $\sin z$, $\sinh z$, $\cos z$, $\cosh z$, etc., and are differentiable at points where the denominators do not become zero.

⁄2.7.2 Short examples. We shall give several examples of functions which are discontinuous at $z = 0$. The discontinuous nature will be demonstrated by showing that different answers are obtained when the limit $z \to 0$ is taken in different ways.

(a) Consider the function

$$f(x,y) = \frac{x^2}{x^2+y^2}, \quad z \neq 0, \tag{2.39}$$

with some value a assigned to $f(0,0)$. The function has the value 0 on the y axis,

$$\forall y \neq 0, \quad f(x,y)|_{x=0} = 0 \quad \Rightarrow \lim_{y\to 0} f(0,y) = 0$$

when the limit is taken along the y axis. On the other hand, along the x axis we get

$$\forall x \neq 0, \quad f(x,y)|_{y=0} = 1 \quad \Rightarrow \lim_{x\to 0} f(x,0) = 1$$

and the limiting value of the function along the x axis is 1. Thus, the limit has two different answers when computed in two different ways, and the function is discontinuous for all possible choices of the value a at the origin.

(b) Next, consider the function

$$f(x,y) = \begin{cases} \frac{xy}{x^2+y^2}, & z \neq 0, \\ 0, & z = 0. \end{cases} \tag{2.40}$$

The function vanishes for all points on the two axes, and hence the limits along the two axes are equal. However, along any other straight line $y = mx$, for $z \neq 0$, the function has the value

$$f(x,y)\Big|_{y=mx} = \frac{m}{1+m^2}$$

and the corresponding value of $\lim_{z\to 0} f(x,y)$ is obviously given by $\frac{m}{1+m^2}$. As different answers will be obtained for different choices of m, the function is discontinuous.

(c) Along all the straight lines $y = mx$, the function

$$f(x,y) = \begin{cases} \frac{x^2y}{x^4+y^2}, & z \neq 0, \\ 0, & z = 0. \end{cases} \tag{2.41}$$

is seen to have the same limiting value 0, but along the parabola $y = ax^2$ the answer for the limit

$$\lim_{z\to 0} \frac{x^2y}{x^4+y^2}\Big|_{y=ax^2} = \frac{a}{1+a^2}$$

depends on a, showing that this function too is not continuous.

Frequently, the nonexistence of the derivative of a function $f(z)$ at a point z_0 can be proven by showing that the expression (2.32) has different limiting values when z approaches z_0 along different curves. If $f(x,y)$ is any one of the three functions considered in ⚡2.7.2(a)–⚡2.7.2(c), taking $g(z) = (x+iy)f(x,y)$ we see that

$$\lim_{z\to 0} \frac{g(z) - g(0)}{z - 0} = \lim_{z\to 0} f(x,y) \tag{2.42}$$

has different values when computed differently, implying that the function $g(z)$ is not differentiable.

▼

☑2.7.3 Problem.

(a) Prove that the function defined by

$$f(z) = \begin{cases} \frac{x^3-y^3+i(x^3+y^3)}{x^2+y^2}, & z \neq 0, \\ 0, & z = 0, \end{cases} \tag{2.43}$$

is continuous.

(b) Show that $f(z)$ satisfies the Cauchy–Riemann equations at $z = 0$ but the function is not differentiable.

(c) Identify which of the four partial derivatives of $f(z)$ are not continuous at $z = 0$.

Solution.

(a) To prove the continuity of the function at $z = 0$ in this case, it is convenient to use polar variables (r, θ). Consider

$$|f(z) - f(0)| = r|(\cos^3\theta - \sin^3\theta) + i(\cos^3\theta + \sin^3\theta)| = r\sqrt{2(\cos^6\theta + \sin^6\theta)} \le 2r^3.$$

Therefore, given arbitrarily small $\epsilon > 0$, if we choose $\delta = (\epsilon/2)^{1/3}$ we would get

$$\big|f(z) - f(0)\big| < \epsilon \quad \forall\, |z| < \delta.$$

Hence, the function is continuous.

(b) The real and imaginary parts of the function are given by

$$u(x, y) = \frac{x^3 - y^3}{x^2 + y^2}, \quad v(x, y) = \frac{x^3 + y^3}{x^2 + y^2}.$$

Computing the partial derivatives at $(0, 0)$, we get

$$\lim_{z\to 0}\frac{\partial u}{\partial x} = \lim_{x\to 0}\frac{u(x,0) - u(0,0)}{x} = \lim_{z\to 0}\frac{x^3}{x^3} = 1, \tag{2.44}$$

$$\lim_{z\to 0}\frac{\partial u}{\partial y} = \lim_{y\to 0}\frac{u(0,y) - u(0,0)}{y} = \lim_{z\to 0}\frac{-y^3}{y^3} = -1, \tag{2.45}$$

$$\lim_{z\to 0}\frac{\partial v}{\partial x} = \lim_{x\to 0}\frac{v(x,0) - v(0,0)}{x} = \lim_{z\to 0}\frac{x^3}{x^3} = 1, \tag{2.46}$$

$$\lim_{z\to 0}\frac{\partial v}{\partial y} = \lim_{y\to 0}\frac{v(0,y) - v(0,0)}{y} = \lim_{z\to 0}\frac{y^3}{y^3} = 1. \tag{2.47}$$

Therefore, the Cauchy–Riemann equations are satisfied at $z = 0$. To show that the function is not differentiable, consider the limit

$$\lim_{z\to 0}\frac{f(z) - f(0)}{z - 0} = \frac{1}{x + iy}\left(\frac{x^3 - y^3}{x^2 + y^2} + i\frac{x^3 + y^3}{x^2 + y^2}\right), \tag{2.48}$$

along the straight line $y = mx$. We get

$$\lim_{z\to 0}\frac{f(z) - f(0)}{z - 0} = \lim\frac{1}{1 + im}\left(\frac{1 - m^3}{1 + m^2} + i\frac{1 + m^3}{1 + m^2}\right), \tag{2.49}$$

an answer depending on the value of m, which will be different for different straight lines. Hence the above limit does not exist and the function is not differentiable at $z = 0$.

(c) In order to check the existence of partial derivatives and of the derivative w.r.t. z at $z = 0$, in part (b), we had to start from the first principles because the usual rules of differentiation, such as $(f/g)' = (f'g - g'f)/g^2$, are *not applicable*, when any one of the terms f' or g' does not exist or g vanishes. These partial differentiation rules *are applicable* in the case where one has $x \neq 0$ and $y \neq 0$ and a straightforward computation gives

$$\frac{\partial u(x,y)}{\partial x} = \frac{x^4 + 3x^2y^2 + 2xy^3}{(x^2+y^2)^2}, \quad \frac{\partial u(x,y)}{\partial y} = \frac{-(3x^2y^2 + 2x^3y + y^4)}{(x^2+y^2)^2}, \tag{2.50}$$

$$\frac{\partial v(x,y)}{\partial x} = \frac{x^4 + 3x^2y^2 - 2xy^3}{(x^2+y^2)^2}, \quad \frac{\partial v(x,y)}{\partial y} = \frac{3x^2y^2 - 2x^3y + y^4}{(x^2+y^2)^2}. \tag{2.51}$$

It is now easily seen that the limit $z \to 0$ of all the four partial derivatives along the straight lines $y = mx$ depends on the value of m, implying that their $z \to 0$ limit does not exist and that they are all discontinuous at $z = 0$. ∎

⊳ Exercise §§2.8, on differentiability.
⊳ Question set §§2.9, on Cauchy–Riemann equations.
⊳ Quiz §§2.10, on differentiability.

§2.8 Analytic Functions

The existence of the differential coefficient at a point by itself does not lead to interesting properties for functions of a complex variable. An analytic function, to be defined below, has several properties which have no analog for functions of one and two real variables. Some remarkable properties are:

- Existence of derivatives of all higher orders;
- Existence of a power series expansion;
- Conformal property when the derivative is nonzero.

We come to the definition of analytic functions.

A function is *analytic in an open set* if it is differentiable at every point of the open set. If a function of a complex variable is differentiable at a point z_0 and at all points inside a circle of nonzero radius and having the center at the point z_0, we say that the function is *analytic at point* z_0. A function being analytic at a point may also be defined as a function which is differentiable at all points of an open set containing the point. A function is called an *entire function* if it is analytic at all points in the complex plane. Clearly, a function is an *entire function* if and only if it has a derivative at all points in the complex plane.

A point z_0 is a *singular point* of a function f if the function is not analytic at that point. A singular point z_0 is called an *isolated singular point* of a function $f(z)$ if there exists a positive number ρ such that the function f is analytic at all points inside the circle $|z - z_0| = \rho$ except at z_0 itself.

§2.8.1 *Properties of analytic functions*

If a function is defined by an algebraic expression involving elementary functions, the following properties are useful for deciding if a function is analytic or singular at a point:

(∗1) The sum $f + g$ of two functions f and g is analytic at all those points where f and g are separately analytic.

(∗2) The product $f \cdot g$ of two functions f and g is analytic at all those points where both f and g are separately analytic.

(∗3) The quotient f/g of two functions f and g is analytic at all those points where f is analytic, g is analytic, and g is not zero. Conversely, f/g is not analytic at all those points where any one of the three properties is not satisfied.

(∗4) If $g(z)$ is analytic at z_0 and $f(z)$ is analytic at $w = g(z_0)$, then the function $F(z) = f(g(z))$ is analytic at z_0.

(∗5) For a composite function $f(g(z))$ the singular points are the values of z such that (i) $f(z)$ is not analytic or (ii) $g(w)$ is not analytic at w, where $w = f(z)$.

(∗6) Let us make contact with the Cauchy–Riemann equations. We know that these equations by themselves do not imply the existence of the derivative w.r.t. z. A sufficient condition for the existence of derivatives is that the Cauchy–Riemann equations be satisfied and all the four partial derivatives be continuous in a neighborhood of the point z_0.

2.8.1 Short examples. Here a set of simple conditions on a function is given which can be used to conclude, by inspection, that a function cannot be analytic. A nonconstant function $f(x, y)$, having any one of the following properties, is not analytic as the Cauchy–Riemann equations cannot be satisfied in an open set.

(a) $f(x, y)$ is purely real;
(b) $f(x, y)$ is pure imaginary;
(c) $f(x, y)$ is independent of $x = \text{Re}\,(z)$;
(d) $f(x, y)$ is independent of $y = \text{Im}\,(z)$;
(e) $f(x, y)$ is independent of r; $r = |z|$;
(f) $f(x, y)$ is independent of $\theta = \arg(z)$;
(g) $f(x, y)$ depends on $\bar{z}$ after the substitution $x = (z + \bar{z})/2,\ y = (z - \bar{z})/2i$ is made.

The last statement is easily understood by noting that the function $\bar{z}$ is not differentiable anywhere, as it does not satisfy Cauchy–Riemann equations at any point.

2.8.2 Short examples. The properties of derivatives on page 39 and sufficiency conditions in §2.6.2 can be used to prove the correctness of the following statements:

(a) A constant function and polynomials are entire functions.
(b) If $P(z), Q(z)$ are polynomials with no common factors, their ratio $P(z)/Q(z)$ is analytic at all those points where $Q(z)$ does not vanish. The singular points of $f(z)$ coincide with roots of the polynomial $Q(z)$.
(c) The functions $\exp(\alpha z)$, $\sin z$, $\cos z$, $\sinh z$, $\cosh z$, etc. can be proven to be differentiable everywhere using the sufficiency conditions in §2.6.2. These functions, therefore, have derivatives for all values of z and are entire functions.
(d) Trigonometric and hyperbolic functions such as $\tan z$ and $\coth z$ can be written as ratios of $\sin z$, $\sinh z$, $\cos z$, $\cosh z$, etc., and are differentiable at points where the denominators do not become zero. Thus, the only singular points of $\cot z$, for example, are $z = n\pi$, where $n \in \mathbb{Z}$.

(e) Composing two entire functions, we obtain more examples of entire functions. Thus, $\exp(z^2)$, $\sin[\exp(\lambda z)]$, $\exp(-\cos^2 z)$, etc. are all entire functions.
(f) The function $\sin(1/z)$ has a singular point at $z = 0$. The singular points of $\sec(1/\sin z)$ are given by $z = n\pi$ and $\sin z = 2/m\pi$, where m is an odd integer.

⧐ Tutorial §§2.11, for checking the analyticity of several functions quickly without a detailed analysis.
⧐ §§2.9, for short questions related to necessary and sufficient conditions for differentiability and analyticity.

§2.8.2 *Power series as an analytic function*

The concept of the analytic function is intimately connected with power series. Therefore, we first introduce the power series and summarize important properties of the power series, and the connection with analytic functions is explained in the end of this section.

A complex power series about a point z_0 is a series

$$\sum_n a_n(z - z_0)^n \tag{2.52}$$

in positive powers of $z - z_0$.

Properties of power series

(◇1) A power series converges inside a circle $|z - z_0| = R$ and diverges outside it. This circle is called the *circle of convergence* of the power series and the radius R is called the *radius of convergence.*
(◇2) Inside the circle of convergence, the power series can be differentiated term by term. This means that if $f(z)$ denotes the sum of the series, then $\frac{df(z)}{dz}$ exists and is given by

$$\frac{df(z)}{dz} = \sum_n n a_n(z - z_0)^{n-1}. \tag{2.53}$$

This series has the same circle of convergence, as that in Eq. (2.52).
(◇3) The radius of convergence R of the power series (2.53) is given by

$$\frac{1}{R} = \overline{\lim}_{n\to\infty}|a_n|^{1/n}, \tag{2.54}$$

where $\overline{\lim}$ is the limit superior of the sequence a_n as $n \to \infty$. For our present purpose, it is sufficient to note that $\overline{\lim}\, a_n$ coincides with $\lim_{n\to\infty} a_n$, whenever latter exists. The radius of convergence can also be computed by using

$$\frac{1}{R} = \lim \frac{a_{n+1}}{a_n}, \tag{2.55}$$

provided that the limit on the right hand side exists.

($\diamond$4) If $R = \infty$, the power series converges for all z and the series represents an entire function analytic everywhere. If R becomes zero, the series has a zero radius of convergence and the series converges only for $z = z_0$.

($\diamond$5) If two power series are given by

$$f(z) = \sum_n a_n(z - z_0)^n, \quad |z - z_0| < R_1, \tag{2.56}$$

$$g(z) = \sum_n b_n(z - z_0)^n, \quad |z - z_0| < R_2. \tag{2.57}$$

then the function $f(z)g(z)$ is given by power series obtained by multiplying the series for $f(z)$ and $g(z)$ and collecting the terms having the same power of z. The circle of convergence of the resulting series is the intersection of the circles of convergence of the the power series in Eqs. (2.56) and (2.57).

($\diamond$6) Power series for the composite function $f(g(z))$ can also be obtained by combining the individual series and rearranging the resulting series.

More details on properties of a power series can be found in several books, such as [Hardy (1971)], [Ahlfors (1996)], and [Titchmarsh (1939)]. As a consequence of the property ($\diamond$2) of power series, we have the following result:

Theorem 2.1 *The function represented by the sum of a complex power series is a differentiable function inside the circle of convergence. Therefore, a power series represents an analytic function inside its circle of convergence.*

The converse result, that every analytic function can be expanded in a power series, is an extremely important result in the theory of functions of a complex variable and will be proven in §5.3.

§2.9 Harmonic functions

We now come to the definition and properties of harmonic functions, which are intimately related to analytic functions.

A function $u(x, y)$ of two real variables is called a harmonic function over a domain D if the following two conditions are satisfied:

(i) The second order partial derivatives

$$\frac{\partial^2 u}{\partial x^2}, \quad \frac{\partial^2 u}{\partial y^2}, \quad \frac{\partial^2 u}{\partial x \partial y}, \quad \frac{\partial^2 u}{\partial y \partial x} \tag{2.58}$$

exist and are continuous in the domain D;

(ii) The function satisfies the Laplace equation

$$\frac{\partial^2 u}{\partial x^2} + \frac{\partial^2 u}{\partial y^2} = 0 \tag{2.59}$$

at every point of the domain D.

Let $u(x, y)$ and $v(x, y)$ be two harmonic functions. These two functions are said to be *conjugate harmonic* if they satisfy the Cauchy–Riemann equations

$$\frac{\partial u}{\partial x} = \frac{\partial v}{\partial y}, \quad \frac{\partial u}{\partial y} = -\frac{\partial v}{\partial x}. \tag{2.60}$$

Theorem 2.2 *A given function $f(z)$ is analytic in a domain f and only if the real part, $u(x, y)$, and the imaginary part, $v(x, y)$, of the function are conjugate harmonic functions.*

☑**2.9.1 Problem.** Find a function which is harmonic conjugate to

$$u(x, y) = \frac{y}{x^2 + y^2},$$

and hence find a function $f(z)$ having $u(x, y)$ as its real part.

Solution. Let $v(x, y)$ denote the harmonic conjugate of $u(x, y)$. Using the Cauchy–Riemann equations, we have the partial derivatives of $v(x, y)$ given by

$$\frac{\partial v}{\partial x} = -\frac{\partial u}{\partial y} = -\frac{(x^2 + y^2) - 2y^2}{(x^2 + y^2)^2} = \frac{y^2 - x^2}{(x^2 + y^2)^2}, \tag{2.61}$$

$$\frac{\partial v}{\partial y} = \frac{\partial u}{\partial x} = \frac{-2xy}{(x^2 + y^2)^2}. \tag{2.62}$$

Integrating Eq. (2.65) w.r.t. y, we get

$$v(x, y) = \frac{x}{x^2 + y^2} + h(x), \tag{2.63}$$

where $h(x)$ is a function of x alone and is to be determined. Making use of Eq. (2.61), we get

$$\frac{\partial h(x)}{\partial x} = 0 \quad \Rightarrow \quad h(x) = C,$$

where C is a constant. This gives

$$f(z) = u(x, y) + iv(x, y) = \frac{y + ix}{x^2 + y^2} + C = \frac{i}{z} + C.$$

▮

⊳ Exercise §§2.12, on harmonic functions.

§2.10 Graphical Representation of Functions

Functions of a complex variable can be represented geometrically in several possible ways. A function of two real variables can be represented by a surface in three

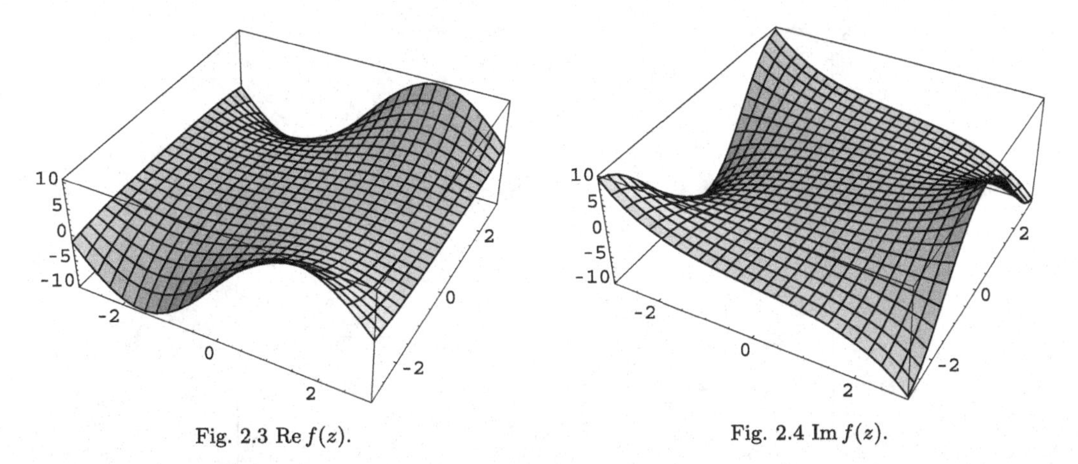

Fig. 2.3 Re $f(z)$.

Fig. 2.4 Im $f(z)$.

dimensions. The height of the surface at any point (x, y) represents the value of the function plotted. Hence, for a complex-valued function $f(z)$, its real and imaginary parts, $u(x, y)$ and $v(x, y)$, given by

$$f(z) = u(x, y) + iv(x, y), \tag{2.64}$$

can be shown as surfaces in three dimensions. As an example, we represent the real and imaginary parts

$$u(x, y) = \sin x \cosh y, \quad v(x, y) = \sinh x \cos y \tag{2.65}$$

of $\sin z$ as surfaces in three dimensions, as shown in Figs. 2.3 and 2.4.

A representation by means of level curves is one of several possible two-dimensional representations of a real-valued function. The level curves of a real function $g(x, y)$ are the curves $g(x, y) =$ const C, for different values of C. Thus, to depict a complex-valued function $f(z)$ by means of level curves, one needs two separate plots. One of these, for example, can be used for the real part, $u(x, y)$, and the other for the imaginary part, $v(x, y)$. So the level curves of the real and imaginary parts of $z^2 = (x^2 - y^2) - 2ixy$ are the hyperbolae

$$x^2 - y^2 = C_1, \quad xy = C_2,$$

which are shown in Fig. 2.5 for a range of values of the constants C_1 and C_2.

Instead of drawing the level curves of the real and imaginary parts of $f(z)$, one may draw the level curves of its argument and modulus, $|f(z)|$ and $\arg f(z)$. So, if we take $f(z) = \frac{z-z_1}{z+z_2}$, the level curves of mod $(f(z))$ and arg $(f(z))$ will be circles of the forms shown in Figs. 1.12 and 1.13.

The Polya representation gives another geometric representation of functions of a complex variable. In this representation, the real and imaginary parts, $u(x, y)$ and $v(x, y)$, are regarded as a vector field in two-dimensions. The vector field $u(x, y), v(x, y)$ is represented by drawing a vector from the point (x, y) with the length proportional to the magnitude $\sqrt{u(x, y)^2 + v(x, y)^2}$. This representation is

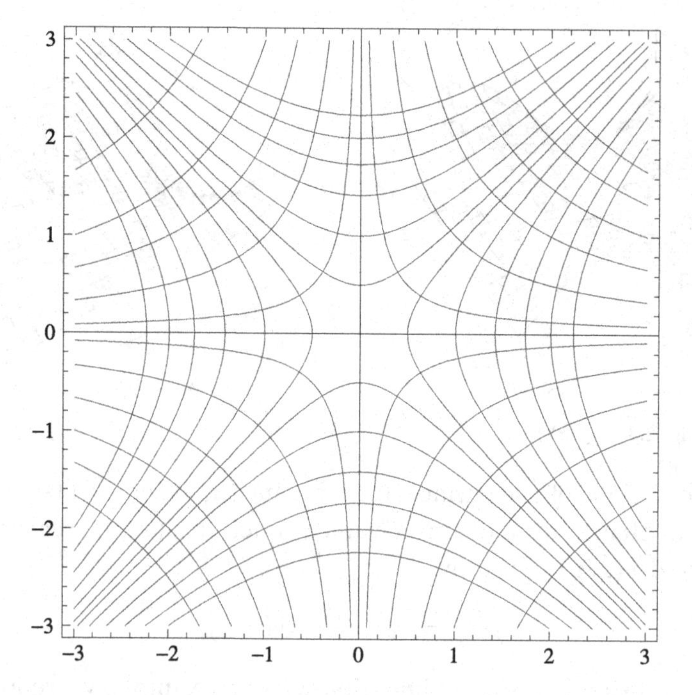

Fig. 2.5 Level curves of Re z^2 and Im z^2.

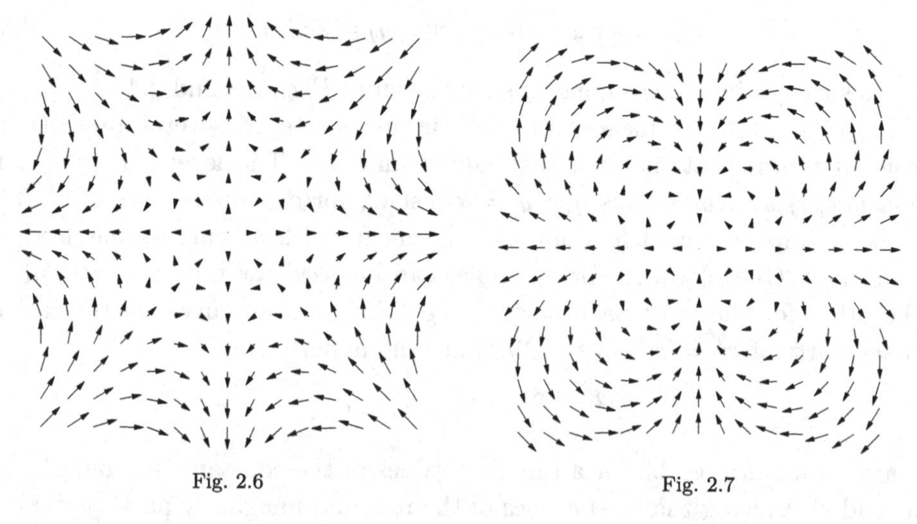

Fig. 2.6

Fig. 2.7

useful for visualization of the flow of fluids and of the electric field in two-dimensional problems. Figure 2.6 shows the Polya field representation of z^3, and the Polya field of $\bar{z}^3$ is shown in Fig. 2.7.

In another frequently used representation of a complex-valued function $f(z)$, two copies of the complex plane are taken and the mapping properties of the function are brought out by showing the image sets of chosen subsets of the complex plane. To do this, for example for the function $f(z) = 1/z$, we may select the four quadrants as the subsets in the z plane — see Fig. 2.8 — and show their images in another

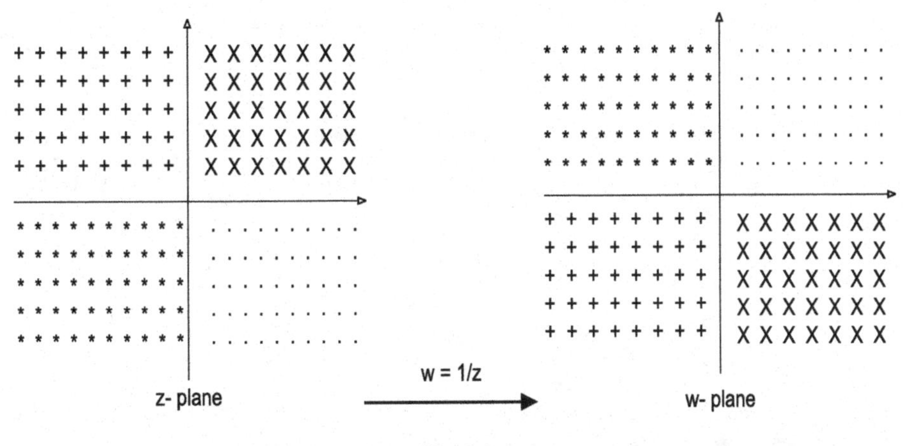

Fig. 2.8

plane, labeled the w plane in the figure. This representation of functions as mappings of the complex plane turns out to be extremely useful for discussion of conformal mappings. Representation of functions will be discussed again, in great detail, in Ch. 9 (on conformal mappings) and will be an essential tool for physical applications in Ch. 10.

§2.11 Notes and References

A wide variety of methods for generating curves and surfaces to represent a given function are available. Here we give a few URLs; while some of these write about examples and the methods, others have a gallery of graphic images.

(◇1) One can have three-dimensional plots of functions of a complex variable on the Riemann sphere; see
http://functions.wolfram.com/ElementaryFunctions/Sech/visualizations/9.html

(◇2) For an example of the representing behavior of a function with colors, see
http://www.andrew.cmu.edu/user/coa/icos-complex.html or
http://home. earthlink.net/~djmp/Cardano3Page.html

Chapter 3

FUNCTIONS WITH BRANCH POINT SINGULARITY

The aim of this chapter is to discuss functions with branch point singularity. The most commonly encountered examples of such functions are the fractional power z^c and the logarithm function $\log z$. Several common functions with branch points arise when we attempt to define inverse functions of many-to-one functions, and we are led to "multivalued functions". Care is needed in defining and handling such functions. In the first few sections of this chapter, important issues concerning the definition of multivalued functions are discussed, while the definition itself is dealt with in §3.4. In later sections, we discuss usage and discontinuity across the branch cut for the functions with branch points. We will mostly concentrate on those multivalued functions which are expressions involving inverses of the functions appearing in ⸗3.1.1. These examples will later appear again in various applications in this book.

As preparation for our discussion of multivalued functions in this section, we summarize important properties of the set of all solutions to simple equations involving elementary transcendental functions.

§3.1 Inverse Functions

In this section we shall start a discussion on functions with branch point singularities. This class of functions requires a careful definition, and we shall consider several examples, such as $\log z, \sqrt{z}, z^\lambda$. In this connection several new concepts, such as the branch point and the branch cut, are needed. Initially these new terms will be introduced and discussed only in the context of a few specific examples. The purpose of this approach is to clarify these concepts to the best possible extent.

The material in this section and the next two sections is preparatory in nature and will be crucial for an understanding of the considerations in §3.4, where the question of defining a few selected functions is dealt with. Since a precise definition of $\log z$ and other examples, as functions, is postponed to §3.4, due to lack of any better nomenclature, we shall call them "multivalued functions" and use the term "single-valued branch" (svb) after a careful definition has been provided.

Many of the multivalued functions arise as solutions to differential equations, and some others are functions defined through integral representations. It will be seen that some of the most common examples of multivalued functions arise in attempts to define "inverse functions" of many-to-one functions. We therefore begin with a few examples of many-to-one functions.

§3.1.1 *Many-to-one functions*

ƒ3.1.1 Short examples. The functions such as z^2, z^N, $\exp(z)$, and $\sin z$ are many-to-one functions from $\mathbb{C}$ to $\mathbb{C}$. To see this we note the following properties:

(a) Under the mapping $z \to z^2$, the complex numbers ξ and $-\xi$ are mapped into the same value, ξ^2.

(b) The complex numbers in the set $\{\xi + 2\pi i N | N \in \mathbb{Z}\}$ are mapped to a single number under the mapping $z \to \exp(z)$.

(c) Let $\Omega_N = \exp(2\pi i/N)$; then $\{\Omega_N^k | k = 0, 1, \ldots, N-1\}$ is the set of N roots of unity. For a complex number ξ, the value of the function $f(z) = z^N$ is the same for all the complex numbers of the set $\{\xi\Omega_N^k | k = 0, 1, N-1\}$.

(d) The trigonometric functions of z are all periodic functions with period 2π, e.g. $\sin(z+2\pi N) = \sin z$. Hence, all trigonometric functions map the points $\{z+2\pi N | N \in \mathbb{Z}\}$ into a single point.

§3.1.2 *Inverse functions*

Given a function $\psi(z)$, many times there is a need to define its inverse function — say, F — having the property that the composite function $\psi \circ F$ is the identity mapping. This means that $\psi(F(z)) = z$. The value, $w = F(z)$, of the function F, inverse to the function ψ, for a given complex number z is to be obtained by solving the equation

$$z = \psi(w). \tag{3.1}$$

For the inverse function F to be well-defined, it is necessary that the function $\psi(z)$ be one-to-one. If the function ψ is a many-to-one function, we will get many solutions for w, and the inverse function will be ill-defined. As has already been noted above, the functions z^2, $\exp(z)$, z^N, and the trigonometric functions are not one-to-one functions. Hence, the corresponding "inverse functions," such as $\sqrt{z}$, $\log z$ and $z^{1/N}$ and inverse trigonometric functions, are multiple-valued correspondences and thus ill-defined as functions. *Due to the lack of a better name, we use the term "multivalued functions," to refer to such examples, always keeping in mind that a function must be single-valued.*

In the first column of Table 3.1, we list some commonly used multivalued functions of a complex variable. With the exception of $\sin^{-1} z$, the other multivalued functions, listed in the first column of the table, will be discussed in this chapter in detail. The second column lists the equations $z = \psi(w)$, which need to be solved in order to define the multivalued functions in the first column. In the third column of

Table 3.1 Examples of multivalued functions arising as solutions to $\psi(w) = z$, where ψ is a many-to-one function. Here $\omega = \exp(2\pi i/3)$ and $\Omega_N = \exp(2\pi i/N)$.

Inverse Function $F(z)$	$z = \psi(w)$	Set of Solutions w	Number of Solutions
$\log z$	$z = \exp(w)$	$w = \ln r + i\theta + 2\pi i m$	Infinite; one for each integer m
$\sqrt{z}$	$z = w^2$	$w = \pm\sqrt{r}\exp(i\theta/2)$	Two
$z^{1/3}$	$z = w^3$	$w_1 = r^{1/3}\exp(i\theta/3)$ $w_2 = \omega r^{1/3}\exp(i\theta/3)$ $w_3 = \omega^2 r^{1/3}\exp(i\theta/3)$	Three
$z^{1/N}$	$z = w^N$	$w_m = \Omega_N^m r^{1/N}\exp(i\theta/N)$	N solutions; one for each $m = 0, 1, \ldots, N-1$
$\sin^{-1} z$	$z = \sin w$	$-i\log(\sqrt{1-z^2} + iz)$	Infinite

the table, we give the set of solutions to equations in the second column. It should be noted that, except for $\sin^{-1} z$, all the other solutions are written in terms of the polar variables r and θ for the complex number z. The solution for $\sin^{-1} z$ is formally expressed in terms of $\log z$ and square root functions. It is seen from the table that in some cases there are a finite number of solutions while in some other cases the number of solutions for w is infinite.

§3.1.3 *Branch of a multivalued function*

In order that the expressions involving the inverse functions can be handled consistently, they must be defined without any ambiguities, and it is important that for every complex number z a unique value, $F(z)$, be assigned to the function, inverse to a given function ψ. A first step toward achieving the uniqueness is to regard each solution to $z = \psi(w)$, for a fixed z, as giving a "different" definition of the inverse function.

When we pick up a solution to $z = \exp(w)$, each such value gives a possible definition of $\log z$ and we will write

$$(\log z)_m \stackrel{\text{def}}{\equiv} \ln r + i(\theta + 2\pi m), \quad m = 0, \pm 1, \pm 2, \ldots. \tag{3.2}$$

✍ *The notation* $\log z$ *will be used for the cases where the argument* z *of the logarithm function is allowed to be a negative real number or a complex number. As very many different definitions are possible for* $\log z$*, we use* $\ln p$ *for the Napierian logarithm of positive real number* p.

Each of the solutions in Eq. (3.2) for a given w is a possible "candidate" for the value of $\log z$. Therefore, we have infinitely many possible loglike functions, $(\log z)_m$, as introduced above. The functions $(\log z)_m$ of Eq. (3.2) are to be treated as separate functions for different values of m. These different functions will be called *branches* of the logarithm of a complex variable. Note that our usage of the term "branch" does not coincide with that in many textbooks.

Similarly, one can define two functions which are equally good candidates for the square root of a complex number. The two functions correspond to the two values listed in the third column of the second row of Table 3.1 and are given by

$$\left[\sqrt{z}\right]_1 \stackrel{\text{def}}{\equiv} +r^{1/2}\exp(i\theta/2), \tag{3.3}$$

$$\left[\sqrt{z}\right]_2 \stackrel{\text{def}}{\equiv} -r^{1/2}\exp(i\theta/2), \tag{3.4}$$

and we say that the square root function has two branches.

Similarly, for the cube root function one has three different branches given by

$$\left[z^{1/3}\right]_1 \stackrel{\text{def}}{\equiv} r^{1/3}\exp(i\theta/3), \tag{3.5}$$

$$\left[z^{1/3}\right]_2 \stackrel{\text{def}}{\equiv} \omega r^{1/3}\exp(i\theta/3), \tag{3.6}$$

$$\left[z^{1/3}\right]_3 \stackrel{\text{def}}{\equiv} \omega^2 r^{1/3}\exp(i\theta/3), \tag{3.7}$$

where $\omega = \exp(2\pi i/3)$ and there are N branches for the function $z^{1/N}$, given by the N solutions in the fourth row of Table 3.1. The function $\sin^{-1} z$ is completely defined once $\sqrt{z^2-1}$ and $\log z$ have been defined.

For reasons to be discussed, even this strategy of regarding each solution to $z = \psi(w)$ as giving rise to a different definition (branch) of the inverse function, F, does not, as yet, lead to an acceptable definition of F. Introducing several branches for a given multivalued function does not completely eliminate the problems of multivaluedness. The expressions for different branches are written in terms of polar variables $r = |z|$ and $\theta = \arg(z)$, and each branch of the multifunctions $\log z, \sqrt{z}$, and $z^{1/N}$ is well-defined a function of r and θ. Although if one knows the polar variables r and θ, and the complex number z is uniquely determined by $z = r(\cos\theta + i\sin\theta)$, specifying a value of z does not give a unique value of (r, θ), but an infinite set of values $(r, \theta + 2m\pi)$, for every integer m. This infinite set of pairs $\{(r, \theta + 2m\pi) | m \in \mathbb{Z}\}$, in general, does not give rise to a unique answer for the value of multifunctions. For example, the values obtained from Eqs. (3.3)–(3.7), etc. are not unique. Thus, the multivalued functions $\log z, \sqrt{z}, z^{1/N}$, etc. *are not single-valued functions* of z even if we focus our attention on a single branch. The source of the multivaluedness is the nonuniqueness of the value of $\arg(z)$ for a given z. For this reason we first discuss the function $\arg(z)$ in the next section.

§3.2 Nature of Branch Point Singularity

§3.2.1 *Multivalued function* arg(z)

A multivalued function has the property that, when the point representing a number z is moved around a closed path, the function does not always return to the initial

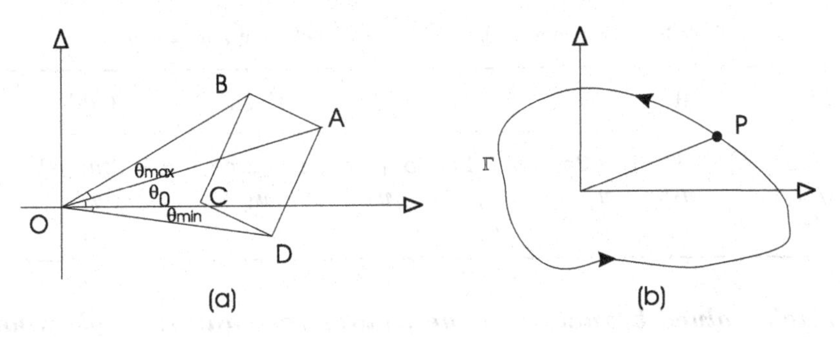

Fig. 3.1

value. A function which always returns to the initial value will be called *single-valued.* Two cases of a closed path are discussed below to illustrate the multivalued property of the function $\arg(z)$:

Case 1: The origin lies outside the closed loop.
Consider a path not enclosing the origin, such as the rectangle ABCD of Fig. 3.1(a). As the point z moves on the rectangle starting from the point A, it is seen that $\arg(z)$ increases from θ_0 to a maximum value, $\theta_{\max}$, at the point B. From B, as one moves toward C, the value of $\arg(z)$ decreases to the minimum value, $\theta_{\min}$, at D, and then it increases so as to become equal to θ_0, the starting value, when the point z reaches the starting point A. This behavior of $\arg(z)$ can easily be verified for every closed path which does not enclose the origin.

Case 2: The closed path encloses the origin.
Next, we consider a closed path Γ as shown in Fig. 3.1(b), which encloses the origin. Assume that, starting from P, a point moves along Γ in the anticlockwise direction and comes back to the starting point after making a full circuit. Let us find the value of $\arg(z)$ as the point moves along the path. The value of $\arg(z)$ increases continuously and changes by 2π when the point z completes one full circuit. In fact, for all such paths $\arg(z)$ increases by $2\pi N$, if N complete loops are made around the path in the anticlockwise direction and decreases by $2\pi N$ when N complete loops are made in the clockwise direction.

The point $z = 0$ is a very special point for the multivalued function $\arg(z)$; the value of $\arg(z)$ does not return to the initial value when a closed circuit is completed in the complex plane around any path enclosing origin. The behavior of $\arg(z)$ discussed here is a reflection of the fact that the values of $\arg(z)$ differing by $2n\pi$ correspond to the same point in complex plane, and that $\arg(z)$ is not a single-valued function of the complex variable z. The property of $\arg(z)$, of not being single-valued, is transmitted to the branches of the multifunctions such as $\log z$, $\sqrt{z}$, $z^{1/3}$, $z^{1/N}$ also. For every closed path encircling the point $z = 0$, the value computed for a branch does not return to the starting value for any of these functions when the point in the z complex plane makes a full loop along a closed path enclosing the origin. This is illustrated in detail in the next subsection.

Table 3.2 Behavior of the square root function around $z = 0$.

L	0	1	2	3	Even	Odd
$\theta = \arg(z)$	α	$\alpha + 2\pi$	$\alpha + 4\pi$	$\alpha + 6\pi$	$\alpha + 2m\pi$	$\alpha + (2m+1)\pi$
$[\sqrt{z}]_1$	u_1	u_2	u_1	u_2	u_1	u_2
$[\sqrt{z}]_2$	u_2	u_1	u_2	u_1	u_2	u_1

§3.2.2 *Single-valued branches of the square root and the cube root*

Some points, called *branch points*, are very special for multivalued functions. As an example to show this special behavior, we shall compare the initial values of different branches of $\sqrt{z}$ and $z^{1/3}$ with the values that one gets for any one, but fixed, branch after completion of a number of loops around $z = 0$.

Let u_1 and u_2 denote the values of the two branches, $[\sqrt{z}]_1$ and $[\sqrt{z}]_2$, of the square root multivalued function at a point $R(\cos\alpha + i\sin\alpha)$. In addition, let us consider the values assumed by the two branches of $\sqrt{z}$ when, starting from the point $r = R, \theta = \alpha$, L complete loops are made around a closed path, encircling the $z = 0$ point. After L full loops the argument of z increases from α to $\alpha + 2L\pi$, and the results for the values for the two branches, $[\sqrt{z}]_1$ and $[\sqrt{z}]_2$, to be computed from Eqs. (3.3) and (3.4), are listed in Table 3.2 for different values of L.

After one loop the finishing value of $\sqrt{z}$ for the first (second) branch coincides with the starting value for the second (first) branch. As more and more loops are completed, one alternately gets u_1 and u_2 for each branch of the square root function. Also note that after an even number of loops, the value for either branch comes back to the starting value, u_1 for the first and u_2 for the second branch. The function $z^{1/N}$ has N branches, defined as follows:

$$\left[z^{1/N}\right]_m \stackrel{\text{def}}{\equiv} r^{1/N} \exp\left(\frac{i\theta}{N}\right) \exp\left(\frac{2\pi i(m-1)}{N}\right), \quad m = 1, 2, \ldots, N. \tag{3.8}$$

If we focus our attention on the values of $z^{1/N}$ for the first branch after completing $1, 2, 3, \ldots, N-1$ loops, we note that these values successively coincide with the starting ($L = 0$) values for the $2^{\text{nd}}, 3^{\text{rd}}, 4^{\text{th}}, \ldots,$ branches, respectively. The value returns to the starting value after $L = N$ loops. A similar pattern is observed for the other branches. This pattern can be described with different words, as follows: The set of all values of $[z^{1/N}]_m$ at a point $|z| = R, \arg(z) = \alpha$, as m is varied, is the same as the set of values when m is fixed and $\arg(z)$ is allowed to take values $\alpha, \alpha \pm 2\pi, \alpha \pm 4\pi, \ldots.$ In Table 3.3 we show this pattern for the cube root multivalued function. In the table v_1, v_2, and v_3 stand for the three values obtained from Eqs. (3.5) and (3.7) for $\theta = \alpha$ and $r = R$. It is to be noted that these three values get repeated successively as the number of loops increases and that the value for every branch returns to the initial ($L = 0$) value after completion of $3m$ loops.

Table 3.3 Behavior of the cube root function around the branch point.

L	0	1	2	3	4	$3m$
θ	α	$\alpha+2\pi$	$\alpha+4\pi$	$\alpha+6\pi$	$\alpha+8\pi$	$\alpha+6m\pi$
$[z^{1/3}]_1$	v_1	v_2	v_3	v_1	v_2	v_3
$[z^{1/3}]_2$	v_2	v_3	v_1	v_2	v_3	v_1
$[z^{1/3}]_3$	v_3	v_1	v_2	v_3	v_1	v_2

In the case of the logarithm function, the value of $\log z$ for every branch $[\log z]_m$ increases by $2\pi i$ after every (anticlockwise) loop and successively combs through the starting values for the $m+1, m+2, \ldots$ branches. There are an infinite number of branches and the value for any fixed branch does not return to the starting value, no matter how many loops are completed around the origin.

It is clear from the above discussion that the point $z = 0$ is a very special point for the functions under consideration, such as $\log z, \sqrt{z}, z^{1/3}$, and $z^{m/n}$. We say that $z = 0$ is a *branch point* of these multivalued functions. In a small-enough neighborhood of a branch point, a multivalued function exhibits a characteristic behavior summarized as follows:

($\star$1) The multivalued functions do not become single-valued in the neighborhood of a branch point even if we consider only a branch;

($\star$2) On traversing closed loops around a branch point, every branch combs through the set of starting values for different branches;

($\star$3) If defined at a branch point, the values for all the branches of a multivalued function coincide at the branch point.

The multivalued functions $\log z$, $\sqrt{z}$, $z^{1/3}$, and $z^{1/m}$ have a branch point at infinity. This can be seen by changing the variable to $t = 1/z$ and examining the behavior of the functions near $t = 0$. For example, the function $z^{1/N}$ under a change of variable from z to $t = 1/z$ becomes $t^{-1/N}$, which shows that $t = 0$ is a branch point and hence that $z^{1/N}$ has a branch point at infinity.

⊳ Questions §§3.1, on this section.

§3.3 Ensuring Single-Valuedness

The multivalued character of $\log z$, $\sqrt{z}$, $z^{1/3}$, and $z^{1/N}$ originates from the multivaluedness of $\arg(z)$, and it makes a consistent discussion on the limit and derivatives for such functions impossible. There are two ways of handling this problem. Firstly, one may regard different values of the pairs (r, θ), for a fixed z, as points in distinct copies of complex plane which are suitably joined to obtain what is known as a *Riemann surface*. The functions such as $\log z$ become single-valued when interpreted as functions on a Riemann surface. The details of the number of copies of the complex plane required and how they are to be joined depend on the function under consideration. We will deal briefly with this topic in §3.10. The interested reader is referred to the textbooks, in particular the book by LePage, cited at the end.

The second approach, which will be needed and will be followed throughout this book, is to restrict values of $\theta = \arg(z)$ to a suitable range such as

$$\theta_0 < \theta < \theta_0 + 2\pi, \tag{3.9}$$

where θ_0 is fixed, but otherwise arbitrary. Each value of θ_0 represents a possible definition of the arg function. Note that both of the values θ_0 and $\theta_0 + 2\pi$ are usually excluded. If we agree to restrict the range of θ as above, the arg function is a well-defined continuous function in the complement of the ray given by $\theta = \theta_0$, also by $\theta = \theta_0 + 2\pi$. *Even if one were to define* $\arg z$ *on the ray itself, the function* $\arg(z)$ *would be discontinuous there.* To see this it is sufficient to note that if we take two points close to, but on the different sides of, the radial line, the values of the arg function differ by 2π as the two points approach the radial line.

In order to further explain some of the above remarks four simple possibilities, out of an infinite number of them, for the restrictions on the range of θ will be considered. These four choices correspond to $\theta_0 = 0, -\pi/2, -\pi$, and $-3\pi/2$ in Eq. (3.9). The corresponding restrictions on the range of θ are as follows:

(a) $0 < \theta < 2\pi$,
(b) $-\pi/2 < \theta < 3\pi/2$,
(c) $-\pi < \theta < \pi$,
(d) $-3\pi/2 < \theta < \pi/2$.

In each of the four cases we want to demonstrate that the function $\arg(z)$ is discontinuous along the radial line $\theta = \theta_0$. The line of discontinuity corresponding to the four ranges of θ is shown in Figs. 3.2(a)–3.2(d), respectively.

The eight points $A_1, A_2, \ldots, A_8$, shown in Fig. 3.3, are all at a distance r from the origin, and make infinitesimally small angles of magnitudes $\delta_1, \ldots, \delta_8$ with the axes, as shown in the figure. We now deal with case (a) and show that the difference in the values of the function $\arg(z)$ just above and just below the line $\theta = 0$, *viz.* the positive real axis, is nonzero. To verify this we consider the two points A_1 and A_2,

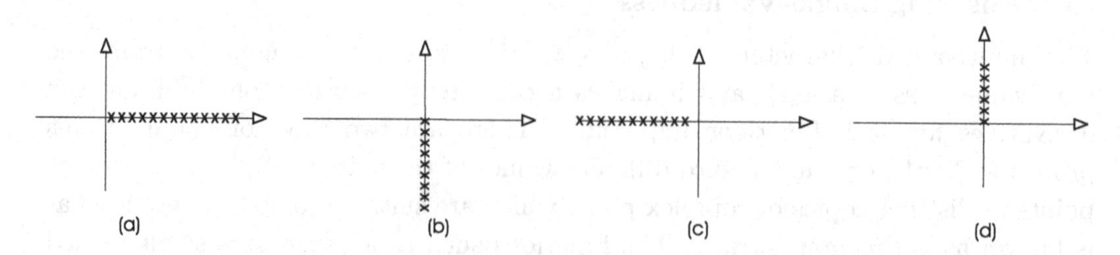

Fig. 3.2

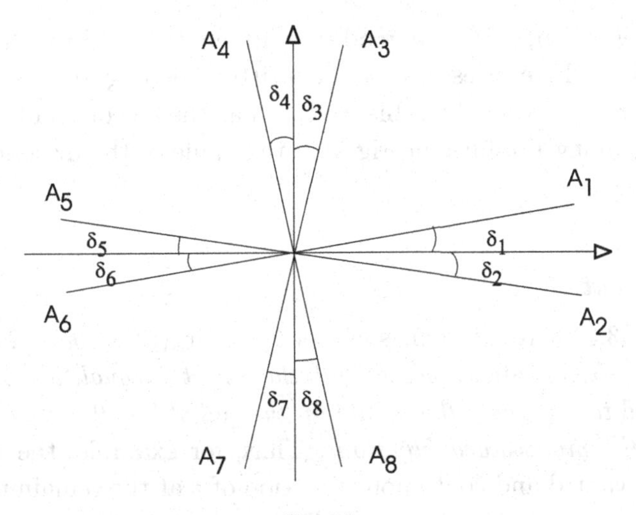

Fig. 3.3

close to the positive real axis, as shown in Fig. 3.3. For the point A_1, just above the positive real axis, $\theta_1 \equiv \arg(z)$ is equal to δ_1. *For the point* A_2, *the value* θ_2, *of* $\arg(z)$ *cannot be taken to be* $-\delta_2$, *but it must be taken to be* $2\pi - \delta_2$ *in accordance with the restriction* $0 < \theta < 2\pi$. We therefore see that the difference, $\theta_2 - \theta_1$, in the values of $\arg(z)$ at A_2 and A_1 becomes 2π when δ_1 and δ_2 go to zero. This shows that, in case (a), the multivalued function $\arg z$ is discontinuous along the line $\theta = 0$.

In Table 3.4, we list values of $\arg(z)$ for other points, $A_3, \ldots, A_8$, and check for the continuity of $\arg(z)$. For cases (b), (c), and (d) it is found to be continuous along the radial line $\theta = 0$ but discontinuous across $\theta = -\pi/2, -\pi, -3\pi/2$, respectively.

Table 3.4 Summary of results on discontinuity of $\arg(z)$ for four different cases on the four radial lines.

	$0 < \theta < 2\pi$	$-\pi/2 < \theta < 3\pi/2$	$-\pi < \theta < \pi$	$-3\pi/2 < \theta < \pi/2$
θ_1	δ_1	δ_1	δ_1	δ_1
θ_2	$2\pi - \delta_2$	$-\delta_2$	$-\delta_2$	$-\delta_2$
$\theta_2 - \theta_1$	2π	0	0	0
θ_3	$\pi/2 - \delta_3$	$\pi/2 - \delta_3$	$\pi/2 - \delta_3$	$\pi/2 - \delta_3$
θ_4	$\pi/2 + \delta_4$	$\pi/2 + \delta_4$	$\pi/2 + \delta_4$	$-(3\pi/2 - \delta_4)$
$\theta_3 - \theta_4$	0	0	0	2π
θ_5	$\pi - \delta_5$	$\pi - \delta_5$	$\pi - \delta_5$	$-(\pi + \delta_5)$
θ_6	$\pi + \delta_6$	$\pi + \delta_6$	$-(\pi - \delta_6)$	$-(\pi - \delta_6)$
$\theta_5 - \theta_6$	0	0	2π	0
θ_7	$3\pi/2 - \delta_7$	$3\pi/2 - \delta_7$	$-(\pi/2 + \delta_7)$	$-(\pi/2 + \delta_7)$
θ_8	$3\pi/2 + \delta_8$	$-(\pi/2 - \delta_8)$	$-(\pi/2 - \delta_8)$	$-(\pi/2 - \delta_8)$
$\theta_7 - \theta_8$	0	2π	0	0

For the other cases, (b)–(d), the reader is urged to go through the entries in Table 3.4 carefully. The entries have to be written keeping in mind the restriction on the range of $\arg(z)$. From the table we see that the discontinuity in the value of $\arg(z)$ is nonzero along the lines in Fig. 3.2; the value of the discontinuity is found to be 2π.

§3.3.1 *Branch cut*

As discussed in §3.3, a restriction, such as $\theta_0 < \arg(z) < \theta_0 + 2\pi$, is needed to make the $\arg(z)$ single-valued. *Note that the strict inequalities on the range of* $\arg(z)$ *correspond to removing the points on the radial line* $\theta = \theta_0$ *from the domain of definition of the multivalued functions.* Thus, for example, the function $\arg(z)$ becomes a single-valued and continuous function of z at the remaining points in the complex plane. The plane, from which a line such as $\theta = \theta_0$ has been removed is called the *cut plane.* This also makes each branch, of the multivalued functions $\log z$, $\sqrt{z}$, $z^{1/3}$, $z^{1/m}$, etc., a well-defined function in the cut plane. A function defined in this manner will be called a *single-valued branch* (svb). The set of points, such as the ray $\theta = \theta_0$, removed to make a multivalued function well-defined and single-valued, is called the *branch cut.* As remarked in the introduction, the location of the branch cut will be discussed for each function separately. Note that very many different choices possible for the branch cut, corresponding to different values of θ_0, will give rise to different svb's of a multivalued function.

§3.4 Defining a Single-Valued Branch

To summarize the discussion in this chapter so far, a complete definition of a multivalued function requires a statement about the following two items:

(1) A definition of the branches of the function and a choice of one of them;
(2) A choice of the branch cut, which is usually specified by imposing restrictions on the range of values of the function $\arg(z)$.

Once these two choices have been specified, each branch of a multivalued function associates a unique value with every complex number in the cut plane and gives rise to a *well-defined function* to be called a *single-valued-branch,* or svb, of the multivalued function. A different choice of value of θ_0, giving a different allowed range of $\arg(z)$, would lead to another possible definition of the function of interest.

As an example, consider the multivalued function $\log z$. One possible choice of svb is

$$\log z = \log r + i\theta + i2\pi m, \quad 0 < \theta < 2\pi, \tag{3.10}$$

with a fixed value of m. Here m specifies the branch of $\log z$ chosen, and the restriction on the range of θ gives a possible location of a branch cut. Note that varying values of m, or selecting a different range of θ, leads to definition of an alternate svb of $\log z$. Thus, there are infinitely many possible definitions of the svb. Once a branch cut has been chosen, the each branch of $\log z$, corresponding to a choice of m, becomes a well defined svb. The same is true of other multivalued functions. *It is thus seen that a family of* svb's *is associated with a multivalued function and it would, therefore, be more appropriate to use the term "multifunction" instead of "multivalued function".*

Although the functions $\log z$, $\sqrt{z}$, $z^{1/3}$, $z^{1/m}$, etc. are not defined in the entire plane, each svb of these multivalued functions represents a continuous function in the cut plane. On the other hand, the familiar functions such as $f(z) = z$, or polynomials in z, are single-valued functions of (r, θ) and they do not have any branch point singularity and introduction of a branch cut is not needed. Such functions will continue to be defined throughout the entire complex plane without the need to introduce a branch cut.

In general, when we want to define some other function we may have to restrict the domain of the function definition in a different fashion, such as by the choice of a range for a set of suitably selected angles other than $\arg(z)$.

§3.4.1 *Principal value of the logarithm and power*

The functions such as the logarithm, square root, and fractional powers are well-defined for positive real numbers. When one is extending the definition to the complex plane, it is useful to select the branch, and the branch cut, so that for positive real values of the argument one recovers the standard definitions. This leads us to define the principal value of the logarithm and other functions, as follows. For the principal value of all the functions below, the branch cut is selected along the negative real axis.

(*1) The principal value of the logarithm, to be denoted by Log z, is defined by selecting the $N = 0$ branch given by

$$\text{Log } z = \ln r + i\theta, \tag{3.11}$$

and taking the branch cut along the negative real axis. The corresponding allowed range of θ is given by $-\pi < \theta < \pi$.

(*2) $[\sqrt{z}]_{\text{PV}} = \sqrt{r}\exp(i\theta/2), \quad -\pi < \theta < \pi.$

(*3) $[z^{1/3}]_{\text{PV}} = r^{1/3}\exp(i\theta/3), \quad -\pi < \theta < \pi.$

(*4) $[z^{m/n}]_{\text{PV}} = r^{m/n}\exp(im\theta/n), \quad -\pi < \theta < \pi.$

(*5) $[z^c]_{\text{PV}} = r^c\exp(ic\theta), \quad -\pi < \theta < \pi.$

(∗6) In a few cases we will also have occassion to take the value of the logarithm function as in the principal value but also defined for $\theta = \pi$, i.e. we shall take $\operatorname{Log} z = \ln r + i\theta$, where θ is restricted to the range $-\pi < \theta \le \pi$.

The principal value for each of the above functions is discontinuous along the negative real axis. We shall verify this in the next section and compute the value of discontinuity for each of the functions discussed here.

⊳ Tutorial §§3.2, on definition and computing the values of $\sqrt{z}$ and of a few other simple functions.

⊳ Exercise set §§3.5 and Mixed Bag §§3.8 have exercises to show that the usual identities, such as $\ln x + \ln y = \ln(xy)$, are not always valid for the logarithm of a complex number.

§3.4.2 *Examples*

Several solved problems of computation of values of multivalued functions are presented. The selection of an svb is to be carefully kept in mind while computing its values at any point.

☑**3.4.1 Problem.** Describe and sketch the image of the real axis under the mapping $w = \operatorname{Log} z$.

Solution. The principal value of the logarithm is taken to be defined as

$$\operatorname{Log} z = \ln r + i\theta, \quad -\pi < \theta \le \pi. \tag{3.12}$$

As r increases from 0 to ∞, $\ln r$ increases monotonically from $-\infty$ to ∞ passing through 0 at $r = 1$. Hence, for the points $P_1, P_2, P_3, P_4, \ldots$ on the positive real axis ($\theta = 0$), Fig. 3.4(a) the image points are on the real axis, as shown in Fig. 3.4(b). Similarly, as z moves on the negative real axis ($\theta = \pi$) away from the origin, passing through the points $N_1, N_2, N_3, N_4, \ldots$, the image moves from the left to right on the line $w = i\pi$ in the w plane, the point $w = i\pi$ being the image of $z = -1$, as shown in the figure.

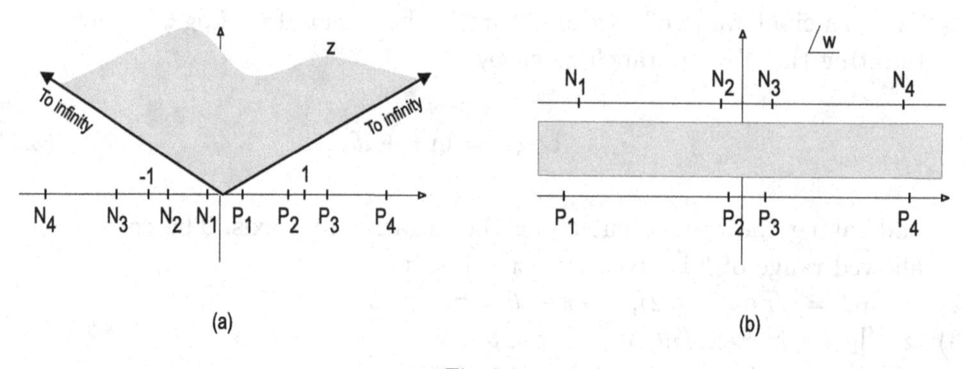

Fig. 3.4

It may be noted that the image of any ray $\theta = \theta_0$ is a straight line parallel to the real w axis in the w plane, the above discussion being a special case of $\theta = 0, \pi$.

∎

☑3.4.2 Problem. For $z = 1+i\sqrt{3}, -1+i\sqrt{3}, -1-i, 1-i$, find the values for the principal branch of $\sqrt{z}$ and $\log z$.

Solution. The principal branches of $\sqrt{z}$ and of $\log z$ are defined by working in polar coordinates and taking

$$(z)^{1/2}_{\text{PV}} = r^{1/2}e^{i\theta/2},$$
$$\text{Log}\, z = \ln r + i\theta,$$

and the range of θ is restricted to being $-\pi < \theta < \pi$. Fig. 3.5 shows the given points in the complex plane and the corresponding values of θ when restricted to the range $-\pi < \theta < \pi$. The values of $r, \theta, (z)^{1/2}_{\text{PV}}$ and Log z are tabulated below for $z = \sqrt{3}, -1+i\sqrt{3}, -1-i, 1-i$.

z	r	θ	$z^{1/2}_{PV}$	$\text{Log}\, z$
$1+i\sqrt{3}$	2	$\pi/3$	$\sqrt{2}e^{i\pi/6}$	$\ln 2 + i\pi/3$
$-1+i\sqrt{3}$	2	$2\pi/3$	$\sqrt{2}e^{i\pi/3}$	$\ln 2 + 2i\pi/3$
$-1-i$	$\sqrt{2}$	$-3\pi/4$	$\sqrt[4]{2}e^{-3i\pi/8}$	$\frac{1}{2}\ln 2 - 3i\pi/4$
$-i$	1	$-\pi/2$	$e^{-i\pi/4}$	$-i\pi/2$
$1-i$	$\sqrt{2}$	$-\pi/4$	$\sqrt[4]{2}e^{-i\pi/8}$	$\frac{1}{2}\ln 2 - i\pi/4$

∎

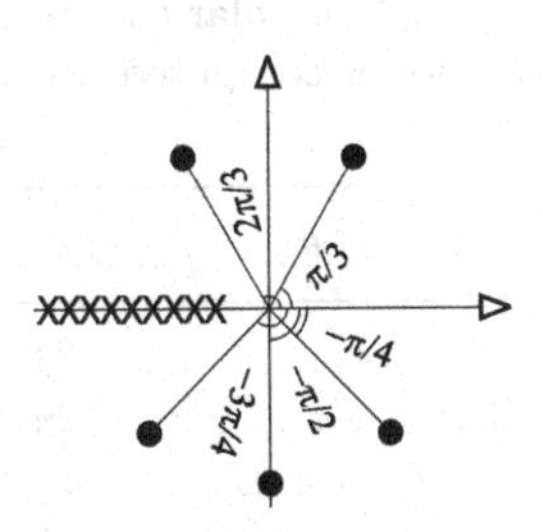

Fig. 3.5

In the previous example, the principal value for the multivalued functions $\log z, z^c$ was introduced by selecting the range

$$-\pi < \theta < \pi.$$

It should be emphasized that the principal value is one of the several possible definitions of the multivalued functions. In the next example, we use a few other possible definitions and compute the values of $z^{1/3}$ for different definitions.

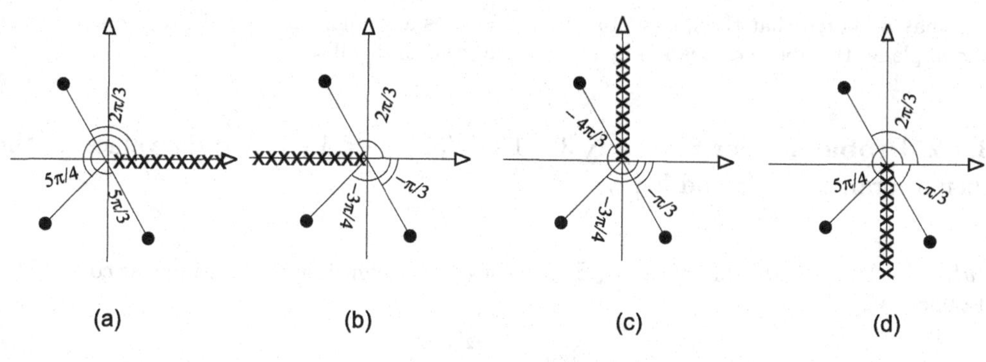

Fig. 3.6

☑**3.4.3 Problem.** Find the values of $z^{\frac{1}{3}}$ for $z = -1 + i\sqrt{3}, 1 - \sqrt{3}i, -\sqrt{2}(1+i)$, taking the expression

$$z^{\frac{1}{3}} = r^{\frac{1}{3}} \exp(i\theta/3)$$

and the branch cut along:
(a) the positive real axis, (b) the negative real axis,
(c) the positive imaginary axis, (d) the negative imaginary axis.

Solution. While computing the values of an svb of a multivalued function, attention must be paid to the the restrictions defining the branch cut. So, in this case, different branch cuts correspond to different restrictions on θ values. The resulting values of θ should be measured without crossing are branch cut and are shown in Fig. 3.6. For the four cases specified in the problem, the ranges of θ are as shown in the third column of the table below. To begin, we first express all the three numbers, where the values of $z^{\frac{1}{3}}$ are required, in polar coordinates with θ restricted as given in the third column. Next, the value of the function is computed using its expression in polar coordinates.

z	r	Range of θ	θ	Value of $z^{\frac{1}{3}}$
$-1+i\sqrt{3}$	2	(a) $0 < \theta < 2\pi$	$2\pi/3$	$\sqrt[3]{2}\exp(2i\pi/9)$
		(b) $-\pi < \theta < \pi$	$2\pi/3$	$\sqrt[3]{2}\exp(2i\pi/9)$
		(c) $-3\pi/2 < \theta < \pi/2$	$-4\pi/3$	$\sqrt[3]{2}\exp(-4i\pi/9)$
		(d) $-\pi/2 < \theta < 3\pi/2$	$2\pi/3$	$\sqrt[3]{2}\exp(2i\pi/9)$
$-\sqrt{2}(1+i)$	2	(a) $0 < \theta < 2\pi$	$5\pi/4$	$\sqrt[3]{2}\exp(5i\pi/12)$
		(b) $-\pi < \theta < \pi$	$-3\pi/4$	$\sqrt[3]{2}\exp(-i\pi/4)$
		(c) $-3\pi/2 < \theta < \pi/2$	$-3\pi/4$	$\sqrt[3]{2}\exp(-i\pi/4)$
		(d) $-\pi/2 < \theta < 3\pi/2$	$5\pi/4$	$\sqrt[3]{2}\exp(5i\pi/12)$
$1-\sqrt{3}i$	2	(a) $0 < \theta < 2\pi$	$5\pi/3$	$\sqrt[3]{2}\exp(5i\pi/9)$
		(b) $-\pi < \theta < \pi$	$-\pi/3$	$\sqrt[3]{2}\exp(-i\pi/9)$
		(c) $-3\pi/2 < \theta < \pi/2$	$-\pi/3$	$\sqrt[3]{2}\exp(-i\pi/9)$
		(d) $-\pi/2 < \theta < 3\pi/2$	$-\pi/3$	$\sqrt[3]{2}\exp(-i\pi/9)$

∎

The properties of multivalued functions, such as $\sqrt{z}\log z$, $\sqrt{z}$, $\sqrt{z-1}$, $\sqrt{z^2-1}$, or $\sqrt{z(z-1)}$, defined by expressions involving a power, or log, of z or $z-\xi$, can be obtained from the properties of the functions discussed above. In general, such functions are expected to have a discontinuity, and hence a branch cut, wherever the individual functions have a discontinuity. A discussion of these type of functions will be taken up in the next section.

§3.5 Multivalued Functions of $z - \xi$

Examples of other multivalued functions which are commonly needed are the log, square root, or other nonintegral powers of $z-\xi$ and expressions involving these multivalued functions. These are defined in terms of $\rho = |z-\xi|$ and $\phi = \arg(z-\xi)$; see Fig. 3.7. The function $\arg(z-\xi)$ exhibits properties similar to those of $\arg(z)$. In this case the value of $\arg(z-\xi)$ changes by 2π in making one full closed circuit enclosing the point ξ. These functions have the branch points at $z=\xi$ and $z=\infty$. The functions such as $(z-\xi)^\lambda$ and $\log(z-\xi)$ will now be defined by specifying the range of ϕ and picking up a branch of the function under consideration, as discussed in §3.4. Using the notation $\omega = \exp(2\pi i/3)$ and $\Omega_N = \exp(2i\pi/N)$, in the examples below, we list branches of some multifunctions involving $z-\xi$. In order to make a branch single-valued, a suitable restriction on ϕ, such as

$$\phi_0 < \arg(z-\xi) < 2\pi + \phi_0,$$

must be imposed on the range of ϕ. For real ξ, and for two possible choices of the allowed range of ϕ, Fig. 3.8 shows the branch cuts for all the functions considered below.

(*1) $\log(z-\xi)$ has the branches

$$[\log(z-\xi)]_N = \ln\rho + i\phi + 2\pi N i.$$

(*2) $[\sqrt{z-\xi}]_1 = +\sqrt{\rho}\exp(i\phi/2)$,
$[\sqrt{z-\xi}]_2 = -\sqrt{\rho}\exp(i\phi/2)$.

(*3) $[(z-\xi)^{1/3}]_1 = \rho^{1/3}\exp(i\phi/3)$,
$[(z-\xi)^{1/3}]_2 = \rho^{1/3}\exp(i\phi/3)\omega$,
$[(z-\xi)^{1/3}]_3 = \rho^{1/3}\exp(i\phi/3)\omega^2$.

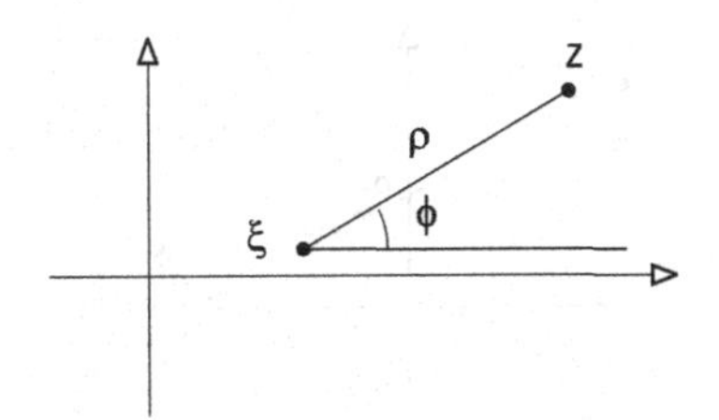

Fig. 3.7

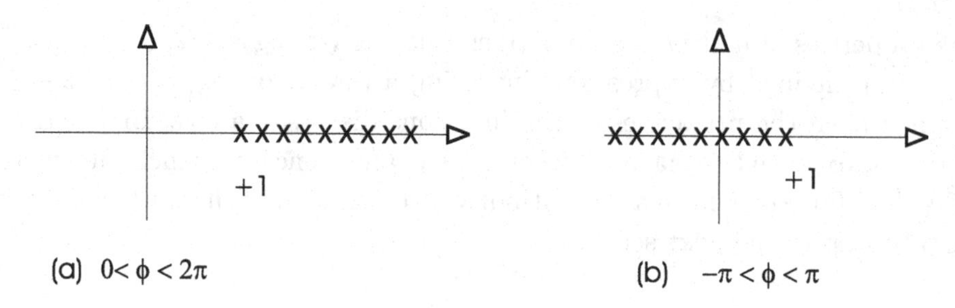

Fig. 3.8

(*4) $[(z-\xi)^{1/N}]_m = \rho^{1/N}\exp(i\phi/N)\Omega_N^m, \quad m = 0, 1, 2, \ldots, N-1,$
(*5) $[(z-\xi)^c]_m = \rho^c \exp(ic\phi)\exp(2\pi icm), \quad m = 0, 1, 2, \ldots.$

§3.5.1 *Sum and product of* $\sqrt{z+1}$ *and* $\sqrt{z-1}$

Many multivalued functions, which will be needed later, are themselves defined in terms of simpler functions already discussed in earlier sections. We shall briefly discuss examples of functions expressed in terms of $\sqrt{z+1}$ and $\sqrt{z-1}$.

¶3.5.1 Short examples. We shall compare and contrast some properties of two such functions, $\sqrt{z+1}+\sqrt{z-1}$ and $\sqrt{z^2-1}$. These functions can be defined, respectively, as the sum and the product of the two multivalued functions $\sqrt{z+1}$ and $\sqrt{z-1}$. The functions $\sqrt{z+1}$ and $\sqrt{z-1}$ are conveniently discussed in terms of the variables ρ_1, ρ_2, ϕ_1, and ϕ_2, defined by (see Fig. 3.9)

$$\rho_1 = |z+1|,$$
$$\rho_2 = |z-1|,$$
$$\phi_1 = \arg(z+1),$$
$$\phi_2 = \arg(z-1).$$

(a) Whereas the multifunction $\sqrt{z+1}+\sqrt{z-1}$ has four branches given by

$$\pm\sqrt{\rho_1}\exp(i\phi_1/2) \pm \sqrt{\rho_2}\exp(i\phi_2/2),$$

there are only two branches of $\sqrt{z^2-1}$, given by

$$\pm\sqrt{\rho_1\rho_2}\exp[i(\phi_1+\phi_2)/2].$$

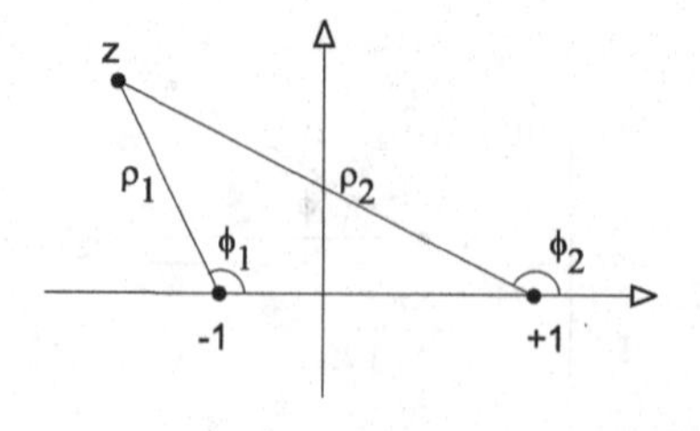

Fig. 3.9

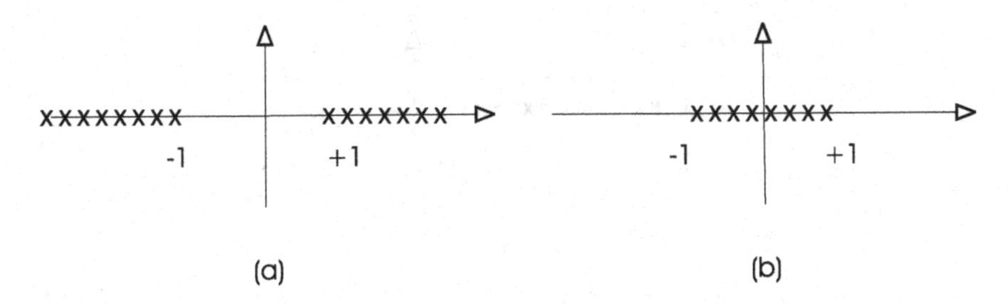

Fig. 3.10

(b) The functions $\sqrt{z-1}$ and $\sqrt{z+1}$ change the sign across their respective branch cuts. It is then easy to see that the value of the product will not change across the real line wherever the cuts for the individual functions, $\sqrt{z-1}$ and $\sqrt{z+1}$, overlap. At these points the sum $\sqrt{z-1}+\sqrt{z+1}$ is discontinuous, but the product function remains continuous.

(c) The union of the cuts for the two functions $\sqrt{z-1}$ and $\sqrt{z+1}$ becomes the branch cut for the sum $\sqrt{z-1}+\sqrt{z+1}$, but not for the product function $\sqrt{z^2-1}$.

(d) For the choice of the ranges

$$0 < \phi_1 < 2\pi, \quad -\pi < \phi_2 < \pi,$$

even though the function $\sqrt{z+1}$ has a branch cut from -1 to ∞ and the function $\sqrt{z-1}$ has a branch cut from $-\infty$ to 1, the function $\sqrt{z^2-1}$ is continuous over the interval $(-1, 1)$ of the real line common to the two branch cuts. Thus, the branch cut for $\sqrt{z^2-1}$ runs from $-\infty$ to -1 and then from 1 to ∞, as shown in Fig. 3.10(a). On the other hand, the cut for the function $\sqrt{z-1}+\sqrt{z+1}$ is a full real line.

(e) For the restrictions,

$$0 < \phi_1 < 2\pi, \quad 0 < \phi_2 < 2\pi,$$

on the angles ϕ_1 and ϕ_2, the branch cut for $\sqrt{z^2-1}$ runs only from -1 to 1 and is shown in Fig. 3.10(b). In contrast, the branch cut for $\sqrt{z+1}+\sqrt{z-1}$ is -1 to ∞.

(f) The function $\sqrt{z+1}$ has a branch point at $z=-1$ and at infinity. The function $\sqrt{z-1}$ has a branch point at $z=1$ and another one at infinity. The function $\sqrt{z^2-1}$ has branch point singularity at $z=\pm 1$ and the point at infinity is not a branch point. On the other hand, $z=\pm 1, \infty$ are the branch points of the function $\sqrt{z+1}+\sqrt{z-1}$.

⊳ §§3.4 Quiz on the branch cut for functions involving powers of $z+1$, $z-1$, etc.
⊳ §§3.7 Mined set on computing values of functions with branch points.

§3.6 Discontinuity Across the Branch Cut

Every branch of a multivalued function is discontinuous on the branch cut. The difference in the limits of the values, as one approaches a point on the branch cut from the two sides of the cut, is called the *discontinuity across the branch cut.*

In this section we shall compute the discontinuity across the branch cut for the principal value of the logarithm, $\log z$, and power, z^c. Recall the definitions

$$\operatorname{Log} z = \ln r + i\theta, \tag{3.13}$$

$$z^c = r^c \exp(ic\theta), \tag{3.14}$$

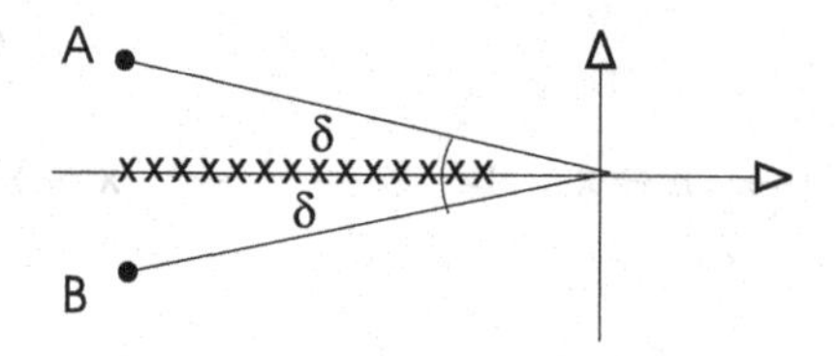

Fig. 3.11

and for the principal value the branch cut is along the negative real axis and the polar angle is restricted to the range $-\pi < \theta < \pi$. Now consider two points A and B — see Fig. 3.11 — just above and below the radial line $\theta = \pi$. For these two points we have $z_A = x + i\epsilon$ and $z_B = x - i\epsilon$, and the polar variables for A and B, respectively, are given by

$$\begin{aligned} z &= x + i\epsilon, \quad & r &= |x|, \quad & \theta &= \pi - \delta(\epsilon), \\ z &= x - i\epsilon, \quad & r &= |x|, \quad & \theta &= -\pi + \delta(\epsilon). \end{aligned}$$

For the branch cut along the negative real axis, the discontinuity is given by

$$\text{disc}\{f\}(x) = \lim_{\epsilon \to 0}[f(x + i\epsilon) - f(x - i\epsilon)], \quad x < 0.$$

Thus, the discontinuity across the cut for $\text{Log}\, z$, for example, is given by

$$\begin{aligned} \text{disc}\{\text{Log}\, z\}(x) &= \lim_{\epsilon \to 0} [\text{Log}(x + i\epsilon) - \text{Log}(x - i\epsilon)] \\ &= \lim_{\delta \to 0}[(\ln|x| + i\pi - i\delta) - (\ln|x| - i\pi + i\delta)] = 2\pi i. \end{aligned}$$

Similarly, for the principal branch of $\sqrt{z} = \sqrt{r}\exp(i\theta/2)$, one has

$$\begin{aligned} \text{disk}\{\sqrt{z}\}(x) &= \lim_{\epsilon \to 0} \left\{ |x|^{1/2} \exp[i(\pi - \delta)/2] - |x|^{1/2} \exp[i(-\pi + \delta)/2] \right\} \\ &= |x|^{1/2} \exp(i\pi/2) - |x|^{1/2} \exp(-i\pi/2) \\ &= 2i|x|^{1/2}. \end{aligned}$$

In Table 3.5 we summarize the computation of discontinuity for the principal value of multivalued functions $\log z$, z^λ. In the second and third columns we list the value

Table 3.5 Computing discontinuity across the branch cut.

SN	$f(z)$	$w_A \equiv f(x + i\epsilon)$	$w_B \equiv f(x - i\epsilon)$	Discontinuity
1	$\sqrt{z}$	$\|x\|^{1/2} \exp(i\pi/2)$	$\|x\|^{1/2} \exp(-i\pi/2)$	$2i\|x\|^{1/2}$
2	$z^{1/3}$	$\|x\|^{1/3} \exp(i\pi/3)$	$\|x\|^{1/3} \exp(-i\pi/3)$	$2i \sin(\pi/3)\|x\|^{1/3}$
3	$z^{1/N}$	$\|x\|^{1/N} \exp(i\pi/N)$	$\|x\|^{1/N} \exp(-i\pi/N)$	$2i \sin(\pi/N)\|x\|^{1/N}$
4	z^c	$\|x\|^c \exp(ic\pi)$	$\|x\|^c \exp(-ic\pi)$	$2i \sin(\pi c)\|x\|^c$
5	$\log z$	$\ln\|x\| + i\pi$	$\ln\|x\| - i\pi$	$2\pi i$

at a point A ($z_A = x+i\epsilon$) just above the cut, and the value at a point B ($z_B = x-i\epsilon$) just below the cut, respectively, in the limit $\epsilon \to 0$. Discontinuity across the branch cut is just the difference $w_A - w_B$ in the limit $\epsilon \to 0$, $\delta \to 0$ and is given in the last column of the table.

§3.7 Examples

3.7.1 Short examples. In the previous section, the principal value for the multivalued functions $\log z$, z^c was introduced by selecting the range

$$-\pi < \theta < \pi,$$

and discontinuity across the branch cut was computed. It should be emphasized that the principal value is one of the several possible definitions of the multiple-valued functions. Here we use another definition by taking the range of θ to be

$$0 < \theta < 2\pi$$

and compute the discontinuity of these functions across the new branch cut along the positive real axis.

(a) Using the definition

$$\log z = \ln r + i\theta, \quad 0 < \theta < 2\pi, \tag{3.15}$$

the function, $\log z$, has a branch cut on the real line for $x > 0$. The values θ_A and θ_B of θ just above and just below the branch cut are given by $\theta_A = 0$ and $\theta_B = 2\pi$, respectively. The discontinuity is computed easily and is given by ($x > 0$)

$$\log(x + i\epsilon) - \log(x - i\epsilon) = \ln x - (\ln x + 2\pi i) = -2\pi i. \tag{3.16}$$

(b) Similarly, the discontinuity of $f(z) = z^c$ across the positive real axis is

$$\begin{aligned} f(x + i\epsilon) - f(x - i\epsilon) &= |x|^c(1 - e^{i2\pi c}) \\ &= -2i \sin c\pi e^{ic\pi}|x|^c. \end{aligned} \tag{3.17}$$

A few problems on discontinuity across the branch cut are presented below. Understanding of these problems is crucial to evaluation of a class of integrals later in Chs. 6 and 7.

☑**3.7.2 Problem.** Select a single-valued branch for the multivalued functions

$$f(z) = z^q, \quad g(z) = (z - 1)^p, \tag{3.18}$$

and locate the branch cut for

$$h(z) = f(z)g(z) = z^q(z - 1)^p, \tag{3.19}$$

and determine the discontinuity of $h(z)$ across its branch cut and briefly discuss the case $p + q = 1$.

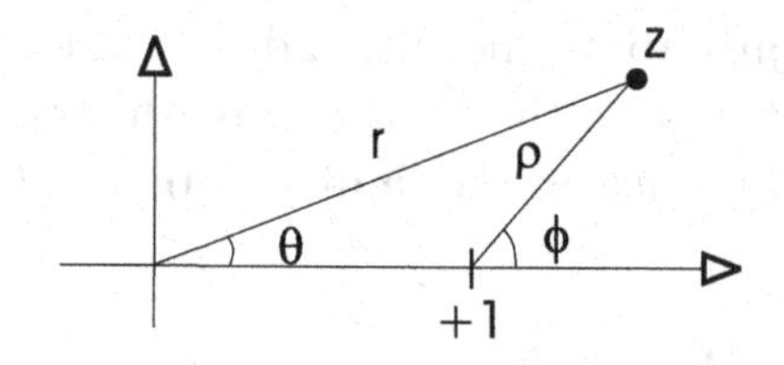

Fig. 3.12

Solution. We define the svb of the functions $f(z)$ and $g(z)$ by taking (Fig. 3.12)

$$f(z) = r^q e^{iq\theta}, \quad 0 < \theta < 2\pi, \tag{3.20}$$

$$g(z) = \rho^p e^{ip\phi}, \quad 0 < \phi < 2\pi, \tag{3.21}$$

where we have used the variables $\rho = |z-1|$ and $\phi = \arg(z-1)$. The branch cut for $f(z)$ is along the positive real axis and that for $g(z)$ is the infinite interval $(1, \infty)$.

For $x < 0$ both of the functions f and g are well-defined and continuous across the real line and the product will be continuous. The product $f(z)g(z)$ is expected to be discontinuous wherever the two individual functions are discontinuous. We therefore compute the discontinuity of $f(z), g(z)$, and $h(z)$ for $x > 0$, considering the cases $x > 1$ and $0 < x < 1$ separately.

Case: $x > 1$

In this case for points A and B, just above and just below the real axis, $z_A = x + i\epsilon, z_B = x - i\epsilon$, the values of the angles θ and ϕ are given by

$$\theta_A = 0, \quad \theta_B = 2\pi; \quad \phi_A = 0, \quad \phi_B = 2\pi. \tag{3.22}$$

This gives the following values for the functions f, g, h and their discontinuities.

Function	Value Above	Value Below	Discontinuity
z^q	x^q	$x^q e^{2iq\pi}$	$x^q(1 - e^{2iq\pi})$
$(z-1)^p$	$\|x-1\|^p$	$\|x-1\|^p e^{2ip\pi}$	$\|x-1\|^p(1 - e^{2ip\pi})$
$z^q(z-1)^p$	$x^q\|x-1\|^p$	$x^q\|x-1\|^p e^{i2(p+q)\pi}$	$x^q\|x-1\|^p(1 - e^{2i(p+q)\pi})$

Form the above table it is clear that the discontinuity of the function $h(z)$ vanishes when $p+q$ is an integer, and hence, in particular for $p+q=1$, the function $h(z)$ is continuous for $x > 1$.

Case: $0 < x < 1$

The values of the angles θ and ϕ for a point A ($z_A = x + i\epsilon$) just above the real axis, and for a point B ($z_B = x - i\epsilon$) just below the real axis, are

$$\theta_A = 0, \quad \theta_B = 2\pi; \quad \phi_A = \pi, \quad \phi_B = \pi. \tag{3.23}$$

The values for the functions f, g, and h in the limit $\epsilon \to 0$ and the corresponding discontinuities are tabulated below:

Function	Value Above	Value Below	Discontinuity
z^q	x^q	$x^q e^{2iq\pi}$	$x^q(1 - e^{2iq\pi})$
$(z-1)^p$	$\|x-1\|^p e^{ip\pi}$	$\|x-1\|^p e^{ip\pi}$	0
$z^q(z-1)^p$	$x^q\|x-1\|^p e^{ip\pi}$	$x^q\|x-1\|^p e^{i(p+2q)\pi}$	$x^q\|x-1\|^p e^{ip\pi}(1 - e^{2iq\pi})$

It is seen that the function $h(z)$ is discontinuous and has a branch cut from $x = 0$ to $x = 1$ only and the value of the discontinuity, with $q = 1 - p$, is given by

$$\text{disk}\{z^{(1-p)}(z-1)^p\}(x) = \begin{cases} 2i\sin(p\pi)x^{(1-p)}(1-x)^p, & x \in (0,1) \\ 0, & x \notin (0,1) \end{cases}. \tag{3.24}$$

In the special case, $p + q = 1$, the product $h(z)$ is well-defined in the cut plane given by the ranges

$$0 \le \theta < 2\pi, \quad 0 \le \phi < 2\pi, \quad \rho + r > 2. \tag{3.25}$$

∎

§3.7.3 Short examples. When a continuous function multiplies a second function with a branch cut, the discontinuity of the product across the branch cut is simply the value of the continuous function multiplied by the discontinuity of the second function across the branch cut. We give several examples of this type:

(a) Let $\chi(z) = \frac{h(z)}{z^2+1}$, with $h(z)$ defined as in Eq. (3.19). The discontinuity of $\chi(z)$ using the answer (3.24) is found to be

$$\chi(x + i\epsilon) - \chi(x - i\epsilon) = 2i\sin p\pi\,\frac{x^{1-p}(1-x)^p}{x^2+1}, \quad 0 < x < 1. \tag{3.26}$$

(b) For the function $f(z) = \frac{\log z}{z^4+a^4}$ with the definition of $\log z$ taken as in Eq. (3.15), we would get

$$f(x + i\epsilon) - f(x - i\epsilon) = \frac{-2\pi i}{x^4 + a^4}. \tag{3.27}$$

(c) Using a result obtained earlier on the discontinuity of z^c — see Eq. (3.17) — we can now write down the answer for the discontinuity of $g(z) = \dfrac{z^c}{z^4 + a^4}$:

$$g(x + i\epsilon) - g(x - i\epsilon) = -2i\sin c\pi\, e^{ic\pi}\,\frac{x^c}{x^4 + a^4}. \tag{3.28}$$

Remarks

($\star$1) If we integrate the Eq. (3.26) result from 0 to 1, we see that $\int_0^1 \frac{x^{1-p}(1-x)^p dx}{x^2+1}$ is related to the integral of the discontinuity across the branch cut.

($\star$2) The integral of the discontinuity across the cut can be converted into an integral over a closed contour and evaluated by the method of contour integration to be discussed in Ch. 7.

($\star$3) Similar remarks apply to the other examples, 7(b)–7(c), considered above.

⊳ Exercise §§3.3, on $\sqrt{\frac{z-a}{z+b}}$. This function has properties similar to those of $\sqrt{z^2-1}$, discussed in §3.5.1.

⊳ Quiz §§3.4, on location of the branch cut.

⊳ Exercise §§3.5, on the logarithm function and its discontinuity.

⊳ Exercise §§3.6, computation of discontinuity.

§3.8 Inverse Trigonometric Functions

The solution to equations, such as $w = \sin(z)$, is easily obtained by writing everything in terms of the exponential function, and it gives a starting point for introducing the inverse trigonometric and inverse hyperbolic functions which will be multivalued. The set of all solutions can be expressed in terms of the logarithm and we have, for example,

$$\sin^{-1} z = -i \log\left(iz + \sqrt{1-z^2}\right), \quad \sinh^{-1} z = \log\left(z + \sqrt{z^2+1}\right), \tag{3.29}$$

$$\cos^{-1} z = -i \log\left(z + \sqrt{z^2-1}\right), \quad \cosh^{-1} z = \log\left(z + \sqrt{z^2-1}\right), \tag{3.30}$$

$$\tan^{-1} z = \frac{i}{2} \log\left(\frac{i+z}{i-z}\right), \quad \tanh^{-1} z = \frac{1}{2} \log\left(\frac{1+z}{1-z}\right). \tag{3.31}$$

We shall now discuss the inverse sine function in detail. It has infinitely many values for every complex number z. Note that by squaring both sides of $iz = -\sqrt{1-z^2}$ we can conclude that the argument of the logarithm is never zero for any complex number z, no matter which definition of $\sqrt{1-z^2}$ is used. *This means that the branch cut of the inverse sine function coincides with the branch cut for* $\sqrt{1-z^2}$. It is apparent that, in general, an svb for $\sin^{-1} z$ can be defined by selecting an svb for the square root $\sqrt{1-z^2}$ and for $\log z$ functions. As an example, we define the *principal value* of the inverse sine function, $\mathrm{Sin}^{-1} z$, by

$$\mathrm{Sin}^{-1} z = -i \mathrm{Log}\,(iz + \sqrt{1-z^2}), \tag{3.32}$$

where $\sqrt{1-z^2}$ is defined in such a way that it reduces to $\sqrt{1-x^2}$ for real values $z = x$ in the range $|x| < 1$. The branch cut for $\sqrt{1-z^2}$ will be on the real axis for $|x| > 1$. The function $\mathrm{Sin}^{-1} x$ has the property that for real x, $|x| < 1$, it becomes

$$\mathrm{Sin}^{-1} x = \mathrm{Arg}(ix + \sqrt{1-x^2}). \tag{3.33}$$

Similar considerations apply to other inverse functions.

↬ Q.[5] in §§3.8 for a proof of Eq. (3.33).

§3.9 Differentiation

We have seen that a multivalued function f frequently appears as an inverse function of a many-to-one function g. The knowledge of the derivative of $g(z)$ can be used to find the derivative of the inverse function.

Assume that the value $f(z)$, of a function f at a point z, is defined to be w obtained by solving the equation $g(w) = z$. Then the derivatives of f and g are related as given by the following theorem:

Theorem 3.1. *Let $g(z)$ be a one-to-one function defined in a domain D. In addition, let S be the range of the function g. Assume that the inverse function $f = g^{-1}$ is continuous on the set S. For every point $w_0 \in D$, such that $g'(w_0) \neq 0$, the derivative of the function f exists at the point z_0 and is given by*

$$\left.\frac{df(z)}{dz}\right|_{z=z_0} = 1 \Big/ \left.\left(\frac{dg(w)}{dw}\right)\right|_{w=w_0},$$

where $w = f(z_0)$.

The above result can be understood by differentiating $g(f(z)) = z$. When $g(z)$ is a many-to-one function, we must restrict the domain of $g(z)$ to make it one-to-one. Thus, for every svb of a multivalued function, the derivative in the cut plane can be computed using the above theorem and taking S to be the cut plane. Knowing about the existence of the derivative w.r.t. the complex variable, it becomes possible to make statements about the analyticity of the function. Alternatively, the analyticity can be investigated using Cauchy Riemann equations and demonstrating the continuity of partial derivatives. For example, one may prove the analyticity of the logarithm and a power of z by first checking the Cauchy Riemann equations in polar variables r, θ and the continuity of the four partial derivatives.

ɇ3.9.1 Short examples. We continue to use the meanings associated with f, g, and w as introduced above.

(a) Let us take $f(z) = \operatorname{Log} z$ to be the principal value, with $g(z) = \exp(z)$; we have in the cut plane

$$\left.\frac{dg(z)}{dz}\right|_{z=w_0} = \exp(w_0), \tag{3.34}$$

$$\therefore \quad \left.\frac{d\operatorname{Log} z}{dz}\right|_{z=z_0} = \exp(-w_0) = \frac{1}{z_0}. \tag{3.35}$$

(b) Let us take $f(z) = \sqrt{z}$ to be the svb which has the branch cut along the negative real axis, and is positive on the real line for $|x| > 0$. This function $f(z)$ is an inverse function of $g(z) = z^2$ and hence $\frac{dg(z)}{dz} = 2z$ gives

$$\frac{d\sqrt{z}}{dz} = \frac{1}{2w} = \frac{1}{2\sqrt{z}}.$$

Note that the same branch of $\sqrt{z}$ appears on both sides of the above equation.

(c) If we consider the function $w = \tan z$ the inverse map, $\tan^{-1} z$ has the derivative given by

$$\frac{d}{dz}\tan^{-1} z = \frac{1}{1+z^2}.$$

Note that the right hand side does not have a branch point singularity.

⇛ Exercise §§3.8, on differentiation of multivalued functions.

⇛ Mined set §§3.7 and locate "mines" in sample answers.

§3.10 Riemann Surface

The simplest multivalued functions arise as the inverse of many-to-one functions which are not one. One way to arrive at single-valued definitions is to introduce a branch cut and to restrict the domain of definition of the function to the cut plane. We shall now describe an elegant, alternate construction which will permit associating a well-defined function with a multivalued function.

We shall explain the basic idea using examples of the square root and the logarithm functions. The square root function has two svb's, which can be taken to be

$$(\sqrt{z})_1 = \sqrt{r}\exp(i\theta), \qquad -\pi < \theta < \pi, \tag{3.36}$$

$$(\sqrt{z})_2 = \sqrt{r}\exp(i\theta), \qquad \pi < \theta < 3\pi, \tag{3.37}$$

where the branch cut is assumed to be along the negative real axis. The values of $\arg(z)$ in the two ranges $(-\pi,\pi)$ and $(\pi,3\pi)$ are represented on two separate copies, also called sheets, of the plane with the negative real axis removed. These two sheets are joined together along the branch cut to form the Riemann surface of the square root function, as shown in Fig. 3.13. The value taken by the square root function on a sheet corresponds to one of the two svb's given above. In an odd number of traversals of closed paths around the origin, a point on the Riemann surface moves from one sheet to another and the value of the function changes sign. In an even number of closed loops around the branch point, the value of $\sqrt{z}$ returns to its starting values. Similar statements hold for $z^{1/n}$ except that the Riemann surface will now have n sheets.

Next, let us consider the function $f(z) = \exp(z)$, which is a many-to-one mapping from the z plane to the w plane, and the inverse mapping $w = \log z = \ln r + i\arg(z)$ is infinite-valued; the "logarithm mapping" gives *distinct answers* $(= \ln r + 2n\pi i)$ for $(r, \theta + 2n\pi i)$ for different n. All the infinite values of the pair $(|z|, \arg z)$ correspond to the *same point* in the z plane. To make a transition to a single-valued mapping, corresponding to the logarithm function, we represent each pair of values

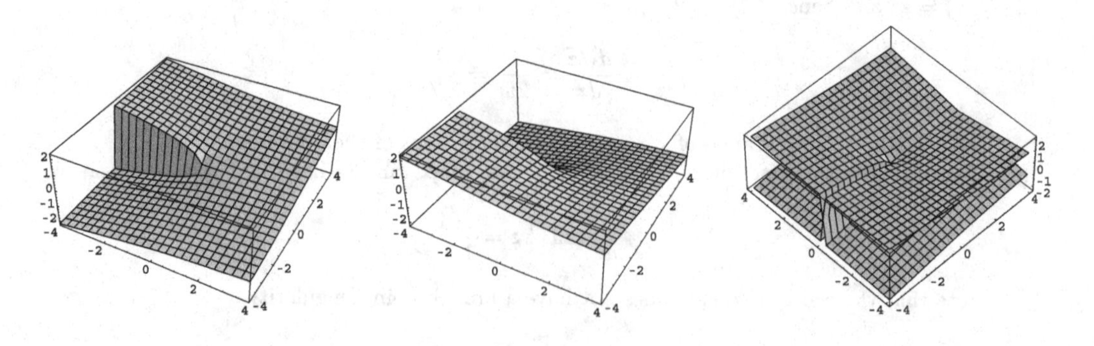

Fig. 3.13

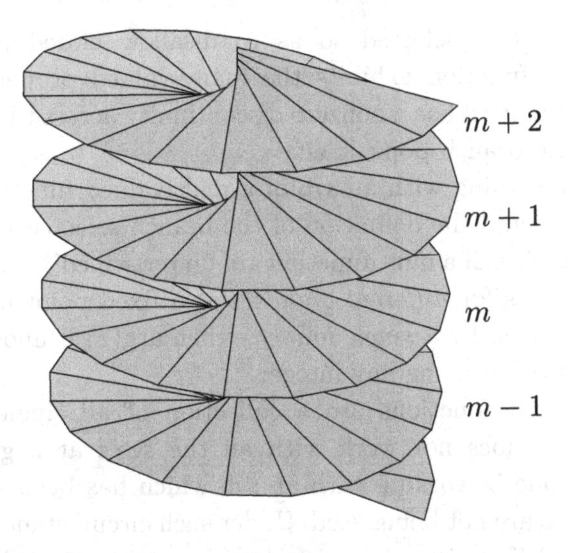

Fig. 3.14

of $|z|$, $\arg(z) = \theta + 2n\pi i$, by a point in a different copy of the complex plane with a branch cut. These different copies, the Riemann "sheets," are glued suitably along the branch cut, giving rise to a *helical* Riemann surface, as shown in Fig. 3.14.

In the Riemann surface description, the logarithm function is viewed as a collection of all the svb's. On a sheet the value of the function is given by one of the svb's—say, $(\log z)_m$. When a point is moved around the origin in a closed path in the anticlockwise direction, the point passes to the corresponding position on the next sheet and there the function assumes the value corresponding to the next branch, $(\log z)_{m+1}$, of the function.

In this book we shall follow the branch cut approach and work with svb's defined in the cut plane and shall not use the Riemann surface description of multivalued functions.

§3.11 Summary

In the following we summarize important general facts about the multivalued functions:

(*1) Multivalued functions are not functions in the usual sense. They denote a collection of functions called single-valued branches.

(*2) Each svb of a multivalued function represents a properly defined single-valued function — the way it should be — in the cut plane.

(*3) The cut plane is the complex plane from which a set of points has been removed, and the set of points removed is called the branch cut.

(*4) The branch cut is a continuous curve joining two branch points, one of which may be at infinity.

(*5) The branch cut is selected so as to disallow closed paths along which a multivalued function exhibits the many-valued property. Each svb of a multivalued function has a nonzero discontinuity across the branch cut except possibly at the branch point itself.

(*6) While one is working with an svb of a multivalued function, the location of the branch cut and the definition of the branch selected must be specified.

(*7) Different branches of a multifunction are "intertwined" in the following sense. The set of values for *different branches* at a fixed point $|z| = R, \arg(z) = \alpha$ can be reproduced by a *single branch*, when $\arg(z)$ is allowed to assume the values $\alpha + 2m\pi$, m being any integer.

(*8) The multivalued functions are a collection of all single-valued branches. Generally, one does not work with all the svb's at a given time. Let us assume that one is working with an svb which has been specified, and that other branches are not being used. Under such circumstances the dividing line between a multifunction and a function becomes very thin. Thus, a phrase, such as the *function* $\log z$, will be commonly used in situations where the use of a term such as the svb *of the multivalued function* $\log z$ would be more appropriate. With this remark it is hoped that the reader is not surprised to find $\log z$, etc. being called functions in this book.

§3.12 Notes and References

There are a wide variety of visuals of Riemann surfaces available on the World Wide Web:

(◇1) For an image of the Riemann surface of the logarithm generated by Matlab, see the following links in MIT OpenCourseWare:
http://ocw.mit.edu/OcwWeb/Mathematics/18-04Fall-2003/CourseHome/index.htm
http://ocw.mit.edu/OcwWeb/Mathematics/18-04Complex-Variables-with-ApplicationsFall1999/StudyMaterials/detail/logsurface.htm

(◇2) Generating pictures of Riemann surfaces can be done with computer packages such as Matlab, Mathematica, and Scilab. For some samples see:

(◇3) Weisstein, Eric W., "Inverse Trigonometric Functions." From MathWorld — A Wolfram Web Resource.

http://www.scilab.org/doc/demos_html/node241.html
http://rsp.math.brandeis.edu/3D-XplorMath/Surface/riemann/riemann.html
http://mathworld.wolfram.com/InverseTrigonometricFunctions.html

This site has a great deal of information and graphics about inverse trigonometric functions. For example, for the function $\sin^{-1} z$ visit the Web page
http://mathworld.wolfram.com/InverseSine.html

(◇4) For some more sample graphics on Riemann surfaces, see:
http://rsp.math.brandeis.edu/3D-XplorMath/Surface/riemann/riemann.html

(◇5) To learn more about using Riemann surfaces in various applications, see the book by Le Page (1980).

Chapter 4

INTEGRATION IN THE COMPLEX PLANE

In this chapter we define line integral in the complex plane and give important results on line integrals. It is assumed that the reader is familiar with elementary results on Riemann integration over a real variable. We begin with a brief introduction to improper integrals. After introducing the line integral in the complex plane, we present Darboux's theorem and Jordan's lemma on bounds of a line integral and Cauchy's fundamental theorem.

In §4.7 of this chapter the relationship between the integrals over a real variable and the line integrals in the complex plane is explored by means of several examples. This relationship is of crucial importance for application of the method of contour integration to computation of proper and improper integrals over a single real variable. The same section includes our first two examples of the use of complex variables in evaluation of improper integrals, and the use of the fundamental theorem will turn out to be sufficient for this purpose. Many more examples, requiring the use of the residue theorem, will appear in a later chapter on contour integration.

Discussion of line integrals of multivalued functions is reserved for the last section of this chapter. The multivalued functions were introduced in Ch. 3 and the reader is urged to revise this chapter before taking the plunge into the last section. In §4.8 several examples of line integrals and definite integrals involving multivalued functions are presented. In particular, we also show how line integrals around a branch cut in the complex plane get related to proper or improper integrals of discontinuity across the branch cut.

§4.1 Improper Integrals

Let f be a real-valued function of a real variable. An integral $\int_a^b f(x)\,dx$ is called an *improper integral* if the range of integration is infinite or if the integrand becomes unbounded in the range of integration. An improper integral has to be defined as a limiting value of a proper integral. Thus, we have the definitions

$$\int_a^{\infty} f(x)\,dx = \lim_{b\to\infty} \int_a^b f(x)\,dx, \tag{4.1}$$

$$\int_{-\infty}^{b} f(x)\,dx = \lim_{a\to-\infty} \int_a^b f(x)\,dx \tag{4.2}$$

whenever the limits exist. The integral over $(-\infty, \infty)$ is defined through a similar double limit of $\int_a^b f(x)\,dx$. If the function $f(x)$ becomes unbounded at a point x_0, the improper integral $\int_{x_0}^b f(x)\,dx$ is defined as

$$\int_{x_0}^{b} f(x)\,dx = \lim_{\varepsilon\to 0} \int_{x_0+\varepsilon}^b f(x)\,dx. \tag{4.3}$$

If the function is unbounded at the upper limit, we define

$$\int_{a}^{x_0} f(x)\,dx = \lim_{\varepsilon\to 0} \int_{a}^{x_0-\varepsilon} f(x)\,dx. \tag{4.4}$$

If the point $x = x_0$, where the function is unbounded falls inside the interval $[a, b]$, several different definite integrals can be defined through limiting procedures, as follows. First, we check if the integrals, over the two ranges (a, x_0) and (x_0, b), exist or not. If the two limits exist their sum gives *a definition* of the improper integral from a to b:

$$\int_{a}^{b} f(x)\,dx = \lim_{\varepsilon\to 0} \int_{a}^{x_0-\varepsilon} f(x)\,dx + \lim_{\varepsilon\to 0} \int_{x_0+\varepsilon}^{b} f(x)\,dx. \tag{4.5}$$

Another possible definition of the improper integral is the Cauchy principle value, defined as

$$\text{PV}\int_{a}^{b} f(x)\,dx = \lim_{\varepsilon\to 0} \left[\int_{a}^{x_0-\varepsilon} f(x)\,dx + \int_{x_0+\varepsilon}^{b} f(x)\,dx\right]. \tag{4.6}$$

The existence of an improper integral implies the existence of the principle value and the two will be equal. However, it is easy to find examples where the improper integral does not exist but the Cauchy principle value is defined. The integral $\int_{-a}^{a} \frac{dx}{x}$ is one such example; it is easily seen that

$$\text{PV}\int_{-a}^{a} \frac{dx}{x} = \lim \left(\int_{-a}^{-\varepsilon} \frac{dx}{x} + \int_{\varepsilon}^{a} \frac{dx}{x}\right) = 0, \tag{4.7}$$

but the separate limits of the two integrals, in the intermediate step, do not exist.

For applications appearing in this book we need improper integrals of complex-valued functions. The above discussion is easily extended to such cases by considering the real and imaginary parts of the integrand separately.

Comparison test. Let $g(x)$ be a positive function.

(i) If M is a constant such that $|f(x)| \leq Mg(x)$ for $a \leq x \leq b$ and if $\int_a^b g(x)$ exists, the integral $\int_a^b f(x)\, dx$ also exists.
(ii) Also, if M is a constant such that $|f(x)| \geq Mg(x)$ for $a \leq x \leq b$ and if $\int_a^b g(x)$ becomes infinite, the integral $\int_a^b f(x)\, dx$ does not exist either.

While the comparison test is a useful tool for checking the existence of improper integrals, there are many important cases where it cannot be applied. For example, the Fresnel integrals $\int_0^\infty \cos x^2\, dx$ and $\int_0^\infty \sin x^2\, dx$ are convergent because the integrand oscillates rapidly for large x and the contribution of large x values to the integral averages out to a small value.

4.1.1 Short examples. Here we give some examples of convergence of improper integrals. The conditions for existence of the improper integral $\int_0^\infty f(x)dx$, where $f(x)$ has a simple form, can be derived by directly integrating $\int_a^b f(x)dx$ and taking the limit $b \to \infty$.

(a) The answer

$$\int x^{c-1}\, dx = \frac{x^c}{c}\bigg|_a^b = \left(\frac{b^c}{c} - \frac{a^c}{c}\right),$$

with $b > a > 0$, shows that:

- The integral $\int_a^\infty x^{c-1}\, dx = \lim_{b\to\infty} \int_a^b x^{c-1}\, dx$ exists only if $c < 0$;
- The integral $\int_0^b x^{c-1}\, dx = \lim_{a\to o} \int_a^b x^{c-1}\, dx$ exists only if $c > 0$.

(b) Next, we consider the integrals of the form $\int_a^b x^c \log x\, dx$. On integrating by parts we get

$$\int_a^b x^{c-1} \log x\, dx = c^{-2}(a^c - b^c) + c^{-1}\,(b^c \,\log b - a^c \,\log a).$$

Taking the limit $b \to \infty$ we see that $\int_a^\infty x^{c-1} \log x\, dx$ exists if $c < 0$, and considering $a \to 0$ we get the result that $\int_0^b x^{c-1} \log x\, dx$ exists if $c > 0$. Similar results for other simple cases are summarized in the table below:

$\int_a^\infty x^{c-1}\, dx$	$\int_a^\infty x^{c-1}(\log x)^n\, dx$	$\int_0^b x^{c-1}\, dx$	$\int_0^b x^{c-1}(\log x)^n\, dx$	$\int_a^\infty e^{cx}\, dx$
$c<0$	$c<0$	$c>0$	$c>0$	$c<0$

In many cases of interest in this book, the functions listed in the table above provide a good choice for applying the comparison test, and convergence of many other integrals can be checked.

(c) On using the inequality $\left|\frac{x^p}{x^4+1}\right| \leq \frac{x^p}{x^4}$, and the comparison test with x^{p-4} as test function, we see that $\int_a^\infty \frac{x^p\, dx}{x^4+1}$ exists if $p < 3$. Similarly, $\left|\frac{x^p}{x^4+1}\right| \leq x^p$ implies that $\int_0^b \frac{x^p\, dx}{x^4+1}$ exists if $p > -1$. Combining these two results we see that $\int_0^\infty \frac{x^p\, dx}{x^4+1}$ exists if p is in the range $-1 < p < 3$.

(d) Using $\exp(-px)$ as test function in the comparison test, we easily see that the integral $\int_{-\infty}^{\infty} \frac{e^{kx}\,dx}{\cosh^2 x}$ exists if $-2 < k < 2$. For this purpose it is sufficient to note the bound

$$\frac{e^{kx}}{\cosh^2 x} = \frac{e^{kx}}{\exp(2x) + \exp(-2x) + 2} \leq e^{(k-2)x} \tag{4.8}$$

for large $x \to \infty$, which shows that $\int_a^{\infty} \frac{e^{kx}\,dx}{\cosh^2 x}$ exists if $k < 2$. Similarly, for $x \to -\infty$, one has the inequality

$$\frac{e^{kx}\,dx}{\cosh^2 x} = \frac{e^{kx}}{\exp(2x) + \exp(-2x) + 2} \leq e^{(k+2)x}, \tag{4.9}$$

which shows that $\int_{-\infty}^{a} \frac{e^{kx}\,dx}{\cosh^2 x}$ exists if $k > -2$. Hence, $\int_{-\infty}^{\infty} \frac{e^{kx}\,dx}{\cosh^2 x}$ exists if $-2 < k < 2$.

⇛ Questions §§4.1, on checking convergence of improper integrals.

§4.2 Definitions

Graphs. Many times, a curve is specified by equations of the type

$$\phi(x, y) = 0. \tag{4.10}$$

For example, any relation linear in x and y, the real and imaginary parts of z, such as

$$ax + by + c = 0, \tag{4.11}$$

represents a straight line. A subset of complex numbers $\{z|\phi(x, y) = 0\}$ will be called a *graph*, to distinguish it from the contour to be defined below. So, for example, the graph of an equation

$$x^2 + y^2 = R^2 \tag{4.12}$$

is a circle.

Curves and contours. A curve in the complex plane is specified by the parametric equations

$$x = \phi(t), \quad y = \psi(t), \quad a \leq t \leq b, \tag{4.13}$$

with

$$z(t) = x(t) + iy(t). \tag{4.14}$$

A *curve* defined as above is called *simple* if distinct values of t correspond to distinct points in the complex plane. If $\phi(a) = \phi(b)$ and $\psi(a) = \psi(b)$ and if no other two values of t correspond to the same point, then the curve is called a *simple closed curve.* If the functions ϕ and ψ are differentiable and the derivatives are continuous and do not vanish simultaneously at any point, the curve is called *smooth curve.* A *contour* is a piecewise continuously differentiable curve in the complex plane.

Orientation. A curve in the complex plane has been defined by means of the parametric equations $x = \phi(t)$ $y = \psi(t)$, where t takes values in a real interval. For example, the curve defined by the parametric equations

$$x(t) = \cos t, \quad y(t) = \sin t, \quad 0 \leq t \leq 2\pi, \tag{4.15}$$

is a circle with radius 1 and the center at $z = 0$.

The above definition of a curve also associates an orientation with the curve. The orientation is given by the direction in which the parameter t increases. Thus, the curve C, defined by the parametric equations (4.15), is a circle with anticlockwise orientation. The parametric equations given below represent the same circle but with clockwise orientation:

$$x(t) = \sin t, \quad y(t) = \cos t, \quad 0 \leq t \leq 2\pi. \tag{4.16}$$

Graphs represented by equations of the type (4.11) carry no information about the orientation of the "curve," as do the equations such as Eqs. (4.15) and (4.16).

§4.2.1 *Integration in the complex plane*

Let f be a complex-valued function of a real variable. In addition, let $u(t)$ and $v(t)$ be the real and imaginary parts of $f(t)$. Integrating the real and imaginary parts separately, the integral of the complex-valued function f is defined as the sum

$$\int_a^b f(t)dt = \int_a^b u(t)dt + i\int_a^b v(t)dt. \tag{4.17}$$

Next, we define the integral of a function of a complex variable along a contour in the complex plane. Let C be a contour given by

$$z(t) = x(t) + iy(t), \tag{4.18}$$

where

$$x(t) = \phi(t), \quad y(t) = \psi(t), \quad a \leq t \leq b, \tag{4.19}$$

and

$$\frac{dz}{dt} = \frac{dx}{dt} + i\frac{dy}{dt} = \frac{d\chi}{dt}, \tag{4.20}$$

where $\chi(t) = \phi(t) + i\psi(t)$. We define the line integral along the contour C by

$$\int_C f(x,y)dz = \int_a^b f(\phi(t), \psi(t))\frac{d\chi}{dt}dt. \tag{4.21}$$

Let $u(x,y)$ and $v(x,y)$ be the real and imaginary parts of $f(x,y)$. Then the integral, defined above, can be put in the form

$$\int_C f(x,y)dz = \int_a^b [u(\phi(t),\psi(t)) + iv(\phi(t),\psi(t))]\left(\frac{d\phi}{dt} + i\frac{d\psi}{dt}\right)dt$$
$$= \int_a^b \left[u(\phi,\psi)\frac{d\phi}{dt} - v(\phi,\psi)\frac{d\psi}{dt}\right]dt$$
$$+ i\int_a^b \left[u(\phi,\psi)\frac{d\psi}{dt} + v(\phi,\psi)\frac{d\phi}{dt}\right]dt. \tag{4.22}$$

Let C and $\tilde{C}$ be two contours consisting of the same set of points but oriented in the opposite directions. Then we define

$$\int_{\tilde{C}} f(z)dz = -\int_C f(z)dz. \tag{4.23}$$

The symbol $\oint$ will be used to denote integration along a closed contour.

§4.3 Examples of Line Integrals in the Complex Plane

We shall now present a simple, but important, line integral as our first example of computation of a line integral. This example utilizes the definition of a line integral in the complex plane to convert it into an integral over a real interval. Later in this book, this process will be reversed when we discuss techniques for evaluating real proper and improper integrals over real variables by relating them to integrals over a closed contour in the complex plane and using the residue theorem.

☑**4.3.1 Problem.** The single most important integral that is evaluated using the definition of the line integral is

$$\oint_C \frac{dz}{(z-z_0)^n} = \begin{cases} 2\pi i, & n = 1, \\ 0, & n \neq 1, \end{cases} \tag{4.24}$$

where n is an integer and C is a positively oriented circle with center z_0 and radius $R > 0$.

Solution. The circle C has the parametric equations

$$(z - z_0) = R\exp(it), \quad 0 < t < 2\pi. \tag{4.25}$$

Therefore,

$$x = x_0 + R\cos t, \quad y = y_0 + R\sin t, \tag{4.26}$$
$$dx = -R\sin t dt, \quad dy = R\cos t\, dt. \tag{4.27}$$

Hence, we have

$$\frac{dx}{dt} + i\frac{dy}{dt} = iR(\cos t + i \sin t) \tag{4.28}$$

$$\frac{1}{z - z_0} = \frac{1}{R}\exp(-it) = \frac{1}{R}(\cos t - i \sin t). \tag{4.29}$$

Therefore, the integral for $n = 1$, when expressed in terms of t, becomes

$$\oint_C \frac{dz}{z - z_0} = \int_0^{2\pi} \frac{1}{R}(\cos t - i \sin t)\left(\frac{dx}{dt} + i\frac{dy}{dt}\right) dt \tag{4.30}$$

$$= \int_0^{2\pi} \exp(-it) i \exp(it) dt = i \int_0^{2\pi} dt \tag{4.31}$$

$$= 2\pi i. \tag{4.32}$$

For the case where $n \neq 1$ we get

$$\frac{1}{(z - z_0)^n} = \frac{1}{R^n}(\cos nt - i \sin nt), \tag{4.33}$$

$$\oint_C \frac{dz}{(z - z_0)^n} = \int_0^{2\pi} \frac{1}{R^n}\Big(\cos nt - i \sin nt\Big)\left(\frac{dx}{dt} + i\frac{dy}{dt}\right) dt \tag{4.34}$$

$$= \int_0^{2\pi} \frac{i}{R^{n-1}}[\cos(n-1)t - i \sin(n-1)t]\, dt$$

$$= 0. \tag{4.35}$$

∎

We continue with examples of computation of line integrals using the definition.

☑4.3.2 Problem. Integrate $z^2 + \lambda\bar{z}$ around the rectangular contour bounded by real and imaginary axes and the lines $\operatorname{Re} z = a$ and $\operatorname{Im} z = b$.

Solution. We denote the four consecutive vertices at the points $0, a, a + ib,$ and ib as A, B, C, and D, and write the integral around the rectangle as a sum of integrals along the four sides to get

$$\oint_{\text{ABCDA}} f(z)\, dz = \int_{\text{AB}} f(z)\, dz + \int_{\text{BC}} f(z)\, dz + \int_{\text{CD}} f(z)\, dz + \int_{\text{DA}} f(z)\, dz, \tag{4.36}$$

with $f(z) = z^2 + \lambda\bar{z}^2$ for the present problem. The four integrals are computed as follows. Along the real axis $z = x$, and hence

$$\int_{\text{AB}} f(z)\, dz = \int_0^a (x^2 + \lambda x) dx = \frac{a^3}{3} + \lambda\frac{a^2}{2}. \tag{4.37}$$

Along the side BC we have $z = a + iy, 0 \leq y \leq b$, and $dz = idy$, so

$$\int_{\text{BC}} f(z)\, dz = \int_0^b [(a + iy)^2 + \lambda(a - iy)]\,(idy) \tag{4.38}$$

$$= \int_0^b [(-2ay + \lambda y) + i(a^2 - y^2 + \lambda y)]\, dy \tag{4.39}$$

$$= \frac{1}{2}(\lambda b^2 - 2ab^2) + \frac{i}{3}(3a^2 b - b^3 + 3\lambda ab). \tag{4.40}$$

Writing $\int_{\rm CD} = -\int_{\rm DC}$ and noting that $z = x + ib, 0 \le x \le a$, along DC, we get

$$\int_{\rm CD} f(z)\,dz = -\int_{\rm DC} f(z)\,dz = -\int_0^a [(x+ib)^2 + \lambda(x-ib)]\,dx \tag{4.41}$$

$$= -\frac{1}{6}(2a^3 - 6ab^2 + 3\lambda a^2) - i(a^2 b - \lambda ab). \tag{4.42}$$

Similarly, $\int_{\rm DA} = -\int_{\rm AD}$, and for the integral along AD we use $z = iy$, $0 \le y \le b$, and therefore

$$\int_{\rm DA} f(z)\,dz = -\int_{\rm AD} f(z)\,dz = -\int_0^b (-y^2 + i\lambda y)(i dy) \tag{4.43}$$

$$= \frac{i}{3}b^3 - \frac{1}{2}\lambda b^2. \tag{4.44}$$

Adding the four contributions we get

$$\oint_{\rm ABCDA} (z^2 + \lambda \bar{z})\,dz = 2i\lambda ab. \tag{4.45}$$

∎

In Chs. 6 and 7 we shall see that evaluation of an integral in the complex plane generally proceeds by making a transition to a closed contour and using the residue theorem. A direct use of the definition, as in the above examples, is of little importance as a method to compute the complex integrals, a notable exception being the class of indefinite integrals appearing in the Schwarz–Christoffel transformation; see §9.8. A few examples belonging to this class of integrals appearing in the Schwarz–Christoffel transformation will be discussed in §4.8.2.

⧐ Tutorial §§4.2, on setting up and computation of complex line integrals in the complex plane.

⧐ Exercises §§4.3, on computation of line integrals.

§4.4 Bounds on Integrals

Theorem 4.1 (Darboux's theorem). *Let f be a function such that $|f(z)| \le M$ for z on a closed contour C of length L. Then*

$$\left|\int_C f(z)dz\right| \le ML. \tag{4.46}$$

This result is known as Darboux's theorem.

An outline of the proof will now be given.

Proof. The line integral in a complex plane

$$\int_C f(z)dz \tag{4.47}$$

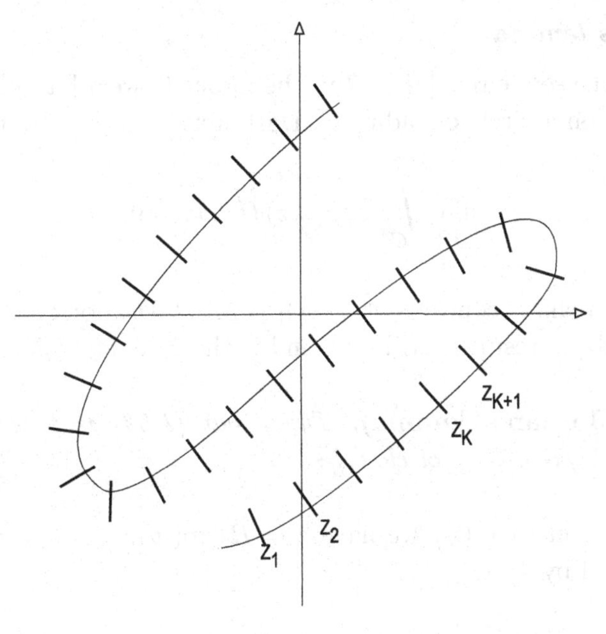

Fig. 4.1

can be defined as a limit of the Riemann sum by dividing the curve into small parts (see Fig. 4.1) and

$$\int_C f(z)dz = \lim \sum_k f(z_k)(z_{k+1} - z_k). \tag{4.48}$$

If M is a bound on $|f(z)|$ for the points on a curve C, i.e. $|f(z)| < M$, then

$$\lim |\sum_k f(z_k)(z_{k+1} - z_k)| \leq M \lim \left|\sum_k (z_{k+1} - z_k)\right| \tag{4.49}$$

$$\leq M \lim \sum_k |(z_{k+1} - z_k)| \tag{4.50}$$

$$\leq ML. \tag{4.51}$$

In writing the last step, we have used the fact that the limit of the sum $\sum_k |(z_{k+1} - z_k)|$ is just the arc length L of the curve C. This gives the desired result for a curve C of length L:

$$\left|\int_C f(z)dz\right| \leq ML. \tag{4.52}$$

∎

§4.4.1 *Jordan's lemma*

Let $C_R^+ (C\bar{R})$ be the semicircle $|z| = R$ in the upper (lower) half plane, see Fig. 4.2. If $|f(z)| \leq M/R^2$ on a circle of radius R, Darboux's theorem implies that

$$\lim_{R\to\infty} \int_{C_R^+} \exp(ikz) f(z) dz = 0. \tag{4.53}$$

If $f(z)$ satisfies a weaker requirement, $|f(z)| \leq M/R$, Darboux's theorem cannot be applied but the above result is still true and is the content of Jordan's lemma.

Theorem 4.2 (Jordan's lemma). *The result (4.53) is true even when $f(z)$ satisfies $|f(z)| < M/R$ on the circle C_R^+.*

Proof. For the semicircle C_R^+ we have $z = R\exp(i\theta)$, $0 < \theta < \pi$. Let the above integral be denoted by I. Then

$$dz = iR\exp(i\theta)d\theta. \tag{4.54}$$

and

$$\begin{aligned} I &= \int_0^\pi \exp\left(ikR\cos\theta - kR\sin\theta\right) f\left(Re^{i\theta}\right) iRe^{i\theta}\, d\theta, \\ |I| &\leq \int_0^\pi \exp\left(-kR\sin\theta\right) |f(Re^{i\theta})| R\, d\theta \\ &\leq M \int_0^\pi \exp(-kR\sin\theta)\, d\theta. \end{aligned} \tag{4.55}$$

We split the above integral into an integral from 0 to $\pi/2$ and another one from $\pi/2$ to π. In the second integral a change of variables is made from θ to $\phi = \pi - \theta$.

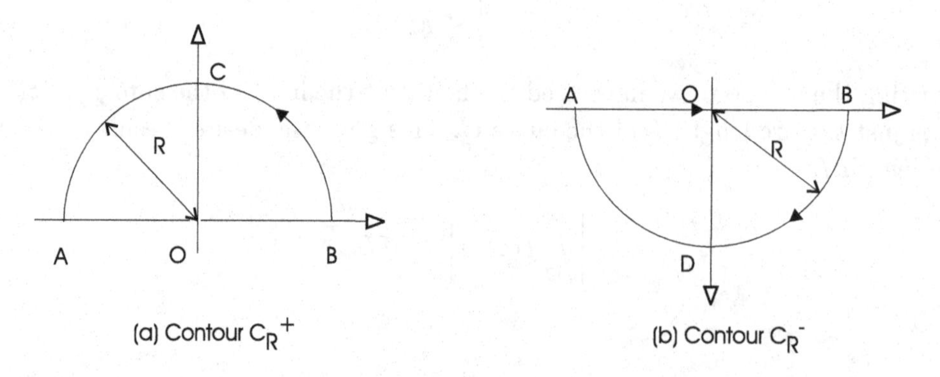

Fig. 4.2

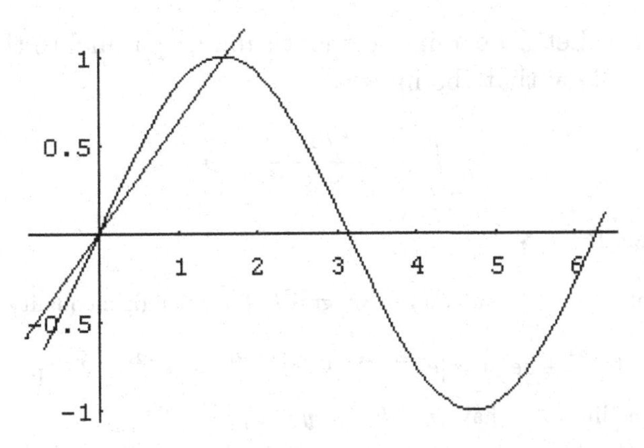

Fig. 4.3

This gives

$$\begin{aligned}\int_0^{\pi} \exp(-kR\sin\theta)\, d\theta &= \int_o^{\pi/2} \exp(-kR\sin\theta)\, d\theta + \int_{\pi/2}^{\pi} \exp(-kR\sin\theta)\, d\theta \\ &= \int_0^{\pi/2} \exp(-kR\sin\theta)\, d\theta + \int_0^{\pi/2} \exp(-kR\sin\phi)\, d\phi \\ &= 2\int_0^{\pi/2} \exp(-kR\sin\theta)\, d\theta. \end{aligned} \tag{4.56}$$

From the plot of $\sin\theta$ in Fig. 4.3 it is seen that we have the inequality $\sin\theta \geq \frac{2\theta}{\pi}$ for the range $0 \leq \theta \leq \pi/2$. Hence, we get

$$|I| \leq 2M \int_0^{\frac{\pi}{2}} \exp(-2kR\theta/\pi)\, d\theta \tag{4.57}$$

$$\leq 2M\frac{\pi}{2kR}[1 - \exp(-2kR)]. \tag{4.58}$$

Hence, for $k > 0$, $\lim_{R\to\infty} I = 0$. For $k < 0$, a similar result holds if instead of C_R^+ the contour C_R^- is used. ∎

§4.5 Examples

In the following examples, and in the rest of this chapter, we shall use C_R to denote the *full circle* $\{z||z| = R\}$, C_R^+ to denote the *semicircular arc* BCA of C_R in the upper half plane given by $\{z = Re^{i\theta}|0 \leq \theta \leq \pi\}$, and by C_R^- we denote the similar *semicircle* BDA in the lower half plane (Fig. 4.2). Also, γ_R will denote an *arbitrary arc* of the circle C_R. As an illustration of the use of Darboux's theorem we will discuss some examples which will be needed in the chapter on applications of contour integration. We begin with a few very simple examples of the use of Darboux's theorem.

☑**4.5.1 Problem.** Let L be a line segment running parallel to the imaginary axis from R to $R+iL$. Show that the integral

$$I = \int_L \frac{e^{pz}\,dz}{e^{-2z}+3e^{5z}}, \quad p < 5,$$

tends to zero when $R \to \infty$.

Solution. We want to put a bound on the integral I. The triangle inequality $|\lambda+\mu| \geq \big||\lambda|-|\mu|\big|$ implies that

$$\left|e^{-2z}+3e^{5z}\right| = \left|e^{-2x-2iy}+3e^{5x+5iy}\right| \geq \left|e^{-2x}-3e^{5x}\right|. \tag{4.59}$$

Since for points on the line L we have $x = R, 0 < y < L$,

$$\left|\frac{e^{pz}}{e^{-2z}+3e^{5z}}\right| = \left|\frac{e^{p(R+iy)}}{3e^{5R+5iy}+e^{-2R-2iy}}\right| \leq \left|\frac{e^{pR}}{3e^{5R}-e^{-2R}}\right| = \left|\frac{e^{(p-5)R}}{3-e^{-7R}}\right|.$$

Hence, Darboux's theorem implies the inequality

$$|I| = \left|\int_0^L \frac{e^{pz}\,dz}{e^{-2z}+3e^{5z}}\right| \leq L\left|\frac{e^{(p-5)R}}{3-e^{-7R}}\right|, \tag{4.60}$$

which tends to zero as $R \to \infty$ for $p < 5$. ∎

☑**4.5.2 Problem.** Using Darboux's theorem show that

$$\lim_{R\to\infty} \int_{C_R^+} \frac{dz}{z^2+1} = 0. \tag{4.61}$$

Solution. Note that, as a simple consequence of the triangle inequality $|\lambda+\mu| \geq \big||\lambda|-|\mu|\big|$, we have

$$|z^2+1| \geq \big||z^2|-1\big|. \tag{4.62}$$

Hence,

$$\frac{1}{|z^2+1|} \leq \frac{1}{\big||z^2|-1\big|}, \tag{4.63}$$

and for points on C_R^+ we have

$$\frac{1}{|z^2+1|} \leq \frac{1}{|R^2-1|}. \tag{4.64}$$

Therefore, using the result (4.51), we get

$$\left|\int_{C_R^+} \frac{dz}{z^2+1}\right| \leq \frac{\pi R}{|R^2-1|}. \tag{4.65}$$

Taking the limit $R \to \infty$ we get the desired result (4.61). A similar result is true if the integration contour is taken to be the semicircle C_R^- or any arc of the circle C_R in the lower half plane. ∎

The result of the above problem is easily extended to several integrals involving rational functions. As preparation, we begin with a result on large $|z|$ behavior of a rational function $Q(z)$.

Let $Q(z) = \frac{p(z)}{q(z)}$ be a rational function of z, with $p(z)$ and $q(z)$ being polynomials of degrees m and n, respectively. Assuming the forms

$$p(z) = a_m z^m + a_{m-1} z^{m-1} + \cdots + a_1 z + a_0, \tag{4.66}$$

$$q(z) = b_n z^n + b_{n-1} z^{n-1} + \cdots + b_1 z + b_0, \tag{4.67}$$

we have the result that the inequality

$$|Q(z)| \leq M R^{m-n} \tag{4.68}$$

holds for the points on a circle $|z| = R$ for large R. The above result and Darboux's theorem will now be used to derive the results in the following examples.

↬ The proof of the above estimates is left as an exercise.

§4.5.3 Short examples. Let $Q(z)$ be a rational function as described above with $n > m + 1$.

(a) $\lim_{R\to\infty} \int_{C_R^\pm} Q(z)\, dz = 0$, because for large R

$$\left| \int_{C_R^\pm} Q(z)\, dz \right| \leq M R^{m-n} \times \pi R = \pi M R^{m+1-n},$$

which tends to zero as $R \to \infty$ if $m + 1 < n$.

(b) Following the above example, one also gets

$$\left| \int_{C_R^\pm} z^p Q(z)\, dz \right| \leq M R^{m+p-n} \times \pi R = \pi M R^{m+p+1-n},$$

which goes to zero for large R if $m + p + 1 - n < 0$.

(c) $\displaystyle\lim_{R\to\infty} \int_{\gamma_R} Q(z) \mathrm{Log}\, z\, dz = 0$, because for $R \to \infty$ we have the bound

$$\left|\mathrm{Log}\, z\right| = \sqrt{(\ln r)^2 + \theta^2} \leq \sqrt{(\ln R)^2 + 4\pi^2}.$$

Therefore, we get

$$\left| \int_{\gamma_R} Q(z) \mathrm{Log}\, z\, dz \right| \leq M \sqrt{(\ln R)^2 + 4\pi^2} \times R^{m-n} \times \pi R,$$

which vanishes as $R \to \infty$ if $m + 1 - n \leq 0$. This result will also hold for the full circle C_R or an arc of C_R.

(d) If $b_0 \neq 0$, the limit $\lim_{|z|\to 0} Q(z)$ exists and equals (a_0/b_0). Therefore, as $\rho = |z|$ approaches 0, $Q(z)$ is bounded by a constant — say, K — on the circle γ_ρ. Hence,

$$\int_{\gamma_\rho} \mathrm{Log}\, z\, Q(z)\, dz \leq K \sqrt{(\mathrm{Log}\, \rho)^2 + 4\pi^2}\, (2\pi\rho) \to 0$$

as $\rho \to 0$.

(e) If we consider the limit $R \to \infty$ of integrals of the type $\int_{C_R^+} e^{ikz} Q(z)\,dz$, the answer depends on the sign of k and we have to be more careful. This is because

$$|\exp(ikz)| = |\exp(ikx - ky)| = \exp(-ky),$$

which for large $|z|$ does not go to zero for $ky < 0$. Thus, we must consider the $k > 0$ and $k < 0$ cases separately. First, consider the case $k > 0$. For the points on the semicircle C_R^+ in the upper half plane, we will have $y \geq 0, ky \geq 0$. This gives

$$\left| \int_{C_R^+} \exp(ikz)\, Q(z)\,dz \right| \leq MR^{m-n} \max(e^{-ky}) \times \pi R \leq MR^{m+1-n} \quad (\because e^{-ky} \leq 1)$$

and the limit exists and is zero for $m + 1 - n < 0$. However, in this case $\int_{C_R^-} e^{ikz}\, Q(z)\,dz$ diverges as $R \to \infty$. It should now be obvious that, in the second case, $k < 0$, $\int_{C_R^-} e^{ikz} Q(z)\,dz$ becomes zero in the limit $R \to \infty$, but $\int_{C_R^+} e^{ikz}\, Q(z)\,dz$ diverges.

▶ The results of the above examples will be useful in Ch. 7, on application of the method of contour integration to improper integrals.

↯4.5.4 Short examples. The following three results can easily be established by the method used in the proof of Jordan's lemma.

(a) If γ_1 is an arc of the circle $|z| = R$ lying in the sector $0 < \theta < \pi/4$, we would have $\lim_{R\to\infty} \int_{\gamma_1} \exp(ipz^2)\,dz = 0$.

(b) Let γ be the arc of the circle $|z| = R$ lying between the radial lines $\theta = 0$ and $\theta = \pi/2N$. Following the proof of Jordan's lemma one can show that $\int_{\gamma_2} \exp(ipz^N)\,dz$, for $N > 1$, vanishes in the limit $R \to \infty$.

(c) The above result, ↯4.5.3(e), is true even when $n - m = 1$.

§4.6 Cauchy's Fundamental Theorem

§4.6.1 *Cauchy's theorem*

We shall first state and prove the result in the form it was originally given by Cauchy.

Theorem 4.3 (Cauchy's theorem). *Let C be a simple closed contour and $f(z)$ be a function such that $\frac{\partial f}{\partial z}$ exists and is continuous everywhere inside and on C. Then*

$$\oint_C f(z)dz = 0. \tag{4.69}$$

Proof. A proof of this theorem follows from Green's theorem for a closed curve in a plane. The existence of the derivative w.r.t. z implies that the real and imaginary parts, $u(x, y)$ and $v(x, y)$, of the function $f(z)$ satisfy the Cauchy Riemann equations

$$\frac{\partial u}{\partial x} = \frac{\partial v}{\partial y}, \quad \frac{\partial v}{\partial x} = -\frac{\partial u}{\partial y}. \tag{4.70}$$

From the conditions of the theorem, it then follows that all the above four partial derivatives are continuous inside and on C. Let D denote the set of points enclosed by C. Green's theorem implies that

$$\oint_C (vdx + udy) = \iint_D \left(\frac{\partial u}{\partial x} - \frac{\partial v}{\partial y} \right) dxdy = 0, \tag{4.71}$$

$$\oint_C (udx - vdy) = -\iint_D \left(\frac{\partial u}{\partial y} + \frac{\partial v}{\partial x} \right) dxdy = 0, \tag{4.72}$$

where the Cauchy–Riemann equations have been used. The above results together with

$$\oint_C f(z)dz = \oint_C (udx - vdy) + i \oint_C (vdx + udy) \tag{4.73}$$

give the desired Cauchy theorem. ∎

It was shown by Goursat that the condition requiring the continuity of derivatives can be dropped. One then has the following result.

Theorem 4.4. *If a function $f(z)$ is analytic inside and on a simple closed contour C then*

$$\oint_C f(z)dz = 0.$$

In the form stated above the Cauchy–Goursat theorem is very useful and is sufficient for most purposes. It is the most important theorem in complex analysis, from which all the other results on integration and differentiation follow. We shall omit the proof of the Cauchy–Goursat theorem, which can be found in most books mentioned in the bibliography.

§4.6.2 *Deformation of contours*

An important consequence of Cauchy's theorem is that, under certain general conditions, the contour of integration can be deformed without changing the value of the integral. This is a powerful tool in applications of the method of contour integration.

Theorem 4.5. *Let z_1 and z_2 be two points in the complex plane. In addition, let Γ_1 and Γ_2 be two arcs connecting the points z_1 and z_2, see Fig. 4.4. If a function f is analytic inside the region enclosed between Γ_1 and Γ_2 and on the two arcs,*

$$\int_{\Gamma_1} f(z)dz = \int_{\Gamma_2} f(z)dz. \tag{4.74}$$

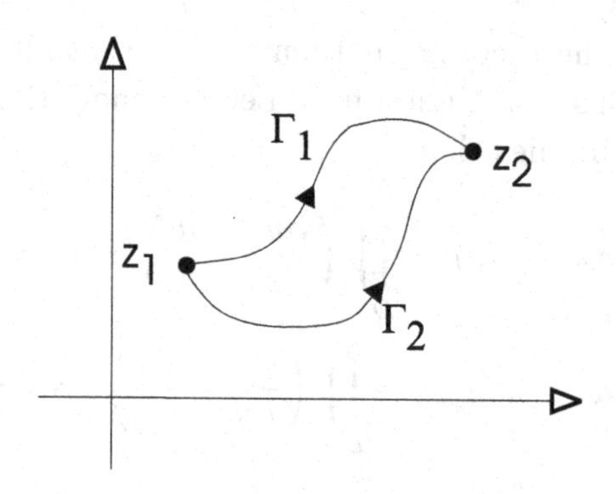

Fig. 4.4

Proof. Under the given conditions the function $f(z)$ is analytic in the domain between Γ_1 and Γ_2. Let $\tilde{\Gamma}_1$ be the countour Γ_1 traversed from z_2 to z_1. Utilizing the Cauchy fundamental theorem the integral $\oint f(z)dz$ along the closed contour consisting of $\tilde{\Gamma}_1$ and Γ_2 vanishes. Hence,

$$\int_{\tilde{\Gamma}_1} f(z)\,dz + \int_{\Gamma_2} f(z)\,dz = 0. \tag{4.75}$$

Since $\int_{\tilde{\Gamma}_1} f(z)\,dz = -\int_{\Gamma_1} f(z)\,dz$, we get the desired result Eq. (4.74). ∎

This result means that the contour Γ_1 can be deformed to a new contour, Γ_2, without changing the value of the integral if the integrand is analytic on the initial and final contours and in the domain enclosed between the two contours.

Theorem 4.6. *Let C be a simple closed curve. In addition, let a function f be analytic everywhere inside C except in a domain D [see Fig. 4.5(a)]. Also, let Γ, be any curve enclosing D. If f is analytic on C, on Γ, and between C and Γ, then*

$$\oint_C f(z)dz = \oint_\Gamma f(z)dz. \tag{4.76}$$

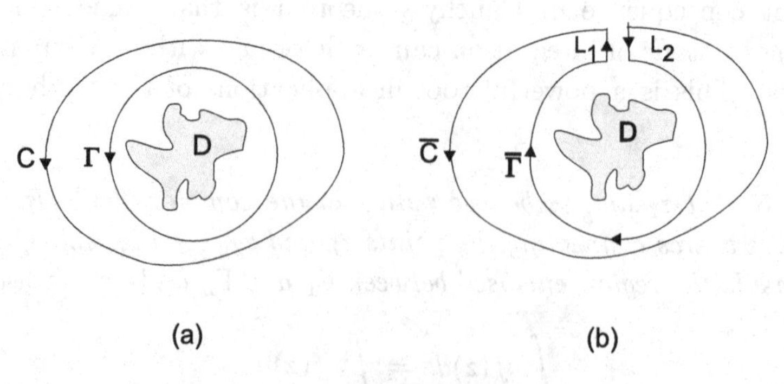

Fig. 4.5

Proof. To prove that the integrals along C and Γ will be equal, consider the integral of $f(z)$ along the closed contour γ of Fig. 4.5(a), consisting of $\bar{C}, \bar{\Gamma}$ and parallel lines L_1, L_2, as shown in Fig. 4.5(b). Here $\bar{C}, \bar{\Gamma}$ are obtained by cutting the contours C, Γ and taking $\bar{\Gamma}$ oriented clockwise. As the function $f(z)$ is analytic in the domain enclosed by the contour γ, $\oint_\gamma f(z)\, dz$ vanishes and hence

$$\int_{\bar{\Gamma}} f(z)\, dz + \int_{L_1} f(z)\, dz + \int_{L_2} f(z)\, dz + \int_{\bar{C}} f(z)\, dz = 0. \tag{4.77}$$

Next, take the limit in which the separation between the two lines L_1, L_2 goes to zero. In this limit $\int_{L_1} f(z)\, dz + \int_{L_2} f(z)\, dz = 0$ and we get

$$\oint_{\bar{\Gamma}} f(z)\, dz + \oint_{\bar{C}} f(z)\, dz = 0, \tag{4.78}$$

implying the result (4.76). ∎

It is obvious that above result in Theorem 4.6 can be generalized as follows. If f is a function which is analytic on a closed curve C and at all points on C except the points in the sets $D_1, D_2, \ldots, D_n$. (Fig. 4.6), then

$$\oint_C f(z)dz = \sum_k \oint_{\Gamma_k} f(z)dz, \tag{4.79}$$

where Γ_k are curves enclosing the domains D_k, as shown in Fig. 4.6.

§4.6.3 *Indefinite integral*

If $f(z)$ is an analytic function in a simply connected domain D, and $z, z_0 \in D$, the function $F(z)$, defined by the integral

$$F(z) = \int_{z_0}^{z} f(\xi)d\xi, \tag{4.80}$$

is called the *indefinite integral* of $f(z)$. We already know from Theorem 4.5 that the value of the integral is independent of the path joining the end points z_0 and z

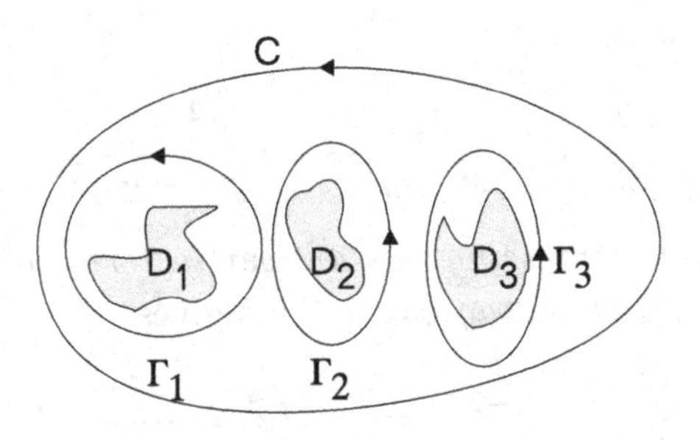

Fig. 4.6

as long as the path lies in the domain D. The indefinite integral has the following properties:

(*1) $F(z)$ is analytic in the domain D.
(*2) The function $f(z)$ is equal to the derivative of $F(z)$ w.r.t. z, i.e.

$$f(z) = \frac{dF(z)}{dz}.$$

(*3) The definite integral of an analytic function can be written down and equals the difference of the values of the indefinite integral, $F(z)$, at the two end points:

$$\int_{z_1}^{z_2} f(\xi)d\xi = F(z_2) - F(z_1). \tag{4.81}$$

(*4) The function $F(z)$, as defined in Eq. (4.80), obeys $F(z_0) = 0$. Changing the lower limit to a point other than z_0 is equivalent to changing $F(z)$ by addition of a constant.
(*5) The above properties are very similar to those of the indefinite integral which one comes across for the functions of a real variable.

An application. As an example of application of the indefinite integral, the Milne–Thomson method of finding a harmonic function conjugate to a given function will be described. The central point of this method is to note that the derivative of an analytic function, $\frac{df}{dz}$, can be written in terms of the partial derivatives of the real, or imaginary, part alone. Let us assume that the real part of an analytic function $f(z)$ is given as

$$u(x,y) = \frac{y}{x^2+y^2},$$

and we have to find the imaginary part, $v(x,y)$, of the function. The derivative of the function $f(z)$ can be written as

$$\frac{df}{dz} = \frac{\partial u}{\partial x} + i\frac{\partial v}{\partial x} = \frac{\partial u}{\partial x} - i\frac{\partial u}{\partial y} = \frac{\partial v}{\partial y} + i\frac{\partial v}{\partial x}. \tag{4.82}$$

We would then get

$$\frac{df}{dz} = -\frac{2xy}{(x^2+y^2)^2} - i\frac{x^2-y^2}{(x^2+y^2)^2}. \tag{4.83}$$

The next step consists in *expressing this answer in terms of the variables* z *and* y (by writing $x = z - iy$), *and setting* $y = 0$. This gives

$$\frac{df}{dz} = -iz^{-2} \quad \Rightarrow \quad f(z) = -\frac{i}{z} + C. \tag{4.84}$$

- ⊳ Quiz §§4.4, on applying Cauchy's theorem.
- ⊳ Exercise §§4.5, on evaluation of line integrals in the complex plane by transformation into integrals over a real variable.
- ⊳ Exercise §§4.6, on computation of line integrals illustrating consequences of Cauchy's theorem and the properties of the indefinite integral.

§4.7 Transforming Integrals over a Real Interval into Contour Integrals

In §4.3 we have seen how a line integral in the complex plane can be transformed into an integral over a real variable. We now present examples to show the *reverse process, i.e. transition from an integral over a real variable to an integral over a contour in the complex plane.* This process usually leads to a line integral over an open contour. However, for applications to computation of improper integrals, one invariably needs to make a transition to a *closed* contour and make use of the residue theorem to be discussed in Ch. 6. As a first step here we show, by means of a few selected examples, how one can make a transition to a closed contour.

§4.7.1 *Adding line segments or circular arcs to close the contour*

The examples of transforming improper integrals over real parameters as contour integrals in the complex plane along a closed contour that appear in this subsection require closing of the contour by adding one or more extra pieces which may be a straight line or a circular arc.

⨍4.7.1 Short examples. An improper integral, such as $\int_{-\infty}^{\infty} f(x)dx$, is a limiting case of a proper integral, $\int_{-R}^{R} f(x)dx$, which in turn can be trivially written as a line integral in the complex plane. The line integral so obtained can be transformed into a closed contour integral by making use of results such as those discussed in ⨍4.5.3.

Let AB denote the line segment on the real axis from $-R$ to R and BCA denote the semicircle C_R^+, in the upper half plane, with AB as diameter, and BDA a similar semicircle C_R^- in the lower half plane, see Fig. 4.2. In addition, let $Q(x) = p(x)/q(x)$ be a rational function with as in short examples ⨍4.5.3.

(a) The integral $\int_{-\infty}^{\infty} Q(x)dx$ can be transformed into a closed contour integral easily by making use of

$$\int_{-\infty}^{\infty} Q(x)dx = \lim_{R\to\infty} \int_{-R}^{R} Q(x)\,dx = \lim_{R\to\infty} \int_{\mathrm{AB}} Q(z)\,dz \tag{4.85}$$

$$= \lim_{R\to\infty} \int_{\mathrm{AB}} Q(z)\,dz + \lim_{R\to\infty} \int_{\mathrm{BCA}} Q(z)\,dz \tag{4.86}$$

$$= \lim_{R\to\infty} \oint_{\mathrm{ABCA}} Q(z)\,dz, \tag{4.87}$$

where the extra second term $\int_{\rm BCA}$, in the second step, is zero by virtue of the result in ∮4.6(a).

(b) A similar scheme works for $\int_{-\infty}^{\infty} e^{ikx}Q(x)dx$. Assuming that $k > 0$,

$$\int_{-\infty}^{\infty} e^{ikx}Q(x)dx = \lim_{R\to\infty}\int_{-R} Re^{ikx}Q(x)\,dx = \lim_{R\to\infty}\int_{\rm AB} e^{ikz}Q(z)\,dz \tag{4.88}$$

$$= \lim_{R\to\infty}\int_{\rm AB} e^{ikx}Q(z)\,dz + \lim_{R\to\infty}\int_{\rm BCA} e^{ikz}Q(z)\,dz \tag{4.89}$$

$$= \lim_{R\to\infty}\oint_{\rm ABCA} e^{ikz}Q(z)\,dz. \tag{4.90}$$

For $k < 0$, we would get

$$\int_{-\infty}^{\infty} e^{ikx}Q(x)dx = \lim_{R\to\infty}\oint_{\rm ABDA} e^{ikz}Q(z)\,dz. \tag{4.91}$$

§4.7.2 *Translation and rotation of the contour*

In this subsection we discuss two examples of Gaussian integrals, which appear frequently in science and engineering applications. Chosen for reasons of simplicity and importance, these examples of improper integrals make use of translation and rotation of contours in the complex plane, a deformation process justified by appealing to Cauchy's fundamental theorem. That the evaluation of the integrals, selected for discussion, makes essential use of Cauchy's fundamental theorem is often not mentioned at all. We present only the essentials, and the details of actual *evaluation* of the improper integrals are left as part of a tutorial workout.

Translation of the contour by a complex number

We consider evaluation of the Fourier transform, $\int_{-\infty}^{\infty}\exp(-x^2 + ikx)dx$, by completing the squares and writing it as

$$\int_{-\infty}^{\infty}\exp(-x^2+ikx)dx = \int_{-\infty}^{\infty}\exp\left[-\left(x-\frac{ik}{2}\right)^2 - \frac{k^2}{4}\right]dx$$

$$= \exp\left(-\frac{k^2}{4}\right)\int_{-\infty}^{\infty}\exp\left[-\left(x-\frac{ik}{2}\right)^2\right]dx. \tag{4.92}$$

At this point one equates the integral on the right hand side with

$$\int_{-\infty}^{\infty}\exp\left(-t^2\right)dt = \sqrt{\pi}, \tag{4.93}$$

giving $\sqrt{\pi}\exp(-k^2/4)$ as the answer for the Fourier transform. An alert reader will ask the following questions:

- *Can a shift in the integration variable x by a complex number, $x \to t = x - \frac{ik}{2}$, be given a meaning?*
- *Is it justified to take limits for t again from $-\infty$ to ∞?*

The answer to both these questions is yes, and a full justification requires the use of Cauchy's fundamental theorem. The shift can be given a meaning by interpreting the last integral in Eq. (4.92) as a contour integral along a line, $\mathrm{Im}\, t = -k/2$, parallel to the real axis in the complex t plane:

$$\int_{-\infty}^{\infty} \exp(-(x - ik)^2)\, dx = \int_{\mathrm{Im}t=-k/2} \exp(-t^2)\, dt. \tag{4.94}$$

Cauchy's theorem can be used to show that it is allowed to shift the contour from the line, $\mathrm{Im}\, t = -k/2$, to the real axis without changing the value. This gives

$$\int_{\mathrm{Im}t=k/2} \exp(-t^2)\, dt = \int_{\mathrm{Im}t=0} \exp(-t^2)\, dt, \tag{4.95}$$

and the right hand side is simply equal to the Gaussian integral (4.93).

Scaling by a complex number and rotation of the contour

We discuss an example of an integral which can be evaluated by justifying scaling of the real integration variable by a complex number. If in the integral

$$\int_0^{\infty} \exp(ix^2)\, dx, \tag{4.96}$$

we change the variable to t by writing $x = \exp(i\pi/4)t$, the integrand $\exp(ix^2)$ becomes equal to $\exp(-t^2)$, and one would like to ask the following questions:

- *Can the scaling of the integration variable x by a complex number, such as $\exp(-i\pi/4)$, be given a meaning?*
- *Can the limits for the integral over t be taken to be from 0 to ∞?*
- *Of the two possible changes of variable $t = x\exp(\pm i\pi/4)$, which is correct and which is incorrect?*

The substitution $x = e^{i\pi/4}t$ and the taking of limits of t from 0 to ∞ is equivalent to a line integral along the ray $\arg(x) = \pi/4$ in the complex x plane. This leads to

$$\int_{\arg(x)=\pi/4} \exp(ix^2)\, dx = e^{i\pi/4} \int_0^{\infty} \exp(-t^2)\, dt = \frac{\sqrt{\pi}}{2}\, e^{i\pi/4}. \tag{4.97}$$

Cauchy's fundamental theorem can be used to justify rotation of the contour from the ray $\arg x = \pi/4$ to the real x axis without changing its value, giving the equality of the two integrals

$$\int_{\arg(x)=\pi/4} \exp(ix^2)\,dx = \int_{\arg(x)=0} \exp(ix^2)\,dx. \tag{4.98}$$

Noting that the right hand side trivially equals the integral (4.96) we started with, we would get

$$\int_0^\infty \exp(ix^2)\,dx = \frac{\sqrt{\pi}}{2}\,e^{i\pi/4}. \tag{4.99}$$

It may emphasized that use of an alternate definition, $x = \exp(-i\pi/4)t$, leads to an integral along the ray $\arg x = -\pi/4$ and that, in this case, the rotation of the contour back to the real positive axis is not permissible.

⊳ Tutorials §§4.7 and §§4.8, on details of justifying scaling and translation by complex numbers and evaluation of Gaussian-like improper integrals.
⊳ Exercise §§4.9, for practice in evaluating integrals related to the Gaussian integral.
⊳ Exercise §§4.10, on evaluation of improper integrals by rotation of contours.

§4.8 Integration of Multivalued Functions

In this chapter, the theory of integration in the complex plane has been developed, culminating in discussions on techniques of contour integration for evaluation of a few classes of improper integrals. Because of its specialized and technical nature, the integration of multivalued functions has not been included as a part of the discussions so far.

The multivalued functions were introduced in Ch. 3. They have branch point singularities and are defined in the cut plane. The definition and usage of multivalued functions require special care, as discussed in Ch. 3. The reader is urged to revise Ch. 3 before attempting to take the plunge into this section. Here, several examples of definite and indefinite line integrals involving multivalued functions are discussed. The examples in §§4.8.4 present the basics of a technique known as integration around a branch cut, and prepare the reader for a complete discussion in Ch. 7 on contour integration.

§4.8.1 *Line integrals*

In this subsection we begin with a simple example of setting up a line integral of a multivalued function.

☑**4.8.1 Problem.** Integrate $\sqrt{z}$ along the unit circle taking the branch cut along (a) The positive real axis, (b) The negative real axis.

Solution.

(a) When the branch cut is along the positive axis, we take

$$z^{1/2} = r^{1/2} \exp\left(\frac{i\theta}{2}\right), \quad 0 < \theta < 2\pi. \tag{4.100}$$

Remembering that along the unit circle $z = e^{i\theta}$, $dz = e^{i\theta} i d\theta$, we get

$$\oint \sqrt{z}\, dz = \int_0^{2\pi} \exp\left(\frac{3i\theta}{2}\right) i d\theta = i \int_0^{2\pi} \left[\cos\left(\frac{3\theta}{2}\right) + i \sin\left(\frac{3\theta}{2}\right)\right] d\theta = -\frac{4}{3}. \tag{4.101}$$

(b) For the branch cut along the negative real axis, if we take svb defined by

$$z^{1/2} = r^{1/2} \exp\left(\frac{i\theta}{2}\right), \quad -\pi < \theta < \pi, \tag{4.102}$$

we will get

$$\oint \sqrt{z}\, dz = \int_{-\pi}^{\pi} \exp\left(\frac{3i\theta}{2}\right) i d\theta = i \int_{-\pi}^{\pi} \left[\cos\left(\frac{3\theta}{2}\right) + i \sin\left(\frac{3\theta}{2}\right)\right] d\theta = -\frac{4i}{3}. \tag{4.103}$$

■

We now present a few examples of setting up the line integral of multivalued functions. Some of them have been selected keeping in mind applications to Schwarz–Christoffel transformations.

§4.8.2 *Indefinite integrals*

☑**4.8.2 Problem.** Find $w(z)$ for real z by integrating

$$w(z) = \int_0^z \xi^{-1/2} (\xi - 1)^{1/2}\, d\xi, \tag{4.104}$$

taking the branch cut for $z^{-1/2}$ along the negative imaginary axis and that for $(z-1)^{1/2}$ along $\arg(z-1) = \frac{3\pi}{2}$.

Solution. We use the definitions

$$z^{-1/2} = r^{-1/2} e^{-i\theta/2}, \quad -\frac{\pi}{2} < \theta < \frac{3\pi}{2}, \tag{4.105}$$

$$(z-1)^{1/2} = |z-1|^{-1/2} e^{i\phi/2}, \quad -\frac{\pi}{2} < \phi < \frac{3\pi}{2}, \tag{4.106}$$

where $\phi = \arg(z-1)$. The branch points of the integrand are at $x = 0, 1$ and hence we shall consider the cases $x < 0, 0 < x < 1$ and $x > 1$ separately.

Case I. First, we consider z real $(= x)$ and $x < 0$. In this case,

$$\sqrt{z} = i\sqrt{|x|} \Rightarrow z^{-1/2} = \frac{-i}{\sqrt{|x|}} = \frac{i\sqrt{|x|}}{x}, \tag{4.107}$$

$$\sqrt{z-1} = i\sqrt{|x-1|}. \tag{4.108}$$

Using the above expressions and substituting $x = -t$, we obtain

$$w(x) = -\int_0^x \frac{1}{x}(|x||x-1|)^{1/2}\,dx = \int_{-x}^0 \frac{\sqrt{t^2+t}}{t}\,dt \tag{4.109}$$

$$= -\sqrt{x^2-x} - \frac{1}{2}\ln\left[1-2x+2\sqrt{x(x-1)}\right] \tag{4.110}$$

$$= -|x|^{1/2}|x-1|^{1/2} - \ln\left(|x|^{1/2}+|x-1|^{1/2}\right), \tag{4.111}$$

where $R = a + bx + cx^2$ and use has been made of the standard results

$$\int \frac{\sqrt{R}}{x}\,dx = \sqrt{R} + a\int\frac{dx}{x\sqrt{R}} + \frac{b}{2}\int\frac{dx}{\sqrt{R}}, \tag{4.112}$$

$$\int\frac{dz}{\sqrt{R}} = \begin{cases} \frac{1}{\sqrt{c}}\ln(2\sqrt{cR}+2cx+b), & c>0, \\ \frac{1}{\sqrt{c}}\cosh^{-1}\frac{2cx+b}{\sqrt{\Delta}}, & c>0,\quad \Delta>0, \\ \frac{-1}{\sqrt{-c}}\sin^{-1}\frac{2cx+b}{\sqrt{-\Delta}}, & c<0,\quad \Delta<0, \\ \ln(2cx+b), & c>0,\quad \Delta=0. \end{cases} \tag{4.113}$$

Case II. Next, we consider z real $(= x)$ and $0 < x < 1$. We now have

$$z^{-1/2} = x^{-1/2}, \quad (z-1)^{1/2} = i|(x-1)|^{1/2} = i(1-x)^{-1/2}. \tag{4.114}$$

Proceeding as above we get

$$w(x) = i\sqrt{x(1-x)} + i\sin^{-1}\sqrt{x}. \tag{4.115}$$

For use below note that $w(1) = \dfrac{i\pi}{2}$.

Case III. Finally, for $x > 1$ we get

$$z^{-1/2} = x^{-1/2}, \quad (z-1)^{1/2} = (x-1)^{1/2} \tag{4.116}$$

and, therefore,

$$w(x) = \int_0^1 z^{-1/2}(z-1)^{1/2}\,dz + \int_1^x z^{-1/2}(z-1)^{1/2}\,dz \tag{4.117}$$

$$= w(1) + \int_1^x t^{-1/2}(t-1)^{1/2}\,dt \tag{4.118}$$

$$= \frac{i\pi}{2} + \sqrt{x(x-1)} - \ln(\sqrt{x}+\sqrt{x-1}). \tag{4.119}$$

Thus, we have the final answer for values of $w(z)$ on the real axis:

$$w(x) = \begin{cases} -|x|^{1/2}|x-1|^{1/2} - \ln(|x|^{1/2}+|x-1|^{1/2}), & x<0, \\ i\sqrt{x(1-x)} + i\sin^{-1}\sqrt{x}, & 0<x<1, \\ \frac{i\pi}{2} + \sqrt{x(x-1)} - \ln(\sqrt{x}+\sqrt{x-1}), & x>1. \end{cases} \tag{4.120}$$

It is not very difficult to check that

$$f(z) = \frac{i\pi}{2} + \sqrt{z(z-1)} - \text{Log}\,(\sqrt{z} + \sqrt{z-1}) \tag{4.121}$$

satisfies $f(x) = w(x)$, for all $x \in \mathbb{R}$. ∎

↬ Proof of the above result is left as an exercise; Q.[13] in §§3.8.

☑**4.8.3 Problem.** Sketch the image of the real axis under the map $z \to w = F(z)$, where

$$F(z) = \frac{h}{\pi} \int_{-1}^{z} (z+1)^{1/2}(z-1)^{-1/2}\, dz + ih,$$

taking the branch cut for the square root in the lower half plane.

Solution. With the branch cuts in the lower half plane, let us select any svb of the two functions $(z+1)^{1/2}$ and $(z-1)^{-1/2}$ so as to have the values

$$(x+1)^{1/2} = \begin{cases} i|x+1|^{1/2}, & x < -1, \\ |x+1|^{1/2}, & x > -1, \end{cases} \tag{4.122}$$

$$(x-1)^{-1/2} = \begin{cases} -i|x-1|^{-1/2}, & x < 1, \\ |x-1|^{-1/2}, & x > 1, \end{cases} \tag{4.123}$$

for points on the real axis. Let $f(z)$ denote the product of svb's of the two functions $(z+1)^{1/2}$ and $(z-1)^{-1/2}$. Then

$$f(x) = \begin{cases} g(x), & x < -1, \\ -ig(x), & -1 < x < 1, \\ g(x), & x > e1, \end{cases} \tag{4.124}$$

where $g(x) = \left|(x+1)/(x-1)\right|^{1/2}$ is a positive function. By suitably splitting the integral, and using $\int_{-1}^{1} g(x)\, dx = \pi$, the function $F(x)$ for real x can be expressed as

$$F(x) = \begin{cases} -\dfrac{h}{\pi}\displaystyle\int_{1}^{|x|} g(-x)\, dx + ih, & x \le -1, \\ -\dfrac{ih}{\pi}\displaystyle\int_{-1}^{x} g(x)\, dx + ih, & -1 \le x \le 1, \\ \dfrac{h}{\pi}\displaystyle\int_{1}^{x} g(x)\, dx, & x \ge 1. \end{cases} \tag{4.125}$$

To trace the point $w = F(x)$ as x assumes real values, we take note of the following properties of $F(x)$:

- As x increases from $-\infty$ to -1, the real part of the function $F(x)$ varies from $-\infty$ to 0 and its imaginary part remains equal to ih.
- For x increasing from -1 to 1, $F(x)$ is pure imaginary and the imaginary part decreases from h to 0.
- For $x > 1$, $F(x)$ is a real positive and monotonically increasing function.

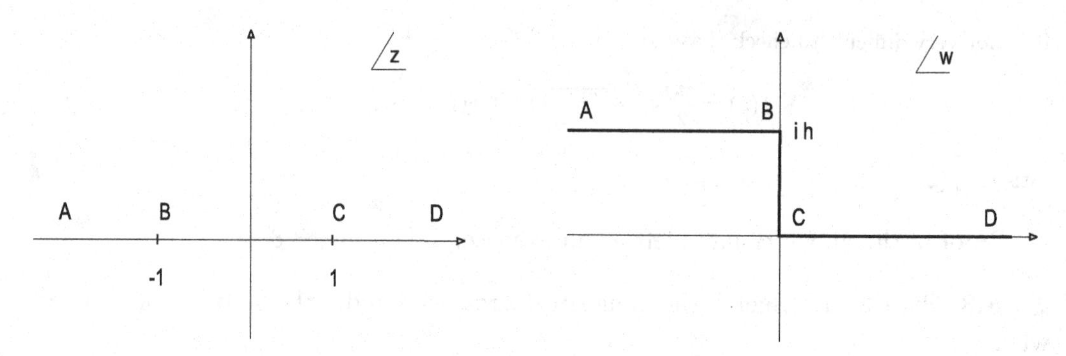

Fig. 4.7

The above results can now be used to trace the image of the real axis and we get the step as shown in Fig. 4.7 as the image in the w plane. ∎

Indefinite integrals of the type discussed above will appear again in §9.8 as Schwarz–Christoffel mapping of the real line onto polygons.

§4.8.3 *Case study of the indefinite integral* $\int \frac{dz}{z}$

The examples considered above involved the integrand of multivalued functions. Once an svb is selected and is used consistently, the computation of the line integral proceeds in a straightforward fashion. Use of indefinite integrals involving a multivalued function requires a lot more care, otherwise one may be led to wrong conclusions. This happens due to biases and prejudices acquired from the results and extensive usage of indefinite integrals of functions of a real variable. As an example, in the case of a real variable, it is correct to write

$$\int_a^b \frac{dx}{x} = \ln b - \ln a. \tag{4.126}$$

The same cannot be said about the indefinite integral in the complex plane. In particular, the result

$$\int_{z_1}^{z_2} \frac{dz}{z} = \log z_2 - \log z_1 \tag{4.127}$$

cannot be used in a *mechanical fashion* without paying attention to the svb selected for $\log z$ and to the path of integration from z_1 to z_2. In the following we present a detailed case study of the integral (4.127) to clarify some of the important issues connected with the indefinite integrals of multivalued functions.

Consider $\oint \frac{dz}{z}$ around a positively oriented rectangle ABCD, Fig. 4.8, with sides of lengths $2a$ and $2b$, and with corners at $a + ib, -a + ib, -a - ib,$ and $-a - ib$, respectively.

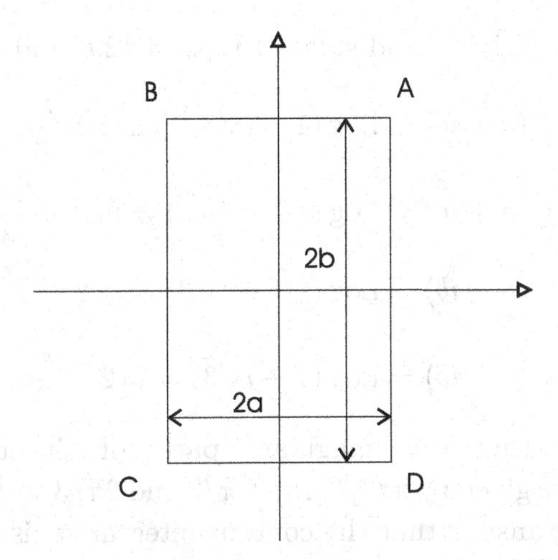

Fig. 4.8

One may attempt to use the indefinite integral to evaluate the contour integral by writing

$$\oint \frac{dz}{z} = I_{\mathrm{AB}} + I_{\mathrm{BC}} + I_{\mathrm{CD}} + I_{\mathrm{DA}}, \tag{4.128}$$

where

$$I_{\mathrm{AB}} = \int_{\mathrm{AB}} \frac{dz}{z} = \log(-a+ib) - \log(a+ib), \tag{4.129}$$

$$I_{\mathrm{BC}} = \int_{\mathrm{AB}} \frac{dz}{z} = \log(-a-ib) - \log(-a+ib), \tag{4.130}$$

$$I_{\mathrm{CD}} = \int_{\mathrm{CD}} \frac{dz}{z} = \log(a-ib) - \log(-a-ib), \tag{4.131}$$

$$I_{\mathrm{DA}} = \int_{\mathrm{DA}} \frac{dz}{z} = \log(a+ib) - \log(a-ib). \tag{4.132}$$

A beginner is likely to conclude that $\oint \frac{dz}{z}$ vanishes because the expressions on the right hand sides of (4.129)–(4.132) when substituted in (4.128) cancel pairwise. This answer is obviously wrong, because we know that the correct value of the integral is $2\pi i$, as can be seen by deforming the rectangle to a circle with the center at the origin and making use of the result for the integral along the circle obtained in ☑4.3.1.

In the above, the logarithm function, $\log z$, has been used without being careful about the choice of svb. One might conclude that consistent use of an svb for the logarithm in (4.129)–(4.132) will may resolve the problem. So, we assume that $a = 1$ and $b = \sqrt{3}$ for the sake of definiteness, and try computing the answers for the above four integrals taking the principle branch $\mathrm{Log}\, z$. An explicit computation of values of different terms on the right hand sides of Eqs. (4.129) and (4.130) gives

$$\text{Log}\,(a+ib) = \text{Log}\,(1+i\sqrt{3}) = \ln 2 + \frac{\pi i}{3}, \tag{4.133}$$

$$\text{Log}\,(-a+ib) = \text{Log}\,(-1+i\sqrt{3}) = \ln 2 + \frac{2\pi i}{3}, \tag{4.134}$$

$$\text{Log}\,(-a-ib) = \text{Log}\,(-1-i\sqrt{3}) = \ln 2 - \frac{2\pi i}{3}, \tag{4.135}$$

$$\text{Log}\,(a-ib) = \text{Log}\,(1-i\sqrt{3}) = \ln 2 - \frac{\pi i}{3}. \tag{4.136}$$

It can now be seen that the imaginary parts of the right hand sides of Eqs. (4.129)–(4.132) are given by $\pi/3, -4\pi/3, \pi/3$ and $2\pi/3$. Adding up these values, we get back the wrong answer that the contour integral vanishes. In fact, it should be obvious that this will be the case for every svb of the logarithm function.

To continue the case study, we will now demonstrate that all the four integrals have positive imaginary parts and, therefore, the sum cannot vanish.

Imaginary part of I_{AB}. We note that the oriented line BA is given by the parametric equation $z = x + ib, -a \le x \le a$ with $dz = dx$. Therefore,

$$I_{\text{AB}} = -I_{\text{BA}} = -\int_{\text{BA}} \frac{\bar{z}dz}{|z|^2} = -\int_{-a}^{a} \frac{x - ib}{x^2 + b^2}\,dx \tag{4.137}$$

$$= -\int_{-a}^{a} \frac{x}{x^2+b^2}\,dx + ib\int_{-a}^{a} \frac{1}{x^2+b^2}\,dx. \tag{4.138}$$

It is now obvious that $\text{Im}\,I_{\text{AB}} > 0$.

Imaginary part of I_{BC}. We note that the oriented line CB is given by the parametric equation $z = -a + iy, -b \le y \le b$ with $dz = idy$. Therefore,

$$I_{\text{BC}} = -I_{\text{CB}} = -\int_{-b}^{b} \frac{-a - iy}{a^2 + y^2}\,(idy) \tag{4.139}$$

$$= ia\int_{-b}^{b} \frac{1}{a^2+y^2}\,dy - \int_{-a}^{a} \frac{y}{a^2+y^2}\,dy. \tag{4.140}$$

It follows that $\text{Im}\,I_{BC} > 0$.

Imaginary part of I_{CD}. We note that the oriented line CD is given by the parametric equation $z = x - ib, -a \le x \le a$ with $dz = dx$. Hence,

$$I_{\text{CD}} = \int_{-a}^{a} \frac{x + ib}{x^2 + b^2}\,dx \tag{4.141}$$

$$= \int_{-a}^{a} \frac{x}{x^2+b^2}\,dx + ib\int_{-a}^{a} \frac{1}{x^2+b^2}\,dx. \tag{4.142}$$

Therefore, $\text{Im}\,I_{\text{CD}} > 0$ follows.

Imaginary part of I_{DA}. Finally, the oriented line DA is given by the parametric equation $z = a + iy, -b \leq y \leq b$ with $dz = idy$. Therefore,

$$I_{\mathrm{DA}} = \int_{-b}^{b} \frac{a - iy}{a^2 + y^2} (idy) \tag{4.143}$$

$$= ia \int_{-b}^{b} \frac{1}{a^2 + y^2} dy + \int_{-a}^{a} \frac{y}{a^2 + y^2} dy, \tag{4.144}$$

giving $\operatorname{Im} I_{\mathrm{DA}} > 0$. Thus, we see that the answer $\operatorname{Log}(-a - ib) - \operatorname{Log}(-a + ib)$ is incorrect for I_{BC} when the principal branch is selected for the logarithm function.

The contradiction between two values — a positive value for I_{BC}, obtained by direct and explicit computation and a negative value obtained by using the indefinite integral $\operatorname{Log} z_C - \operatorname{Log} z_B$ — is resolved by remembering that the result (4.127) is not applicable to I_{BC}, because the svb selected for the logarithm is not analytic on the path of integration BC.

§4.8.4 *Integration around a branch cut*

In §4.7.1 we have seen a few examples of transformation of a class of integrals over a real variable into a contour integral over a closed contour. We now present a couple of examples of an important technique of rewriting *integrals over a real parameter as contour integrals around the branch cut* with the integrand involving multivalued functions. This is usually achieved by integrating the discontinuity of the svb of a suitable multivalued function.

Two simple examples of computation of line integrals over a curve in the complex plane have been given in §4.8 using the definition to set up the line integral. Here we continue to use the same approach and give more examples of reducing a line integral in the complex plane to an integral over a real variable. As will be seen in these examples, frequently an integral of a multivalued function around a branch cut in the complex plane gets related to an integral of the discontinuity across the branch cut. While in our first example in ☑4.12 the branch cut of the integrand is a finite interval, the other two examples involve branch cuts along semi-infinite line segments.

☑4.8.4 Problem. Use the definitions of the multivalued functions $z^q, (1-z)^p$ in terms of the variables variables r, θ, ρ, ϕ defined as in Fig. 3.12, so that the interval $(0, 1)$ becomes the branch cut for the function $\chi(z) = z^{(1-p)}(z-1)^p$. Show that the constant λ in the equation

$$\oint_{\Gamma} z^{(1-p)}(z-1)^p\, Q(z)\, dz = \lambda \int_0^1 x^{(1-p)}(1-x)^p\, Q(x)\, dx \tag{4.145}$$

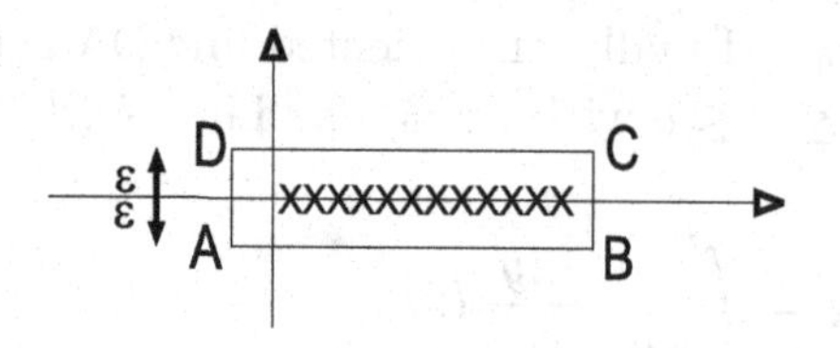

Fig. 4.9

is given by $\lambda = -2i\sin(p\pi)$ where Q is a rational function having singular points outside the interval $[0, 1]$, and Γ is a thin rectangular contour ABCDA — shown in Fig. 4.9 — enclosing the branch cut and excluding singular points of $Q(z)$.

Solution. We introduce the variables r, θ, ρ, and ϕ defined by

$$z = r\exp(i\theta), \qquad z - 1 = \rho\exp(i\phi), \tag{4.146}$$

and restrict the ranges of θ and ϕ as

$$0 < \theta < 2\pi, \qquad 0 < \phi < 2\pi. \tag{4.147}$$

Letting $\chi(z)$ denote $z^{(1-p)}(z-1)^p$, we have

$$\chi(z) = r^{(1-p)}\rho^p \times e^{i(1-p)\theta}e^{ip\phi}, \tag{4.148}$$

and the branch cut for $\chi(z)$ appears in the interval $(0, 1)$ (see ☑3.7.2).

We write the contour integral in Eq. (4.145) as a sum of contributions along the four sides of the rectangle ABCD, and consider the limit in which the sides BC and DA shrink to a point:

$$\oint_\Gamma z^{(1-p)}(z-1)^p\, Q(z)\, dz = \int_{\mathrm{AB}} f(z)dz + \int_{\mathrm{BC}} f(z)dz + \int_{\mathrm{CD}} f(z)dz + \int_{\mathrm{DA}} f(z)dz \tag{4.149}$$

$$= -\int_{\mathrm{DC}} f(z)dz + \int_{\mathrm{AB}} f(z)dz \tag{4.150}$$

$$= -\int_0^1 f(x+i\epsilon)\, dx + \int_0^1 f(x-i\epsilon)\, dx \tag{4.151}$$

$$= -\int_0^1 \left(f(x+i\epsilon) - f(x-i\epsilon)\right) dx. \tag{4.152}$$

Here the first step follows from the fact that the value of the integral on the left hand side remains unchanged as the length of the sides BC and DA is reduced to zero. In this limit the two integrals, along BC and DA, vanish. The integrand in the last line is just the discontinuity across the branch cut from $x = 0$ to $x = 1$. The discontinuity across the branch cut has been computed earlier and we recall the result from Eq. (3.24):

$$f(x+i\epsilon) - f(x-i\epsilon) = 2i\sin(p\pi)\, x^{(1-p)}(1-x)^p\, Q(x). \tag{4.153}$$

Equation (4.153) when substituted in Eq. (4.152) gives us the result (4.145) with $\lambda = -2i\sin(p\pi)$. ∎

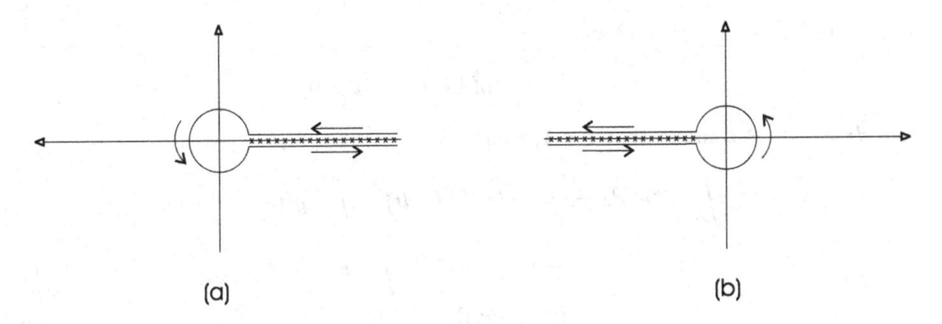

Fig. 4.10

The next two examples furnish two integral representations of Gamma function $\Gamma(\lambda)$ making use of its Euler definition:

$$\Gamma(\lambda) = \int_0^\infty e^{-x} x^{\lambda-1} dx. \tag{4.154}$$

§4.8.5 Short examples. We make use of the discontinuity of z^p found in §3.6.

(a) Let us select the single-valued branch of z^p, defined by

$$z^p = r^p e^{ip\theta}, \qquad 0 < \theta < 2\pi, \tag{4.155}$$

which corresponds to a branch cut along the positive real axis. Setting up the integral of $f(z) = z^p e^{-z}, p > -1$, along the contour C, consisting of two lines running parallel to the positive real axis and a circular arc — shown in Fig. 4.10(a) — in the limit where the radius of the circular arc tends to zero, we get

$$\int_C z^p e^{-z}\, dz = -\int_0^\infty f(t+i\epsilon)\, dt + \int_0^\infty f(t-i\epsilon)\, dt \tag{4.156}$$

The difference, $f(t+i\epsilon)dt - f(t-i\epsilon)$, is just the discontinuity across the branch cut and is easily seen to be

$$f(t+i\epsilon) - f(t-i\epsilon) = f(t) - f(te^{2\pi ip}) = (1 - e^{2p\pi i})t^p. \tag{4.157}$$

Therefore,

$$\int_C z^p e^{-z}\, dz = (e^{2p\pi i} - 1)\int_0^\infty e^{-t} t^p\, dt \tag{4.158}$$

$$= (e^{2p\pi i} - 1)\Gamma(p+1). \tag{4.159}$$

Replacing p with $\lambda - 1$ gives the first integral representation,

$$\therefore \Gamma(\lambda) = \frac{1}{e^{2\lambda\pi i} - 1}\int_C e^{-z}\, z^{\lambda-1}\, dz. \tag{4.160}$$

(b) As our next example we take the principle value of $z^{\lambda-1}$, and repeat the steps of the above example integrating $g(z) = z^{-p}e^z$, and taking the contour C as in Fig. 4.10(b). For the two lines parallel to the cut, we have

$$z = t\exp(\pm i\pi), \qquad t > 0,$$

and $dz = -dt$. Assuming that $p < 1$, we get

$$\int_C z^{-p} e^z\, dz = \int_0^\infty g(te^{i\pi})\,(-dt) - \int_0^\infty g(te^{-i\pi})\,(-dt) \tag{4.161}$$

$$= (-e^{-ip\pi} + e^{ip\pi}) \int_0^\infty e^{-t} t^{-p} dt \tag{4.162}$$

$$= (2i\sin p\pi)\Gamma(1-p). \tag{4.163}$$

Using the identity

$$\Gamma(p)\Gamma(1-p) = \frac{\pi}{\sin \pi p}$$

and writing λ instead of p gives the second integral representation for the Γ function:

$$\frac{1}{\Gamma(\lambda)} = \frac{1}{2\pi i}\int_C e^z\, z^{-\lambda}\, dz. \tag{4.164}$$

§4.9 Summary

In this chapter, after a brief discussion of improper integrals §4.1, integral in the complex plane along a contour has been defined and results on bounds on the contour integrals were discussed in §4.2 and §4.6. Cauchy fundamental theorem was introduced in §4.6 and two important applications, viz. deformation of contours and indefinite integrals have been discussed. This material is basic to all the developments in later chapters.

In §4.7 application to techniques of closing a contour and first examples of contour integration were presented. Note that the Cauchy theorem and other results in the chapter are sufficient for evaluation of these integrals involving Fourier transform of Gaussian and similar integrals.

In §4.8 integration of multivalued functions has been presented. We have shown how contour integrals of multivalued functions can be related to integral over a real variable. Integrals involving multifunctions require care; a case study of $\int \frac{dz}{z}$, recommended for everyone, and several examples are presented in detail to emphasise this point. Some of these examples prepare the reader for later techniques of contour integration of multivalued functions and for Schwarz Christoffel transformation.

Chapter 5

CAUCHY'S INTEGRAL FORMULA

In this chapter we prove Cauchy's integral formula, which is central to the derivation of a number of extremely important results in the theory of functions of a complex variable. Two of the most important results flowing from the integral formula are the existence of derivatives of all orders and Taylor series representation for analytic functions. They are truly remarkable, in the sense that analogous results do not hold for functions of a real variable. It must also be emphasized that the result on the Taylor series follows quite easily from the integral formula, a result in the theory of integration of functions of a complex variable. The same cannot be said about a proof based on differentiation alone and not using any results on integration. The Laurent series expansion theorem, given in §5.6, generalizes the Taylor representation to situations where the function under discussion is analytic only in an annular region centered around a given point.

Besides the results on infinite differentiability, and the two series representation theorems, many other consequences of theoretical importance follow from the integral formula. These results are summarized in §5.9.

§5.1 Cauchy's Integral Formula

Theorem 5.1 (Cauchy's integral formula). *If a function f is analytic inside and on a positively oriented closed contour Γ, and z_0 is a point enclosed inside the contour Γ, we have*

$$\oint_\Gamma \frac{f(\xi)}{\xi - z_0} d\xi = 2\pi i f(z_0). \tag{5.1}$$

Proof. We shall at first show that

$$\oint_\Gamma \frac{f(\xi)}{(\xi - z_0)} d\xi = \oint_\gamma \frac{f(\xi)}{\xi - z_0} d\xi, \tag{5.2}$$

where γ is a circle of radius ρ with the center at the point z_0 and oriented anticlockwise, and is such that γ lies completely inside the closed contour Γ, as shown in Fig. 5.1. If we define $g(z) = \dfrac{f(z)}{z - z_0}$, $g(z)$ is analytic at all points inside Γ

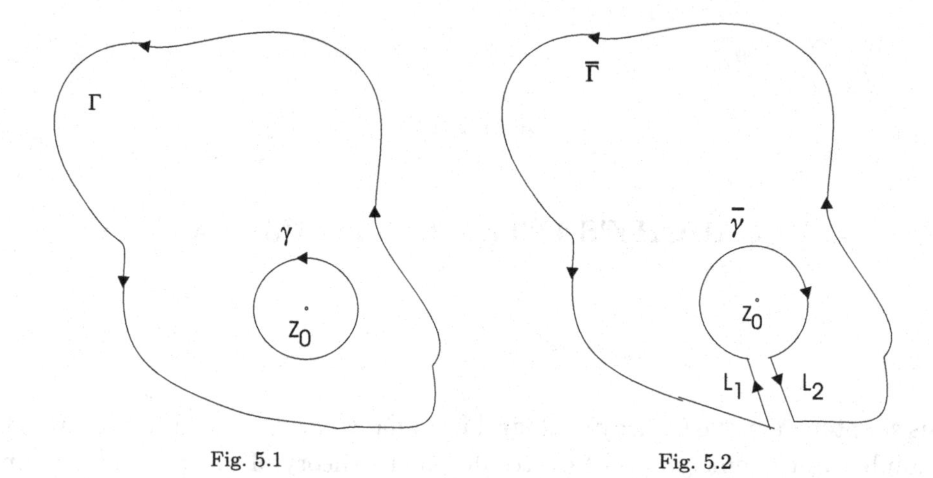

Fig. 5.1 Fig. 5.2

except at the point z_0. Next, form the contours $\bar{\Gamma}$ and $\bar{\gamma}$ by *cutting* the contour Γ and the circle γ and join them by two parallel lines, L_1 and L_2, resulting in a closed contour C of Fig. 5.2.

Then the function g is analytic at all points inside the contour C of Fig. 5.2. Thus,

$$\oint_C \frac{f(z)}{z-z_0}dz = \int_{\bar{\Gamma}} \frac{f(z)}{z-z_0}dz + \int_{L_1} \frac{f(z)}{z-z_0}dz + \int_{L_2} \frac{f(z)}{z-z_0}dz + \oint_{\bar{\gamma}} \frac{f(z)}{z-z_0}dz. \tag{5.3}$$

Since g is analytic on and inside the closed contour C, by Cauchy's theorem we have

$$\oint_C g(z)dz = \oint_C \frac{f(z)}{z-z_0}dz = 0, \tag{5.4}$$

or

$$\int_{\bar{\Gamma}} \frac{f(z)}{z-z_0}dz + \int_{L_1} \frac{f(z)}{z-z_0}dz + \int_{L_2} \frac{f(z)}{z-z_0}dz + \oint_{\bar{\gamma}} \frac{f(z)}{z-z_0}dz = 0. \tag{5.5}$$

In the limit where the parallel lines L_1 and L_2 come very close, the integrals along the two lines cancel,

$$\int_{L_1} \frac{f(z)}{z-z_0}dz + \int_{L_2} \frac{f(z)}{z-z_0}dz = 0, \tag{5.6}$$

and we have

$$\oint_{\bar{\Gamma}} \frac{f(z)}{z-z_0}dz + \oint_{\bar{\gamma}} \frac{f(z)}{z-z_0}dz = 0, \tag{5.7}$$

or

$$\oint_\Gamma \frac{f(z)}{z-z_0}dz = \oint_{\bar{\Gamma}} \frac{f(z)}{z-z_0}dz = -\oint_{\bar{\gamma}} \frac{f(z)}{z-z_0}dz = \oint_\gamma \frac{f(z)}{z-z_0}dz. \tag{5.8}$$

Next, we shall show that the last integral has the value $2\pi i f(z_0)$. For this purpose we note that the function $\phi(z)$, defined by

$$\phi(z) = \frac{f(z)}{z - z_0} - \frac{f(z_0)}{z - z_0} - \left.\frac{df(z)}{dz}\right|_{z=z_0}, \tag{5.9}$$

tends to zero as $z \to z_0$, because the difference of the first two terms tends to the derivative of $f(z)$ at z_0. Hence, for every $\epsilon > 0$, there exists a δ such that

$$|\phi(z)| < \epsilon \quad \text{for} \quad |z - z_0| < \delta. \tag{5.10}$$

We note that the radius ρ of the circle γ can be taken as small as we please; we therefore take $\rho < \delta$ and we get $|\phi(z)| < \epsilon$ everywhere on γ and inside γ. Express the integral in Eq. (5.8) in terms of $\phi(z)$ to get

$$\begin{aligned}\oint_\Gamma \frac{f(z)}{z - z_0} dz = \oint_\gamma \frac{f(z)}{z - z_0} dz &= \oint_\gamma \phi(z) dz + \oint_\gamma \frac{f(z_0)}{z - z_0} dz + f'(z_0) \oint_z dz \\ &= \oint_\gamma \phi(z) dz + f(z_0) \oint_\gamma \frac{dz}{z - z_0} dz.\end{aligned} \tag{5.11}$$

In arriving at the above equation we have used the fact that the constant function, 1, is analytic everywhere, and hence the last integral vanishes using Cauchy's theorem. Using Darboux's theorem and the bound, Eq. (5.10), we get

$$\left| \oint_\gamma \phi(z) dz \right| < 2\pi\rho\,\epsilon.$$

Because ρ can be taken as small as we wish, this shows that the first integral in Eq. (5.11) also vanishes. Thus, Eq. (5.8) implies that

$$\oint_\Gamma \frac{f(z)}{z - z_0} dz = f(z_0) \oint_\gamma \frac{dz}{z - z_0}. \tag{5.12}$$

The integral on the r.h.s. has been computed; see Eq. (4.32). Substituting its value $2\pi i$ we get the desired result

$$\oint_\Gamma \frac{f(z)}{z - z_0} dz = 2\pi i f(z_0), \tag{5.13}$$

keeping in mind that the point z_0 is enclosed inside the contour Γ. ∎

A large number of important results for analytic functions flow from the Cauchy's integral formula. We will first discuss the existence of derivatives of all orders, Taylor and Laurent series expansion theorems. Other important results will be summarized in §5.9.

⊳ Exercise §§5.1, on straightforward application of the integral formula.

§5.1.1 Short examples. The following examples show the use of Cauchy's integral formula:

(a) A straightforward use of Cauchy formula's gives

$$\oint_\gamma \frac{z^2+3}{z-1}\,dz = 8\pi i,$$

where γ is any closed contour enclosing the point $z = 1$.

(b) The partial fraction form of $\frac{1}{z^2-5z+6}$ is

$$\frac{1}{z^3-5z+6} = \frac{1}{z-3} - \frac{1}{z-2}.$$

If γ is taken to be the circle $\{z||z| = 5\}$ of radius 5, an application of Cauchy's integral formula will give

$$\oint_\gamma \frac{e^z}{z^3-5z+6}\,dz = \oint_\gamma \frac{e^z}{z-3}\,dz - \oint_\gamma \frac{e^z}{z-2}\,dz \tag{5.14}$$

$$= 2\pi i(e^3 - e^2). \tag{5.15}$$

(c) If we take γ to be the circle $\{z||z| = 2.5\}$, the answer for the integral in the above example will be given by just the second term of Eq. (5.15):

$$\oint_\gamma \frac{e^z}{z^3-5z+6}\,dz = \oint_\gamma \frac{e^z}{z-3}\,dz - \oint_\gamma \frac{e^z}{z-2}\,dz \tag{5.16}$$

$$= -2\pi i e^2. \tag{5.17}$$

The value of the first integral will be zero, because $z = 3$ lies outside the contour γ and the integrand $e^z/(z-3)$ is analytic inside γ.

§5.2 Existence of Higher Order Derivatives

Theorem 5.2 (higher order derivatives). *If $f(z)$ is analytic at a point, its derivatives to all orders exist at that point and are given by*

$$\frac{d^n}{dz^n} f(z) = \frac{n!}{2\pi i} \oint_C \frac{f(\xi)}{(\xi - z)^{n+1}} d\xi, \quad n = 1, 2, \ldots, \tag{5.18}$$

where C is any contour enclosing the point z_0 and is such that the function f is analytic everywhere inside and on C.

This result is easily proven by starting from Cauchy's integral formula, Eq. (5.13), for the analytic function. It should be emphasized that the above result is quite remarkable. This result on *infinite differentiability* of a function in a region follows from the existence of a derivative of *first order* in the given region. A corresponding result does not hold for functions of a real variable. This result also implies that if $f(z)$ is analytic at a point, all its derivatives, $\frac{df}{dz}, \frac{d^2f}{dz^2}, \ldots, \frac{d^nf}{dz^n}, \ldots,$ are analytic at that point. A function of a real variable may be differentiable at all points in an open set and it may not even have the second order derivative.

§5.3 Taylor Series

We shall now ask when can a function $f(z)$ be expanded in a series $\Sigma a_n(z-z_0)^n$ having positive powers of $z-z_0$. If a series expansion exists, what are the values of z for which the sum of the series coincides with the value of the function? The answer is given by the theorem below:

Theorem 5.3 (Taylor series expansion). *The theorem has two parts:*

(a) *If $f(z)$ is analytic at a point z_0, it can be expanded in a Taylor series,*

$$f(z) = \sum a_n(z-z_0)^n, \tag{5.19}$$

where

$$a_n = \frac{1}{2\pi i}\oint_C \frac{f(\xi)}{(\xi-z_0)^{n+1}}\,d\xi. \tag{5.20}$$

(b) *The power series representation is convergent for all z inside the largest circle, known as the circle of convergence, which can be drawn with the center at z_0 and is such that it does not enclose any singular point of $f(z)$. Thus, the radius of convergence of the power series is given by the distance of z_0 from the nearest singular point.*

Proof. Let $f(z)$ be a function analytic on and inside the contour C, so that the singular points of $f(z)$, marked as $\otimes$ in Fig. 5.3, are outside C. In addition, let R be the distance of the point z_0 from the nearest singular point. Also, Γ denote a circle of radius $\rho < R$ as in Fig. 5.3. The radius ρ can be taken to be as close to R as we may wish. Therefore, using Cauchy's integral theorem we obtain

$$f(z) = \frac{1}{2\pi i}\oint_\Gamma \frac{f(\xi)}{\xi - z}d\xi, \tag{5.21}$$

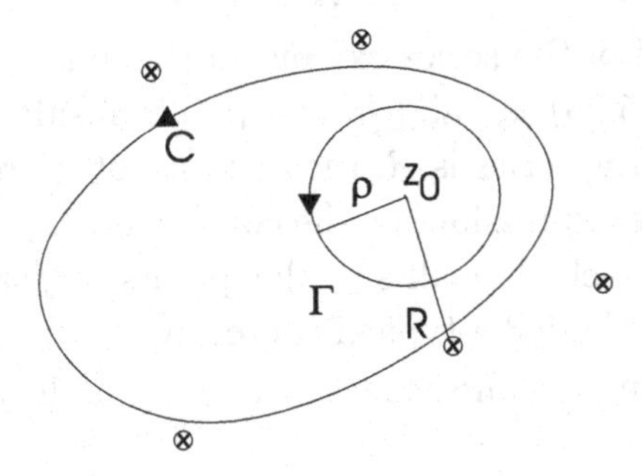

Fig. 5.3

which holds for all points z inside the circle Γ. Next, we write

$$\begin{aligned}
\frac{1}{\xi - z} &= \frac{1}{(\xi - z_0) - (z - z_0)} \\
&= \frac{1}{\xi - z_0} \times \frac{1}{1 - (z - z_0)/(\xi - z_0)} \\
&= \frac{1}{\xi - z_0}\left[1 + \frac{z - z_0}{\xi - z_0} + \frac{(z - z_0)^2}{(\xi - z_n)^2} + \cdots + \frac{(z - z_0)^n}{(\xi - z_0)^n} + \cdots\right] \\
&= \sum_0^\infty \frac{(z - z_0)^n}{(\xi - z_0)^{n+1}}.
\end{aligned} \tag{5.22}$$

The series in square brackets converges for values of z satisfying

$$\left|\frac{z - z_0}{\xi - z_0}\right| < 1. \tag{5.23}$$

Since ξ is a point on the circle Γ and z_0 is the center of the circle $|\xi - z_0| = \rho$, the condition for convergence of the series expansion becomes $|z - z_0| < \rho$, i.e. the point z must lie inside the circle Γ. The expansion for $1/(\xi - z)$ is now substituted in the integral formula. *The integration and the series summation are allowed to be interchanged for all points z inside the circle as the power series is convergent.* Thus, we get

$$\begin{aligned}
f(z) &= \frac{1}{2\pi i}\oint_\Gamma \frac{f(\xi)}{\xi - z} d\xi \\
&= \frac{1}{2\pi i}\sum_0^\infty \oint_\Gamma \frac{(z - z_0)^n}{(\xi - z_0)^{n+1}} f(\xi) d\xi \\
&= \sum_0^\infty a_n (z - z_0)^n,
\end{aligned} \tag{5.24}$$

where a_n are given by

$$a_n = \frac{1}{2\pi i}\int_\Gamma \frac{f(\xi)}{(\xi - z_0)^{n+1}} d\xi. \tag{5.25}$$

This completes the proof of the series expansion theorem.

The representation (5.25) is valid for all points inside the circle Γ. Thus, the radius of convergence of the series is R, the distance of z_0 from the nearest singular point. Note that Γ, in the expression for the coefficients a_n, can be deformed to any closed contour C and is such that all singular points of $f(z)$ are outside C, giving the expression (5.20). In Fig. 5.3 a typical contour C, enclosing the point z_0 but not enclosing any of the singular points, marked with $\otimes$, is drawn. ∎

Combining Eq. (5.25) with Eq. (5.18) one easily gets

$$a_n = \frac{1}{n!}\frac{d^n}{dz^n}f(z)\bigg|_{z=z_0} . \tag{5.26}$$

For $z_0 = 0$, the Taylor series expansion becomes the MacLaurin series

$$f(z) = \sum_{n=0}^{\infty} \frac{1}{n!}\frac{d^n}{dz^n}f(z)\bigg|_{z=0} z^n . \tag{5.27}$$

The representation of a function $f(z)$ in terms of its Taylor series is unique, in the sense that if the sum of a power series $\sum_n b_n(z-z_0)^n$ coincides with the values of $f(z)$ inside the circle of convergence, b_n must coincide with a_n.

⊳ Quiz §§5.2, on the radius of convergence of Taylor series.
⊳ Exercises §§5.3 and §§5.4, on MacLaurin and Taylor series expansions.

§5.4 Real Variable vs. Complex Variable

We present examples of two functions g and f, to show that:

- An infinitely differentiable function of a real variable may not have a Taylor series representation;
- The behavior of a function of a real variable may not give any clue to the radius of convergence of the Taylor series.

We also point out that one can understand the behavior of the Taylor series for two examples, when one regards the functions as functions of a complex variable.

As our first example, we take the function g of a real variable defined by

$$g(x) = \begin{cases} \exp(-1/x^2), & \text{for } x > 0, \\ 0, & \text{if } x \neq 0. \end{cases} \tag{5.28}$$

This function is infinitely differentiable at all points, in particular at $x = 0$ also. A Taylor expansion for the function $g(x)$ in the form

$$\sum a_n x^n$$

can be constructed by writing a_n in terms of the differential coefficients $\frac{d^n g}{dx^n}$ evaluated at $x = 0$. A straightforward computation shows that all the derivatives at $x = 0$ are in fact zero. The Taylor series $\Sigma a_n x^n$ converges to zero for all x but it does not give the value of the function at any point except $x = 0$ and the function cannot be represented by a Taylor series around $x = 0$.

In this example, the function $g(z) = \exp(-1/z^2)$ is not analytic at $z = 0$. Therefore, a series representation in powers of z does not exist. Though the function $g(x)$ is infinitely differentiable at $x = 0$, it is not even continuous at $z = 0$ when considered as a function of a complex variable. This is easily checked by showing that $\lim g(z)$ does not exist as $z \to 0$ along the imaginary axis. In fact, writing $z = ib$ and letting $b \to 0$, we find that

$$g(z) = \exp(-1/z^2) = \exp(1/b^2) \to \infty;$$

the function $g(z)$ is not even continuous at $z = 0$.

As a second example, consider the function

$$f(x) = \frac{1}{1+x^2},$$

which is well-defined and infinitely differentiable for all real x. However, a power series expansion,

$$f(x) = 1 - x^2 + x^4 - x^6 + \cdots, \tag{5.29}$$

does not converge for all values of x.

The function $f(z) = \frac{1}{1+z^2}$ has singular points at $z = \pm i$ in the complex plane. These two singular points determine the radius of convergence of the Taylor expansion about any point in the complex plane, including the points on the real line. Thus, without any further ado, we can say that for the function $f(z)$ in this example:

(a) A series expansion in powers of $z - z_0$ of the form

$$f(z) = \sum a_n (z - z_0)^n$$

exists if $z_0 \neq i$ and $z_0 \neq -i$;

(b) The series expansion in powers of $z - z_0$ converges for $|z - z_0| < R$, where R is the minimum of the two values $|z_0 - i|$ and $|z_0 + i|$.

In the above examples, we have seen that the behavior of a function on the real line does not give us any hint about the behavior of the Taylor series expansions. The only way the peculiar properties of the Taylor series, such as those encountered in the above two examples of functions of a real variable x, can be understood is to replace the real variable x with a complex variable z and to look at the singularities of the function in the complex z plane.

§5.5 Examples

§5.5.1 Short examples. We give a few simple examples of the MacLaurin series:

(a) The MacLaurin expansions for $\exp(z), \sin z, \cos z, \cosh z$, and $\sinh z$ functions are easily obtained by using MacLaurin's theorem. For example,

$$\frac{d^n e^z}{dz^n} = e^z \Rightarrow \left.\frac{d^n e^z}{dz^n}\right|_0 = 1,$$

giving

$$\exp(z) = \sum_n \frac{z^n}{n!} = 1 + z + \frac{z^2}{2!} + \frac{z^3}{3!} + \cdots. \tag{5.30}$$

(b) The MacLaurin series for $\sinh z$ and others can be written in a similar fashion:

$$\frac{d^{2n}}{dz^{2n}}\sinh z = \sinh z, \qquad \frac{d^{2n+1}}{dz^{2n+1}}\sinh z = \cosh z,$$

which gives

$$\frac{d^{2n}}{dz^{2n}}\sinh z\bigg|_{z=0} = 0, \qquad \frac{d^{2n+1}}{dz^{2n+1}}\sinh z\bigg|_{z=0} = 1,$$

and hence

$$\sinh z = z + \frac{z^3}{3!} + \frac{z^5}{5!} + \cdots = \sum_n \frac{z^{2n+1}}{(2n+1)!}. \tag{5.31}$$

The functions $\exp(z), \sin z, \cos z, \cosh z$, and $\sinh z$ are analytic everywhere in the complex plane, and thus their Taylor expansions have an infinite radius of convergence, i.e. the MacLaurin series derived above converge for all values of z.

(c) The MacLaurin expansion for $\tan z$ will have only odd powers of z and therefore we write it in the form

$$\tan z = a_1 + a_3 z^3 + a_5 z^5 + a_7 z^7 + \cdots. \tag{5.32}$$

Next, we rearrange this equation as

$$\sin z = (a_1 + a_3 z^3 + a_5 z^5 + a_7 z^7 + \cdots)\cos z, \tag{5.33}$$

and substitute the series expansions for $\sin z$ and $\cos z$ and compare powers of z. This gives

$$a_1 = 1, \quad a_3 = \frac{1}{3}, \quad a_5 = \frac{2}{15}, \quad a_7 = \frac{17}{315}. \tag{5.34}$$

♯5.5.2 Short examples. We present a few examples of Taylor series expansions and the radius of convergence of Taylor expansion. Recall that a function has Taylor expansion about a point z_0 if it is analytic at that point and the radius of convergence is equal to the distance of z_0 from the nearest singular point of the function.

(a) The functions $\exp(z), \sin z, \cos z, \cosh z$, and $\sinh z$ are analytic everywhere in the complex plane, hence their Taylor expansions, about any point, have an infinite radius of convergence, i.e. the Taylor series for all these functions converge for all values of z. For example,

$$\frac{d^n e^z}{dz^n} = e^z \Rightarrow \frac{d^n e^z}{dz^n}\bigg|_{z_0} = e^{z_0},$$

giving

$$\exp(z) = e^{z_0}\sum_n \frac{(z-z_0)^n}{n!} = e^{z_0}\left[1 + (z-z_0) + \frac{(z-z_0)^2}{2!} + \frac{(z-z_0)^3}{3!} + \cdots\right]. \tag{5.35}$$

This expansion will converge for all values of z.

(b) An expansion of a function about a point z_0 can be conveniently obtained by writing $w = z - z_0$, or $z = w + z_0$, and expanding in powers of w. Thus, for example,

$$\sin z = \sin(w + z_0) = \sin w \cos z_0 + \cos w \sin z_0 \tag{5.36}$$

$$= \cos z_0 \left(w - \frac{w^3}{3!} + \frac{w^5}{5!} - \cdots\right) + \sin z_0 \left(1 - \frac{w^2}{2!} + \frac{w^4}{4!} - \cdots\right) \tag{5.37}$$

$$= \cos z_0 \left[(z - z_0) - \frac{(z - z_0)^3}{3!} + \frac{(z - z_0)^5}{5!} - \cdots\right] \tag{5.38}$$

$$+ \sin z_0 \left[1 - \frac{(z - z_0)^2}{2!} + \frac{(z - z_0)^4}{4!} - \cdots\right]. \tag{5.39}$$

This and other similar expansions of $\exp(z), \cos z, \cosh z$, and $\sinh z$, will converge for all z.

(c) The Taylor expansion of $f(z) = \frac{1}{(z^2+4)(z-5)}$ around a point z_0, where the denominator does not vanish, will have a radius of convergence equal to the minimum of $|z_0 - 2i|, |z_0 + 2i|$, and $|z_0 - 5|$. This expansion can be obtained by writing partial fractions and using the binomial theorem. So, as a special case, an expansion of $f(z)$ in powers of $z + 3$ will have a radius of convergence $|3 + 2i| = \sqrt{13}$.

(d) As our last example, we deal with the function $f(z) = \frac{1}{1+z^2}$. We list the radius and circle of convergence of Taylor expansion in powers of $z - z_0$ for $z_0 = 0, 1/2, 1, -1, 1 - i, 1 + i$.

z_0	0	$1/2$	-1	$1-i$	$1+i$
Nearest singularity	$\pm i$	$\pm i$	$\pm i$	$-i$	i
R	1	$\frac{\sqrt{5}}{2}$	$\sqrt{2}$	1	1
Circle	$\|z\| = 1$	$\|2z - 1\| = \sqrt{5}$	$\|z + 1\| = \sqrt{2}$	$\|z - 1 + i\| = 1$	$\|z - 1 - i\| = 1$

§5.6 Laurent Expansion

In the previous section we have discussed power series representation for functions of a complex variable. A function $f(z)$ has a power series representation of the form

$$\sum_{n=0}^{\infty} a_n (z - z_0)^n, \tag{5.40}$$

if the function $f(z)$ is analytic at the point z_0. The series converges inside a circle in the complex plane with the center at z_0 and the radius given by the distance of z_0 from the nearest singular point. Further, the series converges to $f(z)$ for all z inside this circle of convergence.

For a function of a complex variable, analytic in an annular region between two *concentric circles*, an expansion in powers of $z - z_0$ can be derived. *However, this expansion, known as Laurent expansion, has both positive and negative powers of* $z - z_0$, where z_0 is the common center of the two circles. We shall first state and prove the Laurent expansion theorem and then present some examples.

Theorem 5.4 (Laurent expansion theorem). *Let $f(z)$ be a function which is analytic in a region R which lies between two concentric circles, Γ_1 and Γ_2 with common center at z_0. Then $f(z)$ can be expanded in a Laurent series of positive and negative powers,*

$$f(z) = \sum_{0}^{\infty} a_n(z - z_0)^n + \sum_{1}^{\infty} b_n(z - z_0)^{-n}, \tag{5.41}$$

where a_n and b_n are given by

$$a_n = \frac{1}{2\pi i} \oint_C \frac{f(\xi)}{(\xi - z_0)^{n+1}}\, dz, \tag{5.42}$$

$$b_n = \frac{1}{2\pi i} \oint_C (\xi - z_0)^{n-1} f(\xi)\, d\xi. \tag{5.43}$$

Here C is any positively oriented contour enclosing the point z_0 and lying fully in the region R.

The function may or may not be analytic at the point z_0. In fact, nothing has been said about the nature of z_0. The second part of the Laurent expansion, i.e. the part containing the negative powers of $z - z_0$ is called the *principal part* of the Laurent expansion.

Proof. Let the given function be analytic between two anticlockwise circles, Γ_1 and Γ_2, of radii R_1 and R_2, respectively, with the centers at z_0, as in Fig. 5.4. In addition, let Γ be a contour constructed by cutting the contours Γ_1 and Γ_2 and joining them by parallel lines L_1 and L_2, as shown in Fig. 5.5. Then Γ is a simple closed contour and f is analytic everywhere inside Γ. Using Cauchy's integral theorem we can write

$$f(z) = \frac{1}{2\pi i} \oint_\Gamma \frac{f(\xi)}{\xi - z}\, d\xi, \tag{5.44}$$

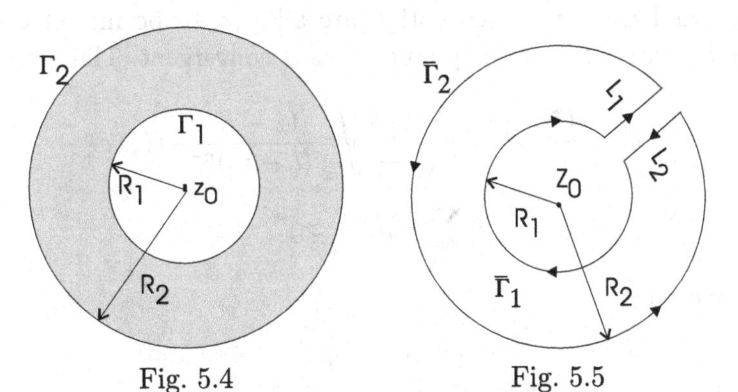

Fig. 5.4 Fig. 5.5

where z is a point in the region enclosed by Γ. Now $\oint_\Gamma = \int_{\bar{\Gamma}_1} + \int_{L_1} + \int_{L_2} + \int_{\bar{\Gamma}_2}$ and in the limit where the two parallel lines L_1 and L_2 coincide $\int_{L_1} + \int_{L_2} \to 0$. Therefore, in this limit we get

$$f(z) = \frac{1}{2\pi i}\oint_{\bar{\Gamma}_1} \frac{f(\xi)}{\xi - z} d\xi + \frac{1}{2\pi}\oint_{\bar{\Gamma}_2} \frac{f(\xi)}{\xi - z} d\xi \tag{5.45}$$

$$= -\frac{1}{2\pi i}\oint_{\Gamma_1} \frac{f(\xi)}{\xi - z} d\xi + \frac{1}{2\pi}\oint_{\Gamma_2} \frac{f(\xi)}{\xi - z} d\xi, \tag{5.46}$$

which holds for all points z in the annular region between the circles Γ_1 and Γ_2. As was done in the case of proof for the Taylor expansion theorem, we write

$$\begin{aligned}
\frac{1}{\xi - z} &= \frac{1}{(\xi - z_0) - (z - z_0)} \\
&= \frac{1}{\xi - z_0} \times \frac{1}{1 - (z - z_0)/(\xi - z_0)} \\
&= \frac{1}{\xi - z_0}\left[1 + \frac{z - z_0}{\xi - z_0} + \frac{(z - z_0)^2}{(\xi - z_0)^2} + \cdots + \frac{(z - z_0)^n}{(\xi - z_0)^n} + \cdots\right] \\
&= \sum_{n=0}^{\infty} \frac{(z - z_0)^n}{(\xi - z_0)^{n+1}}.
\end{aligned} \tag{5.47}$$

The series in square brackets converges for values of z satisfying

$$\left|\frac{z - z_0}{\xi - z_0}\right| < 1. \tag{5.48}$$

For the integral $\int_{\Gamma_2}$, since ξ is a point on the circle Γ_2 and z_0 is the center of the circle, $|\xi - z_0| = R_2$ and hence the condition for convergence of the series expansion becomes $|z - z_0| < R_2$, i.e. the point z must lie inside the circle Γ_2. The expansion for $1/(\xi - z)$ is now substituted in the second term of the integral formula Eq. (5.46). The integration and the series summation are allowed to be interchanged for all points z inside the circle Γ_2, as the power series is convergent. Therefore,

$$\begin{aligned}
\frac{1}{2\pi i}\oint_{\Gamma_2} \frac{f(\xi)}{\xi - z} d\xi &= \frac{1}{2\pi i}\sum_n \oint_{\Gamma_2} \frac{(z - z_0)^n}{(\xi - z_0)^{n+1}} f(\xi)\, d\xi \\
&= \sum a_n (z - z_0)^n,
\end{aligned} \tag{5.49}$$

where a_n are given by

$$a_n = \frac{1}{2\pi i}\oint_{\Gamma_2} \frac{f(\xi)}{(\xi - z_0)^{n+1}}. \tag{5.50}$$

Similarly, for the integral over Γ_1 we use the expansion

$$\begin{aligned}\frac{1}{\xi - z} &= \frac{1}{(\xi - z_0) - (z - z_0)} \\ &= \frac{-1}{z - z_0} \times \frac{1}{1 - (\xi - z_0)/(z - z_0)} \\ &= \frac{-1}{z - z_0}\left[1 + \frac{\xi - z_0}{z - z_0} + \frac{(\xi - z_0)^2}{(z - z_0)^2} + \cdots + \frac{(\xi - z_0)^n}{(z - z_0)^n} + \cdots\right] \\ &= -\sum_{n=1}^{\infty} \frac{(\xi - z_0)^{n-1}}{(z - z_0)^n}. \end{aligned} \tag{5.51}$$

The series in square brackets converges for values of z satisfying

$$\left|\frac{\xi - z_0}{z - z_0}\right| < 1, \tag{5.52}$$

or

$$|z - z_0| > |\xi - z_0|. \tag{5.53}$$

For the integral over Γ_1, the point ξ is on the circle Γ_1 and z_0 is the center of the circle, $|\xi - z_0| = R_1$, and hence the condition for convergence of the series expansion becomes $|z - z_0| > R_1$, i.e. the point z must lie outside the circle Γ_1. The second expansion for $1/(\xi - z)$ from Eq. (5.52) is now substituted in the first term Eq. (5.46). The integration and the series summation are allowed to be interchanged for all points z inside the circle, as the power series is convergent. Thus, we get

$$\begin{aligned}\frac{1}{2\pi i}\oint_{\Gamma_1} \frac{f(\xi)}{\xi - z_0}\,d\xi &= -\frac{1}{2\pi i}\sum_{1}^{\infty}\oint_{\Gamma_1} \frac{(\xi - z_0)^{n-1}}{(z - z_0)^n}\,f(\xi)\,d\xi \\ &= -\sum_{1}^{\infty} b_n (z - z_0)^{-n}, \end{aligned} \tag{5.54}$$

where b_n are given by

$$b_n = \frac{1}{2\pi i}\oint_{\Gamma_1} f(\xi)\,(\xi - z_0)^{n-1}\,d\xi. \tag{5.55}$$

Substituting for the two integrals in Eq. (5.46), we get the required result on the Laurent expansion. By deforming the contours, both the integrals in Eqs. (5.50) and (5.55) giving the coefficients a_n and b_n can be written as integrals over any contour C lying entirely within the two circles Γ_1 and Γ_2, as in Fig. 5.6, and we get the results in (5.42) and (5.43). ∎

With reference to the formula (5.55) for the coefficients b_n in the Laurent expansion, if f is analytic at all points inside Γ_1, including the point z_0, the coefficients b_n all vanish for $n > 1$. This is because the function $f(\xi)(\xi - z_0)^{n-1}$,

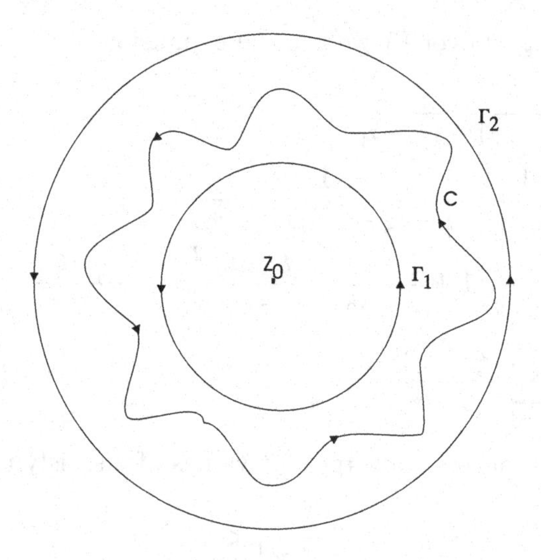

Fig. 5.6

which appears inside the integration, is the product of two functions, $f(\xi)$ and $(\xi - z_0)^{n-1}$, and both are analytic functions of the integration variable ξ on the contour Γ_1 and at all points inside Γ_1. In this case the Laurent expansion coincides with the Taylor expansion.

§5.7 Examples

♯5.7.1 Short examples. If the function $f(z)$ is analytic in the region between two concentric circles, it has a Laurent expansion in positive and negative powers of $z - z_0$, where z_0 is the center of the two circles. Some examples below illustrate this statement.

(a) The singular points of the function $\frac{1}{(z-5)(z^2+4)}$ are $z = 5, \pm 2i$ and is analytic in the annular region $\{z|2 < |z| < 5\}$. Therefore, it will have a Laurent expansion in powers of z convergent in this region.

(b) The function $\frac{1}{z+a}$ has a Taylor expansion *in positive powers of* z,

$$\frac{1}{z+a} = \frac{1}{a}\left(\frac{1}{1+\frac{z}{a}}\right) = \sum_n \left(-\frac{z^n}{a^{n+1}}\right), \tag{5.56}$$

which has the disk $\{z||z| < a\}$ as the region of convergence. The series in Eq. (5.56) can be viewed as a special case of Laurent expansion having no negative powers, and its region of convergence is a limiting case of an annular region $\{z|\rho < |z| < a\}$ as $\rho \to 0$.

(c) The function $f(z) = \frac{1}{(z-2)(z+3)}$ will have three different expansions in powers of z corresponding to three possible annular regions:

(i) $|z| < 2$, (ii) $2 < |z| < 3$, (iii) $|z| > 3$.

Corresponding expansions in powers of z can be easily written using partial fractions and the binomial theorem.

(d) Continuing with the example of the function $\frac{1}{z+a}$, one can expand it *in negative powers of* z using binomial expansion by first writing it as

$$\frac{1}{z+a} = \frac{1}{z}\left(\frac{1}{1+\frac{a}{z}}\right) = \sum_n \left(\frac{(-a)^n}{z^{n+1}}\right). \tag{5.57}$$

This expansion is a Laurent expansion in powers of z converging in the region $\{z||z| > a\}$, which is a limiting case of an annular region $\{z|a < |z| < R\}$ as $R \to \infty$.

(e) The expression $\frac{1}{z+a}$ can also be thought of as a (terminating) Laurent series, *in powers of* $z + a$, and having only one term. Its region of convergence, the entire plane except for $z = -a$, corresponds to a limiting case of an annular region $\{z|\rho < |z+a| < R\}$ as $\rho \to 0$ and $R \to \infty$.

☑5.7.2 Problem. Obtain the first few terms of the Laurent series expansion of

$$f(z) = \frac{5z-3}{z(z^2+z-2)},$$

which would converge in the annular region

(a) $1 < |z| < 2$, (b) $1 < |z-1| < 3$.

Solution. The two specified regions are shown in Fig. 5.7. The first annular region, $\{z|1 < |z| < 2\}$, is between two concentric circles having the center at 0 [Fig. 5.7(a)]. The second region, $\{z|1 < |z-1| < 3\}$, is bounded by a pair of circles with the center at $z = 1$ [Fig. 5.7(b)].

(a) We first write the given function as

$$f(z) = \frac{5z-3}{z(z^2+z-2)} = \frac{5z-3}{3z}\left(\frac{1}{z-1} - \frac{1}{z+2}\right). \tag{5.58}$$

Since $|z| < 2$ the first term $\frac{1}{z+2}$ must be expanded in powers of $z/2$:

$$\frac{1}{z+2} = \frac{1}{2}\cdot\frac{1}{1+\frac{z}{2}} = \frac{1}{2}\left(1 - \frac{z}{2} + \frac{z^2}{4} + \frac{z^3}{8} + \cdots\right). \tag{5.59}$$

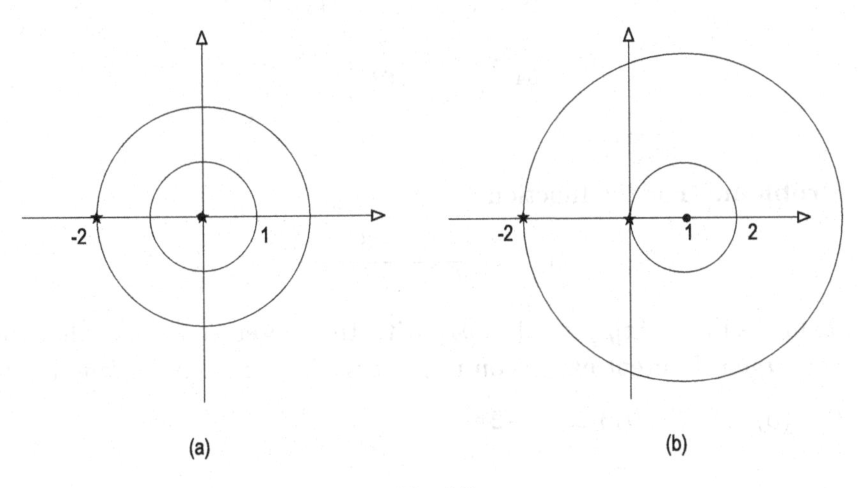

Fig. 5.7

Similarly, in the region of interest $|z| > 1$, and therefore $\frac{1}{z-1}$ should be expanded in powers of $\frac{1}{z}$ to get a convergent series in the required region $1 < |z| < 2$. Thus, we write

$$\frac{1}{z-1} = \frac{1}{z} \cdot \frac{1}{1-\frac{1}{z}} = \frac{1}{z}\left(1 - \frac{1}{z} + \frac{1}{z^2} - \frac{1}{z^3} + \cdots\right). \tag{5.60}$$

Collecting the above results, (5.58)–(5.60), the required expansion becomes

$$f(z) = -\frac{13}{12} + \frac{13}{6}\frac{1}{z} + \frac{2}{3}\frac{1}{z^2} + \cdots \tag{5.61}$$

$$+\frac{13}{24}z - \frac{13}{48}z^2 + \cdots. \tag{5.62}$$

(b) In this case the center of the concentric circles bounding the annular region is at $z = 1$, and the expansion will be in powers of $z - 1 (\equiv w)$. Therefore, we substitute $z = w + 1$ and rewrite the given function as

$$f(z) = \frac{5z-3}{z(z-1)(z+2)} = \frac{5w+2}{w(w+1)(w+3)} \tag{5.63}$$

and the form suitable for the required expansion in powers of w is

$$\frac{5w+2}{w(w+1)(w+3)} = \frac{5w+2}{2w}\left(\frac{1}{w+1} - \frac{1}{w+3}\right). \tag{5.64}$$

In order to get a Laurent series converging in $1 < |w| < 3$ for the above expression, we need to expand $\frac{1}{w+1}$ in powers of $\frac{1}{w}$, and $\frac{1}{w+3}$ in powers of $(w/3)$. These expansions and the corresponding regions of convergence are as follows:

$$\frac{1}{w+1} = \frac{1}{w} \cdot \frac{1}{1+\frac{1}{w}} = \frac{1}{w}\left(1 - \frac{1}{w} + \frac{1}{w^2} - \frac{1}{w^3} + \cdots\right), \quad |w| > 1 \tag{5.65}$$

$$\frac{1}{w+3} = \frac{1}{3} \cdot \frac{1}{1+\frac{w}{3}} = \frac{1}{3}\left(1 - \frac{w}{3} + \frac{w^2}{9} - \frac{w^3}{27} \cdots\right), \quad |w| < 3. \tag{5.66}$$

Thus, we can get the required expansion

$$f(z) = -\frac{13}{18} + \frac{13}{6}\frac{1}{(z-1)} - \frac{3}{2}\frac{1}{(z-1)^2} + \cdots \tag{5.67}$$

$$+\frac{13}{54}(z-1) - \frac{13}{162}(z-1)^2 + \cdots. \tag{5.68}$$

∎

☑**5.7.3 Problem.** For the function

$$f(z) = \frac{z^2 - 3}{z(z^2 - z - 2)},$$

find all annular subsets $\{z | \rho_1 < |z| < \rho_2\}$ with the center at z_0 such that for each region a convergent Laurent expansion in powers of $z - z_0$ may be found, where:

(a) $z_0 = 0$, (b) $z_0 = 2$, (c) $z_0 = -2$.

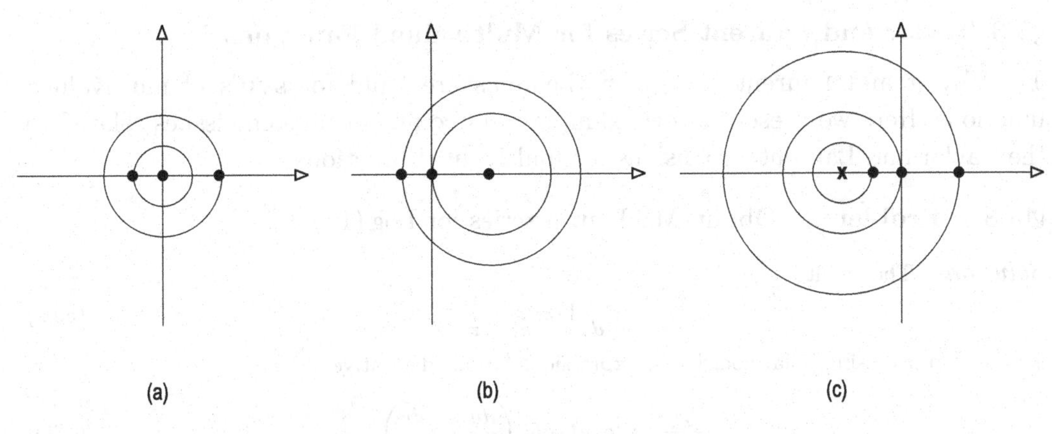

Fig. 5.8

Solution. We should draw concentric circles having centers at z_0 such that the function is analytic inside the annular region bounded by two consecutive circles. So the singular points must be on the circles. The singular points of $f(z)$ are at $z = 0, -1, 2$. These are shown by tiny dark circles in Fig. 5.8.

(a) There are two circles with the centers at $z_0 = 0$ and passing through the singular points. These are $\{z||z| = 1\}$ and $\{z||z| = 2\}$. These circles give three regions — shown in Fig. 5.8(a) — in which Laurent series expansions can be written for the function $f(z)$ in powers of z. The required regions bounded by the two circles are:

(i) $\{z|0 < |z| < 1\}$, (ii) $\{z|1 < |z| < 2\}$, (iii) $\{z||z| > 2\}$.

(b) The concentric circles with the centers at $z_0 = 2$ and passing through the singular points $z = 0, -1$ are drawn in Fig. 5.8(b) and the regions of Laurent expansions are:

(i) $\{z|0 < |z-2| < 2\}$, (ii) $\{z|2 < |z-2| < 3\}$, (iii) $\{z||z-2| > 3\}$.

(c) For $z_0 = -2$, the concentric circles passing through the singular points have radii 1, 2, and 4, and the regions for Laurent expansions are:

(i) $\{z|0 \le |z+2| < 1\}$, (ii) $\{z|1 < |z+2| < 2\}$,
(iii) $\{z|2 < |z+2| < 4\}$, (iv) $\{z||z+2| > 4\}$.

It should be noted that, in case (i), $f(z)$ is analytic at $z = -2$ and hence the Laurent series reduces to the Taylor series and the region of convergence includes the point $z = -2$.

∎

⊳ Tutorial §5.5, on Laurent series expansions to gain an understanding of why negative powers appear in Laurent series and why, in general, one has more than one expansion.
⊳ Quiz §5.7, on the region of convergence of Laurent expansions.
⊳ Exercise §5.8, for practice in getting Laurent series expansions.
⊳ Quiz §§5.9, on the region of convergence for Taylor and laurent series.
⊳ Questions §§5.10, on the uniqueness property of Laurent expansion.

§5.8 Taylor and Laurent Series for Multivalued Functions

The Taylor and Laurent expansion theorems are valid for **svb**'s of multivalued functions. Here we present a few examples and comment on some issues related to the Taylor and Laurent expansions for multivalued functions.

☑5.8.1 Problem. Obtain MacLaurin series for Log $(1+z)$.

Solution. The result

$$\frac{d}{dz}\text{Log } z = \frac{1}{z} \tag{5.69}$$

is easy to prove using polar coordinate expressions for the derivative

$$\frac{df(z)}{dz} = \exp(-i\theta)\left(\frac{du}{dr} + i\frac{dv}{dr}\right). \tag{5.70}$$

Use of the chain rule now gives all the derivatives of Log $(1+z)$,

$$\frac{d}{dz}\log(1+z) = \frac{1}{1+z}, \quad \frac{d^2}{dz^2}\text{Log}(1+z) = -\frac{1}{(1+z)^2}, \ldots, \tag{5.71}$$

needed to construct the MacLaurin series

$$z - \frac{z^2}{2} + \frac{z^3}{3} - \cdots = \sum_n (-1)^{n-1}\frac{z^n}{n}. \tag{5.72}$$

This series converges for $|z| < 1$ and represents the principal branch of Log $(1+z)$. ∎

ℱ5.8.2 Short examples. In this set of examples we comment on the relation between MacLaurin series for different svb's of $\log(1+z)$.

(a) The MacLaruin series $\sum_n (-1)^{n-1}\frac{z^n}{n}$ converges for $|z| < 1$ and represents the principal branch of Log $(1+z)$. In accordance with the Laurent expansion theorem, the radius of convergence is given by the distance of the origin from the nearest singular point, $z = -1$.

(b) If we take the svb of $\log z$, defined by

$$\log z = \ln r + i\theta, \quad 0 < \theta < 2\pi, \tag{5.73}$$

the branch cut of $\log(1+z)$ is along the interval $(-1, \infty)$ and the point $z = 0$ lies on the branch cut. The value of the svb and its derivatives do not exist at $z = 0$ and hence a MacLaurin expansion cannot be written.

(c) For any svb, other those considered above, MacLaurin series may exist and converge for $|z| < 1$, but may not represent values of the function for $|z| < 1$. It may be noted that if the MacLaurin expansion exists for two different svb's, all terms except the first one will be identical because derivatives of all orders will be equal.

ℱ5.8.3 Short examples.

(a) The Taylor series expansion for Log $(1+z)$, in powers of z, converges in a circle of radius 1.

(b) The representation

$$(1+z)^p = 1 + pz + \frac{p(p-1)}{2}z^2 + \cdots, \tag{5.74}$$

in powers of z, would hold for the principal branch of $(1+z)^p$.

(c) The branch cuts for $\log(1+z)$ and $(1+z)^p, p \neq$ integer, run along an infinite interval. Therefore, an annular region $\{z|R_1 < |z+1| < r_2\}$ does not exist in which a Laurent expansion in powers of $z+1$ may be found.

(d) If we define $(z^2-1)^{1/2}$ as a product of the principal values $\sqrt{(z-1)}$ and $\sqrt{(z+1)}$, the branch cut for $\sqrt{z^2-1}$ is along the real interval $(-1,1)$. The function is, therefore, analytic outside the circle $\{z||z|=1\}$. Hence, a Laurent expansion in powers of z and $1/z$ can be found. It is easy to write this expansion,

$$(z^2-1)^{1/2} = z\left(1-\frac{1}{z^2}\right)^{1/2} \tag{5.75}$$

$$= z\left(1-\frac{1}{2}\frac{1}{z^2}-\frac{1}{8}\frac{1}{z^4}-\cdots\right), \tag{5.76}$$

and this represents the function $\sqrt{z^2-1}$ outside the unit circle. Similar results can be seen to hold for $\operatorname{Log}\left(\frac{z-1}{z+1}\right)$ and other functions having a branch cut along a finite line segment.

§5.9 More Results Flowing from the Integral Formula

In this section we will give several other important results which follow from Cauchy's integral formula.

Theorem 5.5 (Cauchy estimates). *Let function $f(z)$ be analytic inside and on a circle C of radius R and center z_0, and M be the maximum value of $f(z)$ on C. Then*

$$\left|f^{(n)}(z_0)\right| \leq \frac{n!M}{R^n}. \tag{5.77}$$

Proof. The integral formula gives

$$f^{(n)}(z_0) = \frac{n!}{2\pi i}\oint \frac{f(z)}{(z-z_0)^{n+1}}\,dz. \tag{5.78}$$

Applying Darboux's theorem gives

$$\left|f^{(n)}(z_0)\right| \leq \left|\frac{n!}{2\pi i}\oint \frac{f(z)}{(z-z_0)^{n+1}},dz\right| \tag{5.79}$$

$$\leq \frac{n!}{2\pi}\frac{M}{R^{n+1}}(2\pi R), \tag{5.80}$$

where M is the maximum value of $f(z)$ on the circle C. Thus, we get the desired result in Eq. (5.77). ∎

Theorem 5.6 (Liouville's theorem). *A function which is analytic everywhere and is bounded for all values of z is a constant.*

Proof. The Cauchy estimate for $n = 1$ implies that

$$|f'(z_0)| \leq \frac{M}{R}, \tag{5.81}$$

for all values of z_0. Since $f(z)$ is analytic everywhere, the radius R is arbitrary and can be taken to be as large as we want. Hence, the inequality (5.81) can hold only if $f'(z_0) = 0$ for all z_0. This implies that $f(z)$ is a constant. ∎

Theorem 5.7 (Fundamental theorem of algebra). *If $P(z)$ is a polynomial,*

$$P(z) = c_0 + c_1 z + c_2 z^2 + \cdots + c_n z^n \quad (c_n \neq 0, n \geq 1), \tag{5.82}$$

of degree 1 or more, the equation $P(z) = 0$ has at least one solution in the complex plane.

Proof. For the sake of obtaining a contradiction, let us assume that $P(z) \neq 0$ for all z. Consider the function $f(z) = 1/P(z)$. It then follows that $f(z)$ is analytic everywhere and is bounded for all z. Liouville's theorem would then require that $f(z)$ be a constant contradicting that $P(z)$ is a polynomial of degree greater than or equal to 1. Hence, $P(z)$ must have at least one zero in the complex plane. ∎

That every polynomial of degree n has exactly n complex zeros requires further development of the theory.

Theorem 5.8 (Morera's theorem). *If $f(z)$ is continuous inside a simply connected domain D and if $\oint_\gamma f(z)dz = 0$ for all closed contours γ lying in D, the function f is analytic throughout D.*

Proof. Under the conditions stated it follows that the indefinite integral

$$F(z) = \int_{z_0}^{z} f(\xi)d\xi \tag{5.83}$$

is analytic and that $f(z) = F'(z)$. Using the result that all higher order derivatives of $F(z)$ exist, we get the result desired that $f(z)$ is analytic in the domain D. ∎

Theorem 5.9 (Maximum modulus theorem). *Let $f(z)$ be analytic inside and on a simple closed curve C and let M be the maximum value of $f(z)$ on C. Then*

$$|f(z)| \leq M \tag{5.84}$$

holds for all z inside the domain bounded by C. In other words, the maximum value is attained on the boundary.

Proof. Since $f(z)$ is analytic, $(f(z))^n$ is also analytic for all n and for all z inside C. Thus, for z_0 inside D, we can write

$$(f(z_0))^n = \frac{1}{2\pi i} \oint_C \frac{(f(\xi))^n}{\xi - z_0} d\xi. \tag{5.85}$$

Let the minimum value of $|\xi - z_0|$, as ξ varies along C, be d. Then by Darboux's theorem

$$|f(z_0)|^n = \frac{1}{2\pi} \frac{M^n L}{d}, \tag{5.86}$$

or

$$|f(z_0)| \leq M \left(\frac{L}{2\pi d}\right)^{1/n}. \tag{5.87}$$

Since n is arbitrary, taking the $n \to \infty$ limit we get

$$|f(z_0)| \leq M. \tag{5.88}$$

∎

Theorem 5.10 (Poisson's integral formula). *If C is a circle $|z| = R$ and $z = re^{i\phi}, r < R$, is a point inside C and if $f(z)$ is analytic on and inside C,*

$$f(re^{i\theta}) = \frac{1}{2\pi} \int_0^{2\pi} \frac{R^2 - r^2}{R^2 - 2Rr\cos(\theta - \phi) + r^2} f(Re^{i\phi})\, d\phi. \tag{5.89}$$

Considering the real and imaginary parts $f(z) = u + iv$, we get Poisson's formula for the harmonic functions $u(r, \theta)$ and $v(r, \theta)$, similar to Eq. (5.89):

$$u(re^{i\theta}) = \frac{1}{2\pi} \int_0^{2\pi} \frac{R^2 - r^2}{R^2 - 2Rr\cos(\theta - \phi) + r^2} u(Re^{i\phi})\, d\phi. \tag{5.90}$$

Proof. Let C be the circle $\{z = Re^{i\phi} | 0 \leq \phi \leq 2\pi\}$ and w be a point inside C. Applying Cauchy's integral formula to the function $f(z)g(z)$, where $g(z) = (R^2 - |w|^2)/(R^2 - \bar{w}z)$, we get

$$\frac{1}{2\pi i} \oint_C \frac{f(\xi)g(\xi)}{\xi - w} d\xi = f(w)g(w). \tag{5.91}$$

Since $g(w) = 1$, the right hand side simplifies to $f(w)$. Poisson's formula then follows, if the substitution $\xi = Re^{i\phi}$ and $w = re^{i\theta}, r < R$, is made on the left hand side of the above equation.

For a circle Poisson's formula expresses the values of a harmonic function *inside the circle* in terms of its values *on the circle.* Similarly, the formula

$$u(x, y) = \frac{1}{\pi} \int_{-\infty}^{\infty} \frac{y u(x', 0)}{(x - x')^2 + y^2} dx' \tag{5.92}$$

called Poisson's formula for a half plane, determines the value of a harmonic function when the values on the boundary, the real axis, are specified.

∎

§5.10 Analytic Continuation

▼

In this section we shall briefly introduce analytic continuation, stating important results without proof. The reader is referred to the books cited in the bibliography for further details and developments. This topic of analytic continuation is not needed later in the book and is included here as it constitutes an extremely important area in the theory of functions of a complex variable. We begin with a result on zeros of an analytic function.

Theorem 5.11 (Zeros of an analytic function). *If $f(z)$ is analytic in a domain D and does not vanish identically in it, the set of zeros of the function in the domain is isolated. In other words, for every zero $z = z_0$, there is a circle $|z-z_0| = \rho(\rho > 0)$ such that within the circle $f(z)$ has no zero other than $z = z_0$ itself.*

As an example, we note that the function $\sin z$ is analytic everywhere and has zeros $z = n\pi, n \in \mathbb{Z}$, which are isolated. This theorem is a powerful statement about the zeros of analytic functions. It implies that if a function $f(z)$ is analytic in a domain D and if the set of points where the function vanishes contains a point which is not isolated, the function vanishes identically everywhere in D. Thus, it follows that if a function analytic in a domain D vanishes on a continuous curve in D, or in a neighborhood of some point in D, or on a set containing a limit point, it must be identically zero in D. Also, if two functions f and g are analytic in D, and if $f = g$ on a curve, or on a neighborhood of some point, or on a set containing a limit point, then $f-g$ must vanish identically in D. Hence, $f = g$ for all points in D.

A useful application of this theorem is that the validity of certain identities for real values implies their validity in a bigger subset of the complex plane. The functions $\sin z$ and $\cos z$ are entire functions and satisfy the identity $\sin^2 x+\cos^2 x = 1$ for all real x. It then follows from Theorem 5.11 that the identity $\sin^2 z+\cos^2 z = 1$ holds for all complex values z.

§5.10.1 *Analytic continuation*

▼

In this section we introduce the concept of analytic continuation of a function and the process of analytic continuation. Though not useful for practical purposes, Taylor series expansion provides a convenient starting point for this discussion.

For a given function, analytic in a domain, several Taylor series expansions, taken about different points z_0, are possible. These series expansions converge in their respective circles of convergence. Thus, the function $F(z) = \frac{1}{a-z}$ has an expansion $\frac{1}{a}\sum_{n=0}^{\infty}(z/a)^n$ converging for $|z| < |a|$. The sum of this series defines a function $f_1(z)$ for $|z| < a$:

$$f_1(z) = \sum_{n=0}^{\infty} \frac{z^n}{a^{n+1}}, \quad |z| < a. \tag{5.93}$$

The functions $f_1(z)$ and $F(z)$ coincide for $|z| < a$, but $f_1(z)$ is not defined for $|z| > a$. The sum of another expansion of $F(z)$, about a different point ξ, defines a function $f_2(z)$,

$$f_2(z) = \sum_{n=0}^{\infty} \frac{(z-\xi)^n}{(a-\xi)^{n+1}}, \quad |z-\xi| < |\xi - a|, \tag{5.94}$$

which coincides with $F(z)$ inside the circle $|z-\xi| = |\xi - a|$. Therefore, we have

$$\begin{aligned} f_1(z) &= F(z), \quad |z| < a, \\ f_2(z) &= F(z), \quad |z-\xi| < |\xi - a|. \end{aligned} \tag{5.95}$$

Notice that a function is defined by specifying the domain of definition of the function and the values it takes on the points in the domain. Here it is natural to think of the functions f_1 and f_2 as parts of the same function $F(z)$, obtained by restricting $F(z)$ over smaller subsets of the plane. The subject of analytic continuation is concerned with a reverse process of extending the domain of definition of a function, such as f_1, and reconstruction of the "whole" function, $F(z)$. *This process, called analytic continuation, is characteristic of complex variables and leads to the concept of an analytic function defined as a single entity without reference to any domain of definition.* We briefly discuss these ideas in the remaining part of this section.

Definition 5.1 If two functions f_1 and f_2, defined on overlapping domains D_1 and D_2, coincide over the points common to the two domains,

$$f_1(z) = f_2, \quad z \epsilon D_1 \cap D_2, \tag{5.96}$$

we say that f_1 and f_2 are analytic continuations of each other. In other words, f_2 is an analytic continuation of f_1 to the domain D_2 and f_1 is an analytic continuation of f_2 to the domain D_1.

If we take $\xi = ia$, the two circles $|z| < a$ and $|z - \xi| < |\xi - a|$ overlap, as shown in Fig. 5.9, and the two functions $f_1(z)$ and $f_2(z)$ in Eqs. (5.93) and (5.94) provide examples of functions which are analytic continuations of each other.

We present two examples to show the process of extending the domain of definition of an analytic function. The first example is analytic continuation by repeated Taylor series expansion. The second example illustrates analytic continuation by repeated integration by parts.

Continuation by Taylor series expansion

As our first example, consider the geometric series

$$1 + z + z^2 + \cdots + z^n + \cdots, \tag{5.97}$$

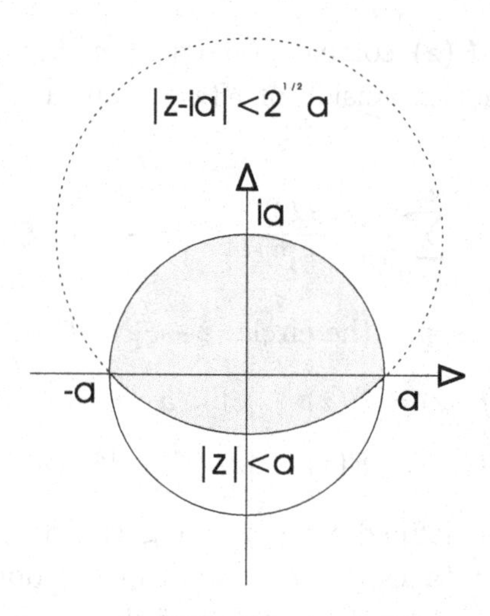

Fig. 5.9

which converges for $|z| < 1$. The sum of (5.97) defines a function $f_1(z)$ analytic inside the unit circle. Though, in this example, the sum of the series is known, the series representation is adequate for finding the value of the function, $f_1(z)$, and all its derivatives for $|z| < 1$. Therefore, we may use Eq. (5.97) and construct its Taylor series representation with $z_0 = -\frac{1}{2}$, and we would then get the series

$$\frac{2}{3} + \left(\frac{2}{3}\right)^2 \left(z + \frac{1}{2}\right) + \left(\frac{2}{3}\right)^3 \left(z + \frac{1}{2}\right)^2 + \cdots . \tag{5.98}$$

in powers of $z + \frac{1}{2}$. The sum of this series gives a function f_2 analytic inside the circle $|z + \frac{1}{2}| < \frac{3}{2}$. This function is an analytic continuation of f_1. The above process can be repeated and one may define a function $f_3(z)$ as the sum of the Taylor series of $f_2(z)$ in powers of $(z + \frac{5}{2})$:

$$f_3(z) = \frac{2}{7} + \left(\frac{2}{7}\right)^2 \left(z + \frac{5}{2}\right) + \left(\frac{2}{7}\right)^3 \left(z + \frac{5}{2}\right)^2 + \cdots . \tag{5.99}$$

The function $f_3(z)$ is analytic inside a bigger circle. $|z + \frac{5}{2}| < \frac{7}{2}$ and provides an analytic continuation of $f_2(z)$.

The three functions f_1, f_2, and f_3 coincide with the sum $\frac{1}{1-z}$ in their respective domains of definition. Here we see how the process of analytic continuation by repeated Taylor expansions can make it possible to reconstruct the "whole" function $\frac{1}{1-z}$ starting from its "part" $f_1(z)$, by extending the original domain of definition of $f_1(z)$.

Continuation by integration by parts

The integral representation:

$$f_1(z) = \int_0^\infty t^{z-1} e^{-t} dt, \tag{5.100}$$

defines function f_1 for $\text{Re}\, z > 0$. An integration by parts allows us to define

$$f_2(z) = \frac{1}{z}\int_0^\infty t^z e^{-t} dt, \tag{5.101}$$

which is well-defined for $\text{Re}\, z > -1$ and coincides with $f_1(z)$ for $\text{Re}\, z > 0$:

$$f_2(z) = f_1(z), \quad \text{Re}\, z > 0. \tag{5.102}$$

In this example $f_2(z)$ is an analytic continuation of $f_1(z)$. Through repeated integrations by parts, we obtain functions $f_3, f_4, \ldots, f_r, \ldots$, where

$$f_r(z) = \frac{1}{z(z+1)\cdots(z+r)}\int_0^\infty e^{-t} t^{z+r} dt \tag{5.103}$$

is defined for $\text{Re}\,(z+r) > 0$. These functions $f_2, f_3, \ldots, f_r$ provide examples of analytic continuations of f_1 to bigger domains and represent the gamma function in their respective domains of definition.

§5.10.2 *Uniqueness of analytic continuation*

It is important to know when the process of analytic continuation does lead to a unique result. Suppose that an analytic function $f(z)$ is defined over a domain D and that $f_1(z)$ and $f_2(z)$ are analytic continuations of $f(z)$ to two domains D_1 and D_2. If D_1 and D_2 have an overlapping part D_3, the two functions $f_1(z)$ and $f_2(z)$ will agree for the points common to two domains, $D_3 = D_1 \cap D_2$,

$$f_1(z) = f_2(z), \quad z \epsilon D_3, \tag{5.104}$$

if D_3 has an overlap with the original domain D, as in Fig. 5.10(a). This can be proven by using the result on zeros of an analytic function. The function f_1 being an analytic continuation of f satisfies $f_1(z) = f(z)$ for points common to D and D_1. Similarly, $f_2 = f(z)$ for $z \epsilon D \cap D_2$. Thus, it follows that $f_1(z) = f_2(z)$ for points in the overlap of all the three domains, D, D_1, and D_2. This ensures that f_1 and f_2 will be equal throughout D_3. If the domain D_3 does not overlap with D, as in Fig. 5.10(b), the values of f_1 and f_2 need not be equal for points in D_3.

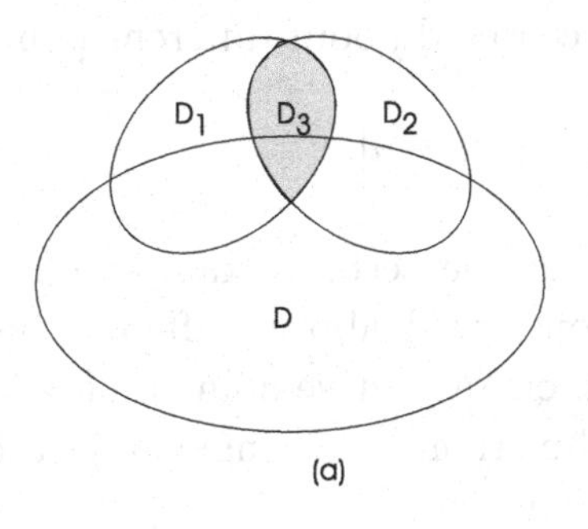

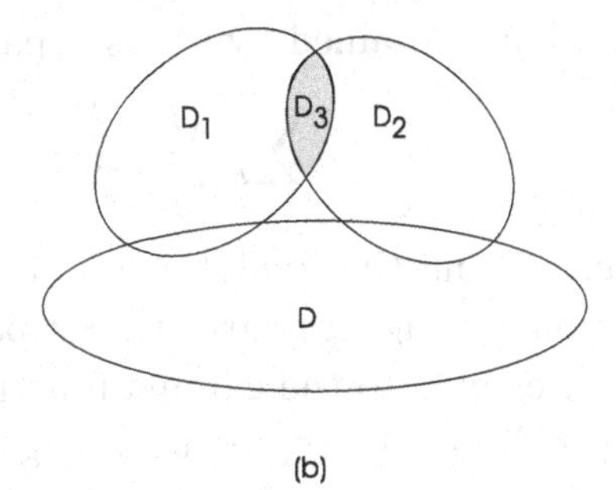

Fig. 5.10

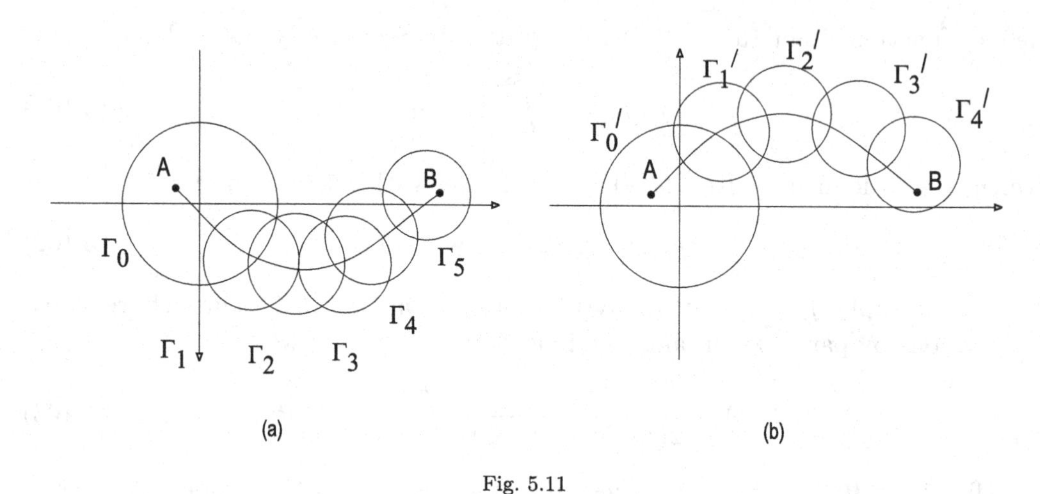

Fig. 5.11

As an example, we consider analytic continuation by Taylor series expansion along two different paths γ_1 and γ_2, connecting two points A and B, as in Fig. 5.11. We would be led to the same answer in the overlap of circles Γ_5 and Γ'_4, if no singular point is enclosed between the two paths. If there are singular points between the two paths, we should expect to get different answers. This is the case when a branch point of a multivalued function lies between the two paths γ_1 and γ_2.

§5.10.3 *Analytic function as a single entity*

▼

A function $f(z)$ is usually defined by specifying values it takes over a subset of complex numbers. The process of analytic continuation allows us to extend the definition to bigger domains of the complex plane, leading to different analytic continuations $f_1, f_2, \ldots$ of the function, as seen in the above examples. All these functions, $\{f_1, f_2, \ldots\}$, obtained by analytic continuation, collectively represent a single entity, an analytic function, now defined without reference to any domain of definition. The process of analytic continuation, starting with an svb of a multivalued function, will lead to all other single-valued branches of the function collectively representing the multivalued function.

In the first example of analytic continuation by Taylor series, in §5.10.1, the set of all functions, obtained by series expansions centered about different points ξ,

$$\sum \frac{(z-\xi)^n}{(a-\xi)^{n+1}}, \quad |z-\xi| < |\xi - a|, \tag{5.105}$$

collectively define the analytic function $\frac{1}{a-z}$. In the second example of analytic continuation through integration by parts, considered above, different functions $f_1, f_2, \ldots$ are equal to the gamma function in their respective domain of definitions and they collectively represent the gamma function as a single object defined without any reference to the domain of definition.

§5.10.4 *Schwarz reflection principle*

Consider a function$f(z)$ which is analytic in a set that contains a point x_0 on the real axis. Then the function can be represented as a Taylor series in powers of x_0:

$$f(z) = \sum_n a_n(z - x_0)^n. \tag{5.106}$$

Let us further assume that the function is real on the real axis. It is then obvious that the coefficients a_n will be real and the function will satisfy $f(\bar{z}) = \overline{f(z)}$. This provides the motivation for the following theorem, which provides an important means of analytic continuation.

Theorem 5.12 (Schwarz reflection principle). *Let a function $f(z)$ be analytic in a domain D in the upper half plane and let an interval on the real line be a part of the boundary of D where the function $f(z)$ is real. If D' is the reflection of D on the real axis, then the function $g(z)$ defined by*

$$g(z) = \begin{cases} f(z), & z \in D, \\ \overline{f(\bar{z})}, & z \in D', \end{cases} \tag{5.107}$$

is analytic in the region $D \cup D'$, see Fig. 5.12.

The Schwarz reflection principle can be generalized to other situations. A function which is pure imaginary on the imaginary axis, $f(z) = -\overline{f(\bar{z})}$, can be analytically continued across the imaginary axis. For a function analytic inside the unit circle and real and continuous on an arc of the unit circle, the relation

$$f(z) = \overline{f\left(\frac{1}{\bar{z}}\right)} \tag{5.108}$$

provides the the analytic continuation across the unit circle. Analytic continuation by reflection in more general curves is possible, if an additional condition of smoothness is met. For a discussion and further references, see Schwarzian reflections in [Needham (1997)].

§5.10.5 *An application*

The uniqueness of analytic continuation can be used to obtain useful conclusions about certain indefinite integrals, especially those appearing in the context of the Schwarz–Christoffel transformation. The integrals under discussion involve multivalued functions and must be computed carefully. One may compute the integral in a convenient fashion and appeal to the uniqueness theorem to prove the correctness of the answer by showing that the answer yields the prescribed value on the real line, when a suitable choice of svb is made for the answer. For instance, in

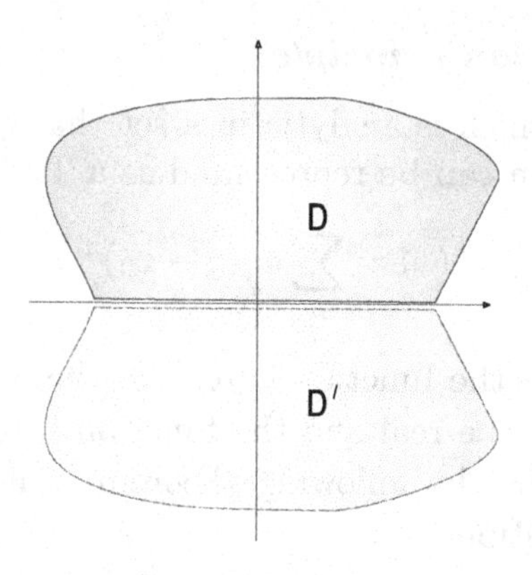

Fig. 5.12

the solved example ☑4.8.2 it was shown that the integral

$$w(z) = \int_0^z \xi^{-1/2}(\xi - 1)^{1/2}\, d\xi \tag{5.109}$$

for real $z(= x)$ has the value

$$w(x) = \begin{cases} -|x|^{1/2}|x-1|^{1/2} - \ln\left(|x|^{1/2} + |x-1|^{1/2}\right), & x < 0, \\ i\sqrt{x(1-x)} + i\sin^{-1}\sqrt{x}, & 0 < x < 1, \\ \dfrac{i\pi}{2} + \sqrt{x(x-1)} - \ln(\sqrt{x} + \sqrt{x-1}), & x > 1. \end{cases} \tag{5.110}$$

It is easy to see that the function $f(z)$,

$$f(z) = \frac{i\pi}{2} + \sqrt{z(z-1)} - \frac{1}{2}\ln(\sqrt{z} + \sqrt{z-1}), \tag{5.111}$$

is analytic in the upper half plane and for $z = x$ coincides with $w(x)$. The uniqueness result allows us to conclude that the indefinite integral $w(z)$ coincides with $f(z)$ for all z in the upper half plane.

§5.11 Summary

The central results of this chapter are Cauchy integral formula, existence of derivatives of all orders for analytic function and Taylor and Laurent series representations. The result that an analytic function is infinitely differentable is a remarkable result with no similar result in case of a function of a real variable. Knowledge of singularities of a function in the complex plane allows us to determine existence of the series expansion and its region of convergence. A Taylor series representation about a point z_0 exists and is a power series in $(z - z_0)$, if the function is analytic at that point and the radius of convergence equals the distance of the point z_0 from nearest singularity in the complex plane. A Laurent series expansion in positive and negative powers of $(z - z_0)$ exists if the function is analytic in an annular region between two concentric circles with centre at z_0.

The Taylor series representation was taken as the starting point of discussion of analytic continuation. The process of analytic continuation allows us to extend the domain of definition of a function and leads to concept of an analytic function of a complex variable without an explicit reference to the domain of definition. In case of a multivalued function, beginning with any one svb analytic continuation leads to all other svb which collectively represent the multivalued function.

Other important results following from integral formula are given in §5.9. The Laurent expansion will form the basis for a classification of isolated singularities. The integral formula will be used to obtain the residue theorem in the next chapter.

Chapter 6

RESIDUE THEOREM

The central result of this chapter is Cauchy's residue theorem. As a preparation for the residue theorem, the definition of an isolated singular point, its classification in terms of a removable singularity, a pole and an essential singular point is described. The residue of a function at an isolated singular point is defined in terms of the Laurent expansion of the function. The residue theorem can also be applied to compute closed contour integrals of functions having a finite number of isolated singularities outside the contour. To make this technique available to the reader, the concept of residue at infinity is introduced. Examples of applications of the residue theorem to proper integrals of trigonometric functions and to the contour integrals of multivalued functions are given.

§6.1 Classification of Singular Points

Singular points of a function of a complex variable are classified as being isolated and nonisolated. Suppose that a function f of a complex variable z is analytic inside a circle $|z-z_0| = R$ except at the point z_0 itself, then z_0 is called an *isolated singular point* of the function f. If a singular point z_0 is such that every circle with center at z_0 has other singular points, then z_0 is not an isolated singular point. Thus, points which are limit points of isolated singular points constitute examples of nonisolated singular points. A singular point z_0 which is not an isolated singular point is called *(nonisolated) essential singular point.*

Isolated singular points can be classified further. Let z_0 be a point such that f is analytic inside the region $0 < |z - z_0| < R$. Assuming f to be single-valued, the Laurent series for the function f can be written as

$$\sum_{n=0}^{\infty} a_n(z-z_0)^n + \sum_{n=1}^{\infty} b_n(z-z_o)^{-n}, \tag{6.1}$$

where the coefficients a_n and b_n are as given by Eqs. (5.42) and (5.43). The series (6.1) converges in the region $0 < |z - z_0| < R$. The following possibilities arise for the Laurent expansion:

Case I: All b's are zero. In this case, only positive powers of $(z - z_0)$ are present in the expansion (6.1). If the sum of the series coincides with the value of the function at $z = z_0$, the function is analytic at the point z_0. If the value of the function at z_0 is different from the sum of the series ($\equiv a_0$), the function is said to have a *removable singularity* at $z = z_0$. The singularity can be *removed* by redefining the function to have the value a_0 at z_0 and it becomes an analytic function.

Case II: Only a finite number of b's are nonzero. In this case, the Laurent expansion contains only a finite number of terms having negative powers of $z - z_0$. Let m be an integer such that $b_m \neq 0$ and $b_n = 0$ for all $n > m$. In this case, we say that the function $f(z)$ has an *order of the pole* at z_0 that is m. The poles of order 1 are also called *simple poles.*

Case III: An infinite number of b's are nonzero. In this case also, we say that the function has an (isolated) *essential singularity* at z_0.

The following two important points are to be noted:

- *An expansion in powers of $z - z_0$ tells us nothing about the nature of the singularity at points other than z_0.*
- *In general, there will be several Laurent expansions about an isolated singular point z_0. The Laurent expansion, which is to be used to decide the nature of a singular point z_0, must be the one which converges in the immediate neighborhood of z_0.*

We explain the above remarks by means of an example. The first remark means, for example, that to fix the nature of the singular point at $z = 3/2$ for the function $f(z)$, given by

$$f(z) = \frac{\operatorname{cosec} \pi z}{2z - 3},$$

one must use the expansion in powers of $z - 3/2$. An expansion in powers of z tells us nothing about the nature of the singular point at $z = 3/2$. The function has different Laurent expansions in powers of z, which are convergent in the annular regions $R_1, R_2, R_3, \ldots$, as shown in Fig. 6.1. However, the expansions in the regions $R_2, R_3, \ldots$ cannot be used to decide the nature of singularity at $z = 0$; only that expansion is useful which is convergent in the region obtained by removing the point $z = 0$ from the interior of the region R_1.

Try §§6.1 and §§6.2 on isolated and nonisolated singular points now.

§6.1.1 *Behavior near an isolated singular point*

It has been noted that there are three types of isolated singularities. We shall briefly summarize the behavior of a function near an isolated singular point z_0.

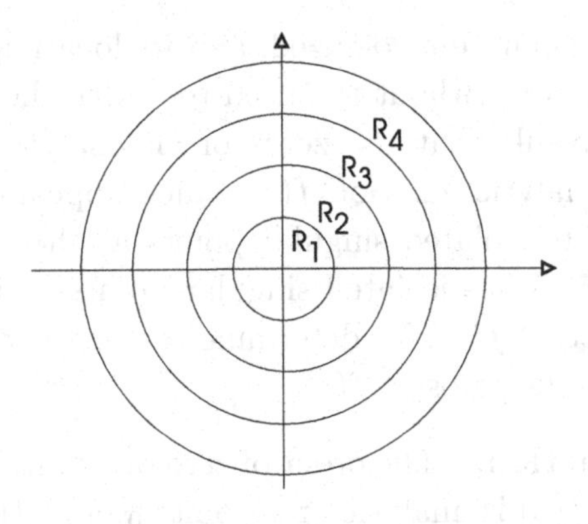

Fig. 6.1

If z_0 is an isolated removable singular point, the function is discontinuous at z_0. If the value of the function $f(z)$ at z_0 is redefined to become equal to $\lim_{z\to z_0} f(z)$, the function becomes analytic at the point. Thus, the singularity can be removed by fixing the value of the function by continuity. If a function $f(z)$ has a pole of order m at a point $z = z_0$, it is clear from the definition that $g(z) = (z - z_0)^m \times f(z)$ is analytic at z_0. Thus, the function $f(z)$ has the form $g(z)/(z - z_0)^m$ with $g(z_0) \neq 0$ and hence $f(z)$ tends to infinity as z tends to z_0. Close to a pole of order m the function is approximated by

$$f(z) \approx \frac{b_m}{(z - z_0)^m}. \tag{6.2}$$

This behavior near a pole helps us in deciding the order of the pole at a point.

Near an essential singularity a function shows more complicated behavior and is given by theorems of Weierstrass and Picard. Briefly, in every neighborhood of an essential singular point z_0, the function assumes every — except possibly one — value in the complex plane. Equivalently, in every neighborhood of z_0 and for every complex number λ — except possibly for at most one complex number — one can always find an infinite subset of complex numbers such that the value of f at each point in the subset is λ. Thus, every neighborhood of an essential singular point is mapped to the entire complex plane — except possibly one point — infinitely many times. For details see the textbooks cited at the end.

§6.2 Finding the Order of Poles and Residues

In order to decide whether a function has a pole at a point, and to find out the order of the pole, it is not always necessary that a Laurent expansion be obtained first. The functions of interest for this purpose will mostly be given by algebraic expressions involving polynomials — the trigonometric, hyperbolic, exponential, and logarithm functions. We shall not be concerned with functions that are, for example, defined by the sum of an infinite series or by an integral representation. For the functions of present interest a pole can be recognized, and its location and order can usually be determined by inspection without the need to derive the Laurent expansion.

It is a well-known result that the zeros of an analytic function are isolated; see Theorem 5.11. An analytic function $f(z)$ which appears in the denominator of an expression gives rise to isolated singular points at the zeros of $f(z)$. Thus, for example, a function $1/f(z)$ has isolated singular points — in fact, poles — at the zeros of an analytic function $f(z)$. To determine the order of the pole in such cases, it is sufficient to look at the zeros of $f(z)$.

Order of zero of a function. The order of zero of an analytic function is defined as follows. If a function $f(z)$ is analytic at a point z_0, and $f(z)$ together with all its derivatives up to order $m-1$ vanish but the mth order derivative does not, we say that for $f(z)$ the *order of zero* at that point is m. For a function $f(z)$ having a zero of order m at a point z_0, the Taylor series expansion begins with $(z-z_0)^m$, i.e. it has the form

$$f(z) = (z-z_0)^m \sum_{n=0}^{\infty} a_n (z-z_0)^n, \quad a_n \neq 0. \tag{6.3}$$

Order of the pole of a function. Noting this fact that the Taylor series of a function having a zero of order m at a point z_0 starts with $(z-z_0)^m$, the following remarks should be helpful in determining whether a singular point is a pole or not, and if it is, then what its order is:

(*1) A function $f(z)$ with a zero of order m has a Taylor series representation of the form (6.3), $1/f(z)$ has a pole of order m at z_0.
(*2) Let $f(z) = \phi(z)/\chi(z)$, with $\phi(z)$ and $\chi(z)$ both being analytic in a domain D. If $\phi(z)$ is nonzero at $z_0 \in D$, and if $\chi(z_0)$ vanishes, the function $f(z)$ has a pole of order equal to the order of zero of $\chi(z)$ at z_0.
(*3) Let $f(z) = \phi(z)/\chi(z)$, with $\phi(z)$ and $\chi(z)$ both being analytic in a domain D. If $\phi(z)$ and $\chi(z)$ have zeros of order p and q, respectively, at a point z_0, then:

   (a) For $p > q$ the function $f(z)$ has a zero of order $p-q$ at the point z_0;
   (b) For $q > p$, the function $f(z)$ has a pole of order $q-p$ at the point z_0;
   (c) $f(z)$ has at most a removable singularity at z_0 if $p = q$.

The above remarks are helpful in determining the order of a pole when the given function is expressible as a ratio of two functions whose zeros can be found easily.

✗6.2.1 Short examples

(a) The polynomial $3z^2+5$ has no singular point and $\exp(1/z)$ is analytic everywhere except at $z=0$. Hence, the product $(3z^2+5)\exp(1/z)$ has only one singular point at $z=0$. This is an isolated singular point.
(b) The function $\cosh \pi z/(z^2+9)$ is the ratio of two entire functions. Such a function can be singular only at the points where the denominator vanishes. Thus, the only singular points are $z = \pm 3i$. These are isolated singular points.

(c) The function $4\sin z + 5$ has no singular points. Hence, the function $1/(4\sin z + 5)$ is singular only at the points for which $4\sin z + 5 = 0$. This equation has the complex solutions

$$z = -\frac{\pi}{2} + 2n\pi - i\ln 2, \quad -\frac{\pi}{2} + 2n\pi + i\ln 2, \tag{6.4}$$

obtainable by expressing $\sin z$ in terms of exponentials. These are all isolated singular points.

(d) The function $(z^3+3)/[\exp(1/z) - 1]$ is the ratio of two functions z^3+3 and $\exp(1/z) - 1$. The numerator is analytic everywhere. The denominator is singular at $z = 0$ and vanishes for $z = i/2m\pi$, where $m = \pm 1, \pm 2, \ldots$. The points $z = 0, i/2m\pi$ are singular points of the given function. Since 0 equals the limit of the singular points $i/2m\pi$, the point $z = 0$ is not an isolated singular point.

(e) In general, for a single-valued branch of a multivalued function, the branch points, as well as all the points on the branch cut, will be singular points. These are nonisolated singular points.

☑**6.2.2 Problem.** Find the singular points and discuss their nature for the following functions:

(a) $z \operatorname{cosec} z$, (b) $\operatorname{cosec} z$, (c) $\operatorname{cosec} z^2$, (d) $\operatorname{cosec}(1/z)$.

Solution.

(a) The given function $f(z) = z\operatorname{cosec} z$ is the ratio of z, and $\sin z$, and both of these have a first order zero at $z = 0$. Therefore, the ratio will have a removable singularity at $z = 0$. The singularity at $z = 0$ can be removed by taking the value of the function at 0 equal to 1 which is the limit of $f(z)$ as z tends to 0.

(b) The function $\sin z$ has first order zeros at $z = n\pi$, and hence $\operatorname{cosec} z$ will have first order poles at $z = n\pi$.

(c) Note that the function $\sin z^2$ vanishes at $z = 0$ and so does its first derivative,

$$\frac{d\sin z^2}{dz} = 2z\cos z^2,$$

but not the second derivative. Hence, $z = 0$ is a pole of order 2. At $z = \sqrt{m\pi}$, $m \neq 0$, the function $\sin z^2$ vanishes but not its derivative. Thus, $z = \sqrt{m\pi}$, $m \neq 0$, are first order zeros of $\sin z^2$ and hence first order poles of $\operatorname{cosec} z^2$.

(d) The function $\sin(1/z)$ is not analytic at $z = 0$ and hence $z = 0$ is a singular point of $\operatorname{cosec}(1/z)$ also. The function $\sin(1/z)$ has zeros of first order at $z = 1/m\pi$ for $m = \pm 1, \pm 2, \ldots$. Therefore, $\operatorname{cosec}(1/z)$ has first order poles at $z = 1/(m\pi)$. The origin, being a limit point of the poles at $1/m\pi$, is an essential singular point which is not isolated.

∎

⊳ Complete the tutorial §§6.2, on classification of isolated singular points.

⊳ Questions §§6.3, to test your understanding of singular points.

§6.3 Residue at an Isolated Singular Point

We now define the residue of a function at an isolated singular point and discuss some useful results on residues.

Definition 6.1. Let $z = z_0$ be an isolated singular point of a function f. In addition, let f be expanded in a Laurent series in a neighborhood of the singular point z_0.

$$f(z) = \sum_{n=0}^{\infty} a_n(z - z_0)^n + \sum_{n=1}^{\infty} b_n(z - z_0)^{-n}. \tag{6.5}$$

The coefficient of $1/(z - z_0)$, b_1, is called the *residue* of the function $f(z)$ at the point z_0 and will be denoted as $\text{Res}\{f(z)\}_{z=z_0}$.

We make several useful remarks about the residue:

(*1) It is important to remember that the Laurent expansion used to compute the residue at a singular point z_0 must be the one which converges in a deleted neighborhood of z_0.

(*2) Using the formula for the coefficients of the Laurent expansion, Eq. (5.43), one gets

$$b_1 = \frac{1}{2\pi i} \oint_C f(z)\, dz, \tag{6.6}$$

where C is any contour such that $f(z)$ is analytic inside and on C except for the point z_0 itself.

(*3) If a function is analytic at a point z_0, the expansion in powers of $z - z_0$ will not contain any negative powers of $z - z_0$. Hence, for a function analytic at a point, the residue at that point is defined to be zero. However, zero residue at a point does not imply that the function is analytic at that point.

(*4) For a function which is the sum of a finite number of functions, the residue at a point is the sum of residues of the individual functions.

§6.3.1 *Residue at a pole*

If a function has a pole of order m at a point z_0, it has an expansion of the form

$$f(z) = \sum_{n=1}^{m} b_n(z - z_0)^{-n} + \sum_{n=0}^{\infty} a_n(z - z_0)^n, \quad b_m \neq 0. \tag{6.7}$$

The function $(z - z_0)^m f(z)$ is an analytic at z_0 and has an expansion in powers of $(z - z_0)$ which can be obtained from the above expansion, Eq. (6.7), by multiplying it by $(z - z_0)^m$:

$$(z - z_0)^m f(z) = \sum_{n=0}^{\infty} a_n(z - z_0)^{n+m} + \sum_{n=1}^{m} b_n(z - z_0)^{m-n}. \tag{6.8}$$

In the above expansion, b_1 is the coefficient of $(z - z_0)^{m-1}$. To determine b_1 differentiate Eq. (6.8) $m - 1$ times w.r.t. z and set $z = z_0$. This gives the following formula for the residue at a pole of order m:

$$b_1 = \frac{1}{(m-1)!} \frac{d^{(m-1)}}{dz^{(m-1)}} \{(z - z_0)^m f(z)\} \bigg|_{z=z_0} \tag{6.9}$$

Special cases of the formula (6.9) for the residue at a pole

The result (6.9) is the basic result for the residue at a pole of order m and it is useful to derive several special cases of Eq. (6.9):

(*1) For a simple pole, $m = 1$, and Eq. (6.9) takes the form

$$b_1 = \lim_{z \to z_0} (z - z_0) f(z). \tag{6.10}$$

(*2) The simplest case of computing the residue is the case where the function $f(z)$ of interest has the form $\phi(z)/(z - z_0)^m$, with $\phi(z_0) \neq 0$, so that $z = z_0$ is a pole of order m. In simple cases of this type, Eq. (6.9) gives the answer in closed form:

$$\text{Res}\,\{f(z)\}_{z=z_0} = \frac{1}{(m-1)!} \frac{d^{m-1}\phi(z)}{dz^{m-1}} \bigg|_{z=z_0}. \tag{6.11}$$

For a first order pole, $m = 1$, the answer for the residue simplifies to $\phi(z_0)$.

(*3) Let the function $f(z)$ be of the form $f(z) = \phi(z)/\chi(z)$. Also, assume that the functions $\phi(z)$ and $\chi(z)$ are analytic at z_0 and that χ has a *first order zero* at z_0 and $\phi(z_0) \neq 0$. Then the function $f(z)$ has a *simple pole* at a point $z = z_0$. In this case the answer for the residue is found to be

$$\text{Res} \left\{ \frac{\phi(z)}{\chi(z)} \right\}_{z_0} = \lim\nolimits_{z \to z_0} \phi(z) \frac{(z - z_0)}{\chi(z)} = \frac{\phi(z)}{\chi'(z)} \bigg|_{z=z_0}, \tag{6.12}$$

where $\chi'(z) = \dfrac{d\chi(z)}{dz}$.

(*4) In those cases where the given function $f(z)$ is of the form $f(z) = \phi(z)/\chi(z)$, where $\phi(z)$ and $\chi(z)$ are analytic at z_0 and assuming that both ϕ and χ vanish at z_0 but z_0 is still a simple pole, one has

$$\text{Res}\,\{f(z)\}_{z=z_0} = \lim_{z \to z_0} (z - z_0) \frac{\phi(z)}{\chi(z)}. \tag{6.13}$$

The limit is most easily computed by repeated use of the analog of l'Hopital's rule, given below.

✍ *Let $g(z)$ and $h(z)$ be analytic at z_0 and both vanish at z_0. Then the limit of the ratio $g(z)/h(z)$ as $z \to z_0$, if it exists, is given by*

$$\lim_{z\to z_0} g(z)\,/h(z) = \lim_{z\to z_0}\left[\frac{dg(z)}{dz}\right] \Big/ \left[\frac{dh(z)}{dz}\right]. \qquad (6.14)$$

(*5) Finally, one can have an approach similar to that in Eq. (6.14) when $f(z) = \phi(z)/\chi(z)$, with $\phi(z)$ and $\chi(z)$ analytic at z_0, and both ϕ and χ vanish at z_0, and are such that the point z_0 is a pole of order greater than 1. However, this is of limited use and, in general, one must fall back on the Laurent expansion to compute the residue.

§6.4 Computing the Residues

We shall briefly discuss a few strategies and give a few tips for computation of residues. In many cases when the Laurent expansion is easy to write down, the residue is easily found by making use of its definition. At an isolated essential singular point, the residue is found by substitution $z = z_0 + w$ followed by Laurent expansion of $f(z_0 + w)$ in powers of w. As the singular point z_0 is assumed to be isolated, an expansion of $f(z_0 + w)$ in powers of w and convergent in a deleted neighbourhood of $w = 0$ exists. We write this expansion and pick up the coefficients of $1/w$.

For residue of a function $f(z)$ at pole of order m, the expression

$$\frac{1}{(m-1)!}\frac{d^{(m-1)}}{dz^{(m-1)}}\{(z-z_0)^m f(z)\}\bigg|_{z=z_0}. \qquad (6.15)$$

and its special cases already discussed in the previous section are useful.

Finally, for sake of completeness, we mention that even Eq. (6.6) on page 150 involving an integration can be used to compute the residues.

⸸6.4.1 Short examples. We shall give several examples of computing the residue.

(a) The coefficient of $(1/z)$ in the expansion

$$\exp(1/z) = 1 + \frac{1}{z} + \frac{1}{2!}\frac{1}{z^2} + \dots$$

gives the residue equal to 1.

(b) The Laurent series for $\sin z^2/z^7$ is easy to write

$$\frac{\sin z^2}{z^7} = \frac{1}{z^5} - \frac{1}{3!z} + \frac{z^3}{5!} + \dots$$

and the residue at $z = 0$ is, therefore, given by -1/6.

(c) The function $\frac{z^2-3}{z^2+7z+12}$ has a first order pole at $z = -3$ and its residue at $z = -3$ is given by

$$\lim_{z\to -3}(z+3)\frac{z^2-3}{z^2+7z+12} = \lim_{z\to -3}(z+3)\frac{z^2-3}{(z+3)(z+4)} = \frac{z^2-3}{(z+4)}\bigg|_{z=-3}$$
$$= 6$$

(d) The residue of $\frac{z\cos z}{z+\pi}$ at $z=-\pi$ is

$$\begin{aligned}\lim_{z\to-\pi}(z+\pi)\frac{z\cos z}{(z+\pi)} &= \lim_{z\to-\pi} z\cos z\\ &= \pi.\end{aligned}$$

(e) The residue of $\frac{1}{(\cosh z-\cosh 5)}$ at $z=5$, where is

$$\lim_{z\to5}\frac{z-5}{\cosh z-\cosh 5} = \frac{1}{\sinh z}\bigg|_{z=\xi_n} = -\frac{1}{\sinh 5}$$

where the limit is computed using l'Hopital rule.

(f) The function $\sin z/(e^z-1)^2$ has a simple pole at $z=0$ and hence we have

$$\text{Residue} = \lim_{z\to0}\frac{z\sin z}{(e^z-1)^2} = \lim_{z\to0}\left(\frac{z}{e^z-1}\right)\lim_{z\to0}\left(\frac{\sin z}{e^z-1}\right) = 1$$

where each limit in the last step is seen to exists separately and is easily computed using l' Hopital rule.

(g) The function $\dfrac{\exp(\alpha z)}{(z+3)^7}$ has a pole of order 7 at $z=-3$. Therefore the residue is

$$\frac{1}{6!}\frac{d^6}{dz^6}(z+3)^7\left[\frac{\exp(\alpha z)}{(z+3)^7}\right]\bigg|_{z=-3} = \frac{1}{6!}\frac{d^6}{dz^6}\exp(\alpha z)\bigg|_{z=-3} = \frac{\alpha^6}{6!}\exp(-3\alpha).$$

☑6.4.2 Problem. Compute the residue of the function $\text{cosec}\pi z^7$ at

(a) at $z=\sqrt[7]{n}$ (b) at $z=0$.

Solution.

(a) The function $\text{cosec}(\pi z^7)$ has a first order pole at $z=\sqrt[7]{n}$ for $n\neq0$. Hence the residue at $z=\sqrt[7]{n}$, using the formula, is

$$\text{Res}\,\{\text{cosec}(\pi z^7)\}z=\sqrt[7]{n} = \lim_{z\to\sqrt[7]{n}}\left(\frac{z-\sqrt[7]{n}}{\sin\pi z^7}\right) = \frac{1}{7\pi z^6\cos\pi z^7}\bigg|_{z=\sqrt[7]{n}} = \frac{(-1)^n}{7\pi\, n^{6/7}}$$

where again the limit is conveniently computed using l'Hopital rule.

(b) The function $\text{cosec}\pi z^7$ has a seventh order pole at $z=0$. The use of the formula for the residue at $z=0$ leads to complicated expressions. It is simpler to derive the Laurent expansion and find the residue as the coefficient of the $1/z$ term. We first derive an expansion for $\text{cosec} z$ which is needed below.

$$\begin{aligned}\sin z &= \left(z-\frac{z^3}{3!}+\frac{z^5}{5!}+\cdots\right)\\ \therefore\quad \frac{1}{\sin z} &= \frac{1}{z}\left(1-\frac{z^2}{3!}+\frac{z^4}{5!}+\cdots\right)^{-1}\\ &= \frac{1}{z}\left[1+\left(\frac{z^2}{3!}-\frac{z^4}{5!}+\cdots\right)+\left(\frac{z^2}{3!}+\frac{z^4}{5!}+\cdots\right)^2+\cdots\right]\end{aligned}$$

The residue of $1/\sin z^7$ at $z = 0$ is most easily computed using the expansion

$$\frac{1}{\sin \pi z^7} = \frac{1}{\pi z^7}\left[1 + \left(\frac{\pi^2 z^{14}}{3!} - \frac{\pi^4 z^{28}}{5!} + \cdots\right) + \left(\frac{\pi^2 z^{14}}{3!} + \frac{\pi^4 z^{28}}{5!} + \cdots\right)^2 + \cdots\right]$$

It is obvious that the term $1/z$ is absent. Hence the residue is zero. ∎

⊳ Tutorial §§6.4, for computation of residues at simple poles.
⊳ Tutorial §§6.5, for computation of residues at poles of order greater than 1.

§6.5 Cauchy's Residue Theorem

Theorem 6.1. *Let f be analytic inside and on a simple closed contour C except at a finite number of singular points, $z_1, z_2, \ldots, z_n$. In addition, let the residue of $f(z)$ at these points be $R_1, R_2, \ldots, R_n$. Then*

$$\oint_C f(z)dz = 2\pi i(R_1 + R_2 + \cdots + R_n). \tag{6.16}$$

Proof. Let $\Gamma_1, \Gamma_2, \ldots, \Gamma_n$ be nonoverlapping circles, each circle Γ_i lying completely inside the contour C and enclosing only one of the singular points. For proof of the residue theorem, the same approach is used as in Theorem 4.6. We form a new contour Γ, as shown in Fig. 6.2(b), by connecting the contour C with Γ_i with parallel lines. Since the function $f(z)$ is analytic inside Γ, we have

$$\oint_\Gamma f(z)dz = 0. \tag{6.17}$$

Taking the limit in which the parallel lines come close, we get

$$\oint_\Gamma f(z)dz = \oint_C f(z)dz - \sum_k \oint_{\Gamma_k} f(z)\,dz \tag{6.18}$$

$$= 0. \tag{6.19}$$

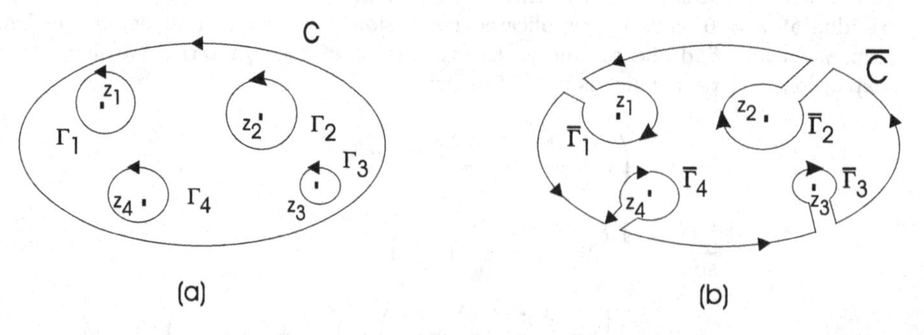

Fig. 6.2

It follows from an earlier property, Eq. (6.6), that integrals $\oint_{\Gamma_k} f(z)\,dz$ equal $2\pi i R_k$. Therefore, we get the residue theorem

$$\oint_C f(z)dz = 2\pi i(R_1 + R_2 + \cdots + R_n). \tag{6.20}$$

This theorem is the most important tool for applications and is essential for computation of integrals. Examples of the use of Cauchy's residue theorem appear repeatedly in problems in subsequent chapters. ▮

⇨ Question set §§6.7, for short questions on the use of the residue theorem.

§6.5.1 *An integral for indented contours.*

When a singular point z_0 of a function $f(z)$ lies on a contour Γ, the contour integral $\int_\Gamma f(z)\,dz$ is not defined. In such cases one may attempt to define the integral by bypassing the singular point in some way. While evaluating such integrals, one has to frequently find the value of the integral $\int_{\gamma_\rho} f(z)\,dz$, where γ_ρ is a circular arc with the singular point z_0 as the center and with radius ρ. If z_0 is a simple pole of $f(z)$, the value of the integral, in the limit where the radius of the arc goes to zero, can be computed and is given by the following result.

Let $f(z)$ be a function having a simple pole at z_0. In addition, let γ be an arc of a circle C of radius ρ and center z_0, and let ρ be chosen to be sufficiently small so that $f(z)$ is analytic inside and on the full circle C except at the center; see Fig. 6.3. Then

$$\lim_{\rho\to 0} \int_\gamma f(z)\,dz = i\alpha \text{Res}\,\{f(z)\}_{z=z_0}. \tag{6.21}$$

This result can be proven by integrating the Laurent series of the function $f(z)$ in powers of $z - z_0$, and the proof is left as an exercise for the reader. This result will later be needed for computation of integrals along indented contours, in §7.3, §§7.10, and §§7.11.

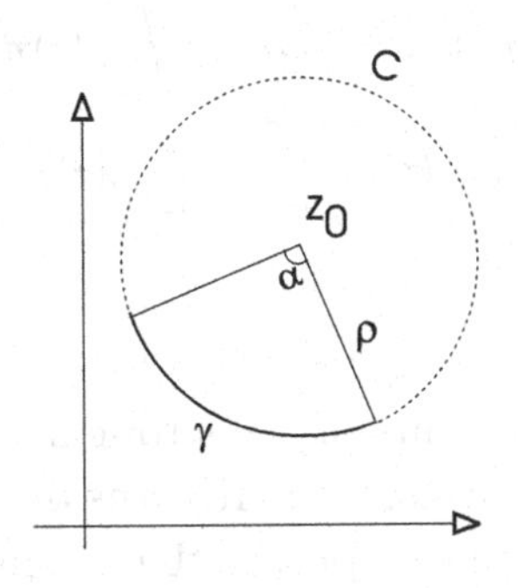

Fig. 6.3

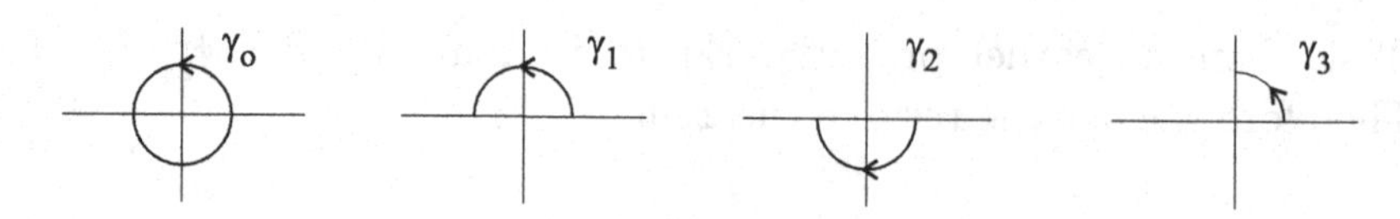

Fig. 6.4

⨍6.5.1 Short examples. It has been noted that if γ is an *open* circular arc, having radius ρ and making an angle α at its center z_0, $\lim_{\rho\to 0}\int_\gamma f(z)\,dz$ exists and is given by Eq. (6.20), when $f(z)$ is analytic or has at most a simple pole at z_0. The integral diverges when $f(z)$ has a pole at z_0 of order 2 or higher.

To illustrate this result a few examples are given below taking integrals of several different functions along three open arcs, γ_1, γ_2, and γ_3 in Fig. 6.4. For comparison we also note the result for a closed contour γ_0.

(a) $\frac{1-e^{iz^2}}{z^2}$ has zero residue $z = 0$, where it has a removable singularity and hence its integral along all the four contours vanishes.

(b) The residue of $\frac{\sin z}{z^2}$ at $z = 0$ is 1 and hence

$$\int_{\gamma_0} \frac{\sin z}{z^2}\,dz = 2\pi i, \quad \int_{\gamma_1} \frac{\sin z}{z^2}\,dz = +\pi i,$$
$$\int_{\gamma_2} \frac{\sin z}{z^2}\,dz = -\pi i, \quad \int_{\gamma_3} \frac{\sin z}{z^2}\,dz = +\frac{\pi i}{2}.$$

(c) The function $z^{-3}\exp(ipz), p \neq 0$, has a third order pole at $z = 0$ with residue $-p^2/2$. Hence,

$$\oint_{\gamma_0} z^{-3}\exp(ipz)\,dz = -\pi i p^2,$$

but the limit of the integral diverges for the other three contours, γ_1, γ_2, and γ_3.

(d) If the first two terms, $z^{-3}(1 + ipz)$, of the Laurent series expansion of $z^{-3}\exp(ipz)$ are subtracted and we consider integrals of

$$g(z) = \frac{\exp(ipz) - 1 - ipz}{z^3},$$

we will get the integrals

$$\oint_{\gamma_0} g(z)\,dz = -\pi i p^2, \quad \int_{\gamma_1} g(z)\,dz = -\frac{\pi i p^2}{2},$$
$$\int_{\gamma_2} g(z)\,dz = \frac{\pi i p^2}{2}, \quad \int_{\gamma_3} g(z)\,dz = -\frac{\pi i p^2}{4}.$$

§6.6 Residue at Infinity

The residue theorem gives the value of a contour integral in terms of the residues at singular points enclosed in the contour. It turns out to be useful to introduce the concept of *residue at infinity* and to rephrase the scheme of evaluation of the contour integral in terms of residues at singular points *outside* the contour. To explain the basic idea, we consider evaluation of an integral

$$I = \oint_C f(z)dz, \tag{6.22}$$

where C is the positively oriented circular contour $|z| = R$. By making a change of variable to $w = 1/z$, the above integral is transformed into a contour integral in the w plane. The image of the circle C under the inversion map, $z \to w = 1/z$, is circle C', given by $|w| = 1/R$ taken in the *clockwise* direction. Thus,

$$I = -\oint_{C'} f(w)\frac{dw}{w^2} = \oint_{C''} g(w)dw, \tag{6.23}$$

where

$$g(w) = w^{-2} f\left(\frac{1}{w}\right), \tag{6.24}$$

and C'' is the positively oriented circle $|w| = 1/R$. The singular points of $g(w)$, lying inside C'' in the w plane, will correspond to the singular points of $f(z)$ *outside the circle* C in the z plane. In addition, the integral in Eq. (6.22) may receive a nonvanishing contribution from the point $w = 0$, which corresponds to the point at infinity in the z plane.

With the above discussion as motivation, we now come to the definition of the residue at infinity.

A function $f(z)$ is *analytic at infinity* if $f(1/w)$ is analytic at $w = 0$, otherwise the function is said to be singular at infinity. For a function $f(z)$ the nature of singularity at infinity is defined to be the same as the nature of singularity of $f(1/w)$ at $w = 0$. It would then be seen that the point at infinity is an isolated singular point of $f(z)$ if there exists a closed contour γ, in the z plane, such that the point at infinity is the only singular point of the function $f(z)$ *outside* γ. For such a function the *residue at infinity* is defined to be

$$\text{Res}\,\{f(z)\}_\infty = -\frac{1}{2\pi i}\oint_\gamma f(z)dz, \tag{6.25}$$

where γ is taken to be in the anticlockwise direction. Apart from the negative sign included in the definition of the residue at infinity, Eq. (6.25) is a straightforward extension of the result given earlier, Eq. (6.6). The minus sign is included because an anticlockwise closed contour has negative orientation w.r.t. the points outside the contour. The residue at infinity is easily computed by changing the variable from z to $w = 1/z$ in Eq. (6.25), giving

$$\text{Res}\,\{f(z)\}_\infty = \frac{1}{2\pi i}\oint_{\gamma'} w^{-2} f\left(\frac{1}{w}\right) dw = -\frac{1}{2\pi i}\oint_{\gamma''} w^{-2} f\left(\frac{1}{w}\right) dw, \tag{6.26}$$

where the image γ' of γ under the inversion map $z \to w = 1/z$ is clockwise and the contour γ'' coincides with the contour γ', but is taken in anticlockwise orientation.

The above discussion shows that one could use the following working recipe for computing the residue at infinity:

- Res $\{f(z)\}_\infty$ *is given by the negative of the coefficient of* w *in expansion of* $f(1/w)$ *in powers of* w*, for small* w*.*
- *For an integral along a closed contour* γ *with the integrand* $f(z)$ *having only a finite number of isolated singular points outside* γ*, we have the result that* $\oint_\gamma f(z)dz$ *equals* $2\pi i$ *times the negative of the sum of residues at all the points outside* γ*.*

The above rule is the simplest method available for a class of integrals when the integrand has a branch cut inside the contour of integration. Examples of such integrals can be found in §§6.10. It is to be noted that the concept of residue at infinity, like the residue at any other point, is defined only when the integrand has an isolated singularity at infinity. While a function analytic at a point in the *finite* complex plane must have a zero residue there, in contrast a function which is not singular at ∞ can have a nonzero residue at infinity. For example, $-\frac{2i}{z}$ is analytic at infinity but has nonzero residue $2i$ there.

§6.7 Illustrative Examples

☑6.7.1 Problem. Compute the integral

$$\oint_C \frac{(3z^2+2)dz}{(z-4)(z^2+9)}$$

for each of the four positively oriented circles C given by:

$$\text{(a) } |3z-2|=1, \quad \text{(b) } |z+1|=4, \quad \text{(c) } |z-2|=3, \quad \text{(d) } |z|=5.$$

Solution. The singular points of the function $(3z^2+2)/(z-4)(z^2+9)$ are at $z = 4, 3i, -3i$. The corresponding residues are easily found to be $2, (3-i)/6$ and $(3+i)/6$. For the above four cases the values of the integrals are computed as follows:

(a) The circle $|3z-2|=1$ has the center at $z=2/3$ and radius 1, and all the three singular points lie outside the circle. Hence, the value of the integral is zero.

(b) The circle $|z+1|=4$ has the center at -1 and radius 4, and it does not enclose the singular point at 4. The integral is then equal to $2\pi i$ times the sum of the residues at the other two points which are inside the circle. The answer is $2\pi i$.

(c) The circle $|z-2|=3$ encloses only one singular point at 4, and the corresponding integral is $2\pi i \times 2 = 4\pi i$.

(d) Finally, the circle $|z|=5$ encloses all the three singular points, and the value of the integral is given by $6\pi i$.

∎

✍6.7.2 Short examples

(a) The function $f(z) = 1/z$ is analytic at ∞, but the residue is -1, a nonzero value.

(b) To find the residue of the function $f(z) = \sqrt{(z-a)(z-b)}$ at infinity, we substitute $z = 1/w$ to get $f = \frac{1}{w}\sqrt{(1-aw)(1-bw)}$, which is expanded in powers of w in the neighborhood of $w = 0$,

$$f = \frac{1}{w}\left(1 - \frac{aw}{2} + \frac{1}{8}a^2w^2 + \cdots\right)\left(1 - \frac{bw}{2} + \frac{1}{8}b^2w^2 + \cdots\right) \tag{6.27}$$

$$= \frac{1}{w} - \frac{a+b}{2} + \frac{1}{8}(a-b)^2 w + \cdots, \tag{6.28}$$

and the residue of $f(z)$ at infinity is given by $-\frac{1}{8}(a-b)^2$.

(c) Consider evaluation of the integral in ☑6.7.1, when C is the circle $|z| = 5$. The integral can be evaluated by making the transformation $z \to w = 1/z$. For $f(z) = \frac{3z^2+2}{(z-4)(z^2+9)}$, $g(w)$ of Eq. (6.23) is

$$g(w) = w^{-1}\frac{3+2w^2}{(1-4w)(1+9w^2)}. \tag{6.29}$$

The residue of $g(w)$ at $w = 0$ is 3 and we recover the value $6\pi i$ for the integral.

(d) Continuing with the above example, the residue at infinity is computed easily by finding the negative of the coefficient of w in the expansion of $f(1/w)$,

$$f\left(\frac{1}{w}\right) = \frac{w(3+2w^2)}{(1-4w)(1+w^2)}, \tag{6.30}$$

and the answer is -3 and the value of the integral used is again $6\pi i$.

☑6.7.3 Problem. Compute the integral

$$\int_0^{2\pi} \frac{d\theta}{7 + 6\cos\theta}. \tag{6.31}$$

Solution. The substitutions $z = \exp(i\theta)$ and

$$\cos\theta = \frac{1}{2}\left(z + \frac{1}{z}\right), \quad \sin\theta = \frac{1}{2i}\left(z - \frac{1}{z}\right) \tag{6.32}$$

in the integral (6.30) give rise to an integral,

$$\int_0^{2\pi} \frac{d\theta}{7 + 6\cos\theta} = -i\oint_C \frac{dz}{3z^2 + 7z + 3}, \tag{6.33}$$

over the unit circle $|z| = 1$ and is easily evaluated with the help of the residue theorem. The integral has two poles at $z_\pm = \frac{-7\pm\sqrt{13}}{6}$. Only one of the two poles, that at $z_+ = \frac{-7+\sqrt{13}}{6}$, lies inside the unit circle, whereas the other pole, at $z_- = \frac{-7-\sqrt{13}}{6}$, lies outside it. Computing the residue,

$$\operatorname{Res}\left\{\frac{dz}{3z^2 + 7z + 3}\right\}_{z=z_+} = \frac{1}{\sqrt{13}}, \tag{6.34}$$

and hence the required integral is found to be

$$\int_0^{2\pi} \frac{d\theta}{7 + 6\cos\theta} = \frac{2\pi}{\sqrt{13}}. \tag{6.35}$$

∎

⊳ Exercise §§6.6, for a practice session on applying the residue theorem.
⊳ Questions §§6.7, to test your understanding of the residue theorem.

- ⊳ Tutorial §§6.8, is a session on the first application of the residue theorem to computation of a class of proper integrals over a real variable with the integrand involving trigonometric functions.
- ⊳ Exercise §§6.9, for practice problems on integrals of type $\int f(\cos\theta, \sin\theta)\, d\theta$.
- ⊳ Quiz §§6.11; computing the residue is an important activity; to test your skills try out this set.
- ⊳ Mined set §§6.12, to see what can happen if one tries to calculate the residue by series expansion ("mechanically") of a function without caring about the convergence properties of the Laurent series.

§6.8 Residue Calculus and Multivalued Functions

The residue theorem is extremely useful for computing integrals along closed contours. In its commonly stated standard form the theorem says that the value of an integral along a closed contour equals $2\pi i$ times the sum of the residues of the integrand at isolated singular points enclosed by the contour. If the contour encloses a branch cut, the residue theorem cannot be applied. However, if the singular points outside the contour, *including the point at infinity*, are isolated, the residue theorem can be applied and the value of the integral can again be written as the *negative* of the sum of residues at all singular points outside the contour, including the point at infinity. We show this by means of two simple examples.

☑**6.8.1 Problem.** Taking the principle value set up integral of $\sqrt{z^2-a^2}$ $(a>0)$, around the branch cut, and show that it is proportional to $\int_{-a}^{a}\sqrt{a^2-x^2}\,dx$. Use the residue theorem to compute the value of the contour integral and hence show that

$$\int_{-a}^{a}\sqrt{a^2-x^2}\,dx = \frac{1}{2}\pi a^2.$$

Solution. For the multivalued function $f(z)=\sqrt{z^2-a^2}$, we select an svb which is the product of the principal value branches of $\sqrt{z+a}$ and $\sqrt{z-a}$, given by

$$\sqrt{z+a} = |z+a|^{1/2}\exp(i\phi_1/2), \tag{6.36}$$

$$\sqrt{z-a} = |z-a|^{1/2}\exp(i\phi_2/2), \tag{6.37}$$

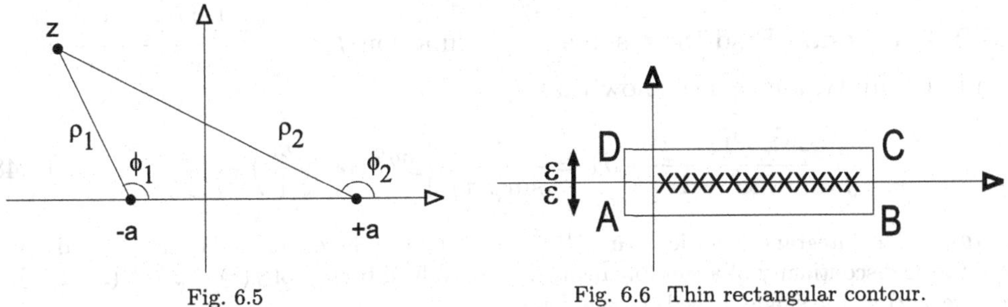

Fig. 6.5

Fig. 6.6 Thin rectangular contour.

where

$$\phi_1 = \arg(z+a), \ -\pi < \phi_1 \le \pi, \tag{6.38}$$

$$\phi_2 = \arg(z-a), \ -\pi < \phi_2 \le \pi, \tag{6.39}$$

and $|z+a| + |z-a| > 2a$, see Fig. 6.5.

It can be easily verified that the svb of $f(z)$ has a branch cut along $(-a, a)$. Noting that just above and just below the cut the values of $\phi_1 + \phi_2$ are given by

$$(\phi_1+\phi_2)_{\mathrm{A}} = \pi, \quad (\phi_1+\phi_2)_{\mathrm{B}} = -\pi, \tag{6.40}$$

we see that the discontinuity of the function across the cut is given by

$$\lim_{\epsilon\to 0}[f(x+i\epsilon) - f(x-i\epsilon)] = 2i\sqrt{a^2-x^2}, \quad x < |a|. \tag{6.41}$$

Next, we write the contour integral around a thin rectangular contour of Fig. 6.6:

$$\lim_{\epsilon\to 0}\oint_{\mathrm{ABCDA}} f(z)\,dz = \int_{\mathrm{AB}} f(z)\,dz + \int_{\mathrm{BC}} f(z)\,dz + \int_{\mathrm{CD}} f(z)\,dz + \int_{\mathrm{DA}} f(z)\,dz \tag{6.42}$$

$$= \int_{\mathrm{AB}} f\,dz + \int_{\mathrm{CD}} f\,dz = \int_{\mathrm{AB}} f\,dz - \int_{\mathrm{DC}} f\,dz, \tag{6.43}$$

because the integrals of $f(z)$ along BC and DA vanish in the limit where $\epsilon \to 0$. Noting that $z = x \pm i\epsilon$ along the two line segments BA and CD, we get

$$\lim_{\epsilon\to 0}\oint_{\mathrm{ABCDA}} f(z)\,dz = -\int_{-a}^{a} f(x+i\epsilon)\,dx + \int_{-a}^{a} f(x-i\epsilon)\,dx = -2i\int_{-a}^{a}\sqrt{a^2-x^2}\,dx; \tag{6.44}$$

the contour integral can be written in terms of the residue at infinity and is equal to the product of $(-2\pi i)\times$ residue of $f(z)$ at infinity. The residue at infinity is obtained by substituting $z = 1/w$, expanding $f(1/w)$ in powers of w, and taking the *negative of the coefficient* of w. Thus,

$$f\left(\frac{1}{w}\right) = \frac{1}{w}(1-a^2w^2)^{1/2} = \frac{1}{w} - \frac{a^2}{2}w + \cdots. \tag{6.45}$$

Therefore,

$$\lim_{\epsilon\to 0}\oint_{\mathrm{ABCDA}} f(z)\,dz = (-2\pi i)\left(\frac{a^2}{2}\right) = -\pi i a^2. \tag{6.46}$$

Hence, we have the answer

$$\int_{-a}^{a}\sqrt{a^2-x^2}\,dx = \frac{i}{2}\lim_{\epsilon\to 0}\oint_{\mathrm{ABCDA}} f(z)\,dz = \frac{\pi a^2}{2}. \tag{6.47}$$

∎

☑**6.8.2 Problem.** Find the residue of the function $f(z) = \dfrac{z^{(1-p)}(z-1)^p}{z^2+1}$ at $z = \pm i$ and at infinity, and hence show that

$$\int_0^1 \frac{x^{(1-p)}(1-x)^p}{x^2+1}\,dx = \frac{\pi}{\sin(p\pi)}\left[2^{p/2}\cos\left(\frac{p\pi}{4}\right) - 1\right]. \tag{6.48}$$

Solution. The integrand is of the form $x^{(1-p)}(1-x)^p Q(x)$, where $Q(x) = (1+x^2)^{-1}$, and can be related to the discontinuity of a suitably defined single-valued branch of $f(z) = z^{(1-p)}(z-1)^p Q(z)$. If we take $f(z)$ as the single-valued branch defined by

$$f(z) = r^{(1-p)}\rho^p\, e^{i[1-p\theta]}e^{ip\phi}Q(z) \tag{6.49}$$

and

$$0 < \theta \le 2\pi, \quad 0 < \phi \le 2\pi, \quad r + \rho > 1. \tag{6.50}$$

Our notation is the same as in ☑4.8.4, where it has been shown that the integral in Eq. (6.56) can be written as a closed contour integral,

$$\int_0^1 x^{(1-p)}(1-x)^p Q(x)\,dx = \left[\frac{i}{2\sin(p\pi)}\right]\oint_\Gamma z^{(1-p)}(z-1)^p Q(z)dz, \tag{6.51}$$

where Γ is a thin rectangular contour of Fig. 6.6 going round the interval $[0,1]$ which does not enclose any poles of $Q(z)$. The value of the integral is given by $-2\pi i$ times the sum of residues of $z^{(1-p)}(z-1)^p\, Q(z)$ at the singular points, $i, -i$, and ∞, lying outside Γ.

For the point at infinity, $\rho = r$ and $\theta = \phi$ and we find that

$$f(z) \to \frac{r\exp(i\theta)}{z^2+1} = \frac{z}{z^2+1} \approx \frac{1}{z}, \tag{6.52}$$

and therefore the residue at ∞ is given by

$$\text{Res}\,\{f(z)\}_\infty = -1. \tag{6.53}$$

For the point $z = i$ one has $r = 1, \rho = \sqrt{2}, \theta = \pi/2, \phi = 3\pi/4$, and hence the residue at this point is

$$\begin{aligned}\text{Res}\left\{\frac{z^{(1-p)}(z-1)^p}{z^2+1}\right\}_{z=i} &= \lim_{z\to i}(z-i)\frac{z^{(1-p)}(z-1)^p}{z^2+1}\\ &= \lim_{z\to i}\frac{\exp[i(1-p)\pi/2]\sqrt{2^p}\exp(3ip\pi/4)}{z+i}\\ &= 2^{p/2-1}\exp(ip\pi/4).\end{aligned} \tag{6.54}$$

Similarly, for the point $z = -i$ one must use $r = 1, \rho = \sqrt{2}, \theta = 3\pi/2$, and $\phi = 5\pi/4$, and the residue is given by

$$\begin{aligned}\text{Res}\left\{\frac{z^{(1-p)}(z-1)^p}{z^2+1}\right\}_{z=-i} &= \lim_{z\to -i}(z+i)\frac{z^{(1-p)}(z-1)^p}{z^2+1}\\ &= \lim_{z\to -i}\frac{\exp[3i(1-p)\pi/2]\sqrt{2^p}\exp(5ip\pi/4)}{z-i}\\ &= 2^{p/2-1}\exp(-ip\pi/4).\end{aligned} \tag{6.55}$$

Thus, the desired answer, Eq. (6.47), is obtained by multiplying the sum of the three residues in Eqs. (6.52)–(6.54) by $-2\pi i$. ∎

⊳ Exercise §§6.10, for computation of integrals using the residue at infinity.
⊳ Mixed Bag §§6.13.

§6.9 Zeros and Poles of a Meromorphic Function

Suppose that a function is analytic at a point z_0, where it has a zero of order m. Then we can write

$$f(z) = (z - z_\circ)^m g(z), \tag{6.56}$$

where $g(z)$ is analytic but not zero at $z_\circ$. Hence,

$$\frac{f'(z)}{f(z)} = \frac{m}{z - z_\circ} + \frac{g'(z)}{g(z)}. \tag{6.57}$$

Thus, $f'(z)/f(z)$ has a simple pole at $z_\circ$ with residue m. Similarly, if $f(z)$ has a pole of order m at $z_\circ$, we can write

$$f'(z) = (z - z_\circ)^{-m} g(z), \tag{6.58}$$

$$\frac{f'(z)}{f(z)} = \frac{-m}{z - z_\circ} + \frac{g'(z)}{g(z)}. \tag{6.59}$$

The function $f'(z)/f(z)$ has a simple pole at $z_\circ$ with residue $-m$. These results imply the following theorem:

Theorem 6.2. *Let $f(z)$ be analytic on and inside a closed contour C except for a finite number of poles. In addition, let $f(z)$ be not zero on C. Then,*

$$\frac{1}{2\pi i} \oint_C \frac{f'(z)}{f(z)} dz = N - P, \tag{6.60}$$

where N is the number of zeros of $f(z)$ enclosed, a zero of order n is to be counted n times, and p is the number of poles of $f(z)$ enclosed inside C, and a pole of order n is to be counted n times.

Noting that

$$\frac{d}{dz} \log f(z) = \frac{f'(z)}{f(z)}, \tag{6.61}$$

and integrating this equation along a closed contour C, we would get

$$\frac{1}{2\pi i} \oint \frac{d}{dz} \log f(z) = N - P. \tag{6.62}$$

Writing $\log f(z) = \log |f(z)| + i \arg f(z)$ and noting that $\log |f(z)|$ is single-valued, the above result, Eq. (6.60), can be restated in the form

$$\frac{1}{2\pi} \Delta_C \phi = N - P, \tag{6.63}$$

where $\phi = \arg f(z)$ and $\Delta_C \phi$ denotes variation of ϕ along C. We thus arrive at the following theorem:

Theorem 6.3 (argument principle). *Let a function $f(z)$ and a contour C be as in Theorem 6.2; then variation in the argument of $\frac{1}{2\pi} \times \log f(z)$ equals $N - P$.*

The argument principle can be used to prove the following important result on zeros of analytic functions:

Theorem 6.4 (Rouche's theorem). *Let two functions $f(z)$ and $g(z)$ be analytic inside and on a closed contour C. Further, let*

$$|f(z)| < |g(z)| \tag{6.64}$$

be satisfied for all points on C; then, inside C, the number of zeros of $f(z) + g(z)$ equals the number of zeros of $f(z)$.

Rouche's theorem imples the fundamental theorem of algebra, which states that every polynomial of degree n has n zeros. For a proof of Rouche's theorem see the textbooks cited.

§6.10 Notes and References

(◇1) The residue theorem has been discussed for simple closed contours. Generalization to cases where a contour need not be simple, and is allowed to cross itself, is easy. For example, a contour having the "figure of eight" can be regarded as consisting of two simple closed contours.

(◇2) A general discussion on the residue theorem requires the use of a winding number. The winding number of a point z_0 w.r.t. a closed curve γ is defined as

$$n(\gamma, z_0) = \frac{1}{2\pi i} \oint_\gamma \frac{dz}{z - z_0}. \tag{6.65}$$

The winding number is zero if the point z_0 lies outside the curve γ and is an integer if γ encloses z_0. To learn more about it, see [Ahlfers (1966)].

(◇3) In many applications, such as solution of differential equations by the method of integral transforms, it is convenient to have closed contours. It therefore turns out to be useful to work with contours on a Riemann sheet when integrating a multivalued function. An example of such a contours is the Pochhammer contour, shown in Fig. 6.7.

(◇4) Given z is a root of

$$z = z_0 + wf(z), \tag{6.66}$$

which equals z_0 when $w = 0$. One wishes to invert the above relation to get z as a power series in w. *Lagrange's method of inversion of series* gives the

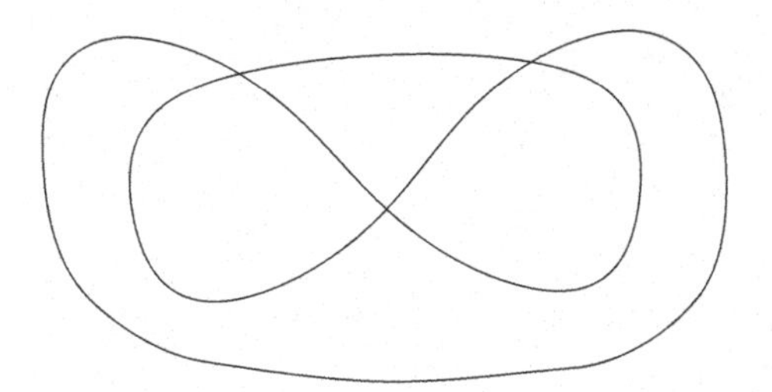

Fig. 6.7

answer for this. Let $f(z)$ be analytic in a circle γ containing w. Then z is given by a series in w:

$$z = z_0 + \sum_{n=1}^{\infty} \frac{w^n}{n!} \frac{d^{n-1}(f(z))^n}{dz^{n-1}}\bigg|_{z=z_0} . \tag{6.67}$$

Also, if $g(z)$ is analytic inside and on the circle γ, $g(z)$ has an expansion in powers of w:

$$g(z) = g(z_0) + \sum_{n=1}^{\infty} \frac{w^n}{n!} \frac{d^{n-1}}{dz^{n-1}} \left\{ g'(z)(f(z))^n \right\}\bigg|_{z=z_0} . \tag{6.68}$$

More details and examples of Lagrange's method of inversion of series can be found in [Polya (1974)].

Chapter 7

CONTOUR INTEGRATION

In this chapter we shall discuss applications of the residue theorem. We first deal with evaluation of definite integrals over a real variable. Several classes of definite integrals, both proper and improper, have already been discussed earlier. Some of the examples, such as in ☑6.7.3, had special properties which made their computation easy. In some other examples the given proper integral was directly related to an integral over a *closed contour* in the complex plane. Several other examples in §4.7.2 included those improper integrals which required the use of Cauchy's fundamental theorem only and *did not require the use of the residue theorem.*

In this chapter we shall first discuss applications of the the method of contour integration, including computation of a variety of *improper* integrals by means of several examples. Later we shall illustrate the use of the residue theorem in series summation and the Mittag–Leffler partial fraction expansion theorem for a meromorphic function.

An improper integral must always be considered as a limiting case of a proper integral which is transformed into a line integral in the complex plane over a contour, and it usually turns out to be an open contour. Next, the contour is suitably closed by adding new pieces, mostly consisting of straight line segments and circular arcs. The evaluation of the contour integral is then completed by a straightforward application of the residue theorem. Several examples of transforming real integrals into line integrals in the complex plane have been discussed in §4.7 and §4.8.4. The process of transforming real definite integrals into contour integrals will invariably require the use of Darboux's theorem; the results of examples ☑6.8.2 in Ch. 4 will be used without a proof. It will therefore be helpful for the reader to recall these examples in the section on bounds where vanishing of some contour integrals in certain limits has been demonstrated. An understanding of these and other similar examples will be needed in making the choice of closed contour for evaluation of improper integrals discussed in this chapter.

Many examples of improper integrals, chosen below, lead to an integrand of the form $g(z) = \frac{f(z)}{z^4+a^4}$. We shall assume that the function $f(z)$ is analytic and nonzero at the points ξ_k given by

$$\xi_k = |a|e^{i\pi/4}e^{2\pi i(k-1)/4}, \qquad k = 1, \ldots, 4, \tag{7.1}$$

where the denominator $z^4 + a^4$ vanishes. Writing explicitly,

$$\xi_1 = |a|e^{i\pi/4}, \quad \xi_2 = |a|e^{3i\pi/4}, \quad \xi_3 = |a|e^{5i\pi/4}, \quad \xi_4 = |a|e^{7i\pi/4}. \tag{7.2}$$

Under the assumed conditions, the function $g(z)$ has poles at the points ξ_k. For our later use we compute the residues of $g(z)$ at these points where the denominator $z^4 + a^4$ vanishes. If ξ denotes any one of these four roots, we have $\xi^4 = -a^4$ and

$$\operatorname{Res}\left\{\frac{f(z)}{z^4+a^4}\right\}_{z=\xi} = \lim_{z\to\xi}\frac{(z-\xi)f(z)}{z^4+a^4} \tag{7.3}$$

$$= \frac{f(\xi)}{4\xi^3} = -\frac{\xi f(\xi)}{4a^4}. \tag{7.4}$$

Here we have used the fact that $f(z)$ is analytic at $z = \xi$ and that

$$\lim_{z\to\xi}\frac{z-\xi}{z^4+a^4} = \frac{1}{4\xi^3}. \tag{7.5}$$

§7.1 Rational and Trigonometric Functions

☑**7.1.1 Problem.** Show that

$$\int_0^\infty \frac{dx}{x^4+a^4} = \frac{\pi}{2\sqrt{2}a^3}. \tag{7.6}$$

Solution. We first write the given integral as

$$\int_0^\infty \frac{dx}{x^4+a^4} = \frac{1}{2}\int_{-\infty}^\infty \frac{dx}{x^4+a^4} = \lim_{R\to\infty}\frac{1}{2}\int_{-R}^R \frac{dx}{x^4+a^4} \tag{7.7}$$

$$= \frac{1}{2}\lim_{R\to\infty}\int_{\mathrm{AOB}}\frac{dz}{z^4+a^4}, \tag{7.8}$$

where AOB is the line segment of the real axis from $-R$ to R; see Fig. 7.1(a). Next, we note that the integral along the semicircular contour BCA vanishes as $R\to\infty$; see ☑4.5.2 and ⚡4.5.3(a).

$$\lim_{R\to\infty}\int_{\mathrm{BCA}}\frac{dz}{z^4+a^4} = 0. \tag{7.9}$$

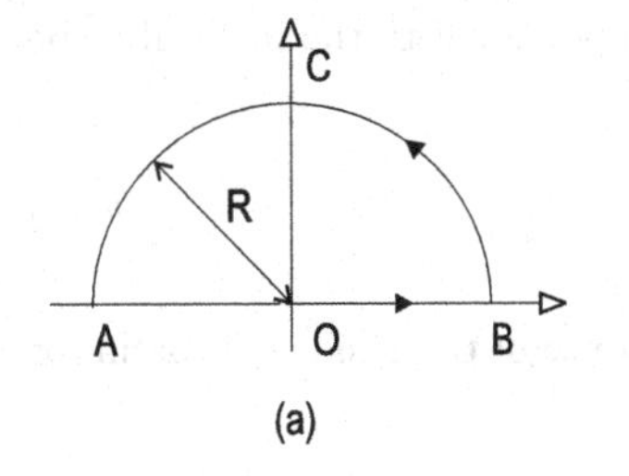

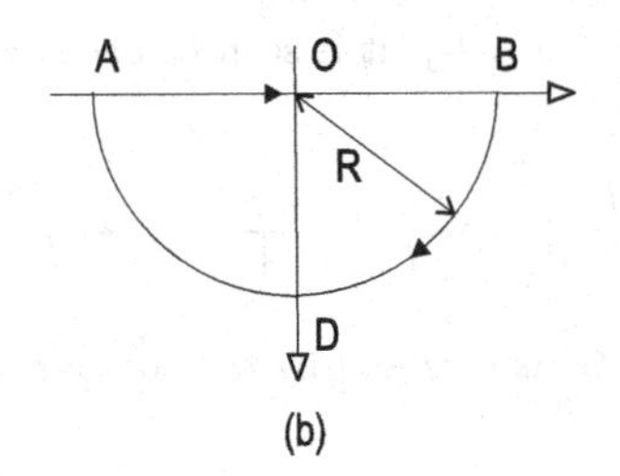

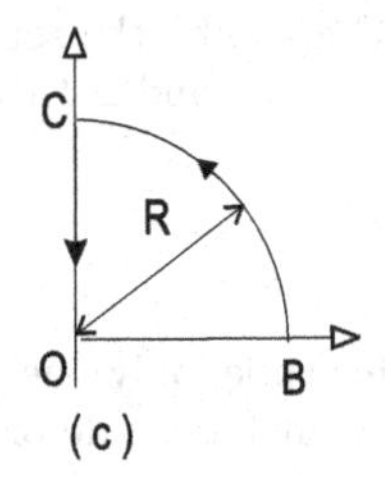

Fig. 7.1

Thus, an addition of the null contribution of the contour BCA to the integral $\int_{\text{AOB}}$ in Eq. (7.8) leads to a closed contour AOBCA and we get

$$\lim_{R\to\infty} \oint_{\text{AOBCA}} \frac{dz}{z^4+a^4} = \lim \int_{\text{AOB}} \frac{dz}{z^4+a^4} + \lim \int_{\text{BCA}} \frac{dz}{z^4+a^4} \tag{7.10}$$

$$= \lim_{R\to\infty} \int_{\text{AOB}} \frac{dz}{z^4+a^4} \tag{7.11}$$

$$= 2\int_0^\infty \frac{dx}{x^4+a^4}. \tag{7.12}$$

The integral along the closed contour AOBCA is easily computed by applying Cauchy's, residue theorem. Of the four poles at ξ_1, ξ_2, ξ_3, and ξ_4, only the poles at ξ_1 and ξ_2 are enclosed in the contour. The residues are computed using (7.4) and given by (7.5). For large R, and the value of the integral on the left hand side of Eq. (7.10) is independent of R. Thus, the large R limit gives

$$\lim_{R\to\infty} \oint_{\text{AOBCA}} \frac{dz}{z^4+a^4} = 2\pi i \times \left(-\frac{\xi_1}{4a^4} - \frac{\xi_2}{4a^4}\right) = \frac{\pi}{\sqrt{2}a^3}, \tag{7.13}$$

which yields the final answer:

$$\int_0^\infty \frac{dx}{x^4+a^4} = \frac{1}{2}\lim_{R\to\infty} \oint_{\text{AOBCA}} \frac{dz}{z^4+a^4} = \frac{\pi}{2\sqrt{2}a^3}. \tag{7.14}$$

∎

We will see that there are several ways of transforming the integral in Eq. (7.6) into a line integral in the complex plane. In all the cases, a closed contour can be obtained by adding an arc of a circle, $|z| = R$, of large radius R. Two alternate choices for closed contours are shown in Figs. 7.1(b) and 7.1(c). The computational details for the semicircle in the lower half plane, AOBDA in Fig. 7.1(b), are very similar to those given above.

We will now write out the steps leading to the result (7.14) using the sector OBCO in Fig. 7.1(c).

Solution. Consider

$$\oint_{\text{OBCO}} \frac{dz}{z^4+a^4} = \int_{\text{OB}} \frac{dz}{z^4+a^4} + \int_{\text{BC}} \frac{dz}{z^4+a^4} + \int_{\text{CO}} \frac{dz}{z^4+a^4}. \tag{7.15}$$

As before, the integral along the arc BC, on the right hand side, vanishes in the limit $R \to \infty$. Noting that the last term of the above equation is the negative of the integral along the line segment

OC given by the set $\{z|z = iy, 0 \le y \le R\}$, it is seen to become proportional to the integral to be evaluated and is therefore

$$\int_{\mathrm{CO}} \frac{dz}{z^4+a^4} = -\int_{\mathrm{OC}} \frac{dz}{z^4+a^4} = -i\int_0^\infty \frac{dy}{y^4+a^4}. \tag{7.16}$$

This time, only one pole, at $\xi = \xi_1$, is enclosed by the contour, and use of the residue theorem for the left hand side of Eq. (7.15) leads to

$$-\frac{2\pi i}{4|a|^3}\exp(i\tfrac{\pi}{4}) = (1-i)\int_0^\infty \frac{dx}{x^4+a^4}, \tag{7.17}$$

which gives the desired result (7.6). ∎

☑7.1.2 Problem. Compute the Fourier transform $\displaystyle\int_{-\infty}^{\infty} \frac{\exp(ibx)}{x^4+a^4}dx$ and hence show that

$$\int_{-\infty}^{\infty} \frac{\cos bx}{x^4+a^4}dx = \left(\frac{\pi}{a^3}\right)e^{-(|b|a/\sqrt{2})}\cos\left(\frac{|b|a}{\sqrt{2}} - \frac{\pi}{4}\right), \tag{7.18}$$

assuming that $a > 0$.

Solution. We write

$$J(b) = \int_{-\infty}^{\infty} \frac{\exp(ibx)}{x^4+a^4}dx = \lim_{R\to\infty}\int_{-R}^{R} \frac{\exp(ibx)}{x^4+a^4}dx \tag{7.19}$$

$$= \int_{\mathrm{AOB}} \frac{\exp(ibz)}{z^4+a^4}dz, \tag{7.20}$$

and examine the possibility of closing the contour in the upper half plane by adding the semicircle BCA [Fig. 7.1(a)].

We first consider the case $b > 0$. We note that the exponential in the integrand, with $z = x+iy$, becomes $\exp(ibz) = \exp(ibx - by)$ which, for $b > 0$, tends to zero on the large semicircle BCA in the upper half plane ($y > 0$) as its radius $R \to \infty$. The absolute value of the integrand, in Eq. (7.20), obeys the inequality

$$\left|\frac{\exp(ibz)}{z^4+a^4}\right| = \left|\frac{\exp(ibx-by)}{z^4+a^4}\right| = \frac{\exp(-by)}{|z^4+a^4|} \le \frac{\exp(-by)}{|R^4-a^4|}, \tag{7.21}$$

on the semicircle BCA. Thus, using Darboux's theorem,

$$\lim_{R\to\infty}\left|\int_{\mathrm{BCA}} \frac{\exp(ibz)}{z^4+a^4}dz\right| \le \lim_{R\to\infty}\frac{\exp(-by)}{|R^4-a^4|}\times\pi R = 0. \tag{7.22}$$

Addition of (7.22) to (7.20) does not change the value of $J(b)$ when $R \to \infty$; it allows us to change the contour to a closed contour, giving

$$J(b) = \lim\left[\int_{\mathrm{AOB}} \frac{\exp(ibz)}{z^4+a^4}dz + \int_{\mathrm{BCA}} \frac{\exp(ibz)}{z^4+a^4}dz\right] = \lim\oint_{\mathrm{AOBCA}} \frac{\exp(ibz)}{z^4+a^4}dz. \tag{7.23}$$

Computing the residues at the two poles, ξ_1 and ξ_2, enclosed by the contour AOBCA [see Eq. (7.4)], and applying the residue theorem, we get

$$J(b) = \left(\frac{\pi}{a^3}\right) e^{-(pa/\sqrt{2})} \cos\left(\frac{ab}{\sqrt{2}} - \frac{\pi}{4}\right), \quad b > 0. \tag{7.24}$$

Next, we assume that $b < 0$. In this case $b = -|b|$ and the exponential in the integrand in (7.20) is seen to diverge,

$$|\exp(ibz)| = |\exp(ibx - by)| = \exp(|b|y) \to \infty, \quad \text{as } R \to \infty,$$

for $y \to +\infty$ in the upper half plane. Hence, the integral over the semicircle BCA will diverge and *the contour cannot be closed in the upper half plane.* On the other hand, it is verified that closing the contour in the lower half plane ($y < 0$), as in Fig. 7.1(b), is permitted and one can compute the integral repeating the steps as in the case of $b > 0$. *Remembering that this contour is clockwise and hence the value of the integral will be the negative of* $2\pi i$ *times the sum of residues at the poles enclosed,* we get

$$J(b) = -2\pi i \left[\frac{\exp(ib\xi_3)\xi_3 + \exp(ib\xi_4)\xi_4}{a^4}\right] \tag{7.25}$$

$$= \left(\frac{\pi}{a^3}\right) e^{-(|b|a/\sqrt{2})} \cos\left(\frac{ba}{\sqrt{2}} + \frac{\pi}{4}\right), \quad b < 0. \tag{7.26}$$

The Fourier cosine transform is the real part of the Fourier transform integral. Collecting the results in (7.23)–(7.26), we get

$$\int_{-\infty}^{\infty} \frac{\cos bx}{x^4 + a^4}\, dx = \left(\frac{\pi}{a^3}\right) e^{-|b|a/\sqrt{2}} \cos\left(\frac{|b|a}{\sqrt{2}} - \frac{\pi}{4}\right). \tag{7.27}$$

It is left as an exercise for the reader to explore what happens if we integrate $\frac{\exp(ibz)}{z^4+a^4}$ around the contour in Fig. 7.1(c). ∎

Remarks. The method for evaluation of improper integrals discussed here is applicable to integrals of the form $\int Q(x) \cos kx\, dx$, where $Q(x)$ is a rational function. The first step consisted of expressing $\cos kx$ as the real part of $\exp(ikx)$ and then closing the contour in the upper half ($k > 0$) or the lower half plane ($k < 0$). This problem could be completed in several different ways. For instance, one may proceed to write $\cos kx$ in terms of the exponentials $\exp(\pm ikx)$ to get

$$\int_{-\infty}^{\infty} Q(x) \cos kx\, dx = \frac{1}{2} \int_{-\infty}^{\infty} Q(x) \exp(ikx)\, dx + \frac{1}{2} \int_{-\infty}^{\infty} Q(x) \exp(-ikx)\, dx. \tag{7.28}$$

The evaluation of the improper integral can be completed by closing the contour in the upper half plane for one of the two terms on the right hand side, and in the lower half plane for the other term.

⊳ Tutorials §§7.1 and §§7.2, for the integral of rational functions.

⊳ Exercises §§7.3 and §§7.4, on integrals of rational functions with or without $\sin x$ or $\cos x$.

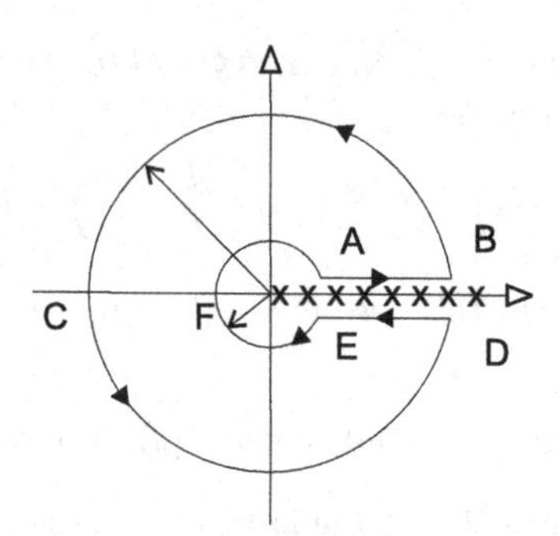

Fig. 7.2

§7.2 Integration Around a Branch Cut

Examples in this section use two circles contour with a cross cut shown in Fig. 7.2 which consists of a large circular arc BCD of radius R, a small circular arc EFA of radius ρ, and two line segments AB and DE running parallel to the real axis and at distances ϵ from it. In the end limits $R \to \infty, \rho \to 0$ and $\epsilon \to 0$ are to be taken. These examples use integration around a branch cut and the reader is urged to revise earlier, related sections on examples of computation of discontinuity in §3.6.

☑**7.2.1 Problem.** Making use of the two circles' contour with a cross cut (Fig. 7.2), prove that

$$\int_0^\infty \frac{x^{p-1}}{x^4 + a^4}\, dx = \frac{\pi |a|^{p-4}}{4\sin(\pi p/4)}, \qquad a \neq 0, \tag{7.29}$$

where $0 < p < 4$, and p is not an integer.

Solution. The required integral is seen to be related to the integral along the two circles' contour by noting the following important steps:

(a) Let the function $\frac{z^{p-1}}{z^4+a^4}$ be defined in the complex plane by taking

$$z^{p-1} = r^{p-1} e^{i(p-1)\theta}, \qquad 0 < \theta < 2\pi, \tag{7.30}$$

as the definition of the function z^{p-1}. The branch cut of z^{p-1}, and hence of $f(z) \equiv \frac{z^{p-1}}{z^4+a^4}$ also, lies along the positive real axis.

(b) The discontinuity of $f(z)$ along the branch cut is proportional to the integrand in Eq. (7.29),

$$\lim_{\epsilon \to 0} [f(x+i\epsilon) - f(x - i\epsilon)] = (1 - e^{2\pi i(p-1)}) \frac{x^{p-1}}{x^4 + a^4}, \qquad x > 0, \tag{7.31}$$

and the integral of the discontinuity (7.31) becomes

$$\int [f(x+i\epsilon) - f(x-i\epsilon)]\, dx = (1 - e^{2\pi i p}) \int \frac{x^{p-1}}{x^4 + a^4}\, dx. \tag{7.32}$$

Thus, in the limit $\epsilon \to 0, R \to \infty$, we have

$$\lim \left[\int_{\mathrm{AB}} f(z) - \int_{\mathrm{ED}} f(z) \right] = (1 - e^{2\pi i p}) \int_0^\infty \frac{x^{p-1}}{x^4 + a^4}\, dx. \tag{7.33}$$

(c) The integral of the function $f(z)$ along the circular contour EFA vanishes as its radius $\rho \to 0$. The integral along the big circle BCD also vanishes in the limit $R \to \infty$.

$$\lim_{\rho\to 0}\int_{\text{EFA}} f(z) = 0, \qquad \lim_{R\to\infty}\int_{\text{BCD}} f(z) = 0. \tag{7.34}$$

Therefore, the integral along the closed contour ABCDEFA is seen to be

$$\begin{aligned}\lim \oint f(z)dz &= \lim \int_{\text{AB}} f(z) + \lim \int_{\text{BCD}} f(z) + \lim \int_{\text{DE}} f(z) + \lim \int_{\text{EFA}} f(z)\\ &= \lim \int_{\text{AB}} f(z)dz - \lim \int_{\text{ED}} f(z)dz\\ &= (1 - e^{2\pi ip}) \int_0^\infty \frac{x^{p-1}}{x^4 + a^4} dx. \end{aligned}\tag{7.35}$$

(d) The closed contour integral appearing on the left hand side is easily evaluated in terms of the sum of all residues, at the poles enclosed. Computing the residues as in Eq. (7.4), we get

$$\sum_{k=1}^{4} \text{Res}\{f(z)\}_{z=\xi_k} = -e^{ip\pi}|a|^{p-4}\cos(p\pi/2)\cos(p\pi/4). \tag{7.36}$$

(e) Collecting the above results and simplifying, we get

$$\begin{aligned}\int_0^\infty \frac{x^{p-1}}{x^4+a^4}dx &= \frac{1}{1-e^{2\pi ip}}\lim\oint f(z)dz = \frac{2\pi i}{1-e^{2\pi ip}}\sum_{k=1}^{4}\text{Res}\{f(z)\}_{z=\xi_k}\\ &= \frac{-2\pi i e^{ip\pi}}{1-e^{2\pi ip}}|a|^{p-4}\cos(p\pi/2)\cos(p\pi/4) \qquad (7.37)\\ &= \pi|a|^{p-4}\frac{\cos(p\pi/2)\cos(p\pi/4)}{\sin p\pi} = \frac{\pi|a|^{p-4}}{4\sin(p\pi/4)}. \qquad (7.38)\end{aligned}$$

■

In the next example, we show how $\int_0^\infty Q(x)$ can be computed using the two circles' contour by means of a method similar to that given in the above example. Although we again take $Q(x) = 1/(x^4 + a^4)$, the method given below is applicable to integrals of general rational functions over the semi-infinite interval $(0, \infty)$ as compared to the method of ☑7.1.1, which applies to the even rational functions only.

☑7.2.2 Problem. Integrate $\frac{\log z}{z^4+a^4}$ around the contour of Fig. 7.2 and prove that

$$\int_0^\infty \frac{dx}{x^4 + a^4} = \frac{\pi}{2\sqrt{2}a^3}. \tag{7.39}$$

Solution. We use $g(z)$ to denote

$$g(z) = \frac{\log z}{z^4 + a^4}. \tag{7.40}$$

(a) Let the integrand $\frac{\log z}{x^4+a^4}$ be defined by taking the branch cut of the logarithm function along the positive real axis and using

$$\log z = \ln r + i\theta, \qquad 0 < \theta < 2\pi, \tag{7.41}$$

as the definition of function $\log z$. The branch cut of $\log z$, and hence of $g(z)$ also, lies along the positive real axis.

(b) The discontinuity of $g(z)$ along the branch cut is proportional to the function $1/(x^4+a^4)$,

$$g(x+i\epsilon) - g(x-i\epsilon) = \frac{-2\pi i}{x^4+a^4}, \quad x>0, \tag{7.42}$$

and the integral of the discontinuity (7.42) becomes

$$\lim_{\epsilon\to 0}\left[\int g(x+i\epsilon) - \int g(x-i\epsilon)\right] dx = \int_\rho^R \frac{-2\pi i}{x^4+a^4}\, dx. \tag{7.43}$$

Thus, in the limit $\epsilon \to 0, \rho \to 0, R \to \infty$ we have

$$\lim \left[\int_{\text{AB}} g(z) - \int_{\text{ED}} g(z)\right] dx = -2\pi i \int_0^\infty \frac{1}{x^4+a^4}\, dx. \tag{7.44}$$

(c) The integral of the function $g(z)$ along the circular contour EFA vanishes as its radius $\rho \to 0$. The integral along the big circle BCD also vanishes in the limit $R \to \infty$:

$$\lim_{\rho\to 0}\int_{\text{EFA}} g(z) = 0, \qquad \lim_{R\to\infty}\int_{\text{BCD}} g(z) = 0. \tag{7.45}$$

Therefore, the value of integral along the closed contour ABCDEFA, in the limit $\rho \to 0, \epsilon \to 0, R \to \infty$, is seen to be

$$\begin{aligned}
\lim \oint g(z)dz &= \lim \int_{\text{AB}} g(z) + \lim \int_{\text{BCD}} g(z) + \lim \int_{\text{DE}} g(z) + \lim \int_{\text{EFA}} g(z) \\
&= \lim \int_{\text{AB}} g(z)dz - \lim \int_{\text{ED}} g(z)dz \\
&= -2\pi i \int_0^\infty \frac{1}{x^4+a^4}\, dx.
\end{aligned} \tag{7.46}$$

(d) The closed contour integral appearing on the left hand side is easily evaluated in terms of the sum of all residues. Computing the residues as in Eq. (7.4), the integral is found to be

$$\lim \oint g(z)dz = 2\pi i \sum_{k=1}^{4} \text{Res}\{g(z)\}_{z=\xi_k}.$$

(e) Computing the residues and collecting the above results, we get the desired answer, Eq. (7.39). ∎

Next, we present another example of integration around a branch cut. We have seen that integrating $\frac{\log z}{z^4+a^4}$ around the two circles' contour of Fig. 7.2 leads to the value of $\int \frac{dx}{x^4+a^4}$ and not $\int_0^\infty \frac{\log x dx}{x^4+a^4}$. The example shows how this last integral can be computed by making use of a different contour of Fig. 7.3 instead of the two circles' contour of Fig. 7.2.

☑**7.2.3 Problem.** Using the closed contour AOBCA of Fig. 7.3, show that

$$\int_0^\infty \frac{\log x}{x^4+1}\, dx = -\frac{\pi^2}{8\sqrt{2}}. \tag{7.47}$$

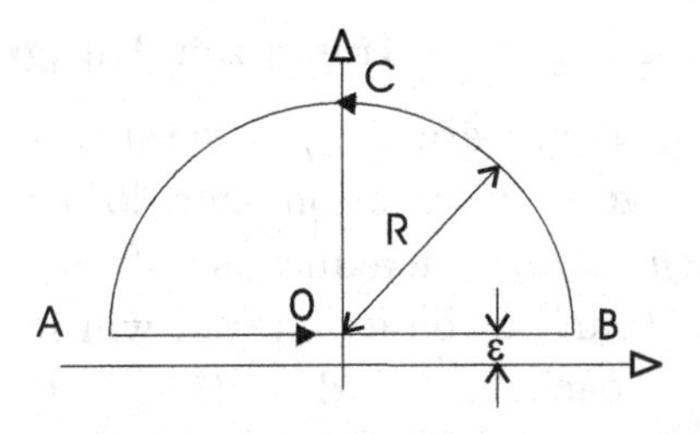

Fig. 7.3

Solution. Let $h(z)$ denote the function

$$h(x) = \frac{\text{Log}\, z}{z^4 + a^4}, \tag{7.48}$$

where $\text{Log}\, z$ is the principal value of the logarithm function. To compute the required integral, we set up the integral of $h(z)$ around the closed contour AOBCA. Therefore, we consider

$$\oint \frac{\text{Log}\, z}{z^4 + a^4}\, dz = \int_{\text{AO}} h(z)\, dz + \int_{\text{OB}} h(z)\, dz + \int_{\text{BCA}} h(z)\, dz. \tag{7.49}$$

Note that the integral along the semicircle BCA vanishes as the radius of the semicircle tends to infinity. In the limit $\epsilon \to 0$ the points on OA are given by $z = re^{i\pi}$ and those on OB are given by $z = r$, and we get

$$\lim_{R\to\infty} \oint \frac{\text{Log}\, z}{z^4 + a^4}\, dz = \lim \left[- \int_{\text{OA}} h(z)\, dz + \int_{\text{AB}} h(z)\, dz + \int_{\text{BCA}} h(z)\, dz \right] \tag{7.50}$$

$$= - \int_0^\infty \frac{\text{Log}\,(re^{i\pi})}{r^4 + a^4}\, e^{i\pi}\, dr + \int_0^\infty \frac{\text{Log}\, r}{r^4 + a^4} dr \tag{7.51}$$

$$= 2 \int_0^\infty \frac{\log r\, dr}{r^4 + a^4} dr + i\pi \int_0^\infty \frac{dr}{r^4 + a^4}. \tag{7.52}$$

Here we have used $\text{Log}\, re^{i\pi} = \log r + i\pi$ for the points on OA. Next, the integral on the left is computed using Cauchy's residue theorem. The two poles at $\xi_1 = |a|e^{i\pi/4}$ and $\xi_2 = |a|e^{i3\pi/4}$ are enclosed in the contour ABCA. Computing the residue as in Eq. (7.4), the integral on the left hand side of Eq. (7.50) is given by

$$\lim_{R\to\infty} \oint \frac{\text{Log}\, z}{z^4 + a^4}\, dz = 2\pi i \Big(\frac{-1}{4a^4}\Big)(\xi_1 \text{Log}\, \xi_1 + \xi_2 \text{Log}\, \xi_2). \tag{7.53}$$

Simplifying and equating the real and imaginary parts of Eqs. (7.52) and (7.53), we get the required answer:

$$\int_0^\infty \frac{1}{r^4 + a^4}\, dr = \frac{\pi}{2\sqrt{2}|a|^3}, \tag{7.54}$$

$$\int_0^\infty \frac{\log r}{r^4 + a^4}\, dr = \frac{\pi}{2\sqrt{2}|a|^3} \log |a| - \frac{\pi^2}{8\sqrt{2}|a|^3}. \tag{7.55}$$

Taking $a = 1$ gives the desired result (7.47). ▮

⇨ Tutorial §§7.6, on integration around a branch cut.
⇨ Exercise §§7.7, for practice problems on integration around a branch cut.

§7.3 Using Indented Contours for Improper Integrals

Indented contours make their appearance in several ways. An integral which has a singular point in the range of integration and does not exist as an improper integral *can be given a definition* by making use of an indented contour. Several choices for an indented contour can be found and will in general lead to different definitions. A different approach to the definition is the "$i\epsilon$ prescription." Yet another, important, definition is the Cauchy principal value. We shall discuss these different ways of defining a singular integrals by means of an example of the *singular integral* $I(x) = \int_{-\infty}^{\infty} \frac{\exp(ikx)}{k^2-k_0^2}\, dk$.

Apart from singular integrals, use of an indented contour may become necessary even for a well-defined improper integral. The process of rewriting the given improper integral as a contour integral along a suitable closed contour may sometimes introduce singular points in the range of integration, which then forces us to use an indented contour. The next example under discussion highlights this point.

§7.3.1 *An example*

First, we present a simple example of a well-defined improper integral, $\int_0^{\infty} \frac{\sin x}{x} dx$, requiring the use of an indented contour. In an earlier example, ☑7.1.2, we switched over from the cosine function to the exponential function [see Eq. (7.20)] while transforming the integral $I(p)$ into a contour integral. This became necessary as the sine and cosine functions, being linear combinations of two the exponentials, such as $\exp(\pm ikx)$, did not vanish on any one of the two possible semicircles in the upper and lower half planes. When we attempt to follow the same procedure for the integral $\int_0^{\infty} \frac{\sin x}{x} dx$, we end up with a singularity in the range of integration and there is a need to use an indented contour, as explained below.

☑7.3.1 Problem. We will make use of an indented contour to show that

$$\int_{-\infty}^{\infty} \frac{\sin x}{x}\, dx = \pi. \tag{7.56}$$

Solution. We begin by noting that the integrand is not singular at $x = 0$ and the integral exists. Following the previous example, ☑7.1.2, we may try to write

$$\begin{aligned}\int_0^{\infty} \frac{\sin x}{x}\, dx &= \frac{1}{2}\int_{-\infty}^{\infty} \frac{\sin x}{x}\, dx \\ &= \frac{1}{2}\,\mathrm{Im} \lim_{R\to\infty} \int_{-R}^{R} \frac{\exp(iz)}{z}\, dz. \end{aligned} \tag{7.57}$$

The original integrand, $\frac{\sin z}{z}$, had a removable singular point at $z = 0$, but the function $\frac{\exp(iz)}{z}$ has a pole in the range of integration and hence the integral in the last step does not exist and none of the contours of Fig. 7.1 is suitable in the present case. We therefore proceed as follows:

$$\int_{-\infty}^{\infty} \frac{\sin x}{x}\, dx = \lim_{R\to\infty}\lim_{\rho\to 0}\left(\int_{-R}^{-\rho} \frac{\sin x}{x}\, dx + \int_{\rho}^{R} \frac{\sin x}{x}\, dx\right) \tag{7.58}$$

$$= \text{Im} \lim_{R\to\infty}\lim_{\rho\to 0}\left(\int_{\text{AB}} \frac{e^{iz}}{z}\, dx + \int_{\text{DE}} \frac{e^{iz}}{z}\, dz\right). \tag{7.59}$$

The last expression is an integral along the contour consisting of two disjoint pieces AB and CD, Fig. 7.4, and we add a semicircle BCD of radius ρ and use the continuous contour $\int_{\text{ABCDE}}$. Next, we close the contour ABCDE by adding a semicircle EFA, and write

$$\oint_{\text{ABCDEFA}} \frac{e^{iz}}{z}\, dz = \int_{\text{AB}} \frac{e^{iz}}{z}\, dz + \int_{\text{BCD}} \frac{e^{iz}}{z}\, dz + \int_{\text{DE}} \frac{e^{iz}}{z}\, dz + \int_{\text{EFA}} \frac{e^{iz}}{z}\, dz. \tag{7.60}$$

The integral on the left hand side vanishes as the integrand is analytic inside and on the contour ABCDEFA:

$$\oint_{\text{ABCDEFA}} \frac{e^{iz}}{z}\, dz = 0. \tag{7.61}$$

The third integral, the one along BCD, can be computed in the limit $\rho \to 0$ making use of the result in §6.5.1, and is given by

$$\lim_{\epsilon\to 0}\int_{\text{BCD}} \frac{e^{iz}}{z}\, dz = -i\pi, \tag{7.62}$$

where the minus sign comes as the contour travels around the origin in clockwise fashion. The last integral along EFA can be seen to vanish in the limit $R \to \infty$:

$$\lim_{R\to\infty}\int_{\text{EFA}} \frac{e^{iz}}{z}\, dz = 0. \tag{7.63}$$

The sum of the imaginary parts of the remaining two integrals, in the limit $R \to \infty$, $\rho \to 0$, is seen from Eq. (7.59) and (7.60) to be just the desired integral $\int_0^\infty \frac{\sin x}{x}\, dx$:

$$\text{Im}\left(\int_{\text{AB}} \frac{e^{iz}}{z}\, dz + \int_{\text{DE}} \frac{e^{iz}}{z}\, dz\right) = \text{Im}\left(\int_{-R}^{-\rho} \frac{e^{iz}}{z}\, dz + \int_{\rho}^{R} \frac{e^{iz}}{z}\, dz\right)$$

$$= \int_{-\infty}^{\infty} \frac{\sin x}{x} dx. \tag{7.64}$$

Using the results in Eqs. (7.60)–(7.64) we get the desired result (7.56). The result (7.56) can also be derived using the contour of Fig. 7.4(b). ∎

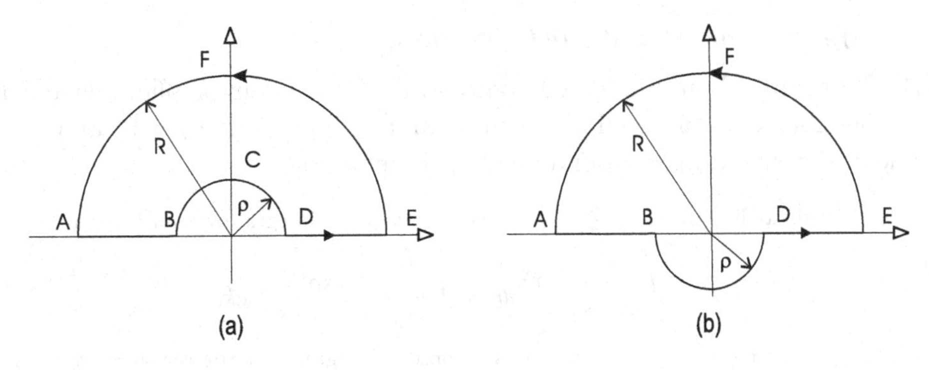

Fig. 7.4

✍ *In general, to compute an integral such as the one considered above, one must remember examples in ↯6.5.1(a) and care must be taken to not have an integrand having a pole of order greater than 1.*

§7.4 Indented Contours for Singular Integrals

In the above example, use of an indented contour became necessary because an intermediate integrand, $\frac{\exp(iz)}{z}$, had a singular point in the range of integration. Though the value of the integral of $\frac{\exp(iz)}{z}$ is different for the two contours in Figs 7.4(a) and 7.4(b), one is led to the same answer for the original integral of $\frac{\sin x}{x}$, which is well defined. However, an integral of the type

$$\int_a^b \frac{f(x)}{x - x_0}\, dx \tag{7.65}$$

is a singular integral and does not exist if, for example, x_0 lies in the range of integration $a < x_0 < b$ and $f(x_0) \neq 0$. A convenient way to define an integral of this type is to interpret the integral (7.65) as an integral in the complex plane and use a suitably indented contour. *Each particular way of indenting the contour is to be regarded as a different definition of the integral.* We illustrate these remarks by means of our next example, which appears as a solution for the Green function for the differential equation

$$\left(\frac{d^2}{dx^2} + k_0^2\right)\phi = 0. \tag{7.66}$$

A formal solution for the Green function of this equation, which satisfies the equation

$$\left(\frac{d^2}{dx^2} + k_0^2\right)G = -2\pi\delta(x), \tag{7.67}$$

can be written as the singular integral $G(x) = \int_{-\infty}^{\infty} \frac{\exp(ikx)dx}{k^2 - k_0^2}$, which is the focus of our discussion in the next few examples.

§7.4.1 *Definition using indented contours*

☑**7.4.1 Problem.** Using indented contours in the complex k plane, give a few different possible ways of defining the integral $\int_{-\infty}^{\infty} \frac{\exp(ikx)}{k^2 - k_0^2}\, dk (k_0 > 0)$ and obtain the value of the integral corresponding to each definition.

Solution. Consider the integral $\int \frac{\exp(ikx)}{k^2 - k_0^2}\, dk$ as a limit of the integral over an interval $(-K, K)$ and write

$$I(x) = \int_{-\infty}^{\infty} \frac{\exp(ikx)}{k^2 - k_0^2}\, dk = \lim_{K \to \infty} \int_{-K}^{K} \frac{\exp(ikx)}{k^2 - k_0^2}\, dk. \tag{7.68}$$

Formally, the integral on the right hand side is a contour integral over the real interval $(-K, K)$ in the complex plane and does not exist because of the singular points at $\pm k_0$ which fall on the

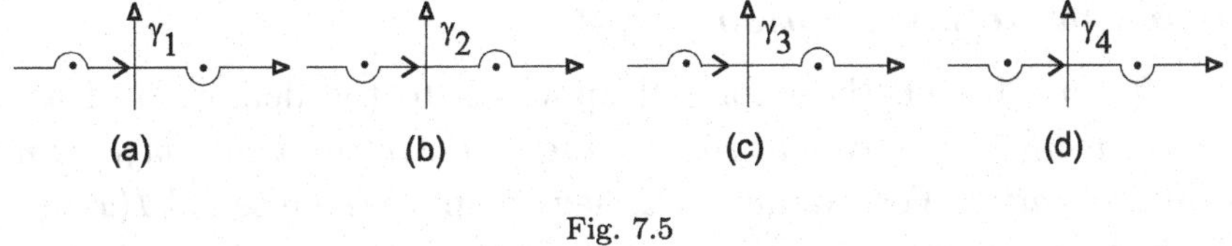

Fig. 7.5

C_R^+ γ_1 (a) $x > 0$ $\qquad$ γ_1 C_R^- (b) $x < 0$

Fig. 7.6

integration contour for large-enough K. The integral can be defined by indenting the integration contour so as to bypass the singular points at $k = \pm k_0$. Several possible contours $\gamma_1, \ldots, \gamma_4$ are shown in Fig. 7.5. Each of these contours γ_k leads to a particular definition I_k of the given integral I. We introduce the contour integrals.

$$J_{\gamma_m}(x) = \int_{\gamma_m} \frac{\exp(ikx)}{k^2 - k_0^2}\, dk, \qquad m = 1, \ldots, 4, \tag{7.69}$$

which in the limit $\epsilon \to 0, K \to \infty$ give four definitions, and four answers $I_m(x)$ for the ill-defined integral I.

As in example ☑7.3.1 of §7.3, each of the integrals J_{γ_k} can be evaluated by closing the contour in the upper half k plane for $x > 0$ and in the lower half k plane for $x < 0$. Considering only J_{γ_1}, the two resulting contours for the integral J_{γ_1} are shown in Figs. 7.6(a) and 7.6(b).

We need the residues at the two singular points $k = \pm k_0$:

$$R_1 = \text{Res}\left\{\frac{\exp(ikx)}{k^2 - k_0^2}\right\}_{k=k_0} = \frac{e^{ik_0x}}{2k_0}, \tag{7.70}$$

$$R_2 = \text{Res}\left\{\frac{\exp(ikx)}{k^2 - k_0^2}\right\}_{k=-k_0} = -\frac{e^{-ik_0x}}{2k_0}. \tag{7.71}$$

For $x > 0$ the contour is closed in the upper half plane, and only the pole at $k = k_0$ is enclosed and hence the value of the integral $J_1(x)$ is $2\pi i R_1$. Similarly, for $x < 0$, the integral J_1 is to be evaluated by closing γ_1 in the lower half plane, and the answer is equal to $-2\pi i R_2$.

The integrals along other possible contours $\gamma_2, \gamma_3, \ldots$ are evaluated in an identical fashion. The resulting values of the integrals $I_m(x)$ for the four definitions are listed in Table 7.1.

∎

Table 7.1 Values of the integrals I_1, I_2, I_3, and I_4.

	$x > 0$	$x < 0$
$I_1(x)$	$2\pi i R_1$	$-2\pi i R_2$
$I_2(x)$	$2\pi i R_2$	$-2\pi i R_1$
$I_3(x)$	0	$-2\pi i(R_1 + R_2)$
$I_4(x)$	$2\pi i(R_1 + R_2)$	0

§7.4.2 *Using the $i\epsilon$ prescription*

Another useful and frequently employed approach to the definition of a singular integral is *to modify the integrand so that the singular point no longer appears in the range of integration.* For example, we may compute the integral $I(x)$ in ☑7.4.1, after making the replacement

$$k^2 - k_0^2 \to k^2 - (k_0 + i\epsilon)^2 \tag{7.72}$$

in the denominator, and take the limit $\epsilon \to 0$. This gives us yet another definition of the integral $I(x)$,

$$\tilde{I}_1(x) = \lim_{\epsilon\to 0} \int_{-\infty}^{\infty} \frac{\exp(ikx)\, dk}{k^2 - (k_0 + i\epsilon)^2}, \tag{7.73}$$

for the singular integral $I(x)$. This replacement, such as in Eq. (7.72), shifts the poles off the real axis, as shown in Fig. 7.7(a). There are several ways of doing this type of replacement (see the last paragraph of this section), and each such replacement amounts to a particular definition of the singular integral $I(x)$.

The integral in Eq. (7.73) is now well defined and can be computed by means of contour integration. As before, we close the contour in the upper k half plane for $x > 0$ as in Fig. 7.7(b) and in the lower half plane for $x < 0$ as in Fig. 7.7(c).

For $x > 0$ this gives

$$\tilde{I}_1(x) = \lim \oint_{\Gamma_1} \frac{\exp(ikx)}{k^2 - (k_0 + i\epsilon)^2}\, dk, \tag{7.74}$$

and for $x < 0$ we have

$$\tilde{I}_1(x) = \lim \oint_{\Gamma_2} \frac{\exp(ikx)}{k^2 - (k_0 + i\epsilon)^2}\, dk. \tag{7.75}$$

Comparing Figs. 7.6(a)–7.6(b) with Figs. 7.7(b)–7.7(c), it is now obvious that, in the limit $\epsilon \to 0$, we would get back the answers in Table 7.1 and corresponding to the choice γ_1 in Fig. 7.6(a) for the indented contour in §7.4.1.

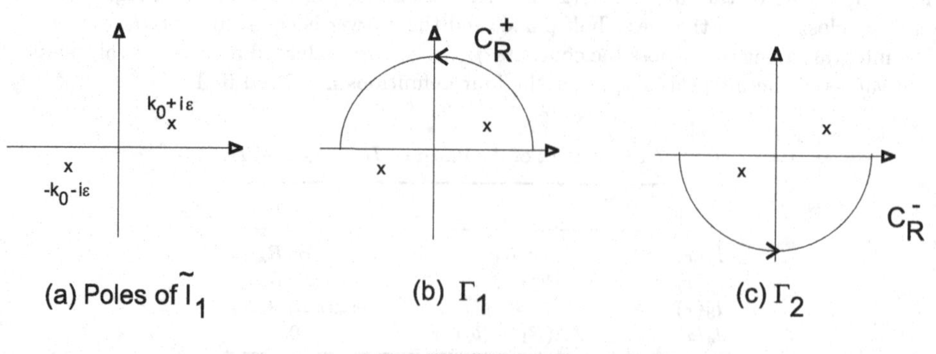

Fig. 7.7

↬ It is left as an exercise for the reader to explore other possible choices for replacement of the denominators such as (a) $k^2-(k_0-i\epsilon)^2$, (b) $k^2-k_0^2+i\epsilon$, (c) $k^2-k_0^2-i\epsilon$, and to discover correspondences of indenting the contour and different "$i\epsilon$ prescriptions."

It may be noted that Eq. (7.67) will have a unique solution only when suitable boundary conditions are specified. Different definitions obtained in the above examples correspond to different possible boundary conditions. In the next example we compute the Cauchy principal value of the integral $I(x)$, which is yet another definition of the integral.

§7.4.3 *Cauchy principal value*

☑**7.4.2 Problem.** Evaluate the Cauchy principal value $I_{\rm PV}(x)$,

$$I_{\rm PV}(x) \equiv PV \int_{-\infty}^{\infty} \frac{\exp(ikx)}{k^2-k_0^2}\, dk,$$

of the integral $I(x)$ of the previous example.

Solution. We note that the integrand is unbounded at $k=\pm k_0$. Therefore, from the definition of the principal value, Eq. (4.5), we have

$$I_{\rm PV}(x) = \mathrm{PV} \int_{-\infty}^{\infty} \frac{\exp(ikx)}{k^2-k_0^2}\, dk$$

$$= \lim \left[\int_{-K}^{-k_0-\rho} \frac{\exp(ikx)}{k^2-k_0^2}\, dk + \int_{-k_0+\rho}^{k_0-\rho} \frac{\exp(ikx)}{k^2-k_0^2}\, dk \right. \tag{7.76}$$

$$\left. + \int_{k_0+\rho}^{K} \frac{\exp(ikx)}{k^2-k_0^2}\, dk \right] = \lim \int_{\gamma} \frac{\exp(ikx)}{k^2-k_0^2}\, dk, \tag{7.77}$$

where lim means that the limits $K\to\infty$ and $\rho\to 0$ are to be taken and the contour γ consists of three disjoint intervals $(-K,-k_0-\rho)$, $(-k_0+\rho, k_0-\rho)$, and $(k_0+\rho, K)$ of the real line, as shown in Fig. 7.8. As in the previous example, we denote the above integral as J_γ and write Eq. (7.77) as

$$I_{\rm PV}(x) = \lim J_\gamma. \tag{7.78}$$

We first make a transition to a continuous contour by adding two little semicircular arcs around the points $\pm k_0$. Several possible choices exist and are shown in Fig. 7.5. *It should be emphasized that all these choices will lead to the same answer because the Cauchy principal value for the integral*

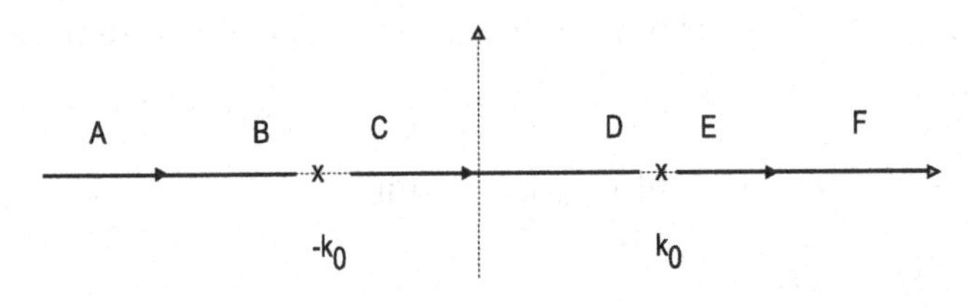

Fig. 7.8

$I(x)$ is well defined. Thus, considering the contour γ_1 of Fig. 7.5(a) we write

$$J_{\gamma_1} = J_\gamma + J_{C_1} + J_{C_2}, \tag{7.79}$$

where C_1 and C_2 are semicircular arcs of Fig. 7.6, of radius ρ each, and with centers at k_0 and $-k_0$, respectively. Therefore,

$$\begin{aligned} I_{\text{PV}}(x) &= \lim J_\gamma \\ &= \lim(J_{\gamma_1} - J_{C_1} - J_{C_2}). \end{aligned} \tag{7.80}$$

The value of $\lim J_{\gamma_1} \equiv I_1$ has already been evaluated in the previous example by closing the contour as in Fig. 7.6, and the value is given in Table 7.1. Also, using the result (6.20) we get

$$\lim_{\rho\to 0} J_{C_1} = \pi i R_1, \qquad \lim_{\rho\to 0} J_{C_2} = -\pi i R_2, \tag{7.81}$$

with R_1 and R_2 being the two residues given by Eqs. (7.70) and (7.71). Note that a negative sign appears in the value of J_{C_2} because the semicircle C_2 has a negative (clockwise) orientation. Collecting the above results we arrive at the required answer for the principal value of the integral $I(x)$. For $x > 0$, $J_\gamma \equiv I_1 = 2\pi i R_1$ and we have, from (7.80),

$$I_{\text{PV}} = \pi i(R_1 + R_2) = -\frac{\pi}{k_0} \sin k_0 x. \tag{7.82}$$

For $x < 0$, $J_\gamma = -2\pi i R_2$ and we get

$$I_{\text{PV}} = -\pi i(R_1 + R_2) = \frac{\pi}{k_0} \sin k_0 x. \tag{7.83}$$

Therefore,

$$I_{\text{PV}}(x) = -\frac{\pi}{k_0} \sin k_0 |x|. \tag{7.84}$$

∎

⊳ Tutorial §§7.10, for use of an indented contour.
⊳ Exercise §§7.11, on the principal value and on use of indented contours.

§7.5 Miscellaneous Contour Integrals

In this section we give a variety of other examples of contour integration. Many of these examples involve integrals of hyperbolic functions which require a rectangular contour.

☑**7.5.1 Problem.** Prove the integration result

$$\int_{-\infty}^{\infty} \frac{e^{kx}}{\cosh^2 x}\, dx = \pi k \operatorname{cosec}\left(\frac{\pi k}{2}\right), \quad -2 < k < 2. \tag{7.85}$$

Solution. Let $f(z)$ denote the integrand $\dfrac{e^{kz}}{\cosh^2 z}$, and we begin by writing the given integral as

$$\int_{-\infty}^{\infty} f(x)dx = \lim_{L\to\infty} \int_{-L}^{L} f(x)dx = \lim_{L\to\infty} \int_{\mathrm{AB}} f(z)dz, \tag{7.86}$$

where AB is the line segment $(-L, L)$ on the real axis; see Fig. 7.9. We note that the integral is related to itself in a simple fashion when x is replaced by $x + \pi i$:

$$\int_{-L}^{L} f(x+\pi i)dx = \exp(ik\pi) \int_{-L}^{L} f(x)dx. \tag{7.87}$$

Noting that $z = x + \pi i$ along DC, we can rewrite Eq. (7.87) as

$$\int_{\mathrm{DC}} f(z)dx = \exp(ik\pi) \int_{\mathrm{AB}} f(z)dz. \tag{7.88}$$

This suggests use of the rectangular contour ABCDA of Fig. 7.9, where the vertical lines BC and DA have been added to close the contour. In fact, it is easy to show that the integrals of $f(z)$ along BC and DA tend to zero as $L \to \infty$. To put all these facts together, consider the integral of $f(z)$ along the closed contour ABCDA of Fig. 7.9:

$$\begin{aligned}\lim_{L\to\infty} \oint f(z)dz &= \lim_{L\to\infty} \left[\int_{\mathrm{AB}} f(z)dz + \int_{\mathrm{BC}} f(z)dz + \int_{\mathrm{CD}} f(z)dz + \int_{\mathrm{DA}} f(z)dz\right] \\ &= \lim_{L\to\infty} \left[\int_{\mathrm{AB}} f(z)dz - \int_{\mathrm{DC}} f(z)dz\right] \\ &= [1-\exp(i\pi k)] \int_{-\infty}^{\infty} f(x)\,dx, \end{aligned} \tag{7.89}$$

$$\therefore \int_{-\infty}^{\infty} f(x)\,dx = \frac{1}{1-\exp(i\pi k)} \lim_{L\to\infty} \oint f(z)dz. \tag{7.90}$$

Therefore, we need to compute only the integral on the right hand side using the residue theorem. The integrand has double poles at $(2n+1)\pi i/2$. Of these, only one pole at $\pi i/2$ is enclosed by the contour and the residue is found to be $-k\exp(ik\pi/2)$. Thus, we get

$$\begin{aligned}\int_{-\infty}^{\infty} f(x)\,dx &= \frac{-2\pi i k \exp(ik\pi/2)}{1-\exp(ik\pi)} \\ &= k\pi \operatorname{cosec}\left(\frac{k\pi}{2}\right). \end{aligned} \tag{7.91}$$

∎

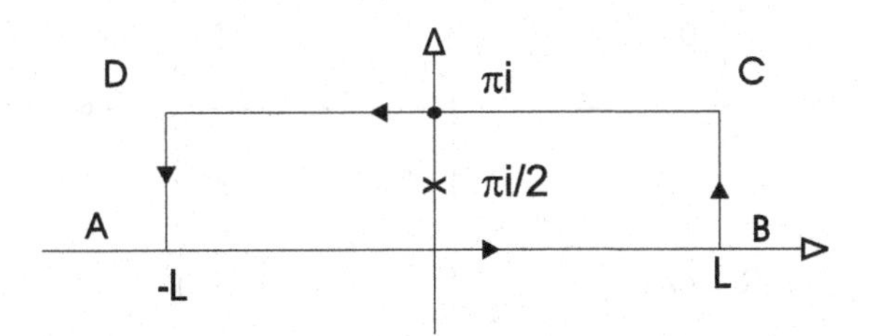

Fig. 7.9

☑**7.5.2 Problem.** By integrating $\exp(i\alpha z^2)\text{cosech}(\pi z)$ around a rectangular contour with corners at $\pm R \pm \frac{i}{2}$, prove that

$$\int_0^\infty \frac{\cos\alpha x^2 \cosh\alpha x dx}{\cosh\pi x} = \frac{1}{2}\cos(\alpha/4), \tag{7.92}$$

$$\int_0^\infty \frac{\sin\alpha x^2 \cosh\alpha x dx}{\cosh\pi x} = \frac{1}{2}\sin(\alpha/4). \tag{7.93}$$

Solution. We first set up the integrals of $\exp(i\alpha z^2)\text{cosech}(\pi z)$ along the four sides of the given rectangle in Fig. 7.10 and take the limit $R \to \infty$. In this limit we have the following expressions for the integrals along the two sides AB and CD. Along the line AB we have $z = x - i/2, -R < x < R$, and therefore

$$\lim_{R\to\infty}\int_{AB}\frac{\exp(i\alpha z^2)}{\sinh\pi z}dz = \lim_{R\to\infty}\int_{-R}^{R}\frac{\exp[i\alpha(x-i/2)^2]}{\sinh\pi(x-i/2)}dx \tag{7.94}$$

$$= i\int_{-\infty}^{\infty}\frac{\exp(i\alpha x^2+\alpha x-i\alpha/4)}{\cosh\pi x}dx. \tag{7.95}$$

Similarly, along DC $z = x + i/2, -R < x < R$ and hence

$$\lim_{R\to\infty}\int_{DC}\frac{\exp(i\alpha z^2)}{\sinh\pi z}dz = \lim_{R\to\infty}\int_{-R}^{R}\frac{\exp[i\alpha(x+i/2)^2]}{\sinh\pi(x+i/2)}dx \tag{7.96}$$

$$= -i\int_{-\infty}^{\infty}\frac{\exp(i\alpha x^2-\alpha x-i\alpha/4)}{\cosh\pi x}dx. \tag{7.97}$$

In the limit $R \to \infty$, the integrals along the remaining two sides, BC and AD, are seen to to be zero,

$$\int_{BC} f(z)dz = \int_{DA} f(z)dz = 0, \tag{7.98}$$

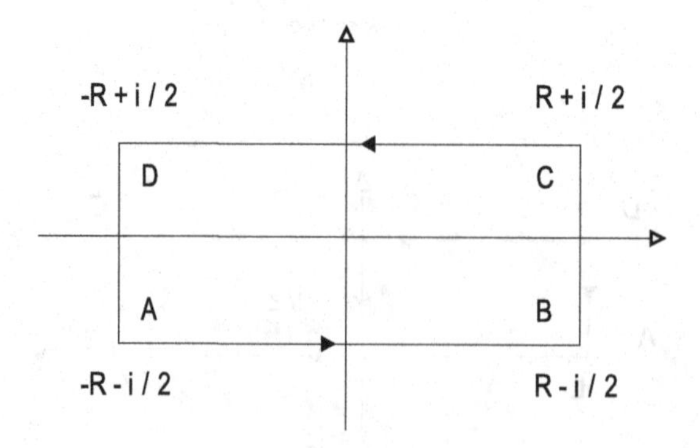

Fig. 7.10

where we have used $f(z)$ to denote the integrand $\frac{\exp(i\alpha z^2)}{\sinh \pi z}$, and hence we have

$$\lim_{R\to\infty} \oint_{\text{ABCDA}} f(z)dz = \int_{\text{AB}} f(z)dz + \int_{\text{BC}} f(z)dz + \int_{\text{CD}} f(z)dz + \int_{\text{DA}} f(z)dz \tag{7.99}$$

$$= \int_{\text{AB}} f(z)dz - \int_{\text{DC}} f(z)dz \tag{7.100}$$

$$= 2ie^{-i\alpha/4} \int_{-\infty}^{\infty} \frac{\exp(i\alpha x^2)\cosh \alpha x}{\cosh \pi x}\, dx. \tag{7.101}$$

For all values of R the contour ABCDA encloses only one singular point of $f(z)$. The residue of $f(z)$ at this point $z = 0$ is

$$\text{Res}\left\{\frac{\exp(i\alpha z^2)}{\sinh \pi z}\right\}_{z=0} = \lim_{z\to 0} \frac{z\exp(i\alpha z^2)}{\sinh \pi z} = \frac{1}{\pi}. \tag{7.102}$$

Therefore, we get

$$\int_{-\infty}^{\infty} \frac{e^{i\alpha x^2}\cosh \alpha x}{\cosh \pi x}\, dx = e^{i\alpha/4} \tag{7.103}$$

and the real and imaginary parts of this equation give the desired answers (7.92) and (7.93). ∎

☑**7.5.3 Problem.** Integrate $\frac{\exp(iaz)}{\exp(2\pi z)-1}$ around a suitably indented rectangular contour with corners of the rectangle at the points $0, R, R+i, i$ and show that

$$\int_0^{\infty} \frac{\sin ax}{\exp(2\pi x) - 1}\, dx = \frac{1}{4}\frac{e^a + 1}{e^a - 1} - \frac{1}{2a}. \tag{7.104}$$

Solution. The integral of $\frac{\exp(iaz)}{\exp(2\pi z)-1}$ around the given rectangular contour ABCDEFA, indented as shown in Fig. 7.11, vanishes,

$$\oint_{\Gamma} \frac{\exp(iaz)}{\exp(2\pi z) - 1}\, dz = 0, \tag{7.105}$$

because the integrand is analytic inside the contour.

Using the notation $f(z) = \frac{\exp(iaz)}{\exp(2\pi z)-1}$, we now write the integrals along different parts of the contour in the limit $R \to \infty$, and $\rho \to 0$.

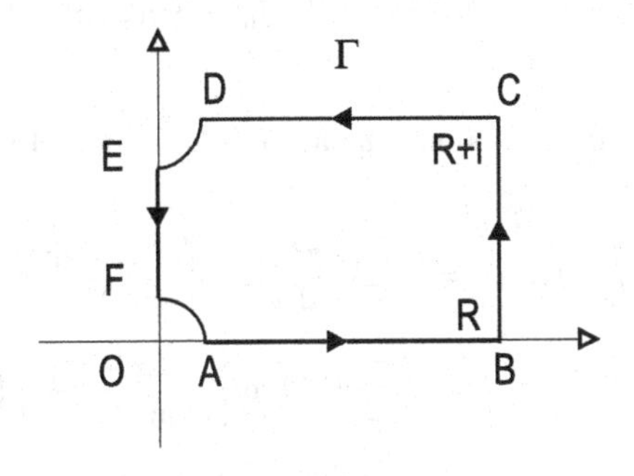

Fig. 7.11

(i) First, the integral along AB is

$$\lim \int_{\mathrm{AB}} f(z)dz = \int_0^\infty \frac{e^{iax}}{\exp(2\pi x) - 1}dx, \qquad \because z = x, \quad dz = dx \text{ along AB.} \tag{7.106}$$

(ii) Next, we shall prove that

$$\lim \int_{\mathrm{BC}} f(z)dz = 0. \tag{7.107}$$

Along BC we have $z = R + iy, dz = idy$, $0 < y < 1$. Hence,

$$\lim \left| \int_{\mathrm{BC}} f(z)dz \right| = \left| \int_0^1 \frac{e^{iaR-ay}}{\exp(2\pi R + 2\pi i y) - 1} idy \right|. \tag{7.108}$$

For large R, use

$$\left| \frac{e^{iaR-ay}}{\exp(2\pi R + 2\pi i y) - 1} \right| \le \frac{\exp(-ay)}{\exp(2\pi R) - 1} \tag{7.109}$$

to get

$$\lim_{R\to\infty} \left| \int_0^1 \frac{e^{iaR-ay}}{\exp(2\pi R + 2\pi i y) - 1} idy \right| \le \lim_{R\to\infty} \frac{1}{\exp(2\pi R) - 1} \tag{7.110}$$

$$= 0. \tag{7.111}$$

(iii) The line segment DC is the set $\{z = x + i | \rho < x < R\}$ and hence

$$\lim \int_{\mathrm{CD}} f(z)dz = -\lim \int_{\mathrm{DC}} f(z)dz \tag{7.112}$$

$$= -\lim \int_\rho^R \frac{\exp(-a + iax)}{\exp(2\pi x) - 1}dx \tag{7.113}$$

$$= -e^{-a} \int_0^\infty \frac{\exp(iax)}{\exp(2\pi x) - 1}dx. \tag{7.114}$$

(iv) The limiting value of the integral along the arc DE is evaluated using Eq. (6.20). This gives

$$\lim \int_{\mathrm{DE}} f(z)dz = (-1)\left(\frac{\pi i}{2}\right) \mathrm{Res} \left\{ \frac{\exp(iaz)}{\exp(2\pi z) - 1} \right\}_{z=i} \tag{7.115}$$

$$= -\frac{\pi i}{2} e^{-a} \lim_{z\to i} \left[\frac{z - i}{\exp(2\pi z) - 1} \right] \tag{7.116}$$

$$= -\frac{i}{4} e^{a}. \tag{7.117}$$

(v) The integral along EF requires more work and we compute

$$\lim \int_{\text{EF}} f(z)dz = -\lim \int_{\text{FE}} \frac{\exp(iaz)}{\exp(2\pi z) - 1} dz \tag{7.118}$$

$$= -\lim \int_{\rho}^{1} \frac{\exp(-ay)}{\exp(2\pi iy) - 1} i dy, \quad \because z = iy \text{ along FE} \tag{7.119}$$

$$= -i \lim \int_{\rho}^{1} \frac{\exp(-ay)\exp(-\pi iy)}{\exp(\pi iy) - \exp(-\pi iy)} dy \tag{7.120}$$

$$= -i \lim \int_{\rho}^{1} e^{-ay} \frac{\cos \pi y - i \sin \pi y}{2i \sin \pi y} dy \tag{7.121}$$

$$= -\frac{1}{2} \lim \int_{\rho}^{1} e^{-ay} \cot \pi y + \frac{i}{2} \int_{0}^{1} e^{-ay} dy \tag{7.122}$$

$$= -\frac{1}{2} \lim \int_{\rho}^{1} e^{-ay} \cot \pi y - \frac{i}{2a}(e^{-a} - 1). \tag{7.123}$$

(vi) Finally,

$$\lim \int_{\text{FA}} f(z)dz = (-1)\left(\frac{\pi i}{2}\right) \text{Res} \left\{ \frac{\exp(iaz)}{\exp(2\pi z) - 1} \right\}_{z=0} \tag{7.124}$$

$$= -\frac{i}{4}, \tag{7.125}$$

where the result (6.20) has been used again.

Noting Eq. (7.105) and adding the results in (i)–(vi) leads to

$$\int_{\text{AB}} f(z)dz + \int_{\text{BC}} f(z)dz + \int_{\text{CD}} f(z)dz + \int_{\text{DE}} f(z)dz + \int_{\text{EF}} f(z)dz$$

$$+ \int_{\text{FA}} f(z)dz = \oint_{\Gamma} f(z)dz = 0. \tag{7.126}$$

We thus get

$$(1 - e^{-a}) \int_{0}^{\infty} \frac{\exp(iax)}{\exp(2\pi x) - 1} dx - \frac{i}{4} e^{-a}$$

$$-\frac{1}{2} \int_{0}^{1} e^{-ay} \cot \pi y dy - \frac{i}{2a}(e^{-a} - 1) - \frac{i}{4} = 0, \tag{7.127}$$

and the required result,

$$\int_{0}^{\infty} \frac{\sin ax}{\exp(2\pi x) - 1} dx = \frac{1}{4} \frac{e^{a} + 1}{e^{a} - 1} - \frac{1}{2a}, \tag{7.128}$$

follows if we equate the imaginary part of the left hand side of (7.127) with zero. Note that evaluation of the integral $\int_0^1 e^{ay} \cot y \, dy$ was never needed. In fact, equating real parts of Eq. (7.127) gives the result

$$\int_{0}^{\infty} \frac{\cos ax}{\exp(2\pi x) - 1} dx = \frac{1}{2} \left(\frac{e^{a}}{e^{a} - 1} \right) \int_{0}^{1} e^{-ay} \cot \pi y \, dy. \tag{7.129}$$

∎

⧐ Exercise §§7.9, on using a rectangular contour.
⧐ Exercise §§7.11, for more problems on use of indented contours.

⊳ Exercise §§7.13, for assorted problems on use of contour integration for improper integrals.

§7.6 Series Summation and Expansion

§7.6.1 *Summation of series*

The method of contour integration can be used to sum a series of the form

$$S = \sum_{n=-\infty}^{\infty} f(n). \tag{7.130}$$

We will assume that $f(z)$ is a function satisfying the following properties:

(*1) $\lim |zf(z)| \to 0$ as $|z| \to \infty$.
(*2) $f(z)$ is analytic except for a finite number of poles at $z_k, k = 1, \ldots, M$.
(*3) The poles of $f(z)$ do not lie at any integer value, otherwise the sum (7.130) would not exist.

The central idea is to note that $\cot \pi z$ has poles at the integers and that

$$\text{Res}\{\pi \cot \pi z f(z)\}_{z=n} = f(n). \tag{7.131}$$

Therefore, if Γ_N is a closed contour such that it encloses the points $z = 0, \pm 1, \pm 2, \ldots, \pm N$, and is such that $f(z)$ is analytic on Γ_N, the integral $\oint_{\Gamma_N} \pi \cot \pi z f(z) dz$ is given by

$$\oint_{\Gamma_N} \pi \cot \pi z f(z) dz = 2\pi i \sum_{k=1}^{N} \text{Res}\{\pi \cot \pi z f(z)\}_{z=\xi_k}, \tag{7.132}$$

where the sum is over all the singular points ξ_k of the integrand, $\pi \cot \pi z f(z)$, enclosed by the contour Γ_N:

$$\oint_{\Gamma_N} \pi \cot \pi z f(z) dz = 2\pi i \left[\sum_{n=-N}^{N} f(n) + \sum_{k} \text{Res}\{\pi \cot \pi z f(z)\}_{z=z_k} \right], \tag{7.133}$$

where the second sum runs over all the poles of $f(z)$ enclosed by Γ_N. By considering a suitable limit in which Γ_N becomes bigger and bigger and encloses all the integers, the sum of the series is obtained by evaluating the integral on the left hand side and the residues at the poles of $f(z)$ appearing on the right hand side.

Choice of contour Γ_N

A useful contour for the summation of series is the square Γ_N (of Fig. 7.12), bounded by the four lines

$$x = \pm \left(N + \frac{1}{2} \right), \qquad y = \pm \left(N + \frac{1}{2} \right). \tag{7.134}$$

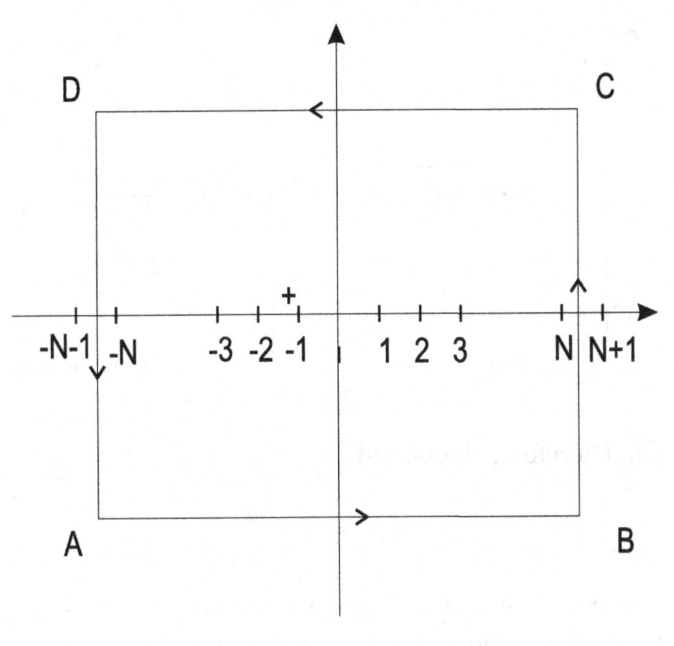

Fig. 7.12

In the limit $N \to \infty$, we will prove that the left hand side of Eq. (7.133) vanishes, i.e.

$$\lim_{N\to\infty} \oint_{\Gamma_N} \pi \cot \pi z f(z) dz = 0. \tag{7.135}$$

Also, when N is sufficiently large, the contour Γ_N will enclose all the poles of $f(z)$ and hence the sum of the series will be given by

$$\sum_{n=-\infty}^{\infty} f(n) = -\sum_{k=1}^{M} \text{Res}\{\pi \cot \pi z f(z)\}_{z=z_k}. \tag{7.136}$$

A proof of Eq. (7.135) will now be given. As a first step we prove that

$$|\cot \pi z| < K \tag{7.137}$$

holds on the contour Γ_N, where the constant K is independent of N. To prove Eq. (7.137) on the horizontal sides of Γ_N, consider

$$|\cot \pi z| = \left|\frac{e^{2\pi i z}+1}{e^{2\pi i z}-1}\right| \le \left|\frac{|e^{2\pi i z}|+1}{|e^{2\pi i z}|-1}\right| = \left|\frac{e^{-2\pi y}+1}{e^{-2\pi y}-1}\right|,$$
$$\therefore |\cot \pi z| \le \coth \pi y. \tag{7.138}$$

The function $\coth \pi y$ blows up at $y = 0$ and is a decreasing function as $|y|$ increases, (see Fig. 7.13), and $y = \pm(N + 1/2)$ on the horizontal sides AB and CD of Γ_N, and

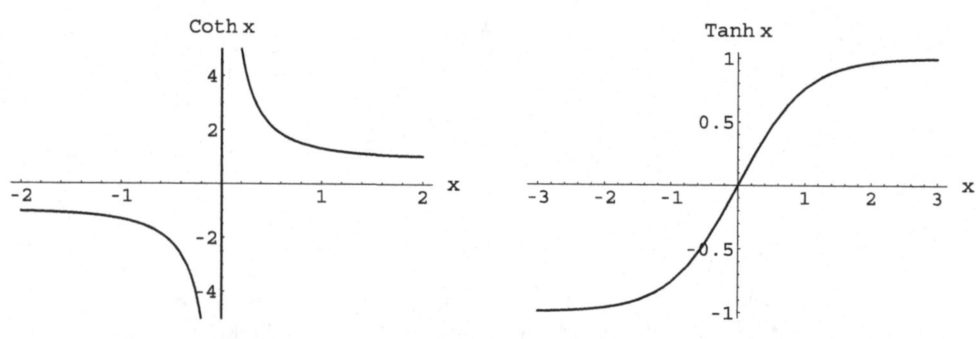

Fig. 7.13 Plot of $\coth x$.

Fig. 7.14 Plot of $\tanh x$.

we have $|y| > 1/2$. Therefore, the bound

$$|\cot \pi z| \leq \coth \frac{\pi}{2} \tag{7.139}$$

holds on the horizontal sides of Γ_N. Next, for the vertical sides of the contour we have $x = \pm(N + 1/2) + iy$, and

$$|\cot \pi z| = |\cot[\pm(N + \tfrac{1}{2})\pi + i\pi y]| = |\tanh \pi y| \leq 1. \tag{7.140}$$

The correctness of the last inequality is seen from the plot of $\tanh x$ in Fig. 7.14. Thus, the result (7.137) has been established on all sides of the contour Γ_N. The result (7.135) now easily follows from

$$\left| \oint_{\Gamma_N} \pi \cot \pi z f(z) dz \right| = \pi K \left| \oint_{\Gamma_N} f(z) dz \right|, \tag{7.141}$$

which tends to zero as $N \to \infty$, because we have $|zf(z)| \to 0$ for $|z| \to \infty$.

§7.6.2 *Mittag–Leffler expansion*

We know that a rational function can be expressed in terms of partial fractions. This result generalizes to meromorphic functions, *viz.* the functions whose singularities are poles only. The result we state and prove here will be applicable when all poles are simple poles. For generalizations refer to the textbooks cited at the end.

Theorem 7.1. *Let $f(z)$ be a function such that its singular points are simple poles at ξ_k, $k = 1, 2, \ldots$, arranged so that $0 \leq |\xi_1| \leq |\xi_2| \leq \cdots \leq |\xi_n| \leq \cdots$, and let ρ_k be the corresponding residues. In addition, let C_n be circles of radii R_n, $n = 1, 2, \ldots$ with $R_n \to \infty$ as $n \to \infty$ and centers at the origin and not passing through any pole of $f(z)$. Also, let $f(z)$ be bounded on the circles C_n,*

$$|f(z)| \leq K \quad \text{for } z \in C_n, \tag{7.142}$$

where K is independent of n. Then $f(z)$ has the representation

$$f(z) = f(0) + \sum_j \left(\frac{\rho_j}{z - \xi_j} - \frac{\rho_j}{z_j} \right). \tag{7.143}$$

Proof. Applying Cauchy's integral formula we have

$$\frac{1}{2\pi i} \oint_{C_n} \frac{f(\xi)}{\xi - z} d\xi = f(z) + \sum_j \frac{\rho_j}{\xi_j - z}. \tag{7.144}$$

This equation gives

$$f(z) - f(0) = \sum_j \left(\frac{\rho_j}{z - \xi_j} - \frac{\rho_j}{\xi_j} \right) + \frac{1}{2\pi i} \oint_{C_n} f(\xi) \left(\frac{1}{\xi - z} - \frac{1}{\xi} \right) d\xi. \tag{7.145}$$

The proof is completed by showing that the last term tends to zero as $n \to \infty$, which follows from

$$\left| \frac{1}{2\pi i} \oint_{C_n} f(\xi) \left(\frac{1}{\xi - z} - \frac{1}{\xi} \right) d\xi \right| \le \frac{K|z|}{R_n(R_n - |z|)} (2\pi R_n) \to 0. \tag{7.146}$$

This completes the proof of the theorem. In the above proof C_n need not be a circular contour; it can be any contour such that, in the limit $n \to \infty$, C_n encloses all the integer values and (7.146) is satisfied. ∎

⊳ Exercise §§7.12, for problems on series summation and partial fraction expansion.

§7.7 A Summary

In this chapter several examples of evaluation of improper integrals by the method of contour integration have been presented. A few examples of application of the method to improper and proper integrals have appeared in earlier chapters. Here we recapitulate these examples along with the different strategies and important points.

(*1) The most important example of a direct evaluation, using the definition of integrals in the complex plane, is the integral $\oint \frac{dz}{(z-z_0)^n}$, which is needed for a proof of the residue theorem.

(*2) Cauchy's fundamental theorem is sufficient for computation of integrals which can be related to Gaussian by a shift or scaling by a complex number. Examples of this type have appeared in §4.7.2.

(*3) Next come a class of proper integrals of trigonometric functions which become integrals over the unit circle after the change of variable $z = \exp(i\theta)$; see ☑6.7.2. These are easily evaluated by making use of the residue theorem.

(*4) A class of integrals such as those appearing in ☑6.8.2 involved fractional powers of the integration variable and could be written as integrals around a branch cut, *extending over a finite interval.* These are most easily computed by applying the residue theorem and considering the singularties outside the contour, especially the point at infinity. This approach is most useful when the singular points lying outside the contour are isolated and are finite in number.

(*5) A large class of improper integrals of rational functions were considered in §7.1, and they are evaluated by using a semicircular contour to close the contour.

(*6) Products of trigonometric functions with rational functions integrated over the real line are also computed by adding a semicircular contour in either the upper or the lower half plane; see ☑7.1.2.

(*7) A class of improper integrals of a rational function, a power or a logarithm, are conveniently computed by turning them into integrals around a branch cut running over the half real line and making use of two circles' contour, see §7.2.

(*8) For a sizable class of integrands involving hyperbolic functions, rectangular contours were found very useful. Examples appear in §7.5.

(*9) Several improper integrals require the use of an indented contour if the contour chosen passes through some of the singular points of the integrand; see §7.3.

(*10) Singular integrals can be given *a definition* using an indented contour for which in general several choices exist and each choice in general leads to different a definition of the singular integral; see §7.4.

In almost all the examples mentioned above and those appearing in this chapter, an application of the contour integration method for a proper or an improper integral $\int f(x)dx$ required identifying a *function* $\phi(z)$ of a complex variable and a *starting contour* γ_0 and a *limiting process.* In an application of the method of contour integration, one or more of the following milestones (*1)–(*7) are easily observed and one is led to several possibilities, indicated as (i), (ii), etc. below.

(*1) The contour γ_0
  (i) is a closed contour;
  (ii) is not a closed contour.

(*2) The function $\phi(z)$ may be
  (i) the function $f(z)$ itself;
  (ii) another function looking different from $f(z)$.

(*3) If the contour γ_0 is not a closed contour, one adds several extra pieces $\gamma_1, \gamma_2, \ldots$ so as to arrive at an integral $\oint_\Gamma \phi(z)dz$ around a closed contour Γ. The integrals $\int_{\gamma_k} f(z)dz$ along some of the pieces may
  (i) become proportional to the integral to be evaluated;
  (ii) combine and produce the required integral $\int f(x)dx$ in a suitable limit.

(∗4) The integrals along the remaining pieces of the contour

(i) vanish in the limit to be taken at the end;
(ii) may not vanish in the limit at the end.

(∗5) The integral $\oint_\Gamma \phi(z)dz$ is evaluated using the residue theorem and the needed limits are taken and the resulting relation

$$\oint_\Gamma \phi(z)dz = \sum \text{residues at singular points inside } \Gamma$$

(i) gives the required answer because all the extra terms vanish or can be easily computed;
(ii) the left hand side contains a new integral coming from line integrals along one or more extra pieces.

(∗6) If there is an extra integral left unevaluated on the left hand side of the final relation obtained by applying the residue theorem, then this extra integral

(i) need not be evaluated separately and gets evaluated along with the original integral by considering real and imaginary parts [see, for example, Eq. (7.52)–(7.55) in ☑7.2.3];
(ii) may require a separate computation;
(iii) does not get evaluated and there is no need to evaluate it either, as is the case in ☑7.5.3 [see comment before Eq. (7.129)];
(iv) cannot be computed by a simple method and remains unknown till the end and the method fails.

(∗7) In the the final limiting situation, the function $\phi(z)$ has

(i) a finite number of poles inside the closed contour;
(ii) an infinite number of poles appear inside the contour and the final answer is, in general, an infinite series for which the sum may or may not be known in a closed form.

⧐ Exercise §§7.14 contains examples of integrals from statistical mechanics.

⧐ Open-Ended §§7.16. Many times answers for two integrals can be obtained by a single contour integration. Try this set for a sample collection of such problems.

⧐ Open-Ended set §§7.17 has some not-so-easy problems, going a little beyond this book.

⧐ Mixed Bag §§7.18 is a collection of assorted problems.

Chapter 8

ASYMPTOTIC EXPANSION

In many applications one needs to describe the behavior of a function $f(z)$ and to compute its values for large $|z|$. If the function is analytic at infinity, it can be represented by a convergent power series of the form

$$a_0 + \frac{a_1}{z} + \frac{a_2}{z^2} + \cdots + \frac{a_n}{z^n} + \cdots . \tag{8.1}$$

Such a representation is not possible if, for example, the function has a branch point at infinity. A series of the form (8.1), even though not convergent, turns out to be a useful representation of $f(z)$ when the partial sum of the first few terms accurately approximates the behavior of the function and the error $|f(z) - \sum_{m=1}^{n} a_m z^{-m}|$, keeping n fixed, becomes smaller and smaller with growing $|z|$. Formally, we say that the series (8.1) is asymptotic to the function $f(z)$, or is an *asymptotic expansion* of $f(z)$, if

$$\lim_{z\to\infty} |z|^n \left| f(z) - \sum_{m=1}^{n} a_m z^{-m} \right| = 0, \tag{8.2}$$

and we write

$$f(z) \sim a_0 + \frac{a_1}{z} + \frac{a_2}{z^2} + \cdots + \frac{a_n}{z^n} + \cdots . \tag{8.3}$$

The limit in Eq. (8.2) will, in general, exist only if $\arg(z)$ is constrained to a range of values (α, β) and the asymptotic expansion will represent the function in the corresponding sector in the complex plane. In many of the examples in this chapter, our discussion will be restricted to the special case of asymptotic expansions of functions of a real variable. As a generalization of Eq. (8.3), we will write

$$f(z) \sim g(z) \left(a_0 + \frac{a_1}{z} + \frac{a_2}{z^2} + \cdots \right) \tag{8.4}$$

if the series $\sum a_m z^{-m}$ is asymptotic to f/g. Here asymptotic, *power series*, expansion of a function for a large argument has been introduced. Although other expansions of the form

$$f(z) \sim \sum_n a_n \phi_n(z) \tag{8.5}$$

can be defined, we shall limit our discussion to expansions of type (8.4). The asymptotic expansions, in powers of $z - z_0$, describing the behavior of the function for $z \to z_0$ will, in principle, require nothing new and can be handled by a change of variable from z to $w = 1/(z - z_0)$, so that as $z \to z_0$ the variable w tends to infinity.

§8.1 Properties of asymptotic expansions

Here we give some important facts about asymptotic expansions; more details and proofs can be found in the books listed at the end.

(1) A function need not have an asymptotic power series expansion. For example, if for $x \to \infty$

$$\exp(-x^2) \sim a_0 + \frac{a_1}{x} + \frac{a_2}{x^2} + \cdots, \tag{8.6}$$

it follows from Eq. (8.2) that the coefficients a_n must all vanish.

(2) Given an asymptotic expansion, a function which it "represents" is not unique. In other words, two different functions may have the same asymptotic expansion. For example, it is obvious that the functions $f(x)$ and $f(x) + \exp(-x^2)$ will have the same asymptotic expansions.

(3) Let $f(z)$ have an asymptotic expansion $\sum a_m z^{-m}$. Then Eq. (8.2) shows that

$$\left| f(z) - \sum_{m=0}^{n} a_m z^{-m} \right| < \frac{\epsilon}{|z|^n} \tag{8.7}$$

holds for arbitrarily small $\epsilon > 0$, when $|z|$ is taken sufficiently large. Splitting the $m = n$ term from the summation, this means that

$$\left| f(z) - \sum_{m=0}^{n-1} a_m z^{-m} \right| < \frac{\epsilon + |a_n|}{|z|^n}. \tag{8.8}$$

Hence, if the first $n - 1$ terms are used to approximate the function $f(z)$, the error is of the same order of magnitude as the nth term, i.e. the absolute value of the first omitted term while making the approximation.

(4) Asymptotic expansion of a function $f(z)$, if it exists, is unique and the coefficients a_n in Eq. (8.2) have the values given by

$$a_n = \lim_{z \to \infty} |z|^n \left[f(z) - \sum_{m=0}^{n-1} a_m z^{-m} \right]. \tag{8.9}$$

(5) The asymptotic expansions of two functions $f(z)$ and $g(z)$ can be combined to give the asymptotic expansion of the linear combination $\lambda f(z) + \mu g(z)$, of the product $h(z) = f(z)g(z)$ and the ratio $f(z)/g(z)$, in the same way as one would do for power series expansion.
(6) The asymptotic expansion of the derivative $f'(z)$ of a function can be obtained by differentiating the asymptotic expansion of the function $f(z)$.

In this chapter we shall discuss the following methods for obtaining asymptotic expansions of functions represented by integrals:

(1) Integration by parts;
(2) Laplace's method and use of Watson's lemma;
(3) Stationary phase approximation;
(4) The method of steepest descent;
(5) The saddle point method.

We shall focus our attention on the working of the methods rather than attempting to give a rigorous mathematical treatment complete with all proofs. It is hoped that the interested reader, who has mastered the details of using the above methods, will be well prepared to read the advanced textbooks where an excellent discussion on the mathematical and technical aspects, not covered here, can be found. The aims of writing this chapter would be served if the reader who has worked through the chapter acquires an appreciation of the applicability, relationship, and working of different methods, as well as their similarities and differences.

§8.2 Integration by Parts

We begin with an example of obtaining asymptotic expansion of integrals using the method of integration by parts.

☑8.2.1 Problem. Obtain the asymptotic expansion of the incomplete error function

$$\operatorname{erfc}(x) = \int_x^\infty \exp(-t^2)\, dt \tag{8.10}$$

as $x \to \infty$ by integrating by parts. Use the result obtained to find asymptotic expansion of $\int_0^x \exp(-t^2)\, dt$ for large x.

Solution. We write

$$\int_x^\infty e^{-t^2}\, dt = \int_x^\infty \left(\frac{-1}{2t}\right) \frac{d}{dt} e^{-t^2}\, dt \tag{8.11}$$

$$= \frac{1}{2x} e^{-x^2} - \int_x^\infty \left(\frac{1}{2t^2}\right) e^{-t^2}\, dt, \tag{8.12}$$

and the second term becomes

$$\int_x^\infty \frac{1}{2t^2}\left(-\frac{1}{2t}\right)\frac{d}{dt}e^{-t^2}\,dt = \frac{e^{-x^2}}{4x^3} - \int_x^\infty \frac{3}{4t^4}e^{-t^2}\,dt. \tag{8.13}$$

Continuing in this way we get the asymptotic expansion

$$\int_x^\infty e^{-t^2}\,dt = \frac{e^{-x^2}}{2x}\left\{1 - \frac{1}{2x^2} + \frac{3}{4x^4} - \frac{15}{8x^6} + \cdots\right\}. \tag{8.14}$$

In order to get the asymptotic expansion of $\int_0^x \exp(-t^2)\,dt$, we first write it as

$$\int_0^x e^{-t^2}\,dt = \int_0^\infty e^{-t^2}\,dt - \int_x^\infty e^{-t^2}\,dt \tag{8.15}$$

$$= \frac{\sqrt{\pi}}{2} - \int_x^\infty e^{-t^2}\,dt \tag{8.16}$$

and then use the result (8.14). ∎

We continue with the above example and compare properties of asymptotic expansion with those of Taylor series expansion. In this example, Taylor series convergent for all x can be obtained as follows.

$$\int_x^\infty e^{-t^2}\,dt = \int_0^\infty e^{-t^2}\,dt - \int_0^x e^{-t^2}\,dt \tag{8.17}$$

$$= \frac{\sqrt{\pi}}{2} - \left(x - \frac{x^3}{3} + \frac{x^5}{10} - \cdots\right). \tag{8.18}$$

In the last step, the e^{-t^2} in the second integral has been expanded in powers of t and has been integrated term by term. Here integration term by term is valid for all x and results in Taylor series expansion of $\operatorname{erfc}(x)$. The first few terms of the Taylor series give good approximation to the numerical value of the function for $x < 1$. Though convergent for all x, the Taylor series has no practical utility for computing values of the function for $x > 1$. For example, the first 20 terms of the Taylor series do not give a reasonable value for $x > 2$. On the other hand, the asymptotic expansion, which is divergent for all x, gives answers with accuracy better than 5% for $x > 2$, when the first two terms are taken into account. The accuracy improves as x increases.

More examples of asymptotic expansions using the method of integration by parts appear in the problem sets. This method has restricted applicability and may fail or even give wrong answers. An excellent discussion and several examples on integration by parts can be found in [Bender and Orsaz (1999)]. Laplace's method, to be discussed in the next section, is a very general and useful method for a class of integrals over a real variable. It is also crucial to the understanding of the method of steepest descents and the saddle point method for integrals in the complex plane.

⧐ Exercise §§8.1, for problems on asymptotic expansion using integration by parts.

§8.3 Laplace's Method

Laplace's method gives asymptotic expansion, for large x, of a function $f(x)$, which can be represented as

$$f(x) = \int_a^b e^{xh(t)} \phi(t)\, dt, \tag{8.19}$$

where $h(t)$ is a real-valued function. Crucial to this method for obtaining asymptotic expansion of $f(x)$ is the observation that for large x the most important contribution comes from a small neighborhood of the point where $h(t)$ is largest. We first present the central idea of Laplace's method to get the *dominant term*, which will be described in several stages. The method to obtain the full expansion and generalizations will be given in the following subsection.

§8.3.1 *Dominant term of asymptotic expansion*

We consider different cases of $h(t)$ being a decreasing function, or having a maximum at the lower limit $x = a$, or a maximum in the interval (a, b).

(i) $h(t)$ is a decreasing function of t. We begin with a simple case where $h(t)$ is largest at $t = a$ and decreases as t increases. Thus, for $t \in (a, b)$

$$h(t) - h(a) < 0, \quad h'(a) < 0. \tag{8.20}$$

Next, we approximate $h(t)$ by a linear function,

$$h(t) = h(a) + (t - a)h'(a) + \cdots, \tag{8.21}$$

and $\phi(t)$ by its value at $t = a$,

$$\phi(t) \approx \phi(a) + \cdots, \tag{8.22}$$

so that

$$f(x) \approx \int_a^{a+\delta} e^{xh(t)} \phi(t)\, dt \tag{8.23}$$

$$= \int_a^{a+\delta} \exp\big(xh(a) + x(t-a)h'(a) + \cdots\big)\big(\phi(a) + \cdots\big)\, dt \tag{8.24}$$

$$= \int_0^{\delta} \exp\big(xh(a) + xuh'(a) + \cdots\big)\big(\phi(a) + \cdots\big)\, du \tag{8.25}$$

$$\sim \frac{e^{xh(a)}\phi(a)}{xh'(a)}\big(e^{x\delta h'(a)} - 1\big). \tag{8.26}$$

Here the integration has been carried out keeping only the first two terms of the exponential. Since $h'(a) < 0$, the exponential term does not contribute to the expansion for large x and one has

$$f(x) \sim -\frac{e^{xh(a)}}{x}\frac{\phi(a)}{h'(a)} = \frac{\phi(a)}{|h'(a)|}\frac{e^{xh(a)}}{x}. \tag{8.27}$$

The last step, leading to Eq. (8.27), is equivalent to taking the upper limit in Eq. (8.25) to infinity.

✍ *The extension in the range of integration from a small neighborhood of a point, such as $t = a$ here, to an infinite range is an important technique which will be used repeatedly.*

(ii) $h(t)$ has a maximum at $t = a$. When $h(t)$ has a maximum at $t = a$, we have $h'(a) = 0$ and $h''(a) < 0$. Instead of Eq. (8.21) we now write

$$h(t) = h(a) + \frac{1}{2}h''(a)(t-a)^2 + \cdots \tag{8.28}$$

and proceed as above to get [with $h''(a) = -|h''(a)|$]

$$f(x) \approx \int_a^{a+\delta} \exp\left[xh(a) - \tfrac{1}{2}x|h''(a)|(t-a)^2 + \cdots\right]\left[\phi(a) + \cdots\right] dt \tag{8.29}$$

$$\sim e^{xh(a)}\,\phi(a)\int_0^{\delta} \exp\left[-\tfrac{1}{2}x|h''(a)|u^2\right] du. \tag{8.30}$$

Extending the upper limit to infinity, we get

$$f(x) \sim \frac{1}{2}\sqrt{\frac{2\pi}{|h''(a)|}}\left(\frac{e^{xh(a)}}{\sqrt{x}}\right)\phi(a). \tag{8.31}$$

(iii) $h(t)$ has a maximum at a point $t_0 (a < t_0 < b)$. When the function $h(t)$ has a maximum at a point inside the range of integration, $h'(t_0) = 0, h''(t_0) \le 0$, we will get

$$f(x) \approx \int_{t_0-\delta}^{t_0+\delta} \exp\left[xh(t_0) - \tfrac{1}{2}x|h''(t_0)|(t-t_0)^2\right]\left(\phi(t_0) + \cdots\right) dt \tag{8.32}$$

$$\sim e^{xh(t_0)}\,\phi(t_0)\int_{-\infty}^{\infty} \exp\left[-\tfrac{x}{2}|h''(t_0)|u^2\right] du \tag{8.33}$$

$$\sim \sqrt{\frac{2\pi}{|h''(t_0)|}}\left(\frac{e^{xh(t_0)}}{\sqrt{x}}\right)\phi(t_0). \tag{8.34}$$

Equation (8.34) differs from the previous case, Eq. (8.31), by a factor of 2 coming from the fact that the range of integration in Eq. (8.33) is extended to $(-\infty, \infty)$.

§8.3.2 *Full asymptotic expansion*

Laplace's method has been outlined above for obtaining the dominant term of the asymptotic expansion of $f(x)$. In order to obtain the full asymptotic expansion, we proceed as follows.

Assuming that $h(t)$ is largest at the lower limit $t = a$, $h(t)$ satisfies Eq. (8.20), and a new variable s is introduced by

$$h(t) = h(a) - s. \tag{8.35}$$

If the function $h(t)$ has a calculus maximum at $t = a$, and hence $h'(a) = 0$, with $h''(a) < 0$, the new variable s is introduced through the equation

$$h(t) = h(a) - s^2. \tag{8.36}$$

In general, if $h'(a) = h''(a) = \cdots = h^{(n-1)}(a) = 0$, but $h^{(n)}(a) \leq 0$, we take

$$h(t) = h(a) - s^n. \tag{8.37}$$

The contribution to the integral, written in terms of the new variable s, comes from a small neighborhood of $s = 0$ and

$$\int_a^b e^{xh(t)}\phi(t)\,dt \sim \int_0^{s(b)} e^{xh(a)-xs^n}\left(\phi\frac{dt}{ds}\right)ds \tag{8.38}$$

$$\sim e^{xh(a)}\int_0^{\delta} e^{-xs^n}\left(\phi\frac{dt}{ds}\right)ds. \tag{8.39}$$

The expression inside the parentheses is written in terms of s and expanded as a series in powers of s. In general, fractional powers of s will appear. The integration range interval $(0, \delta)$ is taken to be small, in which the series expansion converges. The order of integration and series summation are interchanged and the range of integration is extended to infinity by taking δ to infinity. The integrations are then performed and one gets the required asymptotic expansion. The justification for these steps comes from Watson's lemma.

Theorem 8.1 (Watson's lemma). *Let $\phi(t)$ be a function integrable over $(0, b)$ and having asymptotic expansion*

$$\phi(t) \sim \sum_{m=0}^{\infty} a_m t^{pm+q}, \quad p > 0, \ q > -1, \tag{8.40}$$

for $t \to 0$. Then the integral $I(x) = \int_0^b \phi(t)e^{-xt}\,dt$ has large x asymptotic expansion:

$$I(x) \sim \sum_m a_m \Gamma(pm + q + 1)x^{-(pm+q+1)}, \quad x \to \infty. \tag{8.41}$$

In the case where the upper limit b is ∞, $\phi(t)$ must also satisfy the condition $|\phi(t)| < Ke^{ct}$ for large x, where $K, c > 0$ are some constants.

Using Watson's lemma. Laplace's method, as outlined above, requires that the variable t be expressed in terms of the new variable s [see Eq. (8.37)], and this step may become very complicated and tedious. Watson's Lemma eliminates the need for this intermediate step in Laplace's method for obtaining full asymptotic expansion. One may retain the original variable t and, instead of Eq. (8.37), one writes $h(t)$ as

$$h(t) = h(a) + (t-a)^n \frac{h^{(n)}(a)}{n!} + \chi(t), \tag{8.42}$$

if $h'(a) = \cdots = h^{(n-1)}(a) = 0, h^{(n)}(a) \leq 0$. Here $\chi(t)$ is the sum of the remaining terms in the Taylor series of $h(t)$. This time, instead of Eq. (8.39), one gets

$$f(x) = e^{xh(a)} \int_a^b \exp\left[-\frac{x}{n!}(t-a)^n |h^{(n)}(a)|\right] \exp\left[x\chi(t)\right] \phi(t) dt. \tag{8.43}$$

Next, we expand the product $\exp\left[x\chi(t)\right]\phi(t)$ in powers of t, and interchange integration and series summation taking the integral over a small interval $(a, a+\delta)$. Then integrate over a new variable $u = t - a$, taking its range of integration from 0 to ∞. This alternate procedure leads to the full asymptotic expansion. Note that the exponential coming from $\exp(x\chi(t))$ must also be expanded in powers of $t - a$. *It is to be emphasized that merely retaining a few terms in the expansion of $\chi(t)$, and not expanding the exponential, can lead to a wrong answer.* This is because the resulting integrals may not converge or may have spurious maxima.

When the function $h(t)$ has several maxima, the range of integration should be split into suitable subintervals so as to have only one maximum inside, or on, the boundary of a subinterval. Finally, when the given integral has a form different from Eq. (8.1), we must make a suitable transformation to bring it to the desired form, Eq. (8.1). We record some useful identities for later use:

$$\frac{1}{(1+x)^a} = \sum_r \frac{(-1)^r}{r!} \frac{\Gamma(a+r)}{\Gamma(a)} x^r, \tag{8.44}$$

$$\Gamma(2x)\Gamma(\tfrac{1}{2}) = 2^{2x-1}\Gamma(x+\tfrac{1}{2})\Gamma(x), \tag{8.45}$$

$$\int_0^\infty e^{-tx} t^\nu \, dt = x^{-\nu-1}\Gamma(\nu+1). \tag{8.46}$$

☑**8.3.1 Problem.** Use Laplace's method to obtain asymptotic expansion of

$$K_0(x) = \int_1^\infty \frac{\exp(-xt)}{\sqrt{t^2-1}} \, dt. \tag{8.47}$$

Solution. Here $h(t) = -t$ is maximum at $t = 1$ and the change of variable given by $h(t) = h(1) - s$ becomes $t = 1 + s$, giving

$$K_0(x) = e^{-x} \int_0^\infty \frac{e^{-xs}}{\sqrt{s(s+2)}} \, ds = \frac{e^{-x}}{\sqrt{2}} \int_0^\infty e^{-xs} s^{-\frac{1}{2}} \left(1 + \frac{s}{2}\right)^{-\frac{1}{2}} ds. \tag{8.48}$$

Expand $(1+s/2)^{-\frac{1}{2}}$ in powers of s and integrate using Watson's lemma to get

$$K_0(x) \sim \frac{e^{-x}}{\sqrt{2}} \int_0^\infty e^{-xs} s^{-\frac{1}{2}} \left\{1 - \frac{1}{2}\frac{s}{2} + \frac{1}{2}\frac{3}{2}\frac{1}{2!}\left(\frac{s}{2}\right)^2 - \frac{1}{2}\frac{3}{2}\frac{5}{2}\frac{1}{3!}\left(\frac{s}{2}\right)^3 + \cdots\right\} \tag{8.49}$$

$$\sim e^{-x}\sqrt{\frac{\pi}{2x}}\left[1 - \frac{1}{8x} + \frac{9}{32}\left(\frac{1}{2x}\right)^2 - \frac{75}{128}\left(\frac{1}{2x}\right)^3 + \cdots\right]. \tag{8.50}$$

Note that in Eq. (8.48), $\phi(t) = 1/\sqrt{t^2-1}$ becomes infinite at the lower limit $t=1$ but this does not cause any problem. ∎

☑8.3.2 Problem. Find asymptotic expansion of

$$\int_0^{\pi/2} e^{-x\tan t}\, dt \tag{8.51}$$

for large x using Laplace's method.

Solution. In this case $h(t) = -\tan t$ becomes largest at $t = 0$ and the change of variable given by $h(t) = h(0) - s$ becomes $s = \tan t$. This gives

$$\int_0^{\pi/2} e^{-x\tan t}\, dt = \int_0^\infty e^{-sx}(1+s^2)^{-1}\, ds. \tag{8.52}$$

We substitute the power series expansion

$$(1+s^2)^{-1} = 1 - s^2 + s^4 - s^6 + \cdots \tag{8.53}$$

and integrate term by term using Watson's lemma to get

$$\int_0^{\pi/2} e^{-x\tan t}\, dt \sim \frac{1}{x} - \frac{2!}{x^3} + \frac{4!}{x^5} - \frac{6!}{x^7} + \cdots + \frac{(-1)^r(2r)!}{x^{2r+1}} \cdots. \tag{8.54}$$

∎

☑8.3.3 Problem. Obtain the asymptotic expansion

$$\Gamma(x+1) \sim \sqrt{(2\pi x)}e^{-x}x^x\left\{1 + \frac{1}{12x} + \frac{1}{288x^2} + \cdots\right\} \tag{8.55}$$

of the gamma function using the integral representation

$$\Gamma(x+1) = \int_0^\infty e^{-t}t^x\, dt. \tag{8.56}$$

Solution. In order to bring the integral representation to the form (8.19), so that Laplace's method becomes applicable, we first change the variable from t to xt to get

$$\Gamma(x+1) = x^{x+1}\int_0^\infty e^{-xt}t^x\, dt \tag{8.57}$$

$$= x^{x+1}\int_0^\infty e^{x(\log t - t)}\, dt, \tag{8.58}$$

so that $h(t) = \log t - t$. Note that $h'(t) = 0$ gives $t = 1$ and therefore an important contribution comes from a small neighborhood of $t = 1$. Writing

$$\Gamma(x+1) = x^{x+1} I(x) \tag{8.59}$$

and changing the variable to $t = 1 + w$, we get

$$I(x) \sim \int_{1-\eta}^{1+\eta} e^{x \log t - xt}\, dt = e^{-x} \int_{-\eta}^{\eta} \exp\left[x \log(1+w) - xw\right] dw. \tag{8.60}$$

The asymptotic expansion of $I(x)$ can now be obtained in *two different ways.*

First method. We use a change of variable from w to s defined by

$$\log(1+w) - w = -s^2, \tag{8.61}$$

or

$$s^2 = \frac{w^2}{2} - \frac{w^3}{3} + \frac{w^4}{4} - \cdots, \tag{8.62}$$

and write w as a power series in s:

$$w = a_1 s + a_2 s^2 + a_3 s^3 + \cdots. \tag{8.63}$$

The coefficients in Eq. (8.63) are determined by reversion of series in Eq. (8.62) (see §6.10). Substituting Eq. (8.63) in Eq. (8.62) and comparing powers of s on both sides, we get $a_1 = \pm\sqrt{2}$. We take $a_1 = \sqrt{2}$ and determine the other coefficients to get

$$w = \sqrt{2}s + \frac{2}{3}s^2 + \frac{1}{9\sqrt{2}}s^3 - \frac{2}{135}s^4 + \frac{1}{540\sqrt{2}}s^5 + \cdots. \tag{8.64}$$

On extending the integration range to $(-\infty, \infty)$, the integral $I(x)$ becomes

$$I(x) \sim e^{-x} \int_{-\infty}^{\infty} e^{-xs^2} \left(\frac{dw}{ds}\right) ds. \tag{8.65}$$

Substituting the series Eq. (8.64), we get

$$I(x) \sim \sqrt{2} e^{-x} \int_{-\infty}^{\infty} e^{-xs^2} \left\{1 + \frac{1}{6}s^2 + \frac{1}{216}s^4 + \cdots\right\} ds \tag{8.66}$$

$$\sim e^{-x} \sqrt{\frac{2\pi}{x}} \left\{1 + \frac{1}{12x} + \frac{1}{288x^2} + \cdots\right\}. \tag{8.67}$$

Using this result in Eq. (8.59), we see the asymptotic expansion of the gamma function is given by

$$\Gamma(x+1) \sim \sqrt{2\pi x}\, e^{-x} x^x \left\{1 + \frac{1}{12x} + \frac{1}{288x^2} \cdots\right\}. \tag{8.68}$$

Second method. In this method, the variable w is retained as the integration variable and we proceed as explained in §8.3.2. In Eq. (8.60) we first substitute

$$\exp\left[x \log(1+w) - xw\right] = \exp\left(-\frac{xw^2}{2}\right) \exp\left(\frac{xw^3}{3} - \frac{xw^4}{4} + \cdots\right), \tag{8.69}$$

and expand the second exponential and collect powers of x:

$$I(x) = \int_{-\infty}^{\infty} \exp\left(-\frac{xw^2}{2}\right)\left\{1 + x\left(\frac{w^3}{3} - \frac{w^4}{4} + \cdots\right) + \frac{x^2}{2!}\left(\frac{w^3}{3} - \frac{w^4}{4} + \cdots\right)^2 + \cdots\right\} dw. \quad (8.70)$$

Integrating term by term leads to the desired asymptotic expansion of the gamma function. ∎

⊳ Exercise §§8.2, for problems on the dominant term.
⊳ Exercise §§8.3, for full asymptotic expansion using Laplace's method.

§8.4 Method of Stationary Phase

The method of stationary phase applies to integrals of the form of a generalized Fourier integral:

$$f(x) = \int_a^b e^{ixh(t)}\phi(t)\,dt, \quad (8.71)$$

where $h(t)$ is real and the limits are finite. In the limit $x \to \infty$, the integrand oscillates rapidly, resulting in cancelations and a small contribution to the integral. Thus, the dominant contribution comes from the neighborhood of the points where $h(t)$ is stationary, i.e. $h'(t) = 0$. If there are no stationary points, the integral receives the most important contribution from the end points. Integration by parts and a method similar to Laplace's method can be used to isolate the dominant term of the asymptotic expansion. The method for obtaining the dominant term will now be outlined for several cases.

(∗1) If the function $h(t)$ has no stationary point in the interval $(a, b), h'(t) \neq 0$ for $t \in [a, b]$, the large x behavior can be derived by the method of integration by parts. In this case we will get

$$f(x) = \int_a^b \left[\frac{\phi(t)}{ixh'(t)}\right] \frac{d}{dt} e^{ixh(t)}\,dt \quad (8.72)$$

$$\sim \frac{\phi(b)e^{ixh(b)}}{ixh(b)} - \frac{\phi(a)e^{ixh'(a)}}{ixh'(a)} + \cdots. \quad (8.73)$$

(∗2) Assuming that $t = a$ is a stationary point of $h(t)$, we have $h'(a) = 0$ and we proceed as in Laplace's method and substitute

$$h(t) = h(a) + \frac{1}{2}(t-a)^2 h''(a) + \cdots, \quad (8.74)$$

$$\phi(t) = \phi(a) + \cdots \quad (8.75)$$

in the integral and restrict the range of integration to a small interval near $t = a$. Thus, keeping only terms written explicitly in Eqs. (8.74) and (8.75), we get

$$f(x) \approx \int_a^{a+\epsilon} \exp\left[ixh(a) + \frac{ix}{2}(t-a)^2 h''(a)\right]\phi(t)dt \tag{8.76}$$

$$\sim \phi(a)e^{ixh(a)} \int_0^\infty \exp\left[\frac{ix}{2}u^2 h''(a)\right]du, \tag{8.77}$$

and also

$$f(x) \sim \begin{cases} \sqrt{\dfrac{\pi}{2h''(a)x}}\, e^{i\pi/4} e^{ixh(a)}\phi(a), & \text{if } h''(a) > 0, \\ \sqrt{\dfrac{\pi}{2|h''(a)|x}}\, e^{-i\pi/4} e^{ixh(a)}\phi(a), & \text{if } h''(a) < 0. \end{cases} \tag{8.78}$$

When the point $t = b$ is a stationary point, the results are similar to those given above.

(*3) The contribution of a stationary point inside the range of integration has an extra factor of 2 as compared to that coming from a stationary point on the boundary.

(*4) It should be noted that the boundary contributions are of the order $1/\sqrt{x}$ in the case $h'' \neq 0$. If the stationary point t_0 is such that

$$h'(t_0) = h''(t_0) = \cdots = h^{(p-1)}(t_0) = 0, \quad h^p(t_0) \neq 0, \tag{8.79}$$

an interval near a stationary point contributes a term of order $1/\sqrt[p]{x}$. In this case

$$h(t) = h(a) + \frac{1}{p!}(t-a)^p h^p(a) + \cdots \tag{8.80}$$

Further, in this case the dominant term is seen to be given by

$$f(x) \sim \begin{cases} \left[\dfrac{p!}{h^p(a)x}\right]^{1/p} \dfrac{\Gamma(1/p)}{p}\, e^{i\pi/2p} e^{ixh(a)}\phi(a), & \text{if } h^p(a) > 0, \\ \left[\dfrac{p!}{|h^p(a)|x}\right]^{1/p} \dfrac{\Gamma(1/p)}{p}\, e^{-i\pi/2p} e^{ixh(a)}\phi(a), & \text{if } h^p(a) < 0. \end{cases} \tag{8.81}$$

The discussion given here is applicable when $\phi(t)$ does not vanish at the stationary point. The stationary phase approximation is not useful for obtaining terms beyond the dominant term, because the nonstationary points may give an important contribution. In the next section, we shall discuss the method of steepest descent, which is a general method for obtaining the full asymptotic expansion.

§8.5 Method of Steepest Descent

§8.5.1 *Central idea*

The method of steepest descent is a generalization of Laplace's method to integrals in the complex plane, which can be represented as

$$f(x) = \int_C \exp\left[xh(\tau)\right]\phi(\tau)\,d\tau. \tag{8.82}$$

The functions ϕ and h are assumed to be analytic on the contour C running between two points τ_1 and τ_2 in the complex τ plane. *The method of steepest descent consists in deforming the contour C to a new contour C' such that Laplace's method becomes applicable when the integral is computed along C'.* First, we explain the central issues in this method; later, we show its working by means of examples in §8.7.

As in Laplace's method, we note that the integral (8.82) receives important contribution from a neighborhood of a point τ_0 on C where $|\exp\left[xh(\tau)\right]| = \exp(x\mathrm{Re}(h(\tau)))$ is largest.

Choice of a new contour. Among the contours that pass through the point τ_0, we select a new contour along which $\mathrm{Im}[h(\tau)]$ is constant and equals $\mathrm{Im}[h(\tau_0)]$. This choice is made so as to avoid large oscillations as $x \to \infty$, which would otherwise make it difficult to estimate the net contribution to the integral. In addition to demanding $\mathrm{Im}[h(\tau)] = \text{const}$, the new contour must be chosen in such a way that $\mathrm{Re}[h(\tau)]$ falls off as rapidly as possible when one moves away from the point τ_0 where $\mathrm{Re}[h(\tau)]$ is largest.

The steepest path. The real part of $h(\tau)$ can be visualized as a surface in three dimensions, with the height of the surface representing the value of $\mathrm{Re}[h(\tau)]$. A path along which $\mathrm{Re}[h(\tau)]$ varies most rapidly will be called the *steepest path.* Let (u, v) and (ψ, χ) denote the real and imaginary parts of τ and $h(\tau)$, respectively. The direction along which the real part $\psi(u, v) = \mathrm{Re}[h(\tau)]$ varies most rapidly will be along the gradient vector

$$\nabla\psi = \left(\frac{\partial\psi}{\partial u}, \frac{\partial\psi}{\partial v}\right) = \left(-\frac{\partial\chi}{\partial v}, \frac{\partial\chi}{\partial u}\right), \tag{8.83}$$

which is perpendicular to

$$\nabla\chi = \left(\frac{\partial\chi}{\partial u}, \frac{\partial\chi}{\partial v}\right). \tag{8.84}$$

Since the vector (8.84) is along the normal to the curve $\chi(u, v) = \text{const}$, we conclude that if $h'(\tau) \neq 0$ the steepest direction is given by $\chi(u, v) = \mathrm{Im}[h(\tau)] = \text{const}$. When $h'(\tau_0) \neq 0$, there is only one steepest path that passes through the point τ_0 such that $\mathrm{Re}[h(\tau)]$ increases in one direction and decreases in the other direction. If $h'(\tau_0) = 0$

several steep paths cross at the point τ_0 and such a point is called a *saddle point* because the surface representing $\text{Re}[h(\tau)]$ resembles a horse saddle.

§8.5.2 *Local properties of steepest paths*

In order to further discuss the choice of contour through a point τ_0, we write the integral in Eq. (8.82) as

$$f(x) = \exp\left[xh(\tau_0)\right] \int_C \phi(\tau) \exp\left\{x[h(\tau) - h(\tau_0)]\right\} d\tau, \tag{8.85}$$

and examine the behavior of the exponent $h(\tau) - h(\tau_0)$ in a small neighborhood of τ_0. Expanding $h(\tau)$ in a Taylor series gives

$$h(\tau) - h(\tau_0) = \lambda(\tau - \tau_0)^m + \mu(\tau - \tau_0)^{m+1} + \cdots, \tag{8.86}$$

where m is a nonnegative integer. Near τ_0 we introduce local polar coordinates ρ and ϕ by writing $\tau - \tau_0 = \rho e^{i\phi}$ and $\lambda = |\lambda| e^{i\alpha}$; we then have

$$\text{Re}\left[h(\tau) - h(\tau_0)\right] \approx |\lambda| \rho^m \cos(m\phi + \alpha), \tag{8.87}$$

$$\text{Im}\left[h(\tau) - h(\tau_0)\right] \approx |\lambda| \rho^m \sin(m\phi + \alpha). \tag{8.88}$$

We therefore see that in a small neighborhood of τ_0, $\text{Re}\left[h(\tau) - h(\tau_0)\right] \approx 0$ on the set of rays

$$(\mathbf{A}): \quad \cos(m\phi + \alpha) = 0, \quad \text{i.e. } m\phi + \alpha = (n + \tfrac{1}{2})\pi. \tag{8.89}$$

These rays, through τ_0, divide a small neighborhood of τ_0 into sectors such that $\text{Re}[h(\tau) - h(\tau_0)]$ has opposite signs in adjacent sectors. Similarly, Eq. (8.88) shows that the set of rays

$$(\mathbf{B}): \quad \sin(m\phi + \alpha) = 0, \quad \text{i.e. } m\phi + \alpha = n\pi \tag{8.90}$$

has the property that on these rays $\text{Im}[h(\tau) - h(\tau_0)] \approx 0$. These rays can be further divided into two subsets:

$$(\mathbf{B_1}): \quad m\phi + \alpha = 2n\pi, \quad \text{Re}[h(\tau) - h(\tau_0)] > 0; \tag{8.91}$$

$$(\mathbf{B_2}): \quad m\phi + \alpha = (2n+1)\pi, \quad \text{Re}[h(\tau) - h(\tau_0)] < 0. \tag{8.92}$$

Further along the rays in the set (**B2**), the exponent in Eq. (8.85),

$$h(\tau) - h(\tau_0) \approx -|\lambda| \rho^m \tag{8.93}$$

is real and negative and the integrand falls off most rapidly as τ moves away from τ_0. Thus, the rays ($\mathbf{B_2}$) are tangents to the paths of *steepest descent* through the point τ_0. Similarly, the rays ($\mathbf{B_1}$) are tangents to the paths of *steepest ascent* through the point τ_0.

Table 8.1 Direction of steepest paths.

	$m = 1$	$m = 2$	$m = 3$
Set (A)	$0, \pi$	$0, \frac{\pi}{2}, \pi, \frac{3\pi}{2}$	$0, \frac{\pi}{3}, \frac{2\pi}{3}, \pi, \frac{4\pi}{3}, \frac{5\pi}{3}$
Set $(B1)$	$-\frac{\pi}{2}$	$-\frac{\pi}{4}, \frac{3\pi}{4}$	$\frac{\pi}{2}, \frac{7\pi}{6}, \frac{11\pi}{6}$
Set $(B2)$	$\frac{\pi}{2}$	$\frac{\pi}{4}, \frac{5\pi}{4}$	$\frac{\pi}{6}, \frac{5\pi}{6}, \frac{3\pi}{2}$

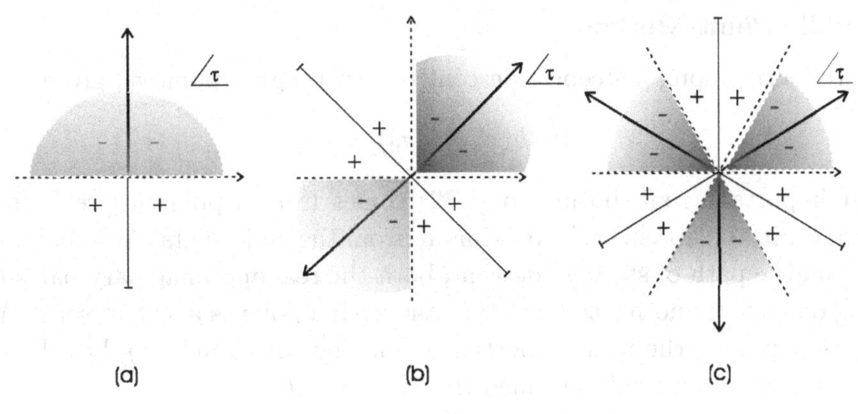

Fig. 8.1

An example. We illustrate the above discussion of local properties by taking the example of $h(\tau) = i\tau^m, \tau_0 = 0$ and three different values of $m = 1, 2, 3$. In Table 8.1 we list the values of ϕ for the sets $(A), (B1)$, and $(B2)$. In Fig. 8.1 we show the sectors where $\text{Re}[h(\tau)] < 0$ and the directions of steepest descent and also the rays in the set $(B2)$ from $\tau_0 = 0$ drawn with bold lines. Also shown are the rays of the set (A) corresponding to $\text{Re}[h(\tau)] = 0$ with broken lines.

It is useful to note that the tangents to the steepest paths, i.e. the rays in the sets $(B1)$ *and* $(B2)$*, bisect the angle, at* τ_0*, between the curves given by* $\text{Re}[h(\tau) - h(\tau_0)] = 0$. Figures 8.1(a)–(c) show the directions of steepest descent by thick lines with arrowheads, and the directions of steepest ascent by thin lines with $\dashv$ as heads, for the three cases $m = 1, 2, 3$, respectively. The level curves of $\text{Re}[h(t)]$, i.e. $\text{Re}[h(\tau) - h(\tau_0)] = 0$ in a small neighborhood of the origin, are shown by thick dashed lines. The positive (or negative) sign in a sector indicates that the value of $\text{Re}[h(\tau) - h(\tau_0)]$ is positive (or negative) in the sector when one is sufficiently close to the origin. In the sectors, shaded in a small neighborhood of the origin, the value of $\text{Re}[h(\tau) - h(\tau_0)]$ decreases as one moves away from the origin, and it increases in the sectors which are not shaded.

§8.5.3 *Change of variable*

Noting Eq. (8.93), we introduce a new integration variable s by means of the equation

$$h(\tau) - h(\tau_0) = -s^m, \tag{8.94}$$

if $h(\tau_0) = h'(\tau_0) = \cdots = h^{(m-1)}(\tau_0) = 0$ and $h^{(m)}(\tau_0) \neq 0$. Along the path of steepest descent $h(\tau) - h(\tau_0)$ is real and Laplace's method becomes applicable when the integral is transformed into an integration along a steepest path, parametrized by the variable s. This procedure for obtaining the asymptotic expansion is called the *method of steepest descents.*

We discuss the saddle point method before taking up examples.

§8.6 Saddle Point Method

When integrating along a steepest descent path through a point τ_0, given by

$$\text{Im}[h(\tau) - h(\tau_0)] = 0, \tag{8.95}$$

the most important contribution to (8.82) comes from a point where $\text{Re}[h(\tau)]$ is maximum or, in the absence of a maximum, from the end points. At a maximum of $\text{Re}[h(\tau)]$, on the path of steepest descent, both the real and imaginary parts have a vanishing derivative and hence $h'(z) = 0$, and such a point is a *saddle point.* Among all the saddle points, the most important is the one where $\text{Re}[h(\tau)]$ has the largest value. This saddle point will be called the *highest saddle point.*

In the *saddle point method* there is considerable flexibility in the choice of a new contour of integration. A useful choice consists of taking *short line segments* in the neighborhood of the saddle points and the end points of the integration range. These straight line segments are taken along the directions of steepest descent, and a new path of integration is constructed by suitably joining them. Of course, the value of the integral should remain unchanged when the contour is deformed from the original one to the new chosen path. The asymptotic expansion is obtained by summing the contribution of linear paths, each of which is computed by *extending the range of integration to infinity.* If there is no saddle point or the contour cannot be deformed to go through a saddle point, only the end points of the contour make an important contribution.

§8.7 Examples

☑**8.7.1 Problem.** Using the method of steepest descent, obtain the asymptotic expansion of the integral

$$f(x) = \int_0^1 \exp[ix(\tau + \tau^2)]\, d\tau. \tag{8.96}$$

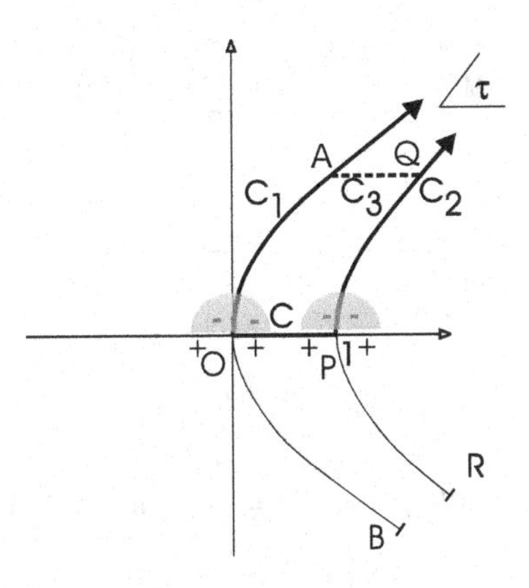

Fig. 8.2

Solution. We begin by noting that $h(\tau) = i(\tau + \tau^2)$ and that the contour for integration is the interval OP from $\tau = 0$ to $\tau = 1$. In order to make use of the method of steepest descent, we must deform the contour OP to a new one consisting of paths of steepest descent through the end points O and P.

Since $h'(0) \neq 0$, there is only one steepest path through the origin. The equation for this path is $\mathrm{Im}[h(\tau) - h(0)] = 0$. This is the curve BOA in Fig. 8.2. The results of §8.5.3 imply that the directions of steepest descent and ascent at the origin are, respectively, given by the rays $\phi = \frac{\pi}{2}$ and $\phi = -\frac{\pi}{2}$. Thus, OA and OB are paths of steepest descent and ascent, respectively, through the origin. Similarly, there is only one steepest path, QPR, through $\tau = 1$ and PQ corresponds to the steepest descent. We can now proceed in two different ways.

First method. We need parametriations of the contours OA and PQ. In general, we introduce a new variable, s, through the equation $h(\tau) - h(\tau_0) = -s$, which in the present case gives the following parametric equations:

$$\mathrm{OA}: \quad \tau = -\frac{1}{2} + \frac{1}{2}\sqrt{1 + 4is}, \quad s > 0; \tag{8.97}$$

$$\mathrm{PQ}: \quad \tau = -\frac{1}{2} + \frac{1}{2}\sqrt{9 + 4it}, \quad t > 0. \tag{8.98}$$

The use of Cauchy's theorem justifies deforming the original contour C, the interval $[0, 1]$, to a new one which goes from $\tau = 0$ to ∞ along C_1 and returns along C_2. The infinite contour is handled by considering it as a limiting case of a finite contour which can be taken to be OAQP, with the points A and Q taken to infinity in the end; see Fig. 8.2. This gives $f(x) = I_1(x) - I_2(x)$, where

$$I_1(x) = \int_{\mathrm{OA}} \exp[ix(\tau + \tau^2)]\, d\tau = \int_0^\infty e^{-sx} \left(\frac{d\tau}{ds}\right) ds, \quad \frac{d\tau}{ds} = i(1 + 4is)^{-\frac{1}{2}}, \tag{8.99}$$

$$I_2(x) = \int_{\mathrm{PQ}} \exp[ix(\tau + \tau^2)]\, d\tau = e^{2ix} \int_0^\infty e^{-tx} \left(\frac{d\tau}{dt}\right) dt, \quad \frac{d\tau}{dt} = i(9 + 4it)^{-\frac{1}{2}}. \tag{8.100}$$

Expanding $\frac{d\tau}{ds}$ and $\frac{d\tau}{dt}$ in power series and making use of Watson's lemma to integrate term by term gives

$$I_1(x) \sim \frac{i}{x} \sum_{m=0}^{\infty} \frac{\Gamma(m + \frac{1}{2})}{\Gamma(\frac{1}{2})} \left(\frac{4}{ix}\right)^m, \tag{8.101}$$

$$I_2(x) \sim \frac{ie^{2ix}}{3x} \sum_{m=0}^{\infty} \frac{\Gamma(m + \frac{1}{2})}{\Gamma(\frac{1}{2})} \left(\frac{4}{9ix}\right)^m. \tag{8.102}$$

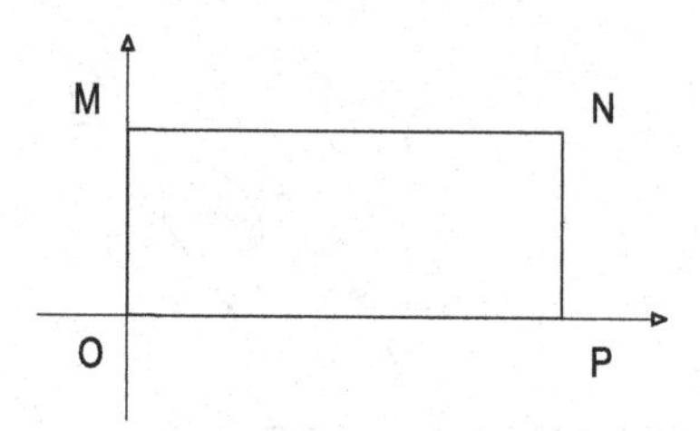

Fig. 8.3

Note that in this case there is a saddle point at $t = -1/2$ and the deformed contour does not pass through the saddle point.

Second method. We now present an alternate method which bypasses integration along the *full path* of steepest descent and where the change of variable is not required. In the limit $x \to \infty$ the contribution of the two integrals along the paths C_1, C_2 comes from a small neighborhood of the points $\tau = 0$ and $\tau = 1$, respectively. Thus, we replace the steepest paths C_1, C_2 by small straight line segments along their tangents and it is sufficient to integrate along a rectangular contour OMNP, shown in Fig. 8.3. Hence,

$$\int_0^1 e^{ix(\tau+\tau^2)}\, d\tau \sim \int_{\text{OM}} e^{ix(\tau+\tau^2)} + \int_{\text{MN}} e^{ix(\tau+\tau^2)} + \int_{\text{NP}} e^{ix(\tau+\tau^2)}, \tag{8.103}$$

with the three integrals given by

$$\int_{\text{OM}} e^{ix(\tau+\tau^2)}\, d\tau = \int_0^{i\eta} e^{ix(\tau+\tau^2)}\, d\tau, \tag{8.104}$$

$$\int_{\text{MN}} e^{ix(\tau+\tau^2)}\, d\tau = \int_{i\eta}^{1+i\eta} e^{ix(\tau+\tau^2)}\, d\tau, \tag{8.105}$$

$$\int_{\text{NP}} e^{ix(\tau+\tau^2)}\, d\tau = -\int_1^{1+i\eta} e^{ix(\tau+\tau^2)}\, d\tau. \tag{8.106}$$

The first integral, along OM, is integrated by extending the range of integration to $(0,\infty)$, and expanding the second exponential, $\exp(ix\tau^2)$, in a series, integrating term by term. This gives

$$\int_0^{i\eta} e^{ix(\tau+\tau^2)}\, d\tau = \int_0^{\eta} e^{-xt-ixt^2}\, idt \sim i\int_0^{\infty} e^{-xt} \sum_{m=0}^{\infty} \frac{(-ix)^m t^{2m}}{m!}\, dt \tag{8.107}$$

$$= \frac{i}{x}\sum_{m=0}^{\infty} \frac{(2m)!}{m!}\left(\frac{1}{ix}\right)^m = \frac{i}{x}\sum_{m=0}^{\infty} \frac{\Gamma(m+\frac{1}{2})}{\Gamma(\frac{1}{2})}\left(\frac{4}{ix}\right)^m. \tag{8.108}$$

It may be noted that the steps here — the power series expansion, use of Watson's lemma, and extending the range of integration to infinity — are very similar to those in Laplace's method. The second integral in Eq. (8.103) is of the order of $\exp(-\eta x)$ for large x

$$\int_{i\eta}^{1+i\eta} e^{ix(\tau+\tau^2)}\, d\tau = e^{-\eta x} e^{-ix\eta^2} \int_0^1 \exp\left[ix(t+t^2+2i\eta t)\right] dt, \tag{8.109}$$

and does not contribute to the asymptotic expansion. The third integral, with $\tau = 1 + it$, gives

$$\int_1^{1+i\eta} e^{ix(\tau+\tau^2)}\, d\tau = \int_0^\eta \exp(2ix)\exp(-3xt)\exp(-ixt^2)\, idt \tag{8.110}$$

$$= i\exp(2ix)\sum_{m=0}^{\infty}\int_0^\eta \exp(-3xt)(-ix)^m \frac{t^{2m}}{m!}\, dt \tag{8.111}$$

$$= i\frac{e^{2ix}}{3x}\sum_{m=0}^{\infty}\frac{(2m)!}{m!}\left(\frac{1}{9ix}\right)^m. \tag{8.112}$$

$$= \frac{ie^{2ix}}{3x}\sum_{m=0}^{\infty}\frac{\Gamma(m+\frac{1}{2})}{\Gamma(\frac{1}{2})}\left(\frac{4}{9ix}\right)^m. \tag{8.113}$$

As expected, we have recovered the result obtained by the first method. ∎

☑8.7.2 Problem. Using the steepest descent method, obtain the asymptotic expansion of

$$I(x) = \int_0^1 \exp\left(ix\tau^2\right) d\tau \tag{8.114}$$

for large x.

Solution. Here, an important contribution comes from the end points. In this case $h(\tau) = i\tau^2$ and $h'(\tau) = 0$ at $\tau = 0$ and two paths of steepest ascent are the rays $\phi = \frac{3\pi}{4}, -\frac{\pi}{4}$. The two rays $\phi = \frac{\pi}{4}, \frac{5\pi}{4}$, drawn from the origin, are the paths of steepest descent.

The steepest path through $\tau = 1$, given by $\text{Im}[h(\tau)] = \text{Im}[h(1)]$, is a branch of the hyperbola $x^2 - y^2 = 1$, the upper part of the hyperbola being the path of steepest descent; see Fig. 8.4. This follows from the fact that, at $\tau = 1$, the direction $\phi = \frac{\pi}{2}$ corresponds to decreasing $h(\tau)$ and $\phi = -\frac{\pi}{2}$ corresponds to increasing $h(\tau)$. Therefore, we write

$$\int_0^1 e^{ix\tau^2}\, d\tau = \int_{C_1} e^{ix\tau^2}\, d\tau - \int_{C_2} e^{ix\tau^2}\, d\tau. \tag{8.115}$$

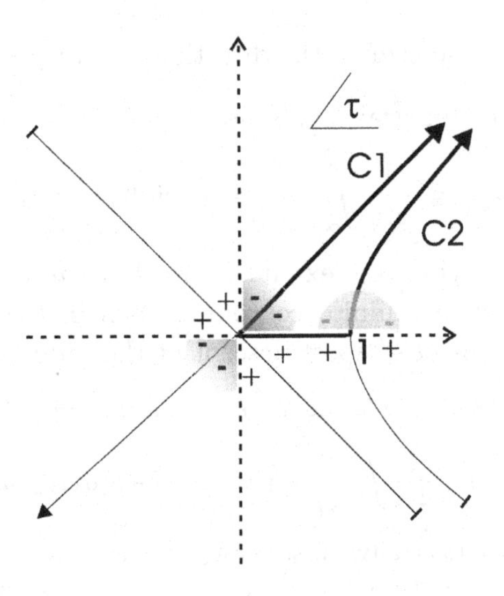

Fig. 8.4

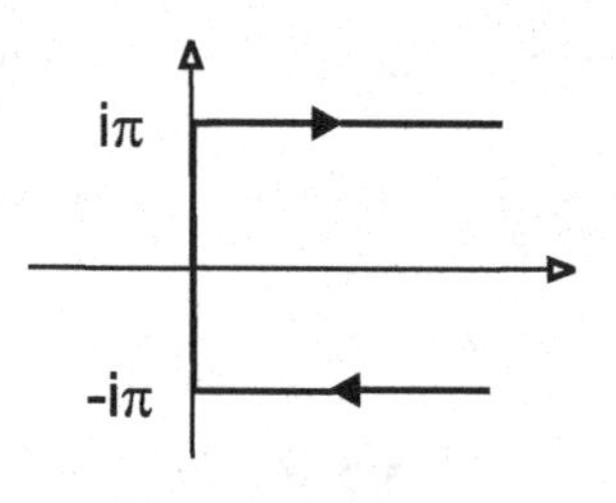

Fig. 8.5

Recalling Eq. (8.94), we choose to parametrize C_1, C_2 by means of new variables s and t, respectively, defined by $i\tau^2 = -s^2$ and $i\tau^2 - i = -t$. Writing the two integrals in terms of the new parameters, we get

$$\int_0^1 e^{ix\tau^2}\, d\tau = \int_0^\infty e^{-xs^2} e^{i\frac{\pi}{4}}\, ds - \int_0^\infty e^{ix-ixt}\left(\frac{d\tau}{dt}\right) dt \tag{8.116}$$

$$= \frac{1}{2}\frac{\sqrt{\pi}}{x} e^{\frac{i\pi}{4}} - \frac{i}{2}e^{ix}\int_0^\infty e^{-xt}(1+it)^{-\frac{1}{2}}\, dt. \tag{8.117}$$

Expanding the square root and integrating term by term gives the desired asymptotic expansion:

$$\int_0^1 \exp(ix\tau^2)\, d\tau = \frac{1}{2}\sqrt{\frac{\pi}{x}}\, e^{\frac{i\pi}{4}} + \frac{1}{2}e^{ix}\sum_{m=0}^{\infty}\frac{\Gamma(m+\frac{1}{2})}{\Gamma(\frac{1}{2})}(ix)^{-m-1}. \tag{8.118}$$

This expansion can also be obtained by the second method of the previous example. ∎

✍ *It is sufficient to know the directions of steepest descent near the end points and near the saddle points, if any. The contribution to the asymptotic expansion comes from small line segments in the directions of the steepest descent in neighborhoods of the end points and the saddle points. In cases where no saddle point exists, only the end points contribute. If the range of integration is infinite, the steepest path must pass through at least one saddle point.*

In the example below we determine the directions of steepest descent.

☑**8.7.3 Problem.** The integral representation

$$J_\nu(x) = \frac{1}{2\pi i}\int_{\infty-\pi i}^{\infty+\pi i} \exp(x\sinh z - \nu z)\, dz \tag{8.119}$$

is to be used to derive asymptotic expansion for large ν and x with $x/\nu = \text{const}$. Here the integral is over an infinite contour, shown in Fig. 8.5. Find the saddle point(s) and the directions of steepest descent at the saddle point(s).

Solution. We first substitute $\nu = x/a$ and $a = \operatorname{sech}\alpha$, and rewrite Eq. (8.119) as

$$J_\nu(x) = \frac{1}{2\pi i}\int_{\infty-\pi i}^{\infty+\pi i} \exp\left[x(\sinh z - z\cosh\alpha)\right] dz. \tag{8.120}$$

Since the contour extends to infinity, the steepest path must pass through a saddle point. Using the notation $h(z) = \sinh z - z\cosh\alpha$, the saddle points are given by $\frac{dh(z)}{dz} = 0$ and are located at

$z = 2n\pi i \pm \alpha$, where n is an integer. The two saddle points, $z = \pm\alpha$, are in the region of interest, $-\pi < \operatorname{Im} z < \pi$. The point $z = \alpha$ is the higher saddle point and is the only one to be considered. Near $z = \alpha$, we introduce the local coordinate w by

$$z = \alpha + w, \quad w = \rho e^{i\phi}, \quad \rho > 0, \tag{8.121}$$

and find the behavior of the function $h(z)$ for small ρ. Thus, we get

$$h(z) - h(\alpha) = (\sinh z - z\cosh\alpha) - (\sinh\alpha - \alpha\cosh\alpha) \tag{8.122}$$

$$= \sinh\alpha(\cosh w - 1) + \cosh\alpha(\sinh w - w) \tag{8.123}$$

$$\approx \tfrac{1}{2}\rho^2 e^{2i\phi}\sinh\alpha + \cdots, \tag{8.124}$$

where only the leading term in series in powers of ρ has been displayed explicitly. To find the directions of steepest paths, one demands $\operatorname{Im}[h(z) - h(\alpha)] = 0$. This gives

$$\phi = 0, \frac{\pi}{2}, \frac{3\pi}{2}, \pi. \tag{8.125}$$

Of these the directions $\phi = 0, \pi$ correspond to steepest ascent, because $\operatorname{Re}[h(z) - h(\alpha)] = \frac{1}{2}\sinh\alpha\rho^2\cos(2\phi)$ is positive along these directions. The other two directions, which bisect the directions of steepest ascent, are $\phi = \frac{\pi}{2}, \frac{3\pi}{2}$ and are the directions of steepest descent. The paths of steepest descent, with $z = x_1 + ix_2$, are given by

$$\operatorname{Im}[h(z) - h(\alpha)] = \cosh x_1 \cos x_2 - x_2\cosh\alpha = 0. \tag{8.126}$$

For $\alpha = 1$, we plot the level curves $\operatorname{Re}[h(z) - h(\alpha)] = 0$, which give the boundary of the shaded region in Fig. 8.6. In this region $\operatorname{Re}[h(z)]$ decreases as one moves away from the saddle point. The steepest descent path must lie in this region and is the middle curve given by $\operatorname{Im}[h(z) - h(\alpha)] = 0$. ∎

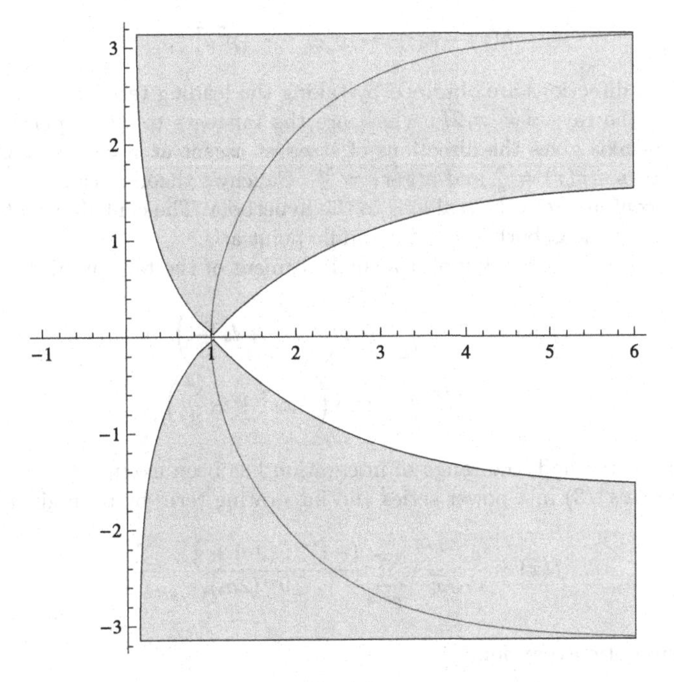

Fig. 8.6

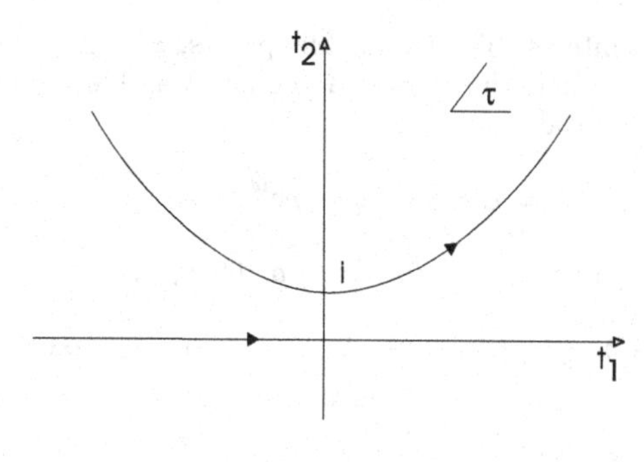

Fig. 8.7

☑**8.7.4 Problem.** Use the saddle point method to find asymptotic expansion of the integral

$$f(x) = \int_{-\infty}^{\infty} \exp\left[ix\left(\tau + \frac{\tau^3}{3}\right)\right] d\tau. \tag{8.127}$$

Solution. The given integral $I(x)$ runs from $-\infty$ to ∞. We want to select a new contour so that the saddle point method may become applicable. Here $h(\tau) = i(\tau + \tau^3/3)$, and the saddle points are $\pm i$. The new path of integration must be an infinite path of steepest descent passing through the saddle point. The steepest path through the saddle point $\tau = i$ is given by $\mathrm{Im}[h(t)] = \mathrm{Im}[h(i)]$. Writing $\tau = t_1 + it_2$, the equation of the steepest path through $\tau = i$ becomes

$$t_1(t_1^2 - 3t_2^2 + 3) = 0. \tag{8.128}$$

This gives the imaginary axis $t_1 = 0$ and the hyperbola $t_1^2 - 3t_2^2 + 3 = 0$. To find the steepest descent directions, we follow §8.3 and write $\tau - i = \rho\exp(i\phi)$ and consider

$$h(\tau) - h(i) = -\rho^2 e^{2i\phi} + \tfrac{i}{3}\rho^3 e^{3i\phi}. \tag{8.129}$$

The steepest descent directions are obtained by taking the leading term, for $\rho \approx 0$, to be real and negative. These are the rays $\phi = \pi, 2\pi$, which are the tangents to the hyperbola. On the other hand, the imaginary axis gives the directions of steepest ascent at $\tau = i$. The asymptotes of the hyperbola are the rays $\arg(\tau) = \frac{\pi}{6}$ and $\arg(\tau) = \frac{5\pi}{6}$. Cauchy's theorem can be used to justify the deformation of the contour from the real axis to the hyperbola. The contribution to the asymptotic expansion comes from a neighborhood of the saddle point at i.

Writing $\tau = i + s$ and integrating over a small segment of the tangent gives

$$f(x) \sim e^{-\frac{2x}{3}} \int_{-\delta}^{\delta} \exp\left(-xs^2 + ix\frac{s^3}{3}\right) ds \tag{8.130}$$

$$\sim e^{-\frac{2x}{3}} \int_{-\infty}^{\infty} \exp\left(-xs^2 + ix\frac{s^3}{3}\right) ds, \tag{8.131}$$

where, as in Laplace's method, the range of integration has been extended to infinity. Expanding the exponential $\exp(ixs^3/3)$ in a power series and integrating term by term gives the result

$$f(x) \sim \frac{e^{-2x/3}}{\sqrt{x}} \sum_{m=0}^{\infty} \frac{(-1)^m \Gamma(3m + \frac{1}{2})x^{-m}}{9^m (2m)!}, \tag{8.132}$$

giving full the asymptotic expansion. ∎

⧐ Exercise §§8.4, on steepest paths.
⧐ Tutorial §§8.6, on the method of steepest descent.
⧐ Tutorial §§8.5, on the saddle point method.
⧐ Exercise §§8.7, on the steepest descent and saddle point methods.

§8.8 Topics for Further Study

Laplace's method is applicable to integrals of the form

$$f(x) = \int_a^b e^{xh(t)}\phi(t)\,dt, \tag{8.133}$$

for $x \to \infty$. Modifications in the method become necessary in a few situations. For example, for the integral

$$\int_0^\infty \exp\left(-xt - \frac{1}{t}\right), \tag{8.134}$$

the important contribution comes from values of t in the neighborhood of $t = 1/\sqrt{x}$, which depends on the variable x. In a situation of this type, known as the case of a *movable maximum*, Laplace's method is not applicable directly. Another situation which requires modification in Laplace's approach is when the function $\phi(t)$ becomes infinity at the maximum. The reader is referred to the books for more examples and for possible modifications in Laplace's method needed to obtain asymptotic expansions.

Laplace's method is applicable when x is real. When x becomes complex, the method of steepest descent can be used and the asymptotic expansion exhibits the *Stokes phenomenon.* In general, the region of the complex x plane where an asymptotic expansion is valid is like a sector bounded by two rays, $\arg(x) = \alpha$ and $\arg(x) = \beta$. On crossing the boundary of a sector, a different expansion will become applicable. The reason for this phenomenon is the a sudden change in the structure of the steepest paths when the argument of x is varied.

For a detailed discussion on these and related matters, see the books cited in the bibliography. In particular, the book [Bender and Orsaz (1999)] contains an excellent discussion and a good number of examples on all the important technical issues concerning asymptotic expansions.

Chapter 9

CONFORMAL MAPPINGS

A function of a complex variable is a mapping from the complex plane to itself. In this chapter we shall be concerned with the properties of mappings defined by functions analytic in a domain except for a few singular points. A useful geometric visualization of a mapping by a function, $w = f(z)$, is to represent different subsets of the complex plane and their images in two different copies of complex planes, to be called the z and the w plane, respectively. In §9.2 conformal mapping is introduced and its properties are discussed. An important class of mappings, bilinear transformation, is discussed in detail in §9.3 and §9.4. Later, in §9.4 and §9.5, the properties of the elementary functions as mappings of the complex plane and the Joukowski map $w = \frac{a}{2}(z + \frac{1}{z})$, a map extremely useful for applications in aerodynamics, are discussed. In §9.7 a few examples and solved problems are taken up. While analytic methods have their role to play, the emphasis of our discussion is on the use of geometric representations of mappings. The Schwarz–Christoffel transformation, giving a mapping of real line onto a specified polygon, is an important tool for applications. The properties and applications of Schwarz–Christoffel transformation are discussed in §9.8 and §9.9. The chapter closes with notes and references for further study.

§9.1 Conformal Mappings

Let γ_1 and γ_2 be two curves intersecting at a point z_0 [see Fig. 9.1], and γ_1' and γ_2' be the images of the two curves under a mapping $w = f(z)$. The mapping is *conformal* if both the magnitude and the sense of the angle between the curves γ_1 and γ_2 equal those for their images, γ_1' and γ_2'. The mapping f is an *isogonal* mapping if the magnitude of the angle between the curves γ_1 and γ_2 is equal to the magnitude of the angle between their images. The images of γ_1 and γ_2 in the z plane, shown in Fig. 9.1(b), give a conformal map, $z \to w$. The map $z \to t$, with images of γ_1 and γ_2 as in Fig. 9.1(c), is not conformal but is only isogonal.

While the conformal mappings preserve the angle between two curves, the arcs of a curve at different places get scaled and rotated by amounts which vary from

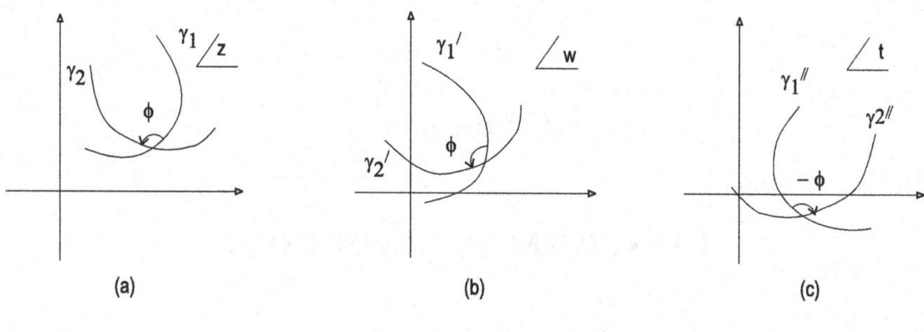

Fig. 9.1

point to point. Though the image of a full curve may appear very different, a small segment of an image curve still looks like, *conforms to*, the original curve.

The linear function functions

$$f(z) = z + \xi, \quad g(z) = e^{i\alpha} z, \quad h(z) = sz$$

represent translation, rotation, and scaling by a fixed amount everywhere. These are the simplest examples of conformal mappings. A combination of all the three, a general linear transformation $w = s\exp(i\alpha)z + \xi$, is also conformal and provides a prototype conformal mapping.

A mapping by an analytic function, having a *nonzero derivative* at a point z_0, can be approximated by a linear function in a small neighborhood of z_0 by keeping the first two terms in its Taylor series expansion:

$$f(z) \approx f(z_0) + (z - z_0) \left. \frac{df}{dz} \right|_{z_0}. \tag{9.1}$$

Therefore, such a mapping is expected to be conformal at z_0. That this result is indeed true is the central result on conformal mapping and will now be discussed. Note that if the derivative $\frac{df}{dz}|_{z_0}$ vanishes, the right hand side of (9.1) reduces to a constant and is not a conformal map.

Theorem 9.1. *A mapping $z \longrightarrow w = f(z)$ is conformal at a point z_0 if the function $f(z)$ is analytic and has a nonvanishing derivative at a point z_0.*

To prove the above result, let a curve γ in a complex plane be represented by the parametric equations

$$\gamma : z(t) = x(t) + iy(t), \qquad t_1 \le t \le t_2. \tag{9.2}$$

The angle α, which the tangent to the curve at a point z_0 $(t = t_0)$ makes with the real axis, is given by

$$\tan\alpha = \left. \left(\frac{dy}{dt} \Big/ \frac{dx}{dt} \right) \right|_{t=t_0} \tag{9.3}$$

$$= \arg \left. \left(\frac{dz}{dt} \right) \right|_{t=t_0}. \tag{9.4}$$

Under a mapping $w = f(z)$, the corresponding angle $\bar{\alpha}$ of the tangent at $w_0 \equiv f(z_0)$ to the image curve $\bar{\gamma}$ in the w plane is given by

$$\bar{\alpha} = \arg\left(\frac{dw}{dt}\right)\bigg|_{t=t_0} \tag{9.5}$$

$$= \arg\left(\frac{dw}{dz} \times \frac{dz}{dt}\right)\bigg|_{t=t_0} \tag{9.6}$$

$$= \arg\left(\frac{dw}{dz}\right)\bigg|_{z=z_0} + \arg\left(\frac{dz}{dt}\right)\bigg|_{t=t_0}. \tag{9.7}$$

Thus, we see that under the mapping f

$$\bar{\alpha} = \alpha + \arg[f'(z_0)] \tag{9.8}$$

and the tangent gets rotated by the angle $\arg[f'(z_0)]$. Note that the angle of rotation is the same for all curves passing through the point z_0.

In Fig. 9.2 two curves γ_1 and γ_2 intersecting at a point z_0 and their tangents having inclinations α and β are shown. Also depicted are the tangents to the image curves $\bar{\gamma_1}$ and $\bar{\gamma_2}$ along with the angles which the tangents make with the real w axis. The angles of rotation of the two tangents given by $\bar{\alpha} - \alpha$ and $\bar{\beta} - \beta$ will be equal for conformal maps, and in that case

$$\bar{\alpha} - \alpha = \bar{\beta} - \beta \Rightarrow \beta - \alpha = \bar{\beta} - \bar{\alpha} \Rightarrow \phi = \bar{\phi}.$$

Thus, the angle $\phi = \beta - \alpha$ between two curves γ_1 and γ_2 is equal to the angle between the image curves given by $\bar{\phi} = \bar{\beta} - \bar{\alpha}$. Therefore, the mapping by the function f is conformal.

A small line segment, Δz, between two neighboring points P and Q, represented by z and $z + \Delta z$ (see Fig. 9.3), is related to the segment Δw between the image

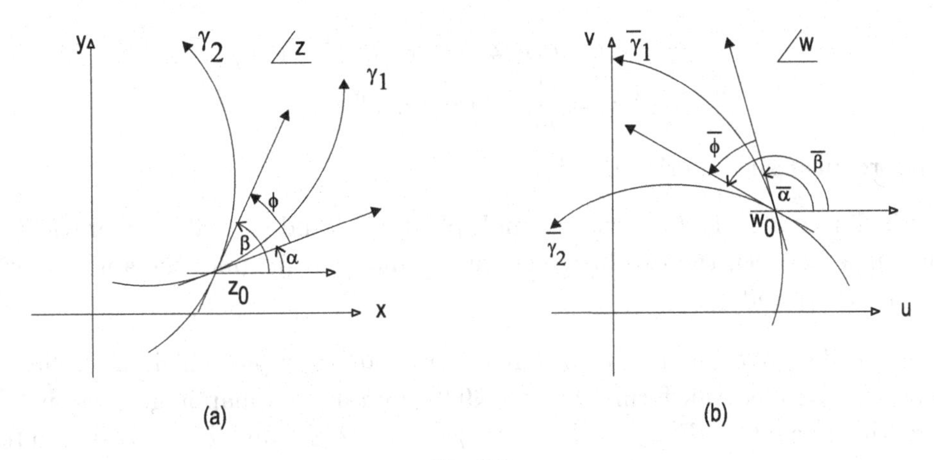

Fig. 9.2

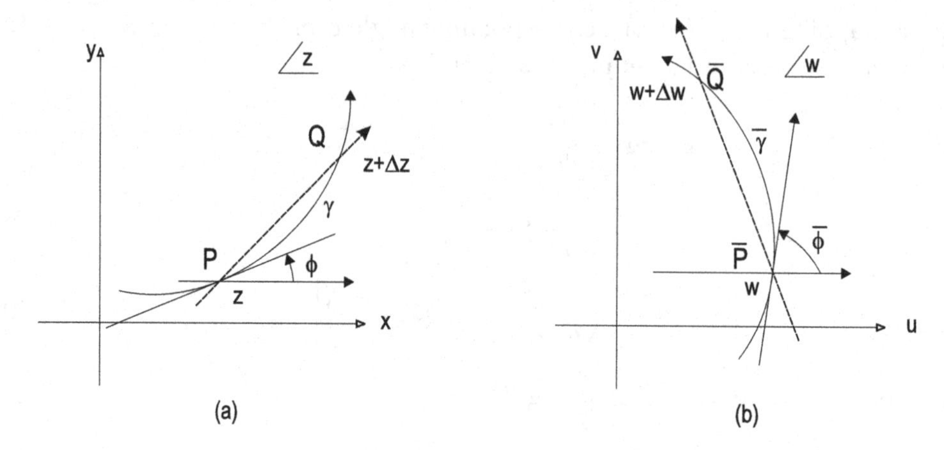

Fig. 9.3

points $\bar{\mathrm{P}}$ and $\bar{\mathrm{Q}}$ by

$$\Delta w = f(z + \Delta z) - f(z), \tag{9.9}$$

$$\Delta w \approx f'(z)\Delta z. \tag{9.10}$$

Thus, the length $|\Delta z|$ of PQ gets scaled by a factor of $|f'(z)|$, and small areas get scaled by $|f'(z)|^2$. Note that, by considering $\arg(w)$, Eq. (9.8) again follows from Eq. (9.10).

A point z_0 is called a *critical point* of a function $f(z)$ if its derivative vanishes at that point. Near a critical point a mapping is not conformal. Let us assume that at a critical point the first $(m-1)$ derivatives vanish, $f'(z_0), f''(z_0), \ldots, f^{(m-1)}(z_0)$, but $f^{(m)}(z_0) \neq 0$. Using the Taylor series we get

$$\Delta w \approx f(z_0 + \Delta z) - f(z_0) \approx \frac{(\Delta z)^m}{m!} f^{(m)}(z_0). \tag{9.11}$$

Thus,

$$\arg(\Delta w) = m \arg(\Delta z) + \arg[f^m(z_0)], \tag{9.12}$$

$$|\Delta w| = \frac{1}{m!}|f^m(z_0)||\Delta z|^m. \tag{9.13}$$

Therefore, near such a critical point,

- The inclination of the tangent gets multiplied by m and is rotated by $\arg[f^m(z_0)]$;
- The angle between the two image curves γ_1' and γ_2' is m times the angle between the curves γ_1 and γ_2.

An example. We shall now briefly discuss an example which underlies the Schwarz–Christoffel transformation (see §9.8), by taking a mapping function $F(z)$, having the derivative $F'(z) = \lambda(z - x_0)^{-b}$, where x_0 and b are real numbers. Assuming that b is not an integer, $F'(z)$ has a branch point at x_0 and, for the

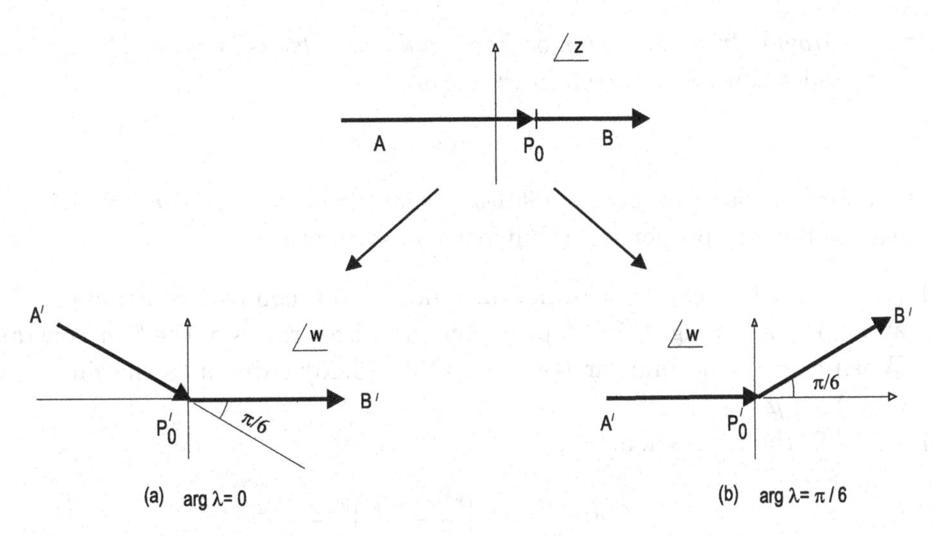

Fig. 9.4

sake of definiteness, we restrict $\arg(z - x_0)$ to the range $(\frac{-\pi}{2}, \frac{3\pi}{2})$. The value of $\arg(z - x_0)$ remains π when z is real and lies to the left of x_0, shown as point P_0 in Fig. 9.4. As the point z moves toward the right and crosses the point x_0, the value of $\arg(z - x_0)$ jumps to 0 and remains constant for all z real and to the right of the point x_0. Using the notation $\Delta z = z - x_0$, $\Delta w = w - w_0$, $w_0 = F(x_0)$, $\lambda = F'(x_0)$, Eq. (9.10) assumes the form

$$\Delta w \approx F'(z)\Delta z = \lambda(z - x_0)^{-b}\Delta z, \tag{9.14}$$

$$\arg(\Delta w) = \begin{cases} \arg(\Delta z) + \arg(\lambda) - b\pi, & \text{if } z \text{ is real and lies to the left of } x_0, \\ \arg(\Delta z) + \arg(\lambda), & \text{if } z \text{ is real and lies to the right of } x_0. \end{cases} \tag{9.15}$$

In Figs. 9.4(a) and 9.4(b) we show the images of the real line as one traverses from left to right for $b = 1/6$ and two values $\arg(\lambda) = 0$ and $\arg(\lambda) = \pi/6$, respectively.

This explains the basic idea behind Schwarz–Christoffel transformation, to be discussed in §9.8, where we will see that by taking $f'(z)$ as a suitable product $\lambda \prod_k (z - x_k)^{-b_k}$, one obtains a mapping $f(z)$ of the real line onto a specified polygon. It is to be noted that the different values $\arg(\lambda)$ produce images with different orientations.

§9.2 Bilinear Transformations

A mapping of the form

$$w = \frac{\alpha z + \beta}{\gamma z + \delta}, \qquad \alpha\delta - \beta\gamma \neq 0, \tag{9.16}$$

is known as *Mobius transformation* or *fractional linear transformation.* The relation between w and z can be rewritten in the form

$$\gamma wz + \delta w - \alpha z - \beta = 0 \tag{9.17}$$

and, therefore, it is also called *bilinear transformation.* In the following, we summarize important properties of bilinear transformations.

(*1) Although a bilinear transformation contains four complex constants, α, β, γ, and δ, the mapping (9.16) depends on only three ratios of the four constants.

(*2) When $\gamma = 0$, the bilinear transformation (9.16) reduces to the linear form $w = \lambda z + \mu$.

(*3) If $\gamma \neq 0$, the expression

$$w = \frac{\alpha}{\gamma} - \left(\frac{\alpha\delta - \beta\gamma}{\gamma}\right)\frac{1}{\gamma z + \delta} \tag{9.18}$$

shows that (9.16) can be thought of as a composition of a linear transformation $u = \gamma z + \delta$, an inversion $v = 1/u$, followed by a linear transformation $w = \frac{\alpha}{\gamma} - \frac{\alpha\delta-\beta\gamma}{\gamma} v$. This decomposition property is extremely useful and will be essential to our understanding of several other properties.

(*4) The bilinear transformation (9.16) maps the point $z = \frac{-\delta}{\gamma}$ to the point at infinity, and the image of the point at infinity in the z plane is the point $\frac{\alpha}{\gamma}$. Equation (9.16) gives a one-to-one mapping of the extended z plane onto the extended w plane. Many of the statements below will need special care when z, or w, is infinity. The reader is urged to be attentive and to explore these special cases separately.

(*5) One of the most important properties of a bilinear transformation is that the image of a circle or a straight line is again a circle or a straight line. The easiest way to see this property is to note that the equation for every circle, and every straight line, can be written in the form

$$az\bar{z} + \bar{\lambda}z + \lambda\bar{z} + b = 0. \tag{9.19}$$

It is now seen that, under a bilinear transformation, Eq. (9.19) is transformed into a equation of the same form in w.

(*6) If we use *generalized circle* as a common term for both circles and straight lines; the result (*5) states that the image of every generalized circle under every bilinear transformation is again a generalized circle.

(*7) It is obvious that every transformation from one parameter family defined by

$$\left(\frac{w - w_1}{w - w_2}\right) = \lambda\left(\frac{z - z_1}{z - z_2}\right), \qquad \lambda \in \mathbb{C}, \tag{9.20}$$

maps z_1 and z_2 into w_1 and w_2, respectively. If, in addition, it is required that the image of a point z_3 ($\neq z_1$ or z_2) be w_3, the value of λ is uniquely fixed. The bilinear transformation is then uniquely determined to be

$$\left(\frac{w - w_1}{w - w_2}\right)\left(\frac{w_3 - w_2}{w_3 - w_1}\right) = \left(\frac{z - z_1}{z - z_2}\right)\left(\frac{z_3 - z_2}{z_3 - z_1}\right). \tag{9.21}$$

(*8) The right hand side of Eq. (9.21) is called the *cross ratio* of complex numbers z, z_1, z_2, z_3, and is denoted as (z, z_1, z_2, z_3). The requirement (9.21) says that the cross ratio remains invariant under a bilinear transformation.

(*9) The bilinear transformation $z \to w = f(z)$, defined by

$$f(z) = (z, z_1, z_2, z_3), \tag{9.22}$$

carries the points z_1, z_2, z_3 into $0, \infty, 1$, respectively. This fact is helpful in remembering the expression for the cross ratio.

(*10) A general, invertible, linear transformation on two complex variables z_1 and z_2 can be written in the form

$$\begin{pmatrix} w_1 \\ w_2 \end{pmatrix} = \begin{pmatrix} \alpha & \beta \\ \gamma & \delta \end{pmatrix} \begin{pmatrix} z_1 \\ z_2 \end{pmatrix}, \qquad \alpha\delta - \beta\gamma \neq 0. \tag{9.23}$$

If we define $z = z_1/z_2$ and $w = w_1/w_2$, the above transformation gives rise to a fractional transformation between z and w.

(*11) It is easy to check that the inverse of a bilinear mapping is again a bilinear mapping and that the result of two successive bilinear transformations is again a bilinear transformation. Also, the mapping composition satisfies the associative property $f \circ (g \circ h) = (f \circ g) \circ h$. These properties, together with the fact that the identity map $w = z$ is also a bilinear map, are summarized by saying that the set of all bilinear transformations of the extended complex plane is a group.

(*12) The image of a point under reflection in a straight line, or a circle, was defined in §1.4. A pair of points was defined to be a symmetric pair w.r.t. a straight line, or a circle, if the two points are reflections of each other. Let Γ be a generalized circle and Γ' be its image under a bilinear transformation. The bilinear maps have the property, known as the *symmetry principle*, that every pair of points P, Q symmetric w.r.t. Γ is mapped into a symmetric pair P′, Q′ w.r.t. the image Γ'. The symmetry principle is very useful for finding bilinear maps.

Quite often we want to find the image of a subset S of the complex plane. For conformal mappings $z \to w = f(z)$, it turns out to be sufficient to find the image of the boundary of S and then use the following rule to determine on which side of the boundary the image region lies.

✍ *A conformal map $z \to w = f(z)$ has the property that it preserves orientation. While moving along an arc of the boundary of the subset S, if S lies on the left, the image region in the w plane will also lie on the left of the image boundary.*

↪The proof of the above rule is left as an exercise for the reader. See [Polya (1974)].

§9.3 Examples

It has been noted that every bilinear transformation is a combination of linear maps and the inversion map $w = 1/z$. In the following we give examples of properties of the inversion map and the general bilinear map.

⚡9.3.1 Short examples. In this set of examples of the inversion map, we use the notation $w = 1/z$ and

$$z = x + iy, \qquad w = u + iv, \tag{9.24}$$

$$z = r\exp(i\theta), \qquad w = \rho\exp(i\phi). \tag{9.25}$$

The relation $w = 1/z$ implies that

$$u = \frac{x}{x^2+y^2}, \qquad v = \frac{-y}{x^2+y^2}, \tag{9.26}$$

$$\rho = 1/r, \quad \phi = -\theta. \tag{9.27}$$

(a) A point in the first quadrant, $x > 0, y > 0$, is mapped onto a point $u > 0, v < 0$. Hence, the image of the first quadrant of the z plane is the fourth quadrant of the w plane. Similarly, a point in the second quadrant is mapped onto a point in the third quadrant; see Fig. 9.5.

(b) A ray drawn from the origin $\theta = \alpha$ is transformed into a ray $\phi = -\alpha$ from the origin, and concentric circles with the center $z = 0$ are transformed into concentric circles with the center again at $w = 0$.

(c) The family of lines $\Gamma_a = \{z|\text{Re}\, z = a,\ a \neq 0\}$, parallel to the y axis under the map $w = 1/z$, is transformed into a family of circles touching at the origin. This can be seen as follows. For the image of Γ_a under the inversion map, using Eq. (9.26) we get

$$\frac{u}{u^2+v^2} = a \quad \text{or} \quad u^2 + v^2 = \frac{u}{a}, \tag{9.28}$$

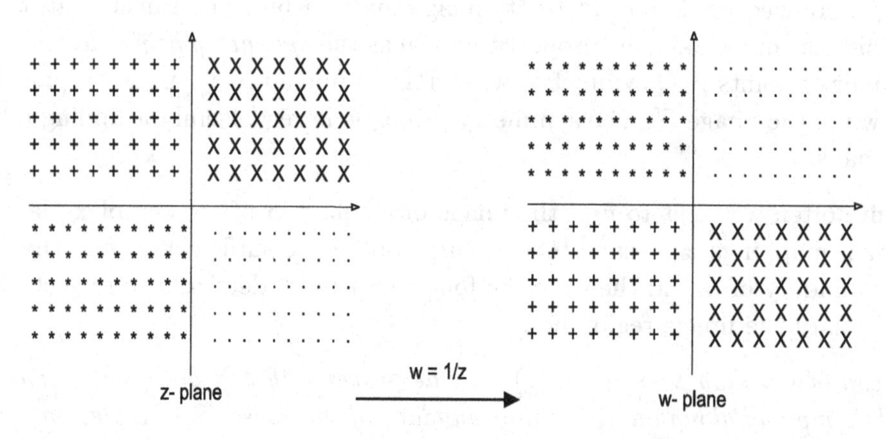

Fig. 9.5

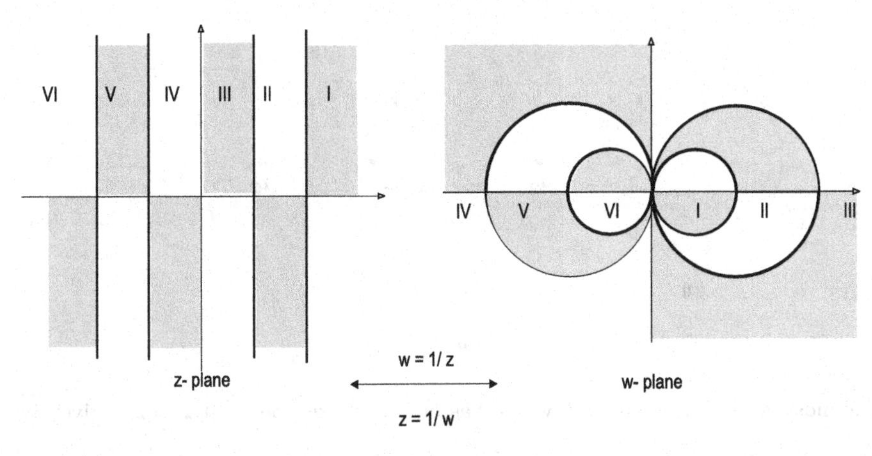

Fig. 9.6

which is the circle

$$\left(u - \frac{1}{2a}\right)^2 + v^2 = \frac{1}{4a^2}, \tag{9.29}$$

with the center at $w = 1/2a$ and with radius $1/4a^2$. The images of several lines parallel to the y axis are circles touching at the origin, as shown in Fig. 9.6.

(d) It is instructive to get the previous result using general properties of bilinear maps. To do this we note that, for the inversion map,
- The image of the real axis in the z plane is again the real axis in the w plane;
- The image Γ'_a must pass through $w = 0$, which corresponds to $z = \infty$;
- The image Γ'_a does not go through $w = \infty$, because the point $z = 0$ is not on Γ_a, and hence Γ'_a must be a circle;
- The map $w = 1/z$ is conformal at $z = a$ and the line Γ_a intersects the real axis at an angle $\pi/2$. Here the image circle Γ'_a and the real axis in the w plane must intersect at $\pi/2$, the point of intersection being $w = 1/a$. This is possible only when the circle Γ'_a has radius $1/2a$ and its center on the real axis itself is at $w = 1/2a$.

(e) The point $z = 0$ is mapped onto a point at infinity, and thus the half plane $\{z|x < a, a > 0\}$ is mapped onto the exterior of the circle Γ'_a. Similar conclusions hold for the lines parallel to the x axis.

⇛ Tutorial §§9.1, on the inversion map.

We give examples on bilinear mappings. It is recommended that the reader may revise §1.14 on bilinear transformation from the first chapter.

ƒ9.3.2 Short examples. We give a few simple examples of finding bilinear transformations:

(a) The inverse of the map $w = \frac{z+1}{z-1}$, obtained by solving for z, is the "same map" $z = \frac{w+1}{w-1}$, whereas the inverse of $w = \frac{z-1}{z+1}$ is the map $z = -\frac{w+1}{w-1}$.

(b) A bilinear map which sends $\{-1, -2, -1+i\}$ into $\{-i, 1-3i, -1+i\}$, using Eq. (9.20), is given by $\frac{w+i}{w-1+3i} = \lambda\frac{z+1}{z+2}$, where λ is obtained by substituting $z = -1+i, w = -1+i$ and solving for λ. This gives $w = \frac{2z+1-i}{z+2-i}$.

(c) The fixed points of the map $w = \frac{2z+3}{3z+2}$ are obtained by demanding $z = \frac{2z+3}{3z+2}$, giving $z = -1, 1$.

(d) A general map with two fixed points 2 and $-3i$ is $\frac{w-2}{w+3i} = \lambda\frac{z-2}{z+3i}$.

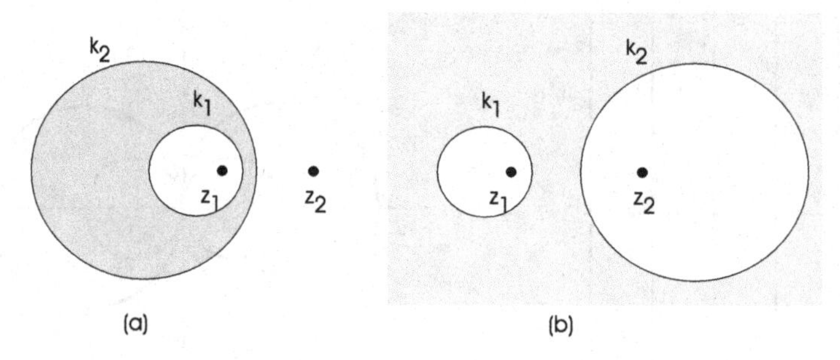

Fig. 9.7

(e) The most general bilinear map which has only one fixed point at $z = 2$ is given by $\frac{1}{w-2} = \frac{\lambda}{z-2} + \mu$.

(f) Two nonintersecting circles, such as those shown in Fig. 9.7(a) and in Fig. 9.7(b), can be thought of as members of the family of circles (see ⚡1.9) given by

$$\left|\frac{z - z_1}{z - z_2}\right| = k, \qquad k \in \mathbb{R}, \tag{9.30}$$

with fixed z_1 and z_2 and the two values $k = k_1, k_2$. Therefore, under a bilinear map $w = \frac{z-z_1}{z-z_2}$ the region between the circles will be transformed into the annular region between the two concentric circles $w = k_1$ and $w = k_2$ in the w plane.

(g) Similarly, a circular arc, with end points at z_1 and z_2, is given by an equation of the form (see ⚡1.7)

$$\arg\left(\frac{z - z_1}{z - z_2}\right) = \alpha. \tag{9.31}$$

Therefore, the map $w = \frac{z-z_1}{z-z_2}$ transforms lens-like regions in Fig. 9.8(a)-(b) into a sector $\alpha_1 < \arg(w) < \alpha_2$ in the w plane.

Our next two examples illustrate explicit construction of these bilinear mappings.

☑**9.3.3 Problem.** Find a bilinear transformation which maps the region between two circles

$$|z| = 1, \quad |z + 1| = \frac{5}{2} \tag{9.32}$$

onto an annular domain between two *concentric circles*. Determine the radii of the two image circles with a common center at the origin.

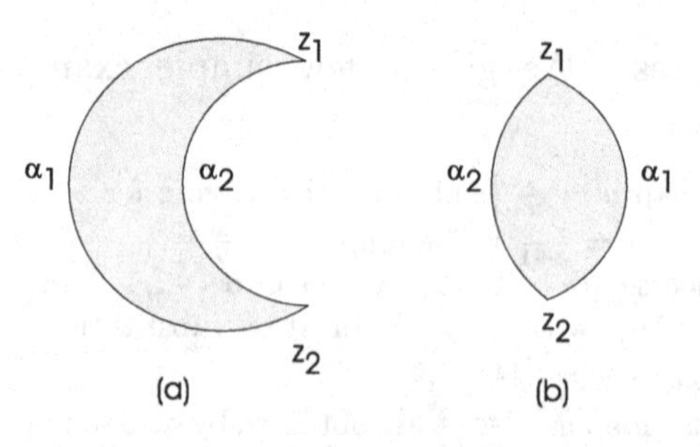

Fig. 9.8

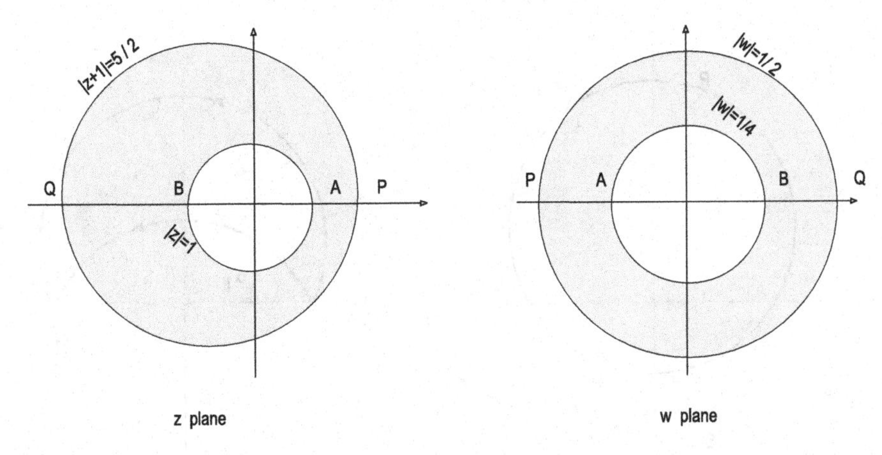

Fig. 9.9

Solution. We shall first find a pair of points a and b which are symmetric w.r.t. both the circles. To get the images to be concentric circles, we choose the map in such a way that a and b are mapped onto $0, \infty$. The common symmetric pair of points, a and b, must lie on the line joining the centers of the two circles, which is the real line. Thus a and b must be real and satisfy

$$ab = 1, \qquad (a+1)(b+1) = \frac{25}{4}. \tag{9.33}$$

This gives two solutions:

$$a = \frac{1}{4}, \quad b = 4 \qquad \text{or} \qquad a = 4, \quad b = \frac{1}{4}. \tag{9.34}$$

Therefore, using Eq. (9.16), a mapping which transforms $1/4, 4$ into $0, \infty$ is

$$w = \frac{z - \frac{1}{4}}{z - 4}. \tag{9.35}$$

To draw the image circles, we note that the conformal property implies that the images of the diameters AB and PQ will again be diameters of the image circles; see Fig. 9.9. Noting that for

$$z_{\rm A} = 1, \quad z_{\rm B} = -1, \quad z_{\rm P} = 1.5, \quad z_{\rm Q} = -3.5, \tag{9.36}$$

we get

$$w_{\rm A} = -\frac{1}{4}, \quad w_{\rm B} = \frac{1}{4}, \quad w_{\rm P} = -\frac{1}{2}, \quad w_{\rm Q} = \frac{1}{2}, \tag{9.37}$$

hence, the image circles have radii $1/4$ and $1/2$.

The map found is not unique; several solutions are possible. A second solution is $w = \frac{z-4}{z-1/4}$. The reader should investigate in what way the two solutions found here are different. ∎

☑9.3.4 Problem. For the bilinear map $w = \frac{1+z}{1-z}$,

(a) Find the images of the circular arcs ABC and ADC of the circle ABCDA in Fig. 9.10(a) passing through the points -1 and 1;
(b) Verify the conformal property of the map at $z = -1$ for the circular arcs ABC and ADC in part (a);

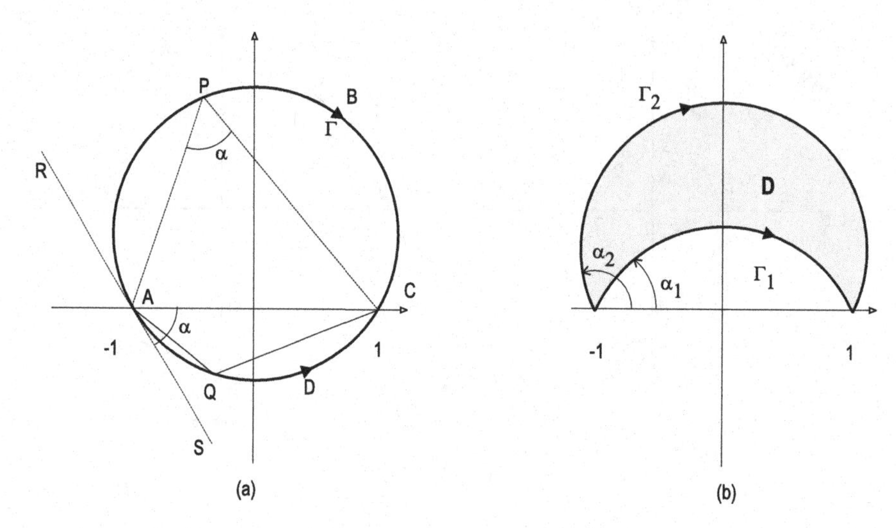

Fig. 9.10

(c) Describe the image of the lens-like domain bounded by two circular arcs through the points -1 and 1, as shown in Fig. 9.10(b).

Solution.

(a) The points $z = \pm 1$ are on the circle Γ and hence the image Γ' passes through the corresponding points $w = 0$ and ∞, and therefore the image will be a straight line passing through $w = 0$. To locate the images of the two arcs, ABC and ADC, note that for a point P on the arc APC ($z = z_{\text{P}}$),

$$\arg w_{\text{P}} = \arg\left(\frac{1+z_{\text{P}}}{1-z_{\text{P}}}\right) = \pi + \arg\left(\frac{z_{\text{P}}+1}{z_{\text{P}}-1}\right) \tag{9.38}$$

$$= \pi + \arg(z_{\text{P}}+1) - \arg(z_{\text{P}}-1) \tag{9.39}$$

$$= \pi + \angle\text{PAC} - (\pi - \angle\text{PCA}) \tag{9.40}$$

$$= \angle\text{CAP} + \angle\text{PCA} \tag{9.41}$$

$$= \pi - \alpha. \tag{9.42}$$

Thus, all points are on the arc ABC are mapped onto the ray $\arg(w) = \pi - \alpha$. Similarly, the points on the arc ADC are mapped onto the ray $\arg w = -\alpha$ in the w plane in the manner shown Fig. 9.11(a). *For circles passing through both of the points ± 1, the image is "tangent to the circle" at the point -1.*

(b) Simple geometrical arguments — a theorem on alternate segments — show that the angles of intersections of the two arcs with the real axis are

$$\angle\text{CAR} = \pi - \alpha, \qquad \angle\text{SAC} = \alpha, \tag{9.43}$$

and are equal to the corresponding angles of the image with the real axis in the w plane; see Fig. 9.11(a). This verifies the conformal property of the map at $z = -1$.

(c) It is now clear that the image of the lens-like domain D in Fig. 9.10(b) will be a sector bounded by the rays $\arg w = \alpha_1$ and $\arg w = \alpha_2$. Since the image of the point $z = -1$ is $w = 0$, the sector D' as shown in Fig. 9.11(b) is the image of the domain D.

∎

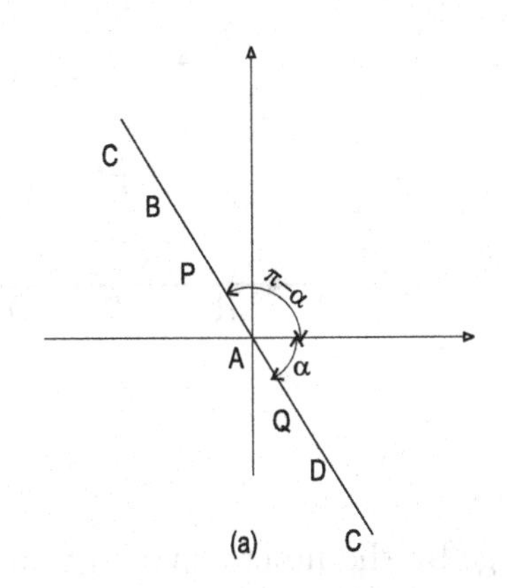

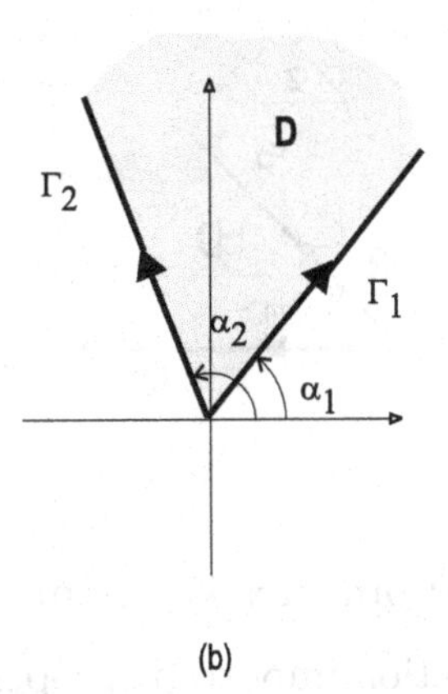

Fig. 9.11

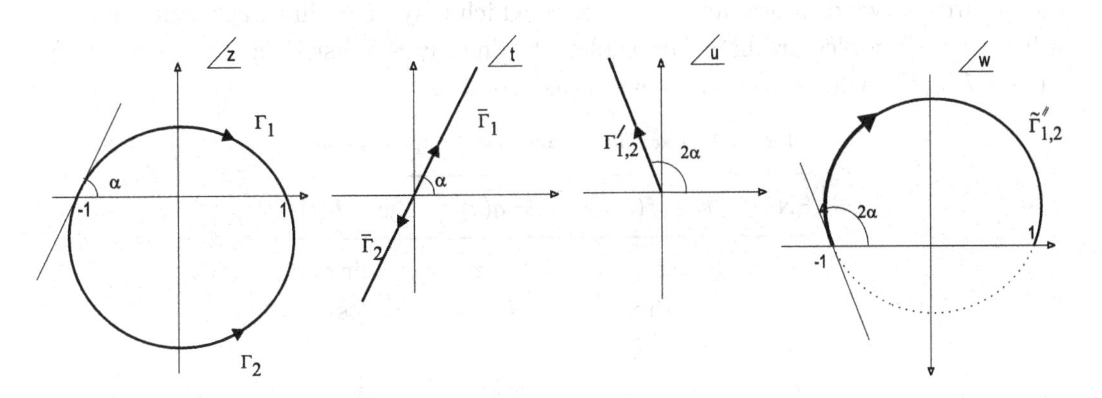

Fig. 9.12

☑**9.3.5 Problem.** Find the image of the arcs Γ_1 and Γ_2 of a circle, passing through the two points $z = \pm 1$ see Fig. 9.12, under composition of the maps

$$t = \frac{1+z}{1-z}, \quad u = t^2, \quad w = \frac{u+1}{u-1}. \tag{9.44}$$

Solution. Recall the result from the above example, ☑9.4.4, and note that circle passing through $z = \pm 1$ is transformed into a straight line through $t = 0$ at angle α with the real axis. The map $u = t^2$ folds the two rays $\arg t = \alpha$ and $\arg t = \pi + \alpha$ into a single ray, $\arg u = 2\alpha$, in the u plane. The map $w = \frac{u+1}{u-1}$ transforms the ray $\arg u = 2\alpha$ back to a circular arc passing through $w = \pm 1$ and intersecting the real axis at angle 2α, as shown in Fig. 9.12. Note that the images of the two arcs Γ_1 and Γ_2 coincide in the u plane, and also in the w plane.

It may be noted here that the composition of maps in Eq. (9.44) is equivalent to the Joukowski map, $w = \frac{1}{2}(z + \frac{1}{z})$, to be discussed in detail in §9.6. ∎

⊳ Exercise §§9.2, on finding images under a bilinear transformation.

⊳ Exercise §§9.3, on bilinear transformations satisfying a given condition.

⊳ Questions §§9.4, on bilinear mappings.

⊳ Exercise §§9.5, on using the symmetry principle for bilinear transformations.

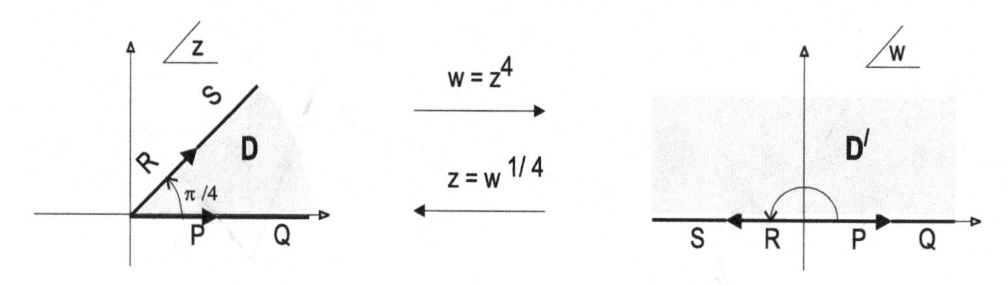

Fig. 9.13

§9.4 Mapping by Elementary Functions

In this section important properties of mappings by the functions z^n, e^z, and $\sin z$, along with their inverse functions, will be described. The mapping properties of other trigonometric and hyperbolic functions can be obtained by regarding them as composites of two or more functions, one of which may be a bilinear transformation; a few such examples are listed in Table 9.1. The maps f listed in the table satisfy $f(z) = h(g(z))$, where h is a bilinear transformation.

Table 9.1 Examples of com-posite maps $f = hog$.

SN	$w = f(z)$	$t = g(z)$	$w = h(t)$
1	$\cos z$	$\frac{\pi}{2} - z$	$\sin t$
2	$\cosh z$	iz	$\cos t$
3	$\coth \frac{z}{2}$	e^z	$\frac{t+1}{t-1}$
4	$\tan^2 \frac{z}{2}$	$\cos z$	$\frac{1-t}{1+t}$

§9.4.1 *Mapping* $w = z^n$

The relation $|w| = |z|^n$ shows that the interior of the circle $|z| < R$ is mapped onto the open disk $|w| < R^n$.

The map $w = z^n$ is conformal at all points except $z = 0$, which is a critical point. The angle at $w = 0$ between the image curves is n times the angle at $z = 0$ between the corresponding curves in the z plane. Thus, a sector of angle α ($\alpha < \frac{2\pi}{n}$) is mapped onto a sector of angle $n\alpha$ in the w plane.

The map $w = z^n$ is not a one-to-one mapping. A sector of angle α is mapped in a one-to-one fashion onto a sector of angle $n\alpha$ if $\alpha < 2\pi/n$. Thus, for example, z^4 maps the sector $0 < \theta < \pi/4$ onto the upper half plane, as shown in Fig. 9.13. The angle at $z = 0$ between the rays PQ and RS gets multiplied by 4. It is obvious that under the inverse map $z = w^{\frac{1}{4}}$ the angle at $w = 0$ will change by a factor of 1/4.

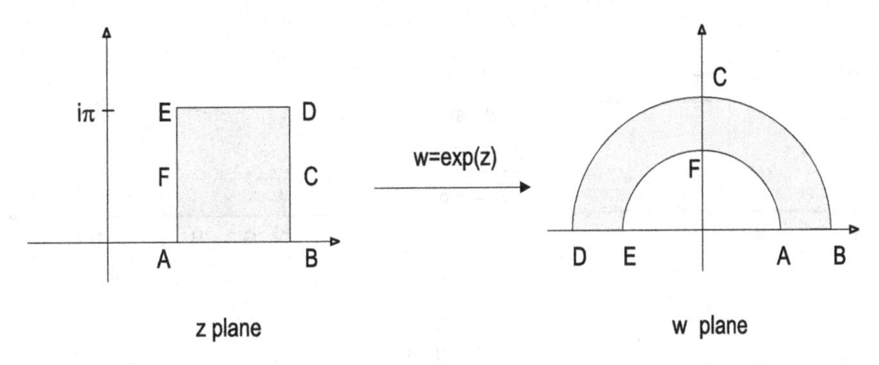

Fig. 9.14

§9.4.2 *Exponential map*

The map by exponential function, $w = e^z$, is analytic everywhere and its derivative is nonzero. Hence, it is conformal everywhere in the complex plane. With $z = x + iy, w = \rho e^{i\phi}$, one has

$$\rho = e^x, \quad \phi = y \qquad \text{mod } 2\pi. \tag{9.45}$$

The family of straight lines $\Gamma_a = \{z|x = a, a \neq 0\}$ is mapped onto a family of concentric circles, $\rho = \text{const}$, with the center at $w = 0$. The images of the lines $y = b$, parallel to the x axis, are the set of rays $\phi = b$ through the origin. Thus, a small rectangular region bounded by ABCDEFA, of height π, is mapped onto a region bounded by ABCDEFA, as shown in Fig. 9.14.

The map $w = e^z$ is not a one-to-one map, because the exponential function is periodic $e^{z+2\pi i} = e^z$. The points with $y \pm 2n\pi i$ (n = integer) have the same image. However, an infinite strip parallel to the x axis and with the width less than 2π is mapped in a one-to-one fashion onto a sector. As an example, the upper half w plane is a one-to-one image of the strip $\{z|0 < y < \pi\}$, as shown in Fig. 9.15.

The image of the left half (right half) of this strip is the interior (exterior) of the semicircle in the upper half plane, as in Fig. 9.16.

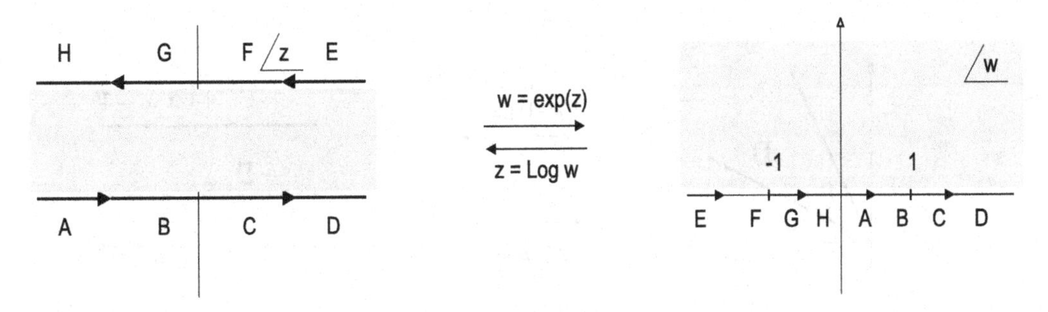

Fig. 9.15

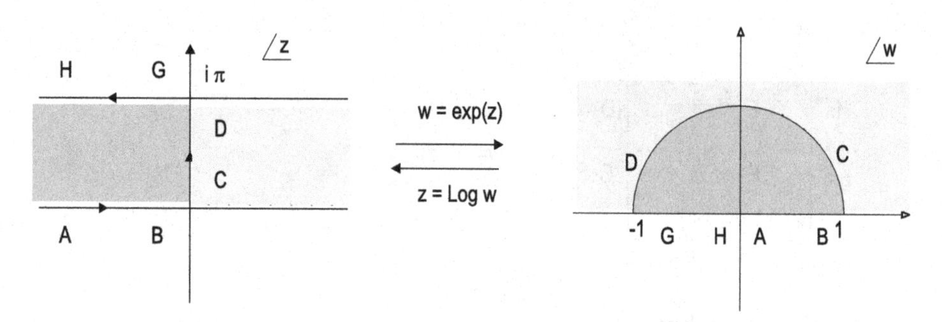

Fig. 9.16

§9.4.3 *Map* $w = \operatorname{Log} z$

The mapping $w = \operatorname{Log} z$ is defined in the complex plane with the cut along the negative real axis:

$$\operatorname{Log} z = \ln r + i\theta, \quad -\pi < \theta \leq \pi.$$

It is the inverse of the exponential map and its properties can be obtained by interchanging the roles of the z and w planes in figures for the exponential map.

An important and useful property of the map $\operatorname{Log} z$ is that a sector $\{z|\alpha_1 < \arg z < \alpha_2\}$ is transformed into an infinite strip $\{w|\alpha_1 < \operatorname{Im} w < \alpha_2\}$, where α_1 and α_2 are in range $(-\pi, \pi)$; see Fig. 9.17. As a special case taking $\alpha_1 = 0$ and $\alpha_2 = \pi$, we see that the function $\operatorname{Log} z$ maps the upper half plane onto a strip bounded by $\operatorname{Im} w = 0, \pi$, as can be inferred from Fig. 9.15.

§9.4.4 *Map* $w = \sin z$

The function $f(z) = \sin z$ is analytic everywhere and hence represents a conformal mapping except at the critical points $z = (2n+1)\pi/2$, where the derivative $f'(z)$ vanishes. Separating the real and imaginary parts of $\sin z$, we get

$$u = \sin x \cosh y, \quad v = \cos x \sinh y. \tag{9.46}$$

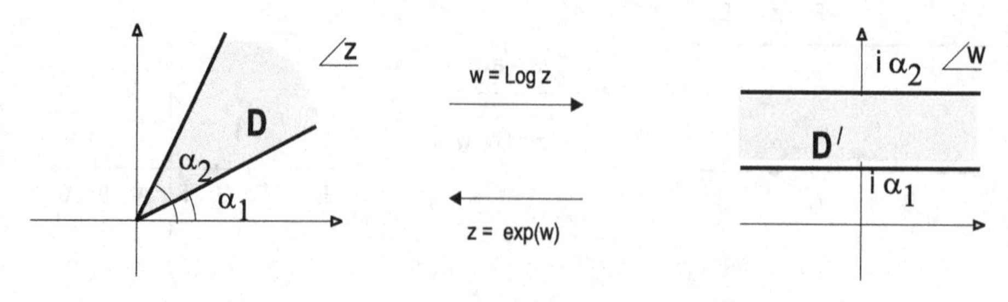

Fig. 9.17

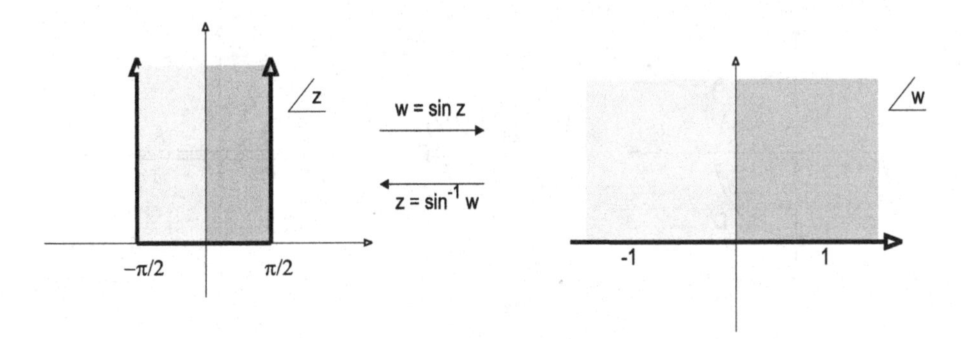

Fig. 9.18

Hence, the line $x = a$ is mapped into a hyperbola,

$$\frac{u^2}{\sin^2 a} - \frac{v^2}{\cos^2 a} = 1, \tag{9.47}$$

and the line $y = b$ is mapped into an ellipse,

$$\frac{u^2}{\cosh^2 b} + \frac{v^2}{\sinh^2 b} = 1. \tag{9.48}$$

Since $\sin(z + 2\pi) = \sin z$ and $\sin(\pi - z) = \sin z$, the $\sin z$ mapping is not a one-to-one mapping. However, on a semi-infinite strip such as $\{-\pi/2 < x < \pi/2, y > 0\}$, the map is one-to-one and the image is the upper half w plane, as shown in Fig. 9.18.

⊳ Exercise §§9.6, on mapping by elementary functions and their composites.

⊳ Exercise §§9.7, on finding maps which transform a given domain into another specified one.

§9.5 Joukowski Map

The map by the function $f(z) = \frac{a}{2}(z + 1/z)$, called the Joukowski function, is conformal everywhere except at $z = 0$ and ± 1; while the point $z = 0$ is a singular point, the points $z = \pm 1$ are critical points of $f(z)$. At the two critical points $f' = 0$ but $f'' \neq 0$, and hence the angles are doubled. [See remarks after Eq. (9.13)].

A useful representation of the Joukowski map is

$$\frac{w + a}{w - a} = \left(\frac{z + 1}{1 - z}\right)^2, \tag{9.49}$$

which shows that it is a result of three consecutive maps: $z \to t = \frac{z+1}{1-z}$, $t \to u = t^2$, and $u \to w = a(\frac{u+1}{u-1})$.

The Joukowski function satisfies $f(z) = f(1/z)$ and hence the map is not one-to-one in any domain that contains both of the point z and $1/z$. It is one-to-one in every domain D, which is such that for every $z \in D$ the point $1/z$ is not in the domain.

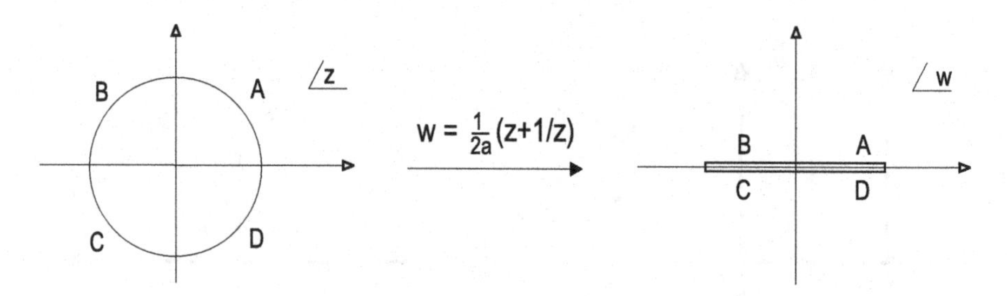

Fig. 9.19

If u and v are the real and imaginary parts of $w = f(z)$, we would get

$$u = \frac{a}{2}\left(r + \frac{1}{r}\right)\cos\theta, \quad v = \frac{a}{2}\left(r - \frac{1}{r}\right)\sin\theta, \tag{9.50}$$

where $z = re^{i\theta}$. These equations can be used to obtain the following properties of the Joukowski map.

The unit circle passes through both of the critical points ± 1. The interior of the unit circle, $\{z = e^{i\theta} | 0 \leq \theta \leq 2\pi\}$, is mapped onto the line segment from $w = a$ to $-a$ in a two-to-one manner, as shown in Fig. 9.19. The family of circles with $|z| = R$ is mapped onto a family of ellipses:

$$\frac{u^2}{a^2(R+\frac{1}{R})^2} + \frac{v^2}{a^2(R-\frac{1}{R})^2} = \frac{1}{4}, \qquad R \neq 1. \tag{9.51}$$

In Fig. 9.20 we show the images of circles for $R > 1$. A circle with the center at the origin is mapped onto an ellipse with the foci at $w = \pm a$. In the limiting case $R \to 1$, the Joukowski map transforms the exterior of the unit circle, in a one-to-one fashion, onto the w plane with a cut from $w = -a$ to $w = a$. Fig. 9.21 shows how the interior and the exterior of the unit circle in the upper half (lower half) plane get mapped onto the cut co plane.

The rays $\theta = 0, \pi$, i.e. the positive and the negative real axis, pass through one of the two critical points, $z = 1$ or $z = -1$, where the angles are doubled. Also, as one travels from 0 to 1 and then to ∞, along the real axis in the z plane, the image

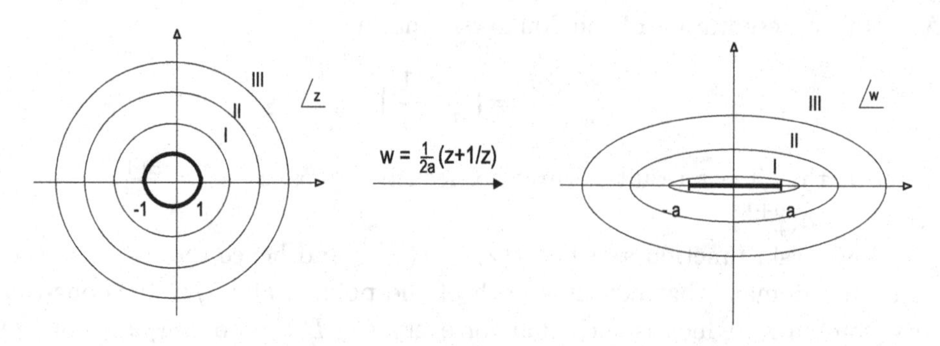

Fig. 9.20

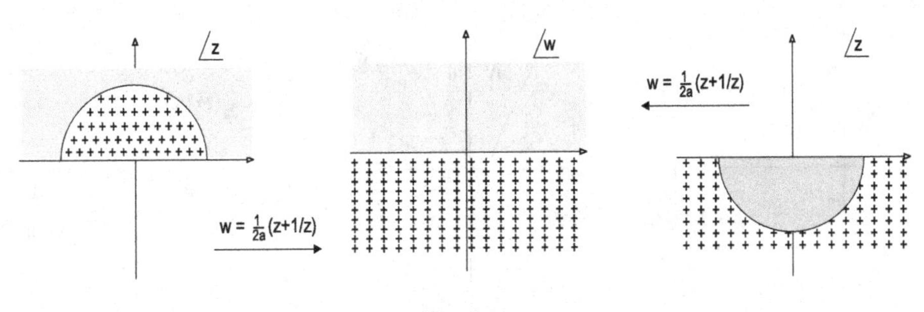

Fig. 9.21

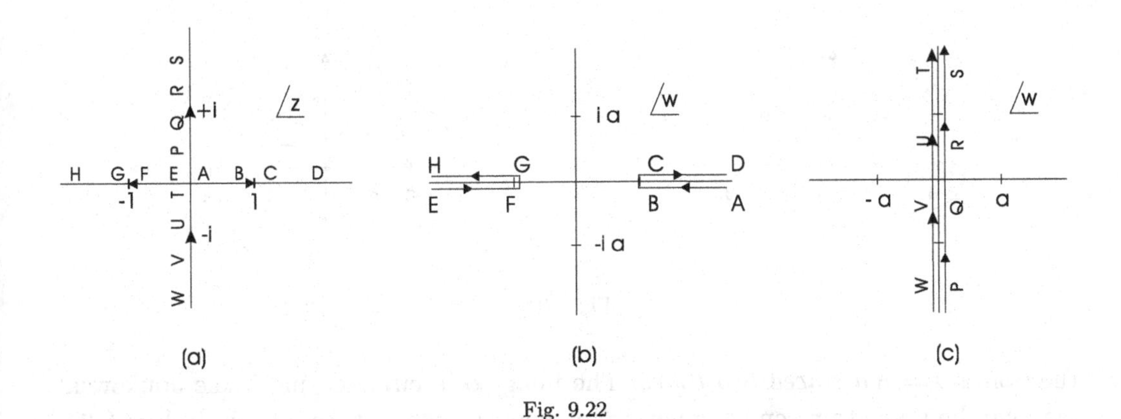

Fig. 9.22

of the point moves from $w = \infty$ to a and back to ∞ along the real axis in the w plane. Similarly, the negative x axis, $\theta = \pi$, is mapped doubly onto the half line segment running from $w = -a$ to $w = -\infty$. For a set of chosen points A, B, C, D, E, F, G, H on the real axis [Fig. 9.22(a)], the image points are shown in Fig. 9.22(b). Different points on the positive imaginary axis, as well as the negative imaginary axis, are mapped in a one-to-one fashion. For a set of selected points P, Q, R, S, W, V, U, T on the imaginary axis in Fig. 9.22(a), the transformed points are shown in Fig. 9.22(c).

A ray $\theta = \alpha$, other than the axes, is transformed into the hyperbola

$$\frac{u^2}{a^2\cos^2\alpha} - \frac{v^2}{a^2\sin^2\alpha} = 1, \qquad \alpha \neq 0, \pm\frac{\pi}{2}, \pi. \tag{9.52}$$

The family of rays $\{z | \arg(z) = \alpha\}$, having $0 < \alpha < \pi$, is mapped onto a family of confocal hyperbolas with the foci at $w = \pm a$ in a one-to-one fashion. The family of rays with $-\pi < \alpha < 0$ is mapped onto the same family of hyperbolas. These mappings are shown in Fig. 9.23. The limiting case $\alpha \to 0$ gives a one-to-one map from the upper half plane onto the entire w plane with slits running from a to ∞ and $-a$ to $-\infty$ along the real axis.

In an earlier example, ☑9.4.5, it has been shown that every circle that passes through both, the critical points $z = \pm 1$ is mapped onto a circular arc connecting

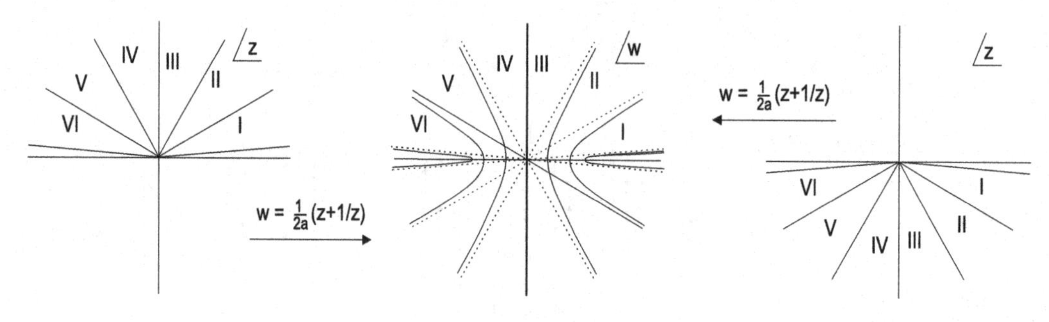

Fig. 9.23

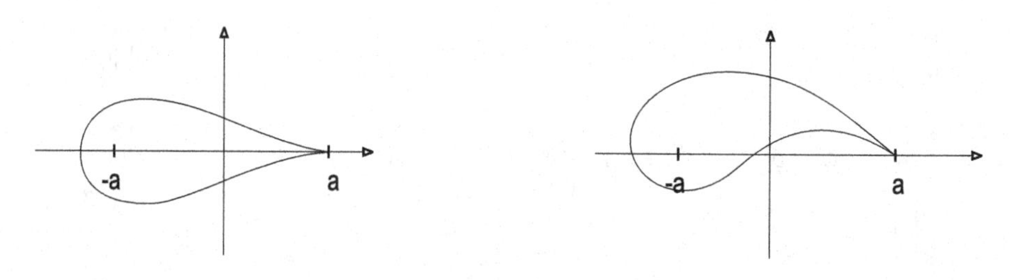

Fig. 9.24

the points $z = \pm a$ *traced two times.* The image of a circle C under the Joukowski map can be made to resemble a variety of shapes, such as those shown in Fig. 9.24, by taking C to be a circle which encloses one of the critical points and passes through the other point.

On solving

$$w = \frac{a}{2}\left(z + \frac{1}{z}\right) \tag{9.53}$$

for z, we get the inverse relation

$$z = \frac{1}{2a}\left(w + \sqrt{w^2 - a^2}\right); \tag{9.54}$$

a suitable svb for the square root must be selected when one is considering an application needing the inverse map.

Note that the Joukoswki map transforms the exterior, as well as the interior, of the unit circle onto the w plane outside the slit from $(-a, a)$. The inverse map (9.54) is double-valued and an svb of (9.54) can be selected to make it one-to-one in two different ways, described below:

- We may take the branch cut for $\sqrt{w^2 - a^2}$ to be the interval $(-a, a)$ and the svb as the one defined by taking $\sqrt{w^2 - a^2}$ as the product of the principal values for $\sqrt{z \pm a}$. In this case the exterior of the slit $(-a, a)$ in the w plane becomes a one-to-one image of the exterior of the unit circle in the z plane.

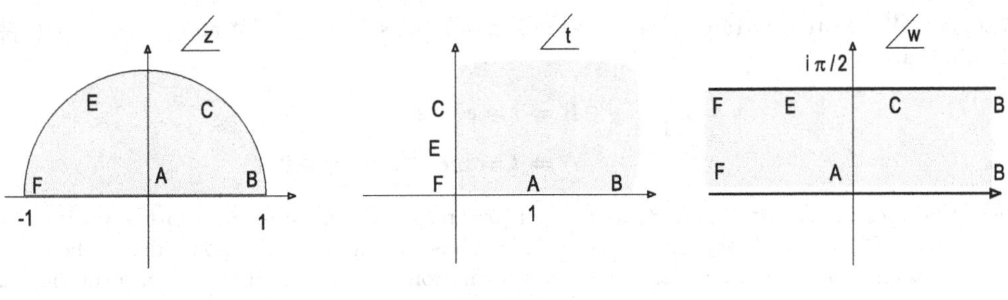

Fig. 9.25

- The Joukowski function maps the upper half plane (and the lower half plane as well) onto the w plane with slits along $(-\infty, -a)$ and (a, ∞). A choice of svb of the inverse mapping, (9.54), can be made which transforms the cut w plane onto the upper half plane (lower half plane) in a one-to-one fashion. In this case the branch cut must be selected along the two slits in the w plane.

✍ *Whenever the image set is covered twice under the Joukowski map, the result can be understood in terms of the property that the points z and $1/z$ have the same images.* ▷ *Q.[4], p. (477).*

✍ *The shapes of aerofoils obtained as images of circles under the Joukowski map can be understood by remembering that the upper and lower arcs of every circle passing through both 1 and -1 are related by the transformation $z \leftrightarrow 1/z$, and are mapped onto the same arc. A slight deformation of a circle will produce an image close to the arc consistent with the fact that angles are doubled at $z = \pm 1$ and that the Joukowski map is conformal at points other than 0 and ± 1.*

⊳ Quiz in §§9.8, on understanding how the Joukowski map gives rise to aerofoils of different shapes.

↬ Q.[5] in §§9.12, on the choice of svb for the inverse map of Joukowski mapping.

§9.6 Examples

We present a few examples of composite maps.

☑**9.6.1 Problem.** Find the image of the half disk bounded by the unit circle and lying in the upper half plane under the map $w = \mathrm{Log}\left(\frac{1+z}{1-z}\right)$.

Solution. The given map can be regarded as a result of two maps $z \to t = \frac{1+z}{1-z}$ and $t \to w = \mathrm{Log}\, t$. Using properties of the map $\mathrm{Log}\, z$ we can easily find the image, as shown in Fig. 9.25 ▮

☑**9.6.2 Problem.** Show that the map $w = \tan z$ transforms the infinite strip $\{z | -\frac{\pi}{4} < \mathrm{Re}\, z < \frac{\pi}{4}\}$ onto the unit disk $\{w | |w| < 1\}$.

Solution. We write $w = \tan z$ as a succession of two maps, $z \to t = e^{2iz}$ and $t \to w = -i(\frac{t-1}{t+1})$, and note that

$$x = \pm\frac{\pi}{4}, \quad y \in \mathbb{R} \Rightarrow t = e^{-2y}e^{\pm i\pi/2} \tag{9.55}$$

$$\Rightarrow t = \pm ie^{-2y}, \qquad y > 0. \tag{9.56}$$

Hence, the lines $\mathrm{Re}\, z = \pi/4$ and $\mathrm{Re}\, z = -\pi/4$ are, respectively, mapped onto the positive and negative parts of the imaginary axis in the t plane. Thus, the image of the boundary of the strip is the full imaginary axes of the t plane. The transformation $t \to w$ is a bilinear transformation and hence the image of the imaginary axis will be a line or circle. It cannot be a line, because it cannot extend to $w = \infty$, which is the image of $t = -1$ not lying on the $\mathrm{Im}\, t$ axis. Hence, the image in the w plane is a circle. That the circle is the unit circle $|w| = 1$ can be easily verified by noting that the images of $t = 0, i, -i$ are $w = -i, 1, -1$. ▮

§9.7 Schwarz–Christoffel Transformation

The Schwarz–Christoffel transformation (SCT) is a mapping of the real axis onto an n-sided polygon giving a one-to-one correspondence between the upper half plane and the interior of the polygon. Let $x_1, x_2, \ldots, x_n$ be points whose images under the map $w = f(z)$ are the vertices $w_1, w_2, \ldots, w_n$ of the polygon arranged in an anticlockwise fashion. Using α_k to denote interior angles, and $\beta_k = \pi - \alpha_k$ to denote exterior angles (measured anticlockwise), see Fig. (9.26), the derivative of the map $f(z)$ is given by

$$f'(z) = \lambda \prod_{k=1}^{n} (z - x_k)^{-c_k}, \tag{9.57}$$

where

$$c_k = \frac{\beta_k}{\pi} = \pi - \frac{\alpha_k}{\pi} \tag{9.58}$$

The interior angles are taken to be in the range $(0, 2\pi)$, and the exterior angles in the range $(-\pi, \pi)$. The points x_k on the real axis are taken to satisfy

$$-\infty < x_1 < x_2 < \cdots < x_n < \infty.$$

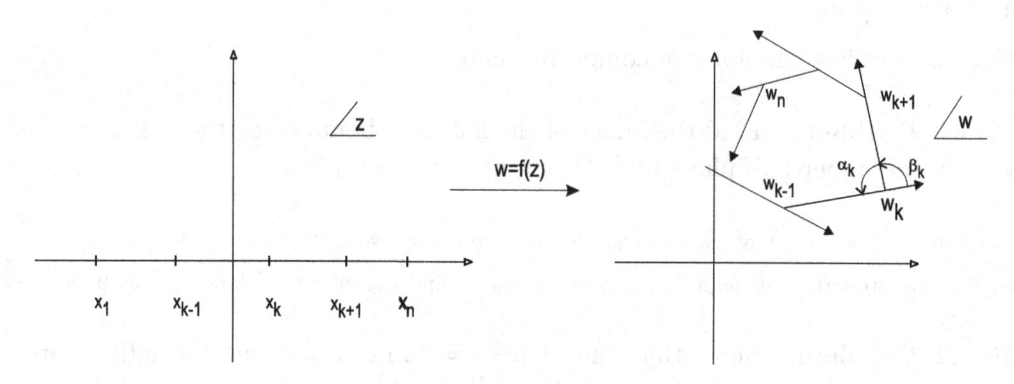

Fig. 9.26

For a closed polygon the angles α_k satisfy

$$\sum \alpha_k = (n-2)\pi, \qquad \sum \beta_k = 2\pi, \tag{9.59}$$

and hence $\sum c_k = 2$. Integrating the expression (9.57) gives the SCT:

$$f(z) = \lambda \int_{z_0}^{z} \prod_{k=1}^{n} (z - x_k)^{-c_k}\, dz + \mu. \tag{9.60}$$

We shall not give a proof and will be content with several explanatory remarks about the formula and its properties and usage.

The function $f'(z)$ becomes analytic in the upper half plane except at the points $x_k, k = 1, 2, \ldots, n$ if the principal value is selected as the svb of $(z - x_k)^{-c_k}$. In fact, one could select the branch cut along any ray in the lower half plane, and a possible choice for the svb, for example, could be

$$(z - x_k)^{-c_k} = |z - x_k|^{-c_k} e^{-ic_k\theta}, \quad -\frac{\pi}{2} < \theta < \frac{3\pi}{2}. \tag{9.61}$$

The integral in (9.60) must be computed taking the contour of integration in the upper half plane and not passing through the points x_k; the function $f(z)$ is analytic for $y \geq 0$ except at the points $z = x_k$, where it turns out to be continuous.

Writing the modulus and argument of the integrand, we get

$$|f'(x)| = |\lambda| \prod_{k=1}^{n} |(x - x_k)|^{-c_k}, \tag{9.62}$$

and

$$\Phi(x) \equiv \arg[f'(x)] = \arg(\lambda) + \sum_{k=1}^{n} \operatorname{Arg}(x - x_k) c_k. \tag{9.63}$$

Next note that $\operatorname{Arg}(x - x_k)$, given by

$$\operatorname{Arg}(x - x_k) = \begin{cases} 0, & x > x_k, \\ \pi, & x < x_k, \end{cases} \tag{9.64}$$

has a jump discontinuity at $x = x_k$ and therefore $\Phi(x)$ will have jump discontinuities at $x = x_k$. Also, $\Phi(x)$ remains constant — say, ϕ_k — in each interval $(x_k, x_k + 1)$. It is now easy to see that the function $f(x)$,

$$f(x) = |\lambda| \int_{x_0}^{x} \prod_{k=1}^{n} e^{i\phi_k} |(x - x_k)|^{-c_k}\, dx, \tag{9.65}$$

assumes the form

$$f(x) = w_k + e^{i\phi_k} \int_{x_k}^{x} \rho(x)\, dx, \quad x_k < x < x_{k+1}, \tag{9.66}$$

in the interval $x_k < x < x_{k+1}$, where $\rho(x)$ stands for the expression on the right hand side of Eq. (9.62), and ϕ_k is given by

$$\phi_k = \arg(\lambda) - \sum_{j=1}^{n} c_j \operatorname{Arg}(x - x_j) = \arg(\lambda) - \sum_{j=k+1}^{n} c_j \pi. \tag{9.67}$$

It is now apparent that as x increases from x_k to x_{k+1}, $w = f(x)$ traces a straight line segment whose inclination with the real axis is ϕ_k. Since the phase $e^{i\Phi(x)}$ jumps when x crosses x_{k+1}, the inclination of the line segment changes abruptly at $w_{k+1} = w(x_{k+1})$.

When the constants λ and μ are varied, keeping x_k and c_k fixed, one gets a set of similar polygons related by a rotation, translation, and scaling transformation.

When mapping of the real axis onto a specific polygon of n sides is desired, not all constants x_k in the formula (9.57) can be chosen arbitrarily. While the constants c_k are determined by the angles of the polygon, there are $n-3$ constraints on x_k coming from the similarity of polygons, and only three real numbers x_k can be selected arbitrarily.

If we write the factor $\lambda(z - x_n)^{-c_n}$ as $\lambda_1(z/x_n - 1)^{-c_n}$ and if x_n is allowed to tend to ∞, with λ_1 kept fixed, the mapping formula assumes the form

$$f(z) = \lambda_1 \int_{z_0}^{z} \prod_{k=1}^{n-1} (z - x_k)^{-c_k} + \mu_1. \tag{9.68}$$

Of the remaining points $x_j, j = 1, 2, \ldots, n-1$, only two can be chosen arbitrarily and λ_1 and μ_1 are constants to be fixed.

While applying the SCT to a boundary value problem, one needs a map of the interior of a polygon onto the upper half plane. The mapping inverse to $f(z)$ of Eq. (9.60), or Eq. (9.68) provides the required transformation. However, the expression for the SCT is obtained in terms of the points x_k and not in terms of the vertices w_k of the polygon. If $n > 3$, some of the points x_k need to be determined in terms of the parameters of the polygon.

The bilinear transformation $w = \frac{z-i}{z+i}$ maps the upper half plane $\operatorname{Im} z > 0$ onto the interior of the circle $|w| < 1$. Let $\xi_1, \ldots, \xi_n$ be defined by

$$\xi_k = \frac{x_k - i}{x_k + i}, \qquad k = 1, 2, \ldots, n, \tag{9.69}$$

be the n points on the unit circle, $|z| = 1$, which are the images of $x_1, x_2, \ldots, x_n$. Performing a change of variable from z to w in the SCT and replacing the integration variable w by z again, we get

$$f(z) = \lambda_2 \int_{0}^{z} \prod_{k=1}^{n} (z - \xi_k)^{-c_k} dz + \mu_2, \tag{9.70}$$

where λ_2 and μ_2 are constants to be fixed. This gives a mapping of the interior of the unit circle onto the interior of a given polygon D.

It is possible to obtain the Schwarz–Christoffel mapping of the upper half plane onto the exterior D' of a given polygon D. Arranging the vertices in the *clockwise direction*, the map can be written in the form

$$f(z) = \lambda_3 \int_{z_0}^{z} \prod_k (z - \xi_k)^{-d_k} \frac{dz}{(z-\zeta)^2(z-\bar{\zeta})^2} + \mu_3. \tag{9.71}$$

Here $d_k = (\pi - \alpha'_k)/\pi$, α'_k denote the interior angles w.r.t. the domain D', and ζ ($\operatorname{Im}\zeta \neq 0$) is the point which is to be transformed into the point at infinity by the mapping. While the form of the first factor inside the integral is the same as in the earlier case of mapping into the interior of the polygon, more work is required to understand the appearance of the factor $(z-\zeta)^2(z-\bar{\zeta})^2$ in the denominator, and the reader is referred to the books cited at the end.

The angles α'_k, written in terms of the angles α_k of the polygon, are $\alpha'_k = 2\pi - \alpha_k$, giving $d^k = -(\pi - \alpha_k)/\pi = -c_k$, and the Schwarz–Christoffel formula for mapping onto the exterior of a polygon takes the form

$$f(z) = \lambda_3 \int_{z_0}^{z} \prod_k (z - \xi_k)^{c_k} \frac{dz}{(z-\zeta)^2(z-\bar{\zeta})^2} + \mu_3, \tag{9.72}$$

where c_k are as given by Eq. (9.58).

We will usually present data needed to set up the SCT in the tabular form given below:

Vertices	P_1	P_2	$\cdots$	P_n
Preimages	x_1	x_2	$\cdots$	x_n
Interior angle	α_1	α_1	$\cdots$	α_n
Exponent	c_1	c_2		c_n

Many unbounded regions, such as an infinite strip, can be considered as limiting cases of the polygon when some of the vertices are moved to infinity. The SCT is applicable to such a region, known as a *degenerate polygon*. It must be emphasized that details of many proofs and usage of the SCT formula require careful treatment beyond the scope of this book. However, in most applications the required mapping can be obtained by a straightforward application of the SCT and it turns out to be easier to verify that the mapping has the desired properties.

§9.8 Examples

☑**9.8.1 Problem.** Find a function $f(z)$ which maps the upper half z plane onto an unbounded region above the step in the w plane shown in Fig. 9.27 such that the points $z = -1$ and $z = 1$ are mapped on the points $w = ih$ and $w = 0$, respectively.

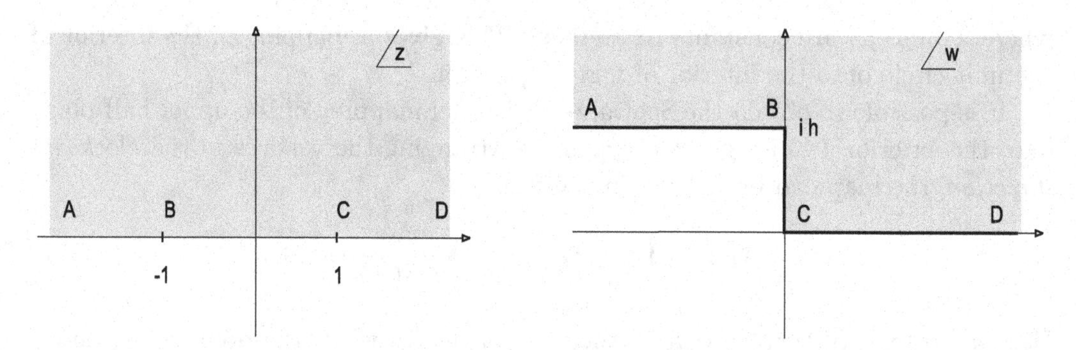

Fig. 9.27

Solution. The Schwarz–Christoffel mapping has the form $z \to w = f(z)$, with

$$f(z) = \lambda \int (z - x_1)^{-c_1}(z - x_2)^{-c_2} \cdots dz + \mu, \tag{9.73}$$

where $x_1, x_2, \ldots$ will have vertices of the polygon as images under the map $f(z)$. We collect those data on the given domain in a tabular form:

Points vertices w_k on the boundary of the polygon	(D, A)	B	C
Points x_k on the real line in the z plane	$-\infty$	-1	1
Interior angles α_k	*	$3\pi/2$	$\pi/2$
Exterior angles β_k		$-\pi/2$	$\pi/2$
Exponents c_k	*	$-1/2$	$1/2$

Recall that one need not assign an exponent for the point at infinity, the corresponding term will be absent from the integrand. Thus, the Schwarz–Christoffel function is given by

$$f(z) = A \int \frac{\sqrt{\xi + 1}}{\sqrt{\xi - 1}}\, d\xi + B. \tag{9.74}$$

We integrate along the real line and determine $f(z)$ as a function of $x = \operatorname{Re} z$. Obtaining $f(x)$, for real x only, is simple and straightforward, and one has

$$f(x) = \lambda \int \frac{\sqrt{x+1}}{\sqrt{x-1}}\, dx + \mu \tag{9.75}$$

$$= \lambda \int \frac{x+1}{\sqrt{x^2 - 1}}\, dx + \mu \tag{9.76}$$

$$= \lambda\left(\sqrt{x^2 - 1} + \cosh^{-1} x\right) + \mu. \tag{9.77}$$

The above expression is analytically continued to the upper half plane and we write

$$f(z) = \lambda\left(\sqrt{z^2 - 1} + \cosh^{-1} z\right) + \mu. \tag{9.78}$$

Now we select a single-valued branch which is analytic in the upper half plane and then fix the constants A and B. The uniqueness of analytic continuation (§5.10) ensures that this will be the desired answer. We write

$$\cosh^{-1} z = \text{Log}\,(z + \sqrt{z^2 - 1}) \tag{9.79}$$

and use the principal value for the functions $\sqrt{z^2 - 1}$ and $\text{Log}\, z$. This also defines a single-valued branch of $\cosh^{-1} z$. Note that the only branch points of the function $\cosh^{-1} z$ are ± 1, which come from $\sqrt{z^2 - 1}$. This result follows from the fact that the expression $z + \sqrt{z^2 - 1}$ never vanishes because

$$z + \sqrt{z^2 - 1} = 0 \Rightarrow z = \sqrt{z^2 - 1} \Rightarrow 1 = 0. \tag{9.80}$$

The principal value of $\sqrt{z^2 - 1}$ is defined by

$$\sqrt{z^2 - 1} = \sqrt{|z^2 - 1|}\exp(i\Phi/2), \tag{9.81}$$

where $\theta_{\pm} = \text{Arg}(z \pm 1)$ and $\Phi = \theta_+ + \theta_-$ should be restricted to the range $-\pi \leq \Phi < \pi$. To fix the constants, we compute the values of $f(z)$ as z approaches ± 1 and

$$f(-1) = i\pi\lambda + \mu, \quad f(1) = \mu.$$

Using $f(-1) = ih$ and $f(1) = 0$, we get

$$\lambda = \frac{h}{\pi}, \quad \mu = 0. \tag{9.82}$$

Thus, we obtain the final answer:

$$f(z) = \frac{h}{\pi}\left(\sqrt{z^2 - 1} + \text{Log}\, z + \sqrt{z^2 - 1}\right). \tag{9.83}$$

▮

↬ Verification that the answer (9.83) indeed gives the correct, required mapping is an elementary exercise, Q.[12](d) in §§3.8 for the reader.

☑**9.8.2 Problem.** Derive the sine function as the inverse mapping of Schwarz–Christoffel mapping from the upper half plane z onto the semi-infinite strip $\{w| - L < \text{Re}\, w < L, \text{Im}\, w > 0\}$.

Solution. For this purpose we regard the semi-infinite strip PABQ, Fig. 9.28, as a (degenerate) triangle with two vertices A and B and one vertex at R which is taken to infinity. The corresponding interior angles are $\pi/2, \pi/2$, and 0. Take w_1, w_2, and w_3 as the images of $-1, 1$, and ∞, respectively. The SCT takes the form $w = f(z)$, with

$$f(z) = \lambda \int_{-1}^{z} (1+z)^{-1/2}(1-z)^{-1/2}\, dz + \mu \tag{9.84}$$

$$= \lambda \sin^{-1} z + \mu. \tag{9.85}$$

Using $w(-1) = -L$ and $w(1) = L$ gives $z = \sin(\pi w/2L)$. ▮

☑**9.8.3 Short examples (Degenerate polygons).** We describe a few examples of domains which when treated as degenerate polygons lead to simple mapping functions $z^{\alpha}, \exp(z)$, and the hyperbolic cosine function.

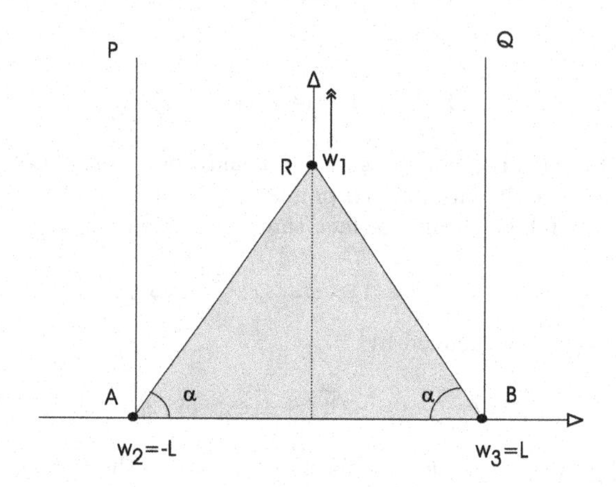

Fig. 9.28

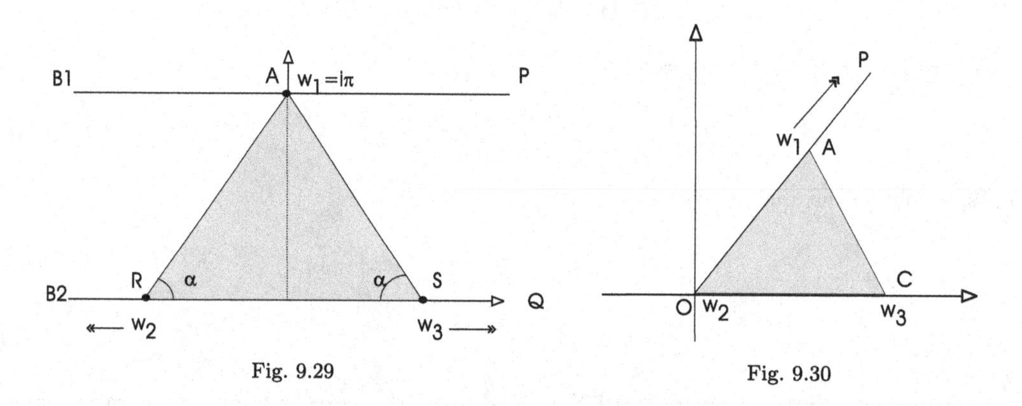

Fig. 9.29

Fig. 9.30

(a) To use the SCT to construct mapping of the upper half plane onto the infinite strip $\{z \,|\, 0 < \operatorname{Im} z < \pi\}$, treat the strip as a limiting case of a triangle ARS, with two vertices w_2 and w_3 going to infinity, as in Fig. 9.29. Taking the vertex w_1 to correspond to $z = 0$, we get

$$w(z) = \lambda \int z^{-1}\, dz + \mu = \lambda \text{ Log } z + \mu, \tag{9.86}$$

and the mapping property of the exponential has been recovered.

(b) We treat the sectorial domain $D = \{w | 0 < \arg w < \alpha\}$ as a "polygon" with two vertices, one of which is at $w = 0$ and the other at $w = \infty$, with the interior angle at $w = 0$ being α. The sector may be regarded as a limiting case of a polygon, as shown in Fig. 9.30. Using the SCT, the function $f(z)$, which maps the upper half plane onto D, we get

$$f(z) = \lambda \int_0^z \xi^{\alpha/\pi - 1} dz = \mu z^{\alpha/\pi}, \tag{9.87}$$

where μ is a constant.

(c) The SCT leads to the function $w = \cosh(\pi z/L)$ as a map such that the image of the interior of a semi-infinite strip,

$$\{z | \operatorname{Re} z > 0, 0 < \operatorname{Im} z < L\},$$

is the upper half plane with $w = 1$ and $w = -1$ as the images of $z = 0$ and $z = iL$, respectively.

(d) Considering the boundary of the infinite strip $\{z | -\frac{\pi}{4} < \operatorname{Re} z < \frac{\pi}{4}\}$ as a two-sided degenerate polygon with the two sides meeting at ∞, at a zero angle, one may construct an SCT $f(z)$ from the unit circle to the strip. Taking the images of the points $z = \pm i$ to be at $w = \infty$, and using Eq. (9.70), with $c_k = -1$, we get

$$f(z) = \lambda_2 \int_0^z \frac{dz}{(z+i)(z-i)} + \mu_2 = \lambda_2 \tan^{-1} z + \mu_2. \tag{9.88}$$

If we use $f(\pm 1) = \pm\frac{\pi}{4}$, we get $f(z) = \tan^{-1} z$.

Many details and proofs in our presentation of the SCT, not needed in this book, have been omitted. The interested reader is referred to the books [Polya (1974); Nehari (1952); Carrier *et al.* (1966)].

⊳ Tutorial §§9.9, on Schwarz–Christoffel mapping.
⊳ Questions §§9.11, on setting up the Schwarz–Christoffel transformation.
⊳ Quiz §§9.10 on finding the images for several Schwarz–Christoffel maps.

§9.9 Notes and References

(◇1) The mapping of the real line onto a rectangle is an SCT which is related to one of the Jacobi elliptic functions; see [Churchill (1964); Ablowitz (2003)].

(◇2) The SCT can be generalized to a map which transforms the real line onto a domain bounded by circular arcs. For a circular triangle, the Schwarz–Christoffel map gets related to hypergeometric functions; see [Nehari (1952); Ablowitz (2003)].

(◇3) To interactively see how the image of a unit circle changes under Joukowsky mapping, visit http://www.diam.unige.it/\~irro/conformi_e.html and http://www.diam.unige.it/~irro/java/conformi1_1.html

(◇4) The SCT maps the real line onto a polygon. A modification can be made to produce the polygon with rounded corners. See [Carrier *et al.* (1966)].

(◇5) The Schwarz–Christoffel mapping can be generalized to give mapping of the upper half plane onto the exterior of a polygon instead of the interior. The details can be found in [Carrier *et al.* (1966); Henricki (1974)].

(◇6) For a description of bilinear transformations on the Riemann sphere, see [Henricki (1974)].

(◇7) Circle packing is a way of constructing conformal mappings proposed in 1985 by William Thurston, a winner of the Fields Medal (the highest honor in mathematics, analogous to the Nobel Prize). This approach does not require complex numbers.

Chapter 10

PHYSICAL APPLICATIONS OF CONFORMAL MAPPINGS

In this chapter, the applications of conformal mapping to two dimensional problems in heat flow, electrostatics and flow of fluids will be presented by means of solved examples. The applications are modelled by Laplace equation

$$\frac{\partial^2 \phi}{\partial x^2} + \frac{\partial^2 \phi}{\partial y^2} = 0 \tag{10.1}$$

in a domain D bounded by a curve C. In a class of problems known as *Dirichlet problems*, the value of the function ϕ is specified on the boundary C. Alternatively, if the value of the normal derivative of ϕ is specified on C, the problem is known as a *Neumann problem*. There are other cases where a mixed type of boundary condition is imposed on the solution. In the conformal mapping approach, we attempt to find a function $\Omega(z)$ analytic in D such that its real (or imaginary) part gives us the desired solution by mapping the given problem onto another, simpler problem. When the given domain D is unbounded, the behavior of the solution, or equivalently of $\Omega(z)$, must be specified at infinity.

In many examples of interest, $\Omega(z)$ is required to be singular at some points of the domain D. We shall assume that the function $\Omega(z)$ has at most a logarithmic singularity at infinity and at some chosen points inside D. This specification of behavior at infinity and at singular points inside D is essential for obtaining a unique solution. Occasionally, in a few idealized examples, such as those involving uniform flows or dipoles, the complex functions $\Omega(z)$ may exhibit a more singular behavior.

§10.1 Model problems

In this section solutions to some very simple boundary value problems will be presented as the real and imaginary parts of a few chosen analytic functions. These problems serve as model problems for more complicated, realistic situations which will be dealt with later.

First, consider the problem of obtaining the solution to the Laplace equation $\nabla^2 \phi = 0$ in an infinite strip subject to the boundary conditions indicated in Fig. 10.1(a). It is obvious that the function $\phi(x, y) = c_1 + c_2 x$, being the real part

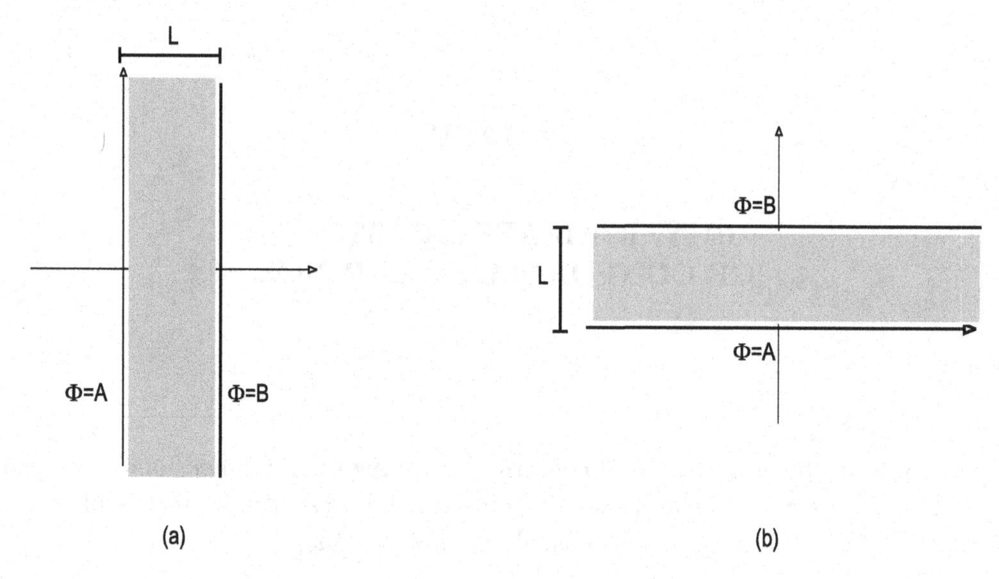

Fig. 10.1

of $f(z) = c_1 + c_2 z$, is harmonic and that the real constants c_1 and c_2 can be chosen to satisfy the required boundary conditions $\phi(0, y) = A$ and $\phi(L, y) = B$. Similarly, the imaginary part of $f(z)$,

$$\operatorname{Im} \Omega(z) = c_1 + c_2 y, \tag{10.2}$$

has the correct form to be used as the solution for the boundary value in Fig. 10.1(b). The answers to these two problems are given by

$$\text{(a)} \quad \phi(x, y) = A + (B - A)\frac{x}{L}, \tag{10.3}$$

$$\text{(b)} \quad \phi(x, y) = A + (B - A)\frac{y}{L}. \tag{10.4}$$

The solution to the problems in Fig. 10.2 can be written in terms of the real and imaginary parts of $g(z) = c_1 + c_2 \operatorname{Log}(z)$. Thus, the real part, $c_1 + c_2 \log r$, is appropriate for the annular domain in Fig. 10.2(a) and the imaginary part, $c_1 + c_2\theta$, solves the boundary value problem for the domains in Figs. 10.2(b) and 10.2(c). Fixing the constants c_1 and c_2 by using the specified boundary conditions, the answers to the three problems in Fig. 10.2 are

$$\text{(a)} \quad \phi(x, y) = A + (B - A)\frac{\ln(r/r_1)}{\ln(r_2/r_1)}, \tag{10.5}$$

$$\text{(b)} \quad \phi(x, y) = A + (B - A)\frac{\theta}{\alpha}, \tag{10.6}$$

$$\text{(c)} \quad \phi(x, y) = A + (B - A)\frac{\theta}{\pi}. \tag{10.7}$$

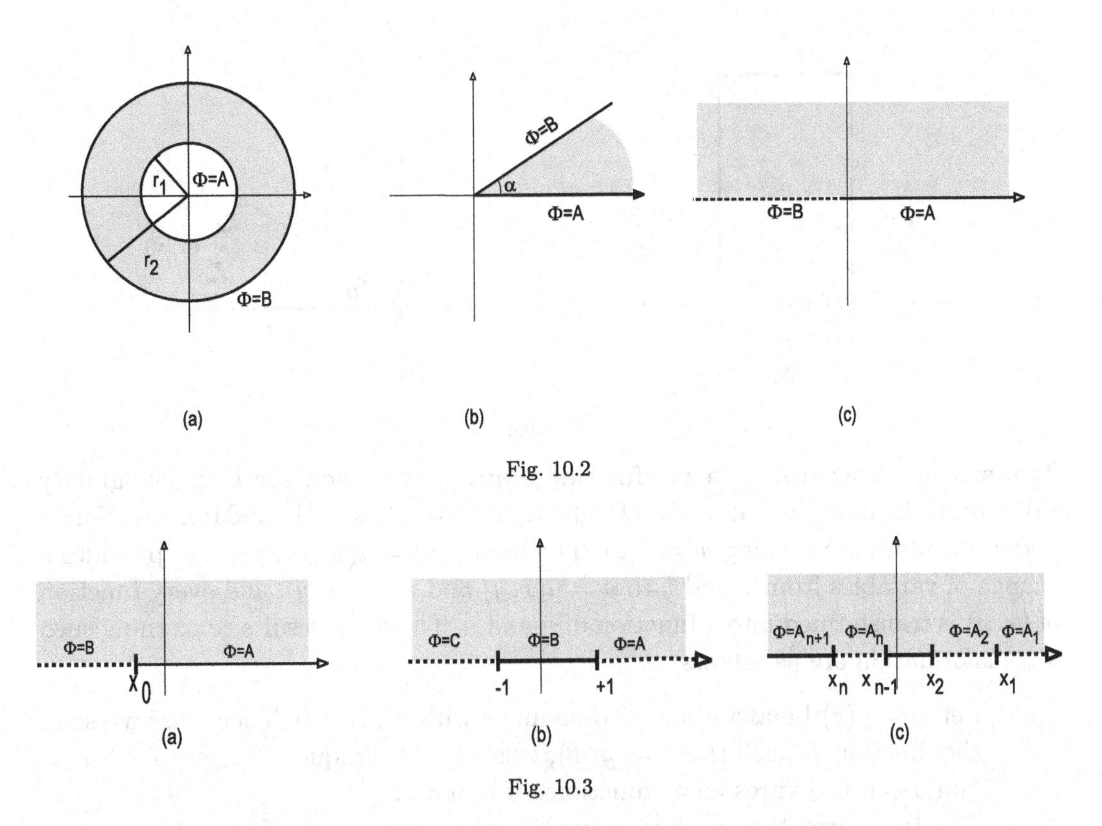

Fig. 10.2

Fig. 10.3

An observation crucial to the solution of this problem is that $\theta = \arg z$ assumes different values on the real axis to the left and to the right of $x = 0$. *Note that in the boundary value problem (c), the upper half plane has been treated as a sector* $0 < \theta < \pi$. The shifted polar variable $\phi = \arg(z - x_0)$ will have a similar property and should be used for the solution of the model problem of Fig. 10.3(a). For the problems in Figs. 10.3(b) and 10.3(c), several such variables are needed. The final answers to the three problem are

$$\text{(a)} \quad \phi = A + \frac{B-A}{\pi} \arg(z - x_0), \tag{10.8}$$

$$\text{(b)} \quad \phi = A + \frac{B-A}{\pi} \arg(z-1) + \frac{C-B}{\pi} \arg(z+1), \tag{10.9}$$

$$\text{(c)} \quad \phi = A_1 + \frac{A_2 - A_1}{\pi}\theta_1 + \frac{A_3 - A_2}{\pi}\theta_2 + \cdots + (A_{n+1} - A_n)\theta_n, \tag{10.10}$$

where $\theta_k = \arg(z - x_k), k = 1, \ldots, n$. *The part of the boundary marked by crosses, as in Fig. 10.4, will be used to denote the Neumann boundary condition of the vanishing normal derivative.* The solution to these problems is easily seen to be given by (10.3) and (10.6), respectively. The answer (10.3) for the model problem in Fig. 10.1(a) has normal derivative zero on the x axis, and hence it is also a solution to the model problem in Fig. 10.4(a). Similarly, the answer (10.6) has normal derivative zero on a circular arc, as required for Fig. 10.4(b).

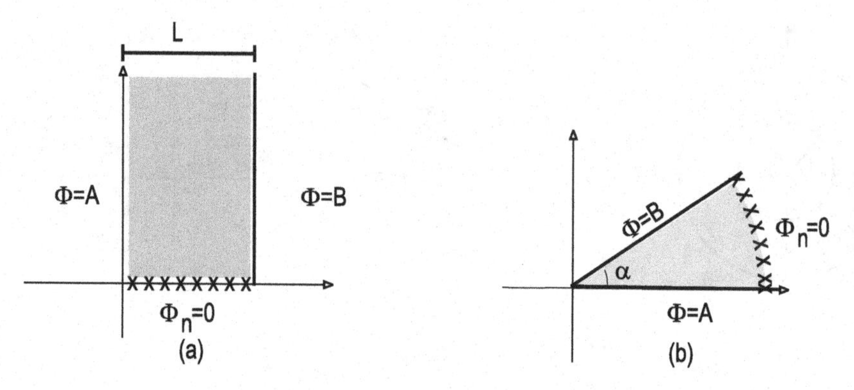

Fig. 10.4

Transformations under a conformal map. Let us now see how a boundary value problem in a given domain D, having a curve C as its boundary, transforms under a conformal mapping $w = f(z)$. The map $f(z) = u(x, y) + iv(x, y)$ provides a change of variables from x and y to $u = u(x, y)$ and $v = v(x, y)$, and every function of (x, y) is transformed into a function of u and v. The main results concerning such a transformation are as follows:

(*1) Let $w = f(z)$ be analytic in a domain D with $f'(z) \neq 0$. Then the inverse of the function f, such that $z = g(w)$, exists and is unique. This ensures that x and y can be expressed as functions of u and v.

(*2) The transformation $z \to w = f(z)$ carries every harmonic function $\phi(x, y)$ into a harmonic function of the variables u and v. Thus, the function $\Phi(u, v) = \phi(x(u, v), y(u, v))$ is a harmonic function of u and v.

(*3) If $\phi(x, y)$ is *constant* on a part of the curve C, the transformed function $\Phi(u, v)$ is also constant along the corresponding part of the boundary in the w plane.

(*4) If the normal derivative of $\phi(x, y)$ is zero along a subset γ of C, the transformed function Φ has a zero normal derivative on the image of γ. It must be emphasized that if the normal derivative of ϕ has a nonzero constant value, the transformed function Φ need not have a constant normal derivative on the corresponding subset of the boundary in the w plane.

(*5) In order that under a conformal mapping the boundary of the domain can be transformed to the boundary of the image domain, it is necessary that the mapping be one-to-one. Therefore, it is important to use only one-to-one mapping while attempting to solve a boundary value problem.

↬The proof of statement (*4) is left as exercise in Q.[1] in §§10.11.

§10.2 Physical Applications

Almost all problems of interest in science and engineering applications are formulated in three dimensions. Two-dimensional problems arise naturally whenever there is a symmetry under translations perpendicular to a plane. Thus, for example, a circular boundary in a two-dimensional problem arises from an infinite cylinder in three dimensions, and a "point" charge in two dimensions is a line charge of a three-dimensional problem.

The method of conformal mappings is applicable for solutions to the Laplace equation

$$\frac{\partial^2 \phi}{\partial x^2} + \frac{\partial^2 \phi}{\partial y^2} = 0 \tag{10.11}$$

in two dimensions only. The solution ϕ is harmonic in the domain of interest and, frequently, it is useful to introduce the conjugate harmonic function $\psi(x, y)$, which also has a physical interpretation. These two functions, $\phi(x, y)$ and $\psi(x, y)$, are the real and imaginary parts of an analytic function to be denoted by Ω:

$$\Omega(z) = \phi(x, y) + i\psi(x, y). \tag{10.12}$$

Many quantities of interest can be expressed in terms of Ω and its use simplifies computations.

We shall now briefly describe the boundary value problems arising in physical applications and relate ϕ and ψ to the quantities of interest.

§10.3 Steady State Temperature Distribution

In problems of heat conduction, the function $\phi(x, y)$ represents the temperature at a point (x, y). In the steady state ϕ satisfies the Laplace equation in domains that do not contain sources and sinks of heat.

The boundary of D, or a part of it, may be held at a constant temperature or may be insulated. The flow of heat across a boundary is determined by the temperature gradient or the value of the normal derivative of ϕ. An insulated part means that there is no flow of heat across that subset and the normal derivative of ϕ must vanish on the insulated part of the boundary. Thus, we are led to the Dirichlet, Neumann, and mixed boundary value problems. The level curves $\phi(x, y) =$ const, are known as *isothermal curves.* The direction of the heat flow is along the normals to the isothermal curves, which are the same as the tangents to the level curves of the conjugate function $\psi(x, y)$. The quantity $\Omega(z) = \phi(x, y) + i\psi(x, y)$ will be called the *complex temperature.*

The solutions to model boundary value problems in §10.1 were found by inspection and did not require the use of conformal mapping. Here we deal with the solution of a few simple heat conduction problems using conformal mapping by mapping the region of interest to that in one of the model problems of §10.1.

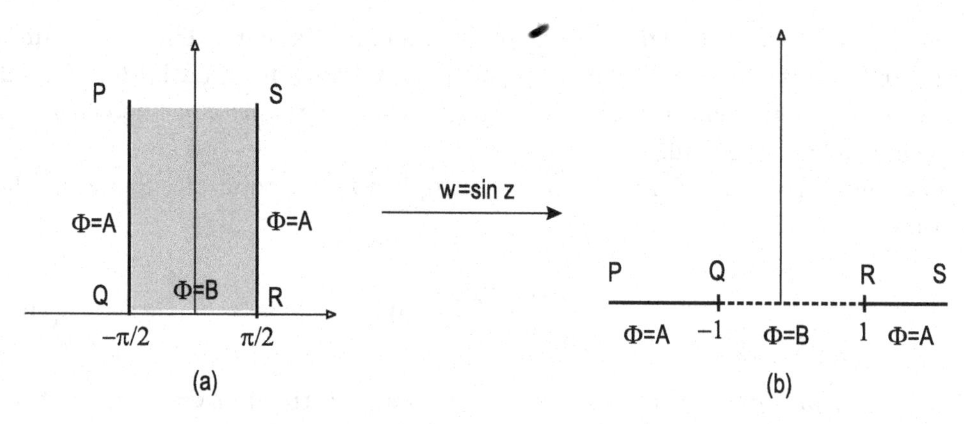

Fig. 10.5

☑**10.3.1 Problem.** Find the solution to the Laplace equation

$$\frac{\partial^2 \phi}{\partial x^2} + \frac{\partial^2 \phi}{\partial y^2} = 0 \tag{10.13}$$

in a semi-infinite strip [Fig. 10.5(a)] subject to the boundary conditions

$$\phi(x, y)\Big|_{x=\pm\frac{\pi}{2}} = A, \quad \text{for all } y > 0, \tag{10.14}$$

$$\phi(x, y)\Big|_{y=0} = B, \quad \text{for all } x, \tag{10.15}$$

such that $\frac{\pi}{2} < x < \frac{\pi}{2}$ and ϕ is bounded as $y \to \infty$.

Solution. Use of the map $w = \sin z$ transforms the given boundary value problem into a problem in the upper half plane with boundary conditions, as shown in Fig. 10.5(b). This problem is recognized as the model problem in Fig. 10.3(b) and the solution is

$$\phi = A + \frac{A-B}{\pi}\text{Arg}(w+1) + \frac{B-A}{\pi}\text{Arg}(w-1) \tag{10.16}$$

$$= A + \frac{B-A}{\pi}\arg\left(\frac{w-1}{w+1}\right) \tag{10.17}$$

$$= A + \frac{B-A}{\pi}\text{Arg}\left(\frac{2v}{u^2+v^2-1}\right), \tag{10.18}$$

where $u = \text{Re}\sin z = \sin x \cosh y$ and $v = \text{Im}\sin z = \cos x \sinh y$. The above answer can be shown to be equal to

$$\phi = A + \frac{2(B-A)}{\pi}\tan^{-1}\left(\frac{\cos x}{\sinh y}\right). \tag{10.19}$$

∎

Next, we present an example of mixed boundary conditions. The reader should compare this example with the previous one.

☑**10.3.2 Problem.** Find the temperature distribution in the upper half plane subject to the boundary conditions

$$T(x,0) = \begin{cases} A, & x < -1, \\ B, & x > 1, \end{cases} \tag{10.20}$$

and where part of the boundary between -1 and 1 is insulated, i.e.

$$\frac{\partial T(x,y)}{\partial y} = 0 \quad \text{if } -1 < x < 1. \tag{10.21}$$

Solution. We use the map $w = \sin^{-1} z$, which transforms the upper half plane onto an infinite strip $\left\{-\frac{\pi}{2} < u < \frac{\pi}{2}\right\}$ in the w plane. The new problem is solved trivially by $T = C_1 + C_2 u$. Fixing the constants C_1 and C_2, and using the boundary conditions at $u = \pm\frac{\pi}{2}$, we get

$$T = \frac{A+B}{2} + \frac{B-A}{\pi} u. \tag{10.22}$$

Using the result in Q.[14] in §§2.13,

$$2 \sin u = \sqrt{(x+1)^2 + y^2} - \sqrt{(x-1)^2 + y^2}, \tag{10.23}$$

we get the final answer,

$$T(x,y) = \frac{A+B}{2} + \frac{B-A}{\pi} \sin^{-1}\left[\frac{1}{2}\sqrt{(x+1)^2+y^2} - \frac{1}{2}\sqrt{(x-1)^2+y^2}\right],$$
$$x > 0, y > 0, \tag{10.24}$$

where the inverse sine function takes values between $-\frac{\pi}{2}$ and $\frac{\pi}{2}$. For x and y in the second quadrant, the solution is obtained by making the replacement $x \to -x$.

It is straightforward to verify that the answer (10.24) does satisfy the boundary conditions (10.20) and (10.21) if one remembers that, for $y = 0$, the positive value of the two square roots must be taken. ∎

⧐ Tutorial §§10.1, for simple heat conduction problems.
⧐ Exercise §§10.2, for several heat conduction problems.

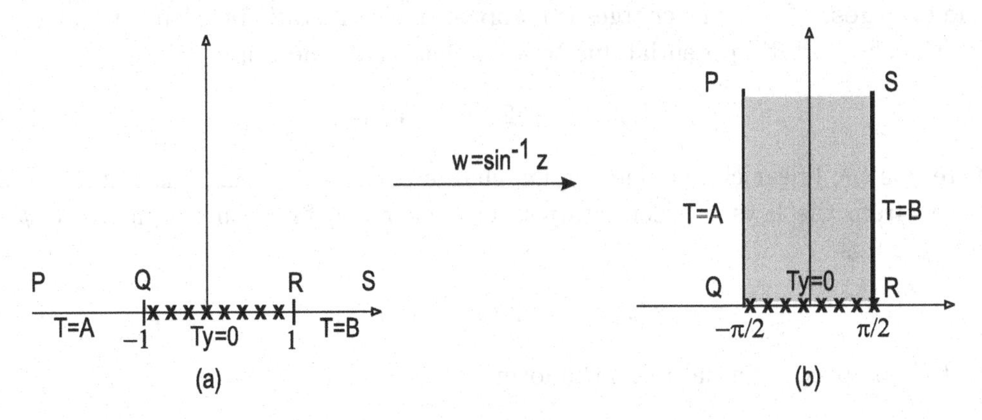

Fig. 10.6

§10.4 Electrostatic Potential

Complex potential. In the case of electrostatic problems, the electric potential is required in the presence of charges and perfect conductors. In mathematical terms, the presence of conductors means that the potential is constant on such a conductor. In domains free of charges the potential satisfies the Laplace equation and one is led to Dirichlet boundary value problems for domains free of charges. The level curves $\phi(x, y) = \text{const}$ are called *equipotential curves.* Thus, *all conductors are equipotential curves.* If $\psi(x, y)$ is a conjugate harmonic function, *the level curves of* ψ, $\psi = $ *const, give the direction of flux lines or the electric field.* The values of the electric field components at a point (x, y) are easily found if the potential function $\phi(x, y)$ or the conjugate function $\psi(x, y)$ is known. Thus,

$$E_x = -\frac{\partial \phi}{\partial x} = -\frac{\partial \psi}{\partial y}, \tag{10.25}$$

$$E_y = -\frac{\partial \phi}{\partial y} = \frac{\partial \psi}{\partial x}. \tag{10.26}$$

If we introduce the complex potential $\Omega(z) = \phi(x, y) + i\psi(x, y)$, whose real and imaginary parts are the two harmonic functions ϕ and ψ, it is easy to see that

$$\mathcal{E} = -\overline{\left(\frac{d\Omega}{dz}\right)}, \tag{10.27}$$

where $\mathcal{E}$ is the *complex electric field* $\mathcal{E} = E_x + iE_y$. The magnitude of the electric field has a simple expression in terms of the complex potential:

$$|E| = |\mathcal{E}| = \left|\frac{d\Omega}{dz}\right|. \tag{10.28}$$

We will be using the complex potential and complex electric field, which greatly simplifies computations when a solution is obtained using conformal mapping.

Line charges. Since line charges will appear in electrostatic problems, we mention that the electrostatic potential due to a line charge at the origin is

$$\phi(x, y) = -2q \ln r + \text{const}, \tag{10.29}$$

where q is the linear charge density, i.e. charge per unit length. This result can be derived from the laws of electrostatics. The harmonic function conjugate to ϕ in Eq. (10.29) is

$$\psi(x, y) = -2q \arg z + \text{const}, \tag{10.30}$$

and the complex potential takes the form

$$\Omega(z) = -2q \log z + \text{const}. \tag{10.31}$$

For several line charges at z_k, $k = 1, \ldots, n$, the complex potential is given by superposition of potentials due to individual charges and one has

$$\Omega(z) = -2\sum_{k=1}^{n} q_k \log(z - z_k) + \text{const.} \tag{10.32}$$

The constant in Eqs. (10.30)–(10.32) is usually fixed by setting the potential equal to zero at some reference point.

Work and flux in electrostatics. Let C be a curve and let E_t and E_n denote the tangential and normal components of the electric field. The work done when a unit positive charge is moved along C is given by

$$\text{work done} = \int_C E_t \, ds. \tag{10.33}$$

Another important quantity, the flux of the electric field through C, is given by the line integral

$$\text{flux} = \int_C E_n \, ds. \tag{10.34}$$

Since there are two directions of the tangent and normal vectors, we must fix our conventions before proceeding further. We assume that the tangent vector is obtained from the normal vector by anticlockwise rotation of $\frac{\pi}{2}$, see Fig. 10.7. With this convention the results on the work and flux can be summarized as

$$\text{work} + i \text{ flux} = -\int_C \left(\frac{d\Omega}{dz}\right) dz. \tag{10.35}$$

↬ Proof of Eq. (10.36) is left as an exercise, Q.[2], in §§10.11.

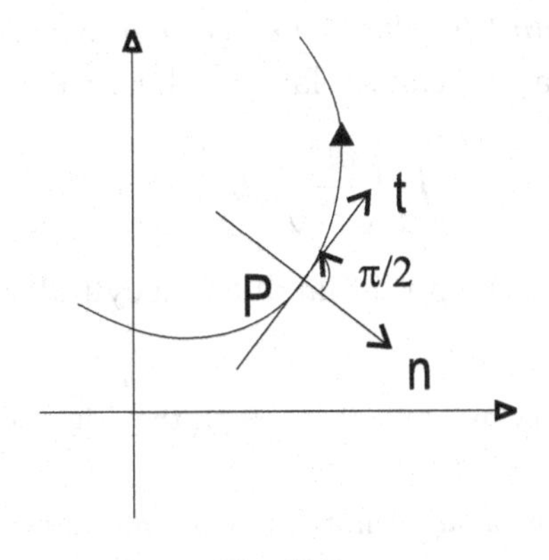

Fig. 10.7

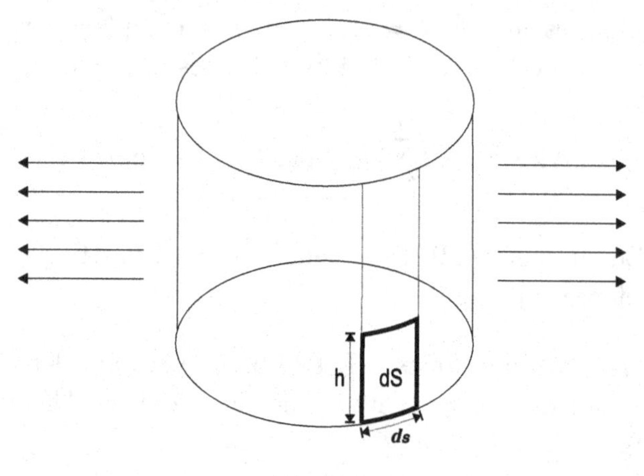

Fig. 10.8

Gauss' law. In electrostatics, Gauss' law states that the flux of the electric field through a closed surface S, defined as the surface integral of the electric field

$$\Phi_S = \iint_S \mathbf{E} \cdot \mathbf{n} \, dS, \tag{10.36}$$

is equal to 4π times the net charge enclosed by S. We rewrite Gauss' law for two-dimensional problems. Assuming translation symmetry perpendicular to the x–y plane, and taking S to be a cylinder of height h, with a closed curve C as its base (with $dS = hds$) — see Fig. 10.8 — one gets

$$\Phi_S = \iint_S E_n \, dS = h \oint_C E_n \, ds. \tag{10.37}$$

The integral $\oint_C E_n \, ds$ will be called the *flux, per unit length of the cylinder, through the closed curve* C and using Gauss' law, and is given by

$$\oint_C E_n \, ds = 4\pi q, \tag{10.38}$$

where q *is the charge per unit length of the cylinder.* Since the work done in a closed path vanishes in electrostatics, Gauss' law can be written as

$$i \oint_C \left(\frac{d\Omega}{dz} \right) dz = 4\pi q. \tag{10.39}$$

☑**10.4.1 Problem.** Find the potential due to a cylinder of radius R having charge q per unit length.

Solution. If the axis of the cylinder passes through the origin and no other charge is present, then

- the complex potential must be analytic in the whole complex plane excluding the point z_0;
- due to symmetry, concentric cylindrical surfaces must be equipotentials.

Thus, on $|z| = R$ the potential must be a constant. This is the model problem in Fig. 10.1(a), with $r_2 \to \infty$, and the complex potential is given by

$$\Omega(z) = K \log z + C. \tag{10.40}$$

To fix the constant, we compute the electric field and use Gauss' law. The complex electric field is

$$\bar{\mathcal{E}} = -\frac{K}{z}, \tag{10.41}$$

and its flux must be equal to $4\pi q$, giving $K = -2q$; see Eq. (10.40). The constant C can be fixed by specifying the potentials at some point $z = z_0$, giving

$$\Omega(z) = -2q \log z + C. \tag{10.42}$$

∎

§10.5 Flow of Fluids

The Laplace equation in two dimensions also appears in ideal fluid flow problems, whenever the following conditions are met:

(*1) The velocity of the fluid is assumed to be parallel to the x–y plane. It is described by two components, v_x and v_y.

(*2) The flow is assumed to be irrotational, which means that the velocity can be derived from a *velocity potential* function $\phi(x, y)$, i.e.

$$v_x = \frac{\partial \phi}{\partial x}, \quad v_y = \frac{\partial \phi}{\partial y}. \tag{10.43}$$

(*3) The fluid is assumed to be incompressible and nonviscous.

(*4) There are no sources and sinks of the fluid in the volume of interest. The fluid enters and leaves through the boundary only.

(*5) It is also assumed that the flow is steady and the velocity is independent of time.

(*6) There are no external forces acting on the fluid.

§10.5.1 *Stream function and stream lines*

In the case of two-dimensional ideal flow of fluids, the *velocity potential* function is a harmonic function and its conjugate harmonic function $\psi(x, y)$ is called the *stream function*. The curves $\phi(x, y) = \text{const}$ and $\psi(x, y) = \text{const}$ are called, respectively, the *equipotential curves* and the *stream lines* of the flow. We list some important facts about the fluid flow:

(*1) The stream lines and the equipotential curves, being the level curves of conjugate harmonic functions, are orthogonal.

(*2) The velocity vectors are normal to the equipotential $\phi = \text{const}$, and hence are along the tangents to the stream lines, which represent the path followed by the fluid particles.

(*3) In the problems of fluid flows in channels and across obstacles, the normal component of the velocity along the boundaries is zero. Since the velocity must be tangential to the boundaries, these boundaries must be stream lines and the stream function must be constant on the boundary.

(*4) The velocity, using Cauchy–Riemann equations, is also given by

$$v_x = \frac{\partial \psi}{\partial y}, \quad v_y = -\frac{\partial \psi}{\partial x}. \tag{10.44}$$

The complex velocity $\mathcal{V} = v_x + iv_y$ is obtained from the complex potential $\Omega = \phi + i\psi$ using the relations

$$\mathcal{V} = \overline{\left(\frac{d\Omega}{dz}\right)}, \tag{10.45}$$

$$|\boldsymbol{v}| = \left|\frac{d\Omega}{dz}\right|. \tag{10.46}$$

(*5) The flux of the fluid through a cylindrical surface of unit height and with the base on a curve γ (Fig. 10.9) is equal to the difference in the values of the stream function at the end points of γ. This can be seen as follows. If v_n is the normal component of the velocity, the flux is

$$\text{flux} = \int_\gamma v_n\, ds = \int_\gamma \left(\frac{\partial \phi}{\partial x}\, dy - \frac{\partial \phi}{\partial y}\, dx\right) \tag{10.47}$$

$$= \int_\gamma \left(\frac{\partial \psi}{\partial x}\, dx + \frac{\partial \psi}{\partial y}\, dy\right) \tag{10.48}$$

$$= \psi(x_2, y_2) - \psi(x_1, y_1). \tag{10.49}$$

(*6) The line integral $\oint_\gamma v_t\, ds$ along a closed curve γ, where v_t is the tangential component, is called the *circulation* along γ.

(*7) Let z_0 be a point such that the flux through the circles $|z - z_0| = R$ is equal to a constant k for all sufficiently small R. The point z_0 is called a *source of strength* k if $k > 0$ and a *sink of strength* k if $k < 0$. A point z_0 is called a

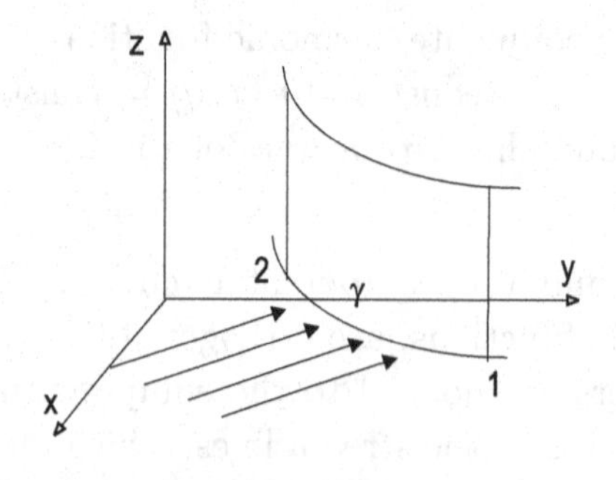

Fig. 10.9

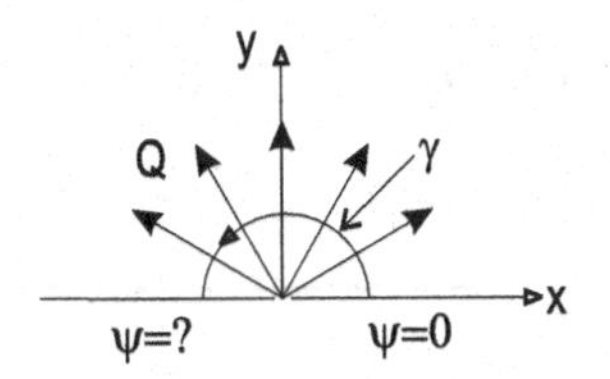

Fig. 10.10

vortex if the circulation along the circles $|z - z_0| = R$ is nonzero for sufficiently small values of R.

(*8) The circulation and the flux can be related to the complex potential in a manner similar to that for the electrostatics, and we have

$$\text{circulation} + i \text{ flux} = \int_C \frac{d\Omega}{dz} dz. \tag{10.50}$$

(*9) *Bernoulli's theorem.* If p is the pressure exerted by the fluid, v the speed and ρ the density, then along a stream line

$$p^2 + \frac{1}{2}\rho v^2 = \text{const.} \tag{10.51}$$

(*10) Using Bernoulli's theorem one can derive expressions for the force and torque exerted by the fluid on an obstacle.

↬ Q.[4], for an expression for force and torque on an obstacle placed in the flow of a fluid; and Q.[5], for an exercise on the lift on a cylinder.

The examples below show how Eq. (10.50) could be used to determine the boundary conditions on the stream function in the presence of sources on the boundary of the domain of interest.

§10.5.1 Short examples. The following short examples show uses of the flux argument.

(a) For the flow of a fluid entering the upper half plane through a slit at $z = 0$, the positive and negative parts of the real axis will be stream lines. If the quantity of the liquid flowing into the upper half plane per unit height of the slit is Q, the boundary condition on the stream function can be determined by using Eq. (10.50). Assuming that $\psi = 0$ on the positive portion of the real axis, taking a semicircular arc as γ, as in Fig. 10.10, we see that the value of the stream function for $x < 0$ is given by

$$\psi(x, 0) = \begin{cases} 0 & x > 0, \\ Q & x < 0. \end{cases} \tag{10.52}$$

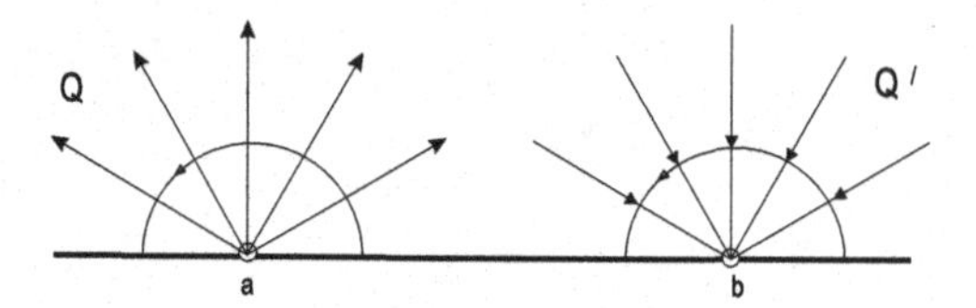

Fig. 10.11

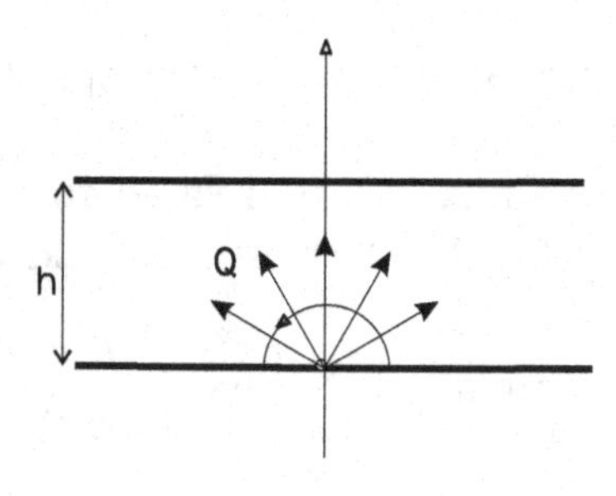

Fig. 10.12

Since the stream function is harmonic in the upper half plane, using Eq. (10.8) we get

$$\psi(x,y) = \frac{Q}{\pi}\theta = \frac{Q}{\pi}\tan^{-1}\left(\frac{y}{x}\right) \tag{10.53}$$

and the complex potential is seen to be $\Omega(z) = (Q/\pi)\text{Log}\, z$. One can easily verify that Eq. (10.53) is satisfied for the curve γ.

(b) The fluid enters the upper half plane from a source of strength Q at $z = a$ and leaves through a sink of strength Q' at $z = b$; see Fig. 10.11. The complex potential for this problem would be

$$\Omega(z) = Q\,\text{Log}\,(z-a) - Q'\,\text{Log}\,(z-b). \tag{10.54}$$

(c) Consider an infinitely long channel, of uniform cross section, having a source of strength Q on the boundary, as in Fig. 10.12. The fluid flowing into the channel per unit time at $z = 0$ is $Q/2$. It follows from conservation of flux that the speed of the leaving at $x \to \pm\infty$ is $Q/4h$, where h is the width of the channel.

(d) The strength of an image source, or image sink, at the image point w_0, under a conformal mapping $z \to w = f(z)$, is equal to the strength of the source, or the sink, at the point corresponding to w_0, i.e. at the point z_0 given by $f(z_0) = w_0$.

⊳⊳ §§10.10, for more problems on sources and sinks.

⊳⊳ §§10.10: Q[3], Q[4], and Q[5], for use of simple arguments in fluid flow to fix an extra constant in Schwarz–Christoffel mapping.

§10.6 Solutions Described by Simple Complex Potentials

⨍10.6.1 Short examples. We will now describe the flow of fluids represented by several complex potentials and draw stream lines and equipotentials.

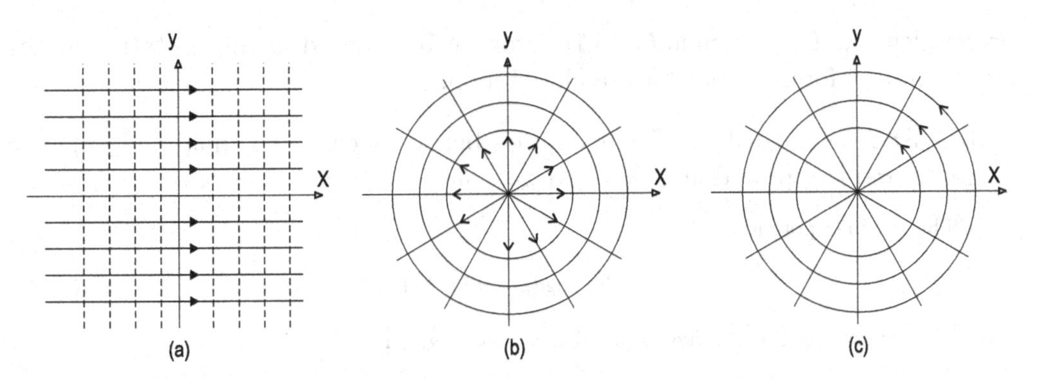

Fig. 10.13

(a) For the complex potential $\Omega(z) = v_0 z, v_0 =$ real, the real and imaginary parts are

$$\phi(x,y) = v_0 x, \quad \psi(x,y) = v_0 y. \tag{10.55}$$

Hence, the velocity is given by $v = (v_0, 0)$. It represents a uniform flow parallel to the x axis. The equipotentials are lines parallel to the y axis and the stream lines are lines parallel to the x axis [Fig. 10.3(a)].

(b) The complex potential $\Omega(z) = k \operatorname{Log} z$ gives

$$\phi = k \ln r, \quad \psi = k\theta. \tag{10.56}$$

Thus,

$$v_r = \frac{\partial \phi}{\partial r} = \frac{k}{r}, \quad v_\theta = \frac{\partial \phi}{\partial \theta} = 0, \quad v = \frac{k}{r}(\cos\theta, \sin\theta). \tag{10.57}$$

The above expression for velocity shows that the flow is radially outward for $k > 0$ (inwards for $k < 0$) and the stream lines are radial lines [Fig. 10.13(b)]. The equipotential curves are concentric circles with the center at $z = 0$. There is a source at the origin having strength σ, given by

$$\sigma = -i \oint \Omega'(z)\, dz = 2\pi k. \tag{10.58}$$

(c) For $\Omega(z) = ik \operatorname{Log} z$ we have

$$\phi = k\theta, \quad \psi = k \ln r. \tag{10.59}$$

This gives

$$v_r = \frac{\partial \phi}{\partial r} = 0, \quad v_\theta = \frac{\partial \phi}{\partial \theta} = k, \quad v = \frac{k}{r}(-\sin\theta, \cos\theta). \tag{10.60}$$

This complex potential describes the flow of a liquid around a *vortex at the origin* with constant angular velocity k. The stream lines are concentric circles and the equipotentials are radial lines. The equipotentials and the stream lines for the three cases are shown in Fig. 10.13.

§10.6.2 Short examples. We shall now describe the electric fields corresponding to the complex potentials in §10.6.1. In each case the equipotential is the same as described in §10.6.1. The electric field is given by $E = -\nabla\phi$; the expressions for the electric field are obtained by changing the sign of the velocity in §10.6.1. The flux lines have the direction opposite to that of the stream lines. Thus, §10.6.1(a) corresponds to the uniform electric field along negative x axis. The example §10.6.1(b) corresponds to the field of a uniform line charge of density $-2\pi k$

per unit length. The example ⚡10.6.1(c) cannot be realized in electrostatics, as the electric lines of force cannot form a closed loop.

⚡10.6.3 Short examples. The examples below concern two equal and opposite charges, q and $-q$, placed at $z = a$ and $z = -a$.

(a) The complex potential is

$$\Omega = -2q \operatorname{Log}(z - a) + 2q \operatorname{Log}(z + a). \tag{10.61}$$

(b) The equipotential curve having a value $\phi = \phi_0$ is given by

$$\left|\frac{z-a}{z+a}\right| = \exp\left(-\frac{\phi_0}{2q}\right). \tag{10.62}$$

This is a circle whose center z_0 and radius R can be found using ☑1.8 (p. 28) and is given by

$$z_0 = -a \coth\left(\frac{\phi_0}{2q}\right), \quad R = a\left|\operatorname{cosech}\left(\frac{\phi_0}{2q}\right)\right|. \tag{10.63}$$

(c) The stream function $\psi(x, y) = \operatorname{Im}\Omega(z)$ is

$$\psi(x, y) = 2q \arg\left(\frac{z+a}{z-a}\right). \tag{10.64}$$

This equation represents a circle with the center at $z = -2ia \coth(\psi_0/2q)$ and radius $R = a|\operatorname{cosec}(\psi_0/2q)|$.

☑10.6.4 Problem. Two coaxial cylindrical shells of radii r_1 and r_2 are held at constant potentials V_1 and V_2, respectively. Find the potential between the shells.

Solution. The cross section of the shells consists of two concentric circles, and the corresponding two-dimensional problem is just the model problem shown in Fig. 10.2(a). The function $A \ln r + B$ is harmonic for $r_1 < r < r_2$ and assumes constant values on the two circles. Thus, the solution is $\phi = A \ln r + B$; see the answer for the model problem in Fig. 10.3(a). Fixing the constants by demanding $\phi(r_1) = V_1, \phi(r_2) = V_2$, we get

$$\phi(r) = \begin{cases} V_1 + (V_2 - V_1)\frac{\ln(r/r_1)}{\ln(r_2/r_1)}, & r_1 < r < r_2, \\ 0, & r > r_2, \end{cases} \tag{10.65}$$

and the electric field is given by $\mathbf{E} = -\nabla\phi$. ∎

§10.7 Using Conformal Mappings

☑10.7.1 Problem. Find the complex potential for flow in a quadrant $x \geq 0, y \geq 0$ and show that the stream lines are rectangular hyperbolas.

Solution. Using the map $w = z^2$ the quadrant is mapped onto the upper half plane. The complex potential for the upper half plane is $\Omega = Aw$ and hence for the quadrant we have

$$\Omega = Az^2, \qquad A \in \mathbb{R}. \tag{10.66}$$

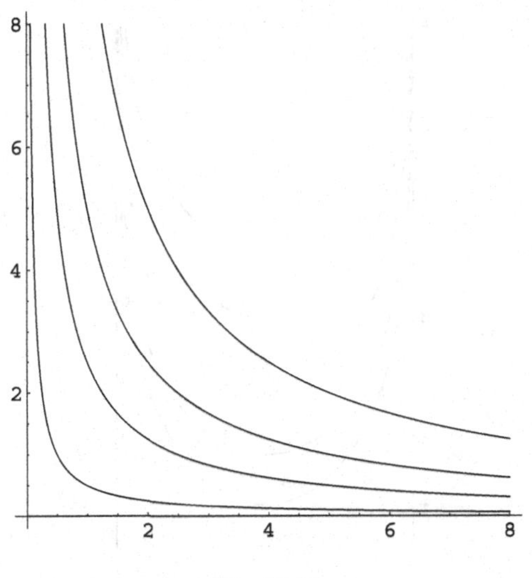

Fig. 10.14

Taking the imaginary part we get the required stream function

$$\psi = 2Axy. \tag{10.67}$$

Thus, the stream line $\psi(x, y) = \psi_0$ is a branch of a rectangular hyperbola; see Fig. 10.14. ∎

☑10.7.2 Problem. Obtain the complex potential and the stream function for flow between two semi-infinite walls.

Solution. We take the boundary of the flow region to be at $x = 0, x = \pi$, and $y = 0$. This domain can be mapped onto the upper half plane by the mapping $w = \sin(z - \pi/2) = -\cos z$. The complex potential for flow in the upper half plane is $\Omega = Aw$, giving

$$\Omega(z) = A\cos z. \tag{10.68}$$

for flow in the strip D. Thus, the stream function is

$$\psi(x, y) = \operatorname{Im}\Omega(z) = -A\sin x \sinh y. \tag{10.69}$$

The stream lines which are level curves of $\psi(x, y)$ are shown in Fig. 10.15. The fluid velocity is along the tangent to the stream lines. ∎

☑10.7.3 Problem. A conducting grounded cylinder of radius R is placed in an electric field which becomes uniform at infinity. Find the resulting field everywhere in the presence of the conducting cylinder.

Solution. We use the map

$$w = \frac{1}{2}\left(z - \frac{R^2}{z}\right) \tag{10.70}$$

to transform the given problem in Fig. 10.16(a) into the model problem in Fig. 10.16(b) in the w plane. The circular boundary of the cylinder is transformed into a line segment from $-iR$ to iR. This transformation makes the imaginary axis an equipotential surface with $\phi = 0$. The solution

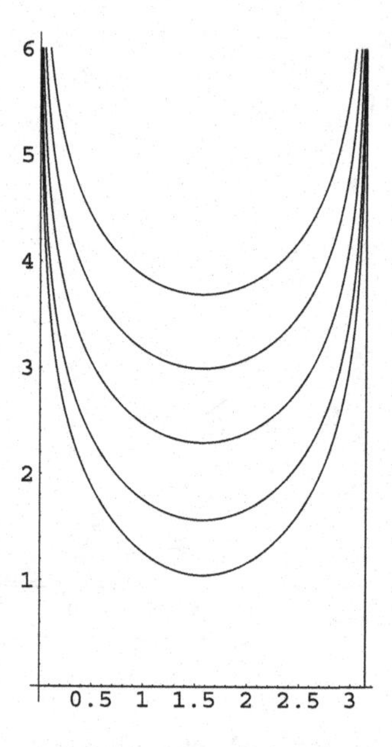

Fig. 10.15

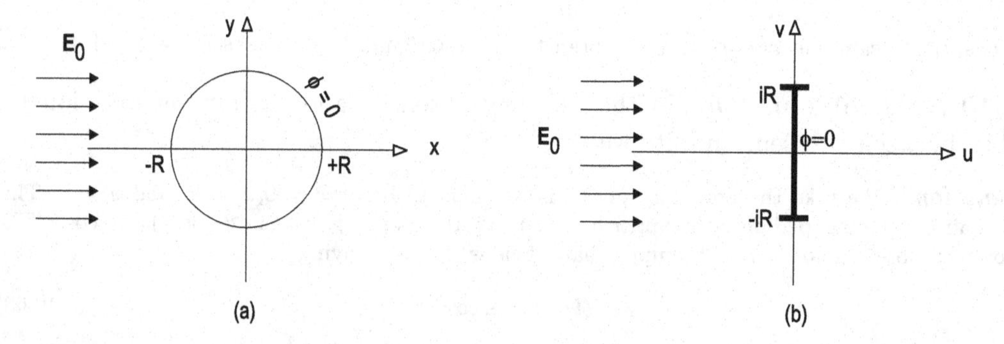

Fig. 10.16

to the model problem is obviously of the form $Aw + B$. Hence, the complex potential $\Omega(z)$ is given by

$$\Omega(z) = \frac{A}{2}\left(z - \frac{R^2}{z}\right) + B \tag{10.71}$$

and the complex electric field is

$$\mathcal{E} = -\frac{\bar{A}}{2}\left(1 + \frac{R^2}{\bar{z}^2}\right) \tag{10.72}$$

$$= -\frac{\bar{A}}{2}\left(1 + \frac{R^2}{r^2}e^{2i\theta}\right). \tag{10.73}$$

Demanding that as $r \longrightarrow \infty$, the electric field should tend to $(E_0, 0)$; we get

$$-\frac{\bar{A}}{2} = E_0 \tag{10.74}$$

and hence

$$\mathcal{E} = E_0\left(1 + \frac{R^2}{r^2}e^{2i\theta}\right). \tag{10.75}$$

This gives the components of the electric field at any point (r, θ):

$$E_x = E_0\left(1 + \frac{R^2}{r^2}\cos 2\theta\right), \tag{10.76}$$

$$E_y = E_0\frac{R^2}{r^2}\sin 2\theta. \tag{10.77}$$

The electrostatic potential is then given by

$$\phi(r, \theta) = -\mathrm{Re}\left[E_0\left(z - \frac{R^2}{z}\right) + B\right] \tag{10.78}$$

$$= -E_0\left(r - \frac{R^2}{r}\right)\cos\theta + \mathrm{Re}\,B. \tag{10.79}$$

This expression shows that the boundary of the cylinder, $r = R$, is an equipotential surface. Since $\phi = 0$ on the circle $|z| = R$, we get $\mathrm{Re}\,B = 0$ and the final expression for the potential is

$$\phi(r, \theta) = -E_0\left(r - \frac{R^2}{r}\right)\cos\theta. \tag{10.80}$$

A case where the electric field at infinity is along a given direction, making an angle α with the x axis, is left as an exercise for the reader. It should be noted that the imaginary part of the axis outside $|z| = R$, i.e. along the rays $\theta = \frac{\pi}{2}, \frac{3\pi}{2}$, is also at zero potential, the same as that of the conductor. This could have been guessed from the symmetry of the problem. ∎

☑10.7.4 Problem. For a fluid flowing past a cylindrical obstacle, find the complex potential, sketch the stream lines and find the velocity at any point.

Solution. In this problem the boundary of the cylinder is a stream line. Use the transformation

$$w = \frac{1}{2}\left(z + \frac{R^2}{z}\right) \tag{10.81}$$

to map the given problem, Fig. 10.17(a), into a problem of flow, as in Fig. 10.17(b). In this case the circle is mapped onto a line segment on the real axis. The complex potential for the new problem

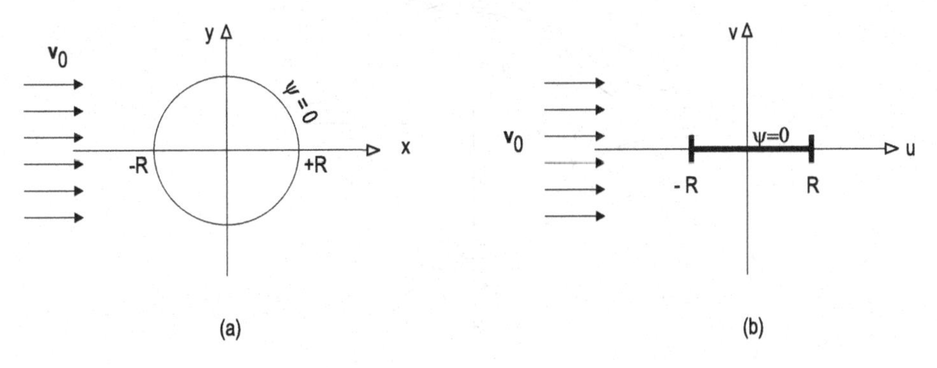

Fig. 10.17

is $Aw + B$, giving

$$\Omega(z) = \frac{A}{2}\left(z + \frac{R^2}{z}\right) + B. \tag{10.82}$$

Assuming A and B to be real, the potential and the stream functions are

$$\phi = \operatorname{Re}\Omega = \frac{A}{2}\left(r + \frac{R^2}{r}\right)\cos\theta + B, \tag{10.83}$$

$$\psi = \operatorname{Im}\Omega = \frac{A}{2}\left(r - \frac{R^2}{r}\right)\sin\theta. \tag{10.84}$$

The complex velocity at a point (r, θ) is given by

$$\mathcal{V} = \overline{\left(\frac{d\Omega}{dz}\right)} = \frac{A}{2}\left(1 - \frac{R^2}{\bar{z}^2}\right). \tag{10.85}$$

Now we use $v \longrightarrow v_0$ as the condition on the velocity at large distances to get $A/2 = v_0$. Therefore,

$$\mathcal{V} = v_0\left(1 - \frac{R^2}{\bar{z}^2}\right) = v_0\left(1 - \frac{R^2}{r^2}e^{2i\theta}\right). \tag{10.86}$$

Thus, the components of the velocity at any point are

$$v_x = \operatorname{Re}\mathcal{V} = v_0\left(1 - \frac{R^2}{r^2}\cos\theta\right), \tag{10.87}$$

$$v_y = \operatorname{Im}\mathcal{V} = -v_0\frac{R^2}{r^2}\sin 2\theta. \tag{10.88}$$

The potential and the stream functions are given by

$$\phi = v_0\left(r + \frac{R^2}{r}\right)\cos\theta, \quad \psi = v_0\left(r - \frac{R^2}{r}\right)\sin\theta. \tag{10.89}$$

The real axis is thus a stream line $\psi = 0$. The stream lines are plotted in Fig. 10.18. Note that the speed becomes zero at $z = \pm R$. These points are called *stagnation points*. ∎

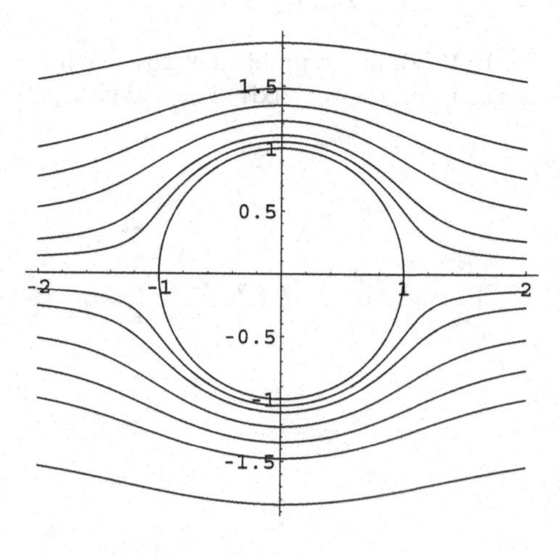

Fig. 10.18

⊳ Exercise §§10.3, for boundary value problems in electrostatics on using conformal maps.
⊳ Exercise §§10.5, for fluid flow problems using conformal maps.

§10.8 Method of Images

The method of images is an extremely useful method for solving boundary value problems. When combined with conformal mapping it provides a convenient and easy solution to many problems. Suppose that we need to find the electrostatic potential in a domain D in the presence of conductors and charges. Further, assume that the potentials of the conductors are known. This problem is equivalent to another problem in which some of the conductors have been removed and additional charges have been placed *outside* D. Placing the charges outside the domain is necessary in order to ensure that the charge density in D remains the same in the second problem. The value and location of additional charges, called *image charges*, must be selected so that the curves, corresponding to the boundaries, *representing the conductors in the first problem but absent in the second*, remain equipotential curves. In addition, the potential on each of these curves must coincide with the value of the potential on the corresponding conductor in the original problem.

While one is applying the method of images to problems of the flow of fluids, the stream function is constant over the boundary of the fluid flow and the stream function takes over the role of the equipotentials. The value of the stream function must not change on the boundary when the image sources are considered.

⁄10.8.1 Short examples. We give a few simple examples to show the method of images:

(a) Figure 10.19(a) shows an infinite plane conductor at $y = 0$ and a line charge q located in the upper half plane at z_0 The second problem, sketched in Fig. 10.19(b), contains the original charge at z_0 and an image charge $-q_0$ at $\bar{z}_0$ but the conductor has been removed. In Fig. 10.19(b) the x axis remains an equipotential curve at zero potential. Also the charge distribution in the upper half plane for the two problems coincides. The method of images then tells us that the potential at all points in the upper half plane will be the same for the

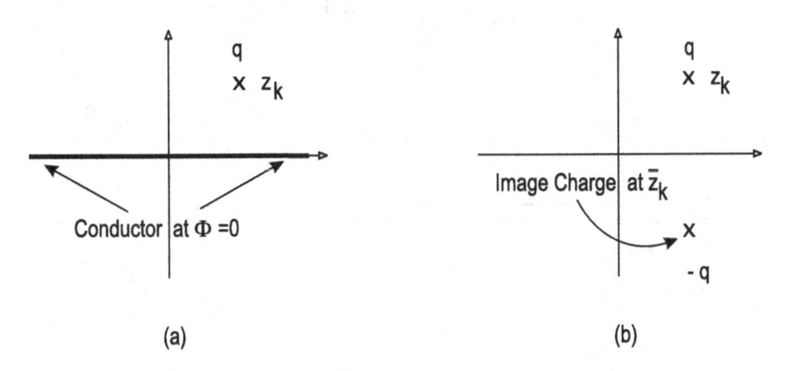

Fig. 10.19

two problems. Since the potential for Fig. 10.19(b) is known, one easily gets the solution to the problem in Fig. 10.19(a) and the required answer is

$$\phi(x,y) = -2q\text{Re}\left[\text{Log}\,(z-z_0) - \text{Log}\,(z-\bar{z}_0)\right]. \tag{10.90}$$

(b) We shall now find the complex potential for fluid flow from a point source of strength k at $z = ia$ in the presence of an infinite plane boundary at $y = 0$. This solution to this problem runs parallel to the previous electrostatic case. In this case the boundary condition is $\psi = 0$, or $\frac{\partial \phi}{\partial n} = 0$, and the *image source must be of the same sign,* $+k$, placed at $-ia$. The resulting complex potential is

$$\Omega(z) = 2k\left[\text{Log}\,(z-ia) + \text{Log}\,(z+ia)\right] \tag{10.91}$$

and its imaginary part vanishes on the real axis, i.e. $\psi = 0$ for $y = 0$.

(c) As another simple example of the method of images applied to the flow of fluids, recall that the radial lines are stream lines for a point source at $z = 0$. If we replace the rays $\theta = 0$ and $\theta = \pi$ with boundary, the new problem that we get is that of a fluid flow into the upper half plane through a narrow slit at $z = 0$. The complex potential for this problem has the same form,

$$\Omega(z) = k\,\text{Log}\,z, \tag{10.92}$$

as for a point source alone. The constant k has to be determined in terms of the rate of flow of the fluid through the slit, see ↯10.4(a).

☑10.8.2 *Problem.* *Field and capacitance of two parallel cylindrical capacitors* Use the method of images to find the capacitance of two cylinders of radii r_1 and r_2 with distance $d > r_1 + r_2$ between their axes and having potentials $-V_0$ and V_0. Compute the electric field in domain D outside both the cylinders.

Solution. We shall replace the two cylinders by two image charges q and $-q$, with the centers located at $z = a$ and $z = -a$, and determine the values of q and a such that the potentials on the two circular boundaries coincide with $\pm V_0$. There are no other charges in the problem and hence the complex potential is given by

$$\Omega(z) = -2q\log(z-a) + 2q\log(z+a) + C. \tag{10.93}$$

A constant C has been included to adjust the reference point of the potential. Thus, the electric potential is given by

$$\phi(x,y) - C = 2q\log\frac{(x+a)^2+y^2}{(x-a)^2+y^2}. \tag{10.94}$$

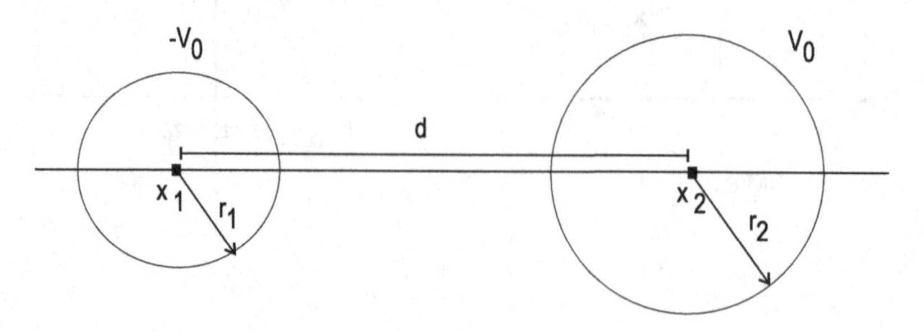

Fig. 10.20

An equipotential curve with potential $\pm V_0$ given by

$$\left|\frac{z+a}{z-a}\right| = \exp\left(\frac{\pm V_0 - C}{2q}\right) \tag{10.95}$$

is a circle. If x_1 and x_2 are the centers of the two circles, using the results of Q.[11], §§1.9, we have

$$x_1 = a\coth\left(\frac{-V_0 - C}{2q}\right), \quad r_1 = a\left|\operatorname{cosech}\left(\frac{V_0 + C}{2q}\right)\right|, \tag{10.96}$$

$$x_2 = a\coth\left(\frac{V_0 - C}{2q}\right), \quad r_2 = a\left|\operatorname{cosech}\left(\frac{V_0 - C}{2q}\right)\right|. \tag{10.97}$$

Since $2V_0$ is the potential difference between the two conductors, the capacitance is $q/2V_0$ and we must express it in terms of d, r_1, and r_2. Equations (10.97) and (10.98) give

$$\cosh\left(\frac{V_0 + C}{2q}\right) = -\frac{x_1}{r_1}, \quad \cosh\left(\frac{V_0 - C}{2q}\right) = \frac{x_2}{r_2}. \tag{10.98}$$

Using the identity for $\cosh(\alpha + \beta)$ with $\alpha = (V_0 + C)/2q$ and $\beta = (V_0 - C)/2q$, we get

$$\cosh\left(\frac{V_0}{q}\right) = \frac{-x_1 x_2 + a^2}{r_1 r_2} \tag{10.99}$$

$$= \frac{d^2 - r_1^2 - r_2^2}{2r_1 r_2}, \tag{10.100}$$

where we have used $x_1 x_2 = (r_1^2 + r_2^2 + 2a^2 - d^2)/2$, which follows from the relations

$$d = |x_1 - x_2|, \quad x_1^2 - r_1^2 = a^2, \quad x_2^2 - r_2^2 = a^2.$$

Thus, the capacitance is given by

$$\text{capacitance} = \frac{q}{2V_0} = \frac{1}{2\cosh^{-1}\left(\frac{d^2 - r_1^2 - r_2^2}{2r_1 r_2}\right)}. \tag{10.101}$$

▮

⊳⊳ Tutorial §§10.8, on boundary value problems.

⊳⊳ Exercise §§10.6, on boundary value problems in electrostatics and flow of fluids using the method of images.

§10.9 Using the Schwarz–Christoffel Transformation

☑**10.9.1 Problem.** A semi-infinite channel consists of two parallel half lines, $L_1 = \{z|y = d, x \leq 0\}$ and $L_2 = \{z|y = -d, x \leq 0\}$. Find the complex potential, stream lines, and velocity at any point for the flow of fluid through the channel.

Solution. Figure 10.21(a) shows the given semi-infinite channel. This problem can be solved by mapping it onto the model problem of flow in an infinite channel in Fig. 10.21(b), in the w plane. For the model problem, the complex potential is given by $\Omega = Aw + B$.

Due to the symmetry of the given problem, it is sufficient to consider the mapping of the domain of interest in the upper half plane only. Thus, we restrict our attention to the domain D consisting of the upper half z plane with a slit at $\operatorname{Im} z = id$ [see Fig. 10.22(a)]. The solution will

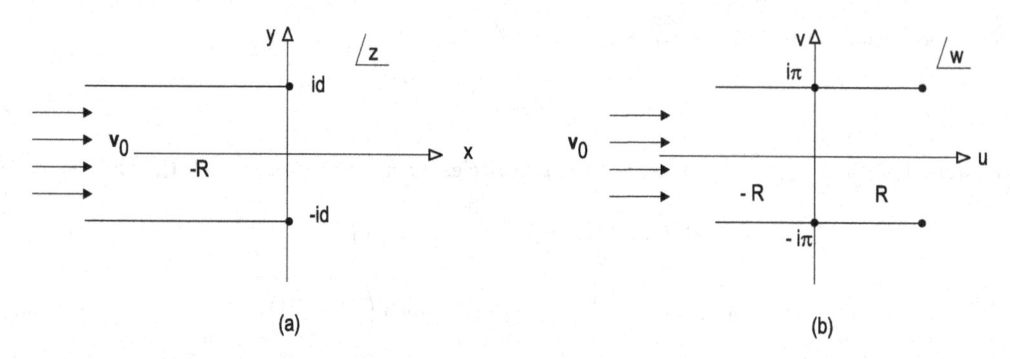

Fig. 10.21

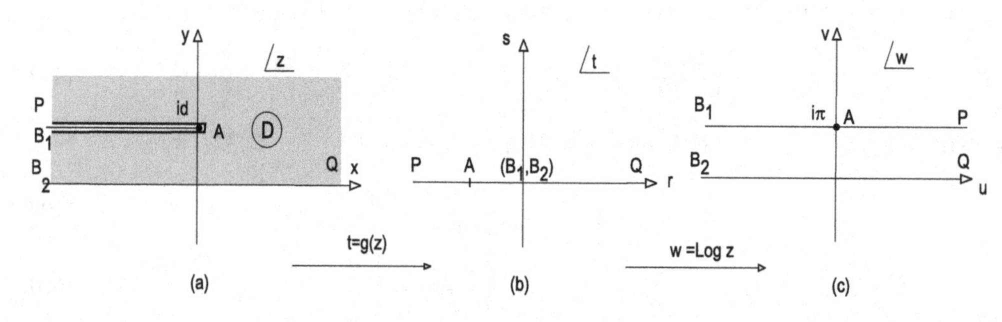

Fig. 10.22

be obtained by finding a map from the z plane to the w plane as the composite of two maps. The first map, $z \to t = g(z)$, transforms D onto the upper half t plane. The second map, $t \to w = w(t)$, transforms the upper half plane of Fig. 10.22(b) onto the problem of flow in a channel shown in Fig. 10.22(c).

To construct the mapping $g(z)$, we use the Schwarz–Christoffel transformation to find the map $z = f(t)$ of the upper half t plane onto the domain D and get

$$f(t) = \frac{d}{\pi}(t + \operatorname{Log} t + 1). \tag{10.102}$$

The required mapping $g(z)$ is then just the inverse mapping of $f(t)$. Finally, $w = \operatorname{Log} t$ maps the given problem onto the problem of flow in a channel in the w plane. Hence, we get

$$z = \frac{d}{\pi}\left(w + e^{w} + 1\right). \tag{10.103}$$

The velocity at any point is obtained from

$$v_x - iv_y = \bar{\mathcal{V}} = \frac{d\Omega}{dz} \tag{10.104}$$

$$= A \bigg/ \frac{dz}{dw} \tag{10.105}$$

$$= \frac{\pi A}{d} \cdot \frac{1}{1 + e^{w}}. \tag{10.106}$$

Note that $v_x \to \frac{\pi A}{d}$ as $x \to -\infty$ ($|y| < d$) because $\operatorname{Re} w < 0$, and $\exp(w) \to 0$. On the other hand, $v \to 0$ for $x \to \infty$ because $\operatorname{Re} w > 0$ and $\exp(w) \to \infty$.

∎

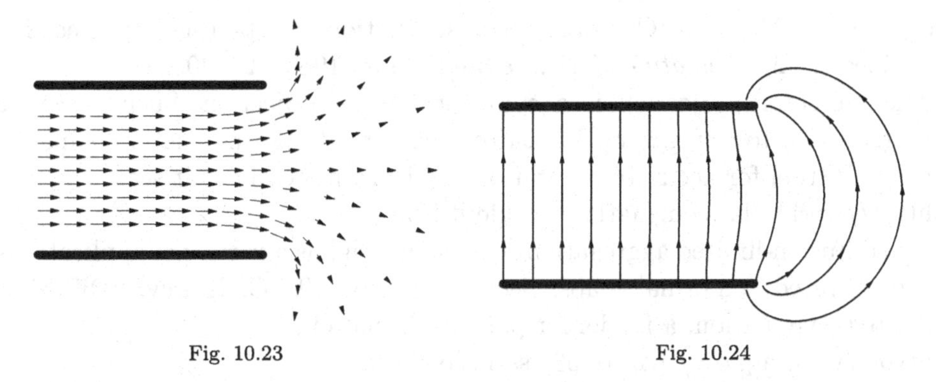

Fig. 10.23 Fig. 10.24

The stream lines are the images of $v = \text{const}$ and become parallel to the x axis as $x \to -\infty$, and are shown in Fig. 10.23. In the above fluid flow problem, the boundary condition used is that the stream function, $\psi(x, y) = \operatorname{Im} \Omega(z)$, has constant values on the channel boundary. The same solution with a different boundary condition gives the electric field inside a capacitor with semi-infinite plates. In Fig. 10.24 we give the plot of the electric field lines for this electrostatic problem. The details of this problem are left as an exercise for the reader.

It may be noted that, while using the Schwarz–Christoffel transformation, it is easier to compute the complex velocity and complex electric field directly using the complex potential, rather than calculating the real and imaginary parts.

Exercise §§10.9, for several exercises using the Schwarz–Christoffel transformation.
Quiz §§10.4, on nine problems with solutions using one conformal mapping.
Mixed Bag §§10.11, for a variety of problems.

§10.10 Notes and References

(◇1) Conformal mapping has been used to design high-tech metamaterials which can make an object invisible. To learn more about it see: J. B. Pendry, D. Schurig, and D. R. Smith (23 June 2006), *Science* **312** (5781), 1780 [DOI: 10.1126/science.1125907].

(◇2) There are several cases of flows where the boundary is not given in advance, but is to be determined after the problem has been solved. See discussion of the hodograph in [Polya (1974); Carrier *et al.* (1966)].

(◇3) One of the important applications of conformal mappings is in design of wing shaping. An article for nonexperts in aerodynamics is: R. S. Burlington, "On the use of conformal mapping in shaping wing profiles," *The American Mathematical Monthly*, **47**(6) (Jun.–Jul., 1940), 362–373.

(◇4) The Smith chart is a useful graphical representation of bilinear transformations used by electrical engineers. To learn more about it see: H. F. Mathis, "Conformal transformation charts used by electrical engineers," *Mathematics Magazine* **36**(1) (Jan. 1963), 25–30.

($\diamond$5) The Mercator projection is a cylindrical map projection. Such a map is of great use to navigators. Mercator's map made the magnetic compass a practical tool for accurate navigation. To learn more about it see: http://worldatlas.com/aatlas/worldout.htm

($\diamond$6) The diffusion-limited aggregation scheme, which has a variety of applications, uses iterated conformal maps. See for example: T. C. Halsey, "Diffusion-limited aggregation: a model for pattern formation," http://www.aip.org/ pt/vol-53/iss-11/p36.html

Part II

Problem Sessions

Chapter 1

COMPLEX NUMBERS

§§1.1 Exercise

Polar Form of Complex Numbers

[1] Which of the following equations are correct? If an equation is incorrect, correct it by changing the right hand side only.

(a) $\exp(i\pi/2) = -1$, (b) $\exp(i\pi) = -i$,

(c) $\exp(-i\pi/2) = i$, (d) $\exp(21\pi i) = 1$,

(e) $\frac{1}{2}(\sqrt{3} + i) = \exp(i\pi/3)$, (f) $\frac{1}{2}(1 - \sqrt{3}i) = \exp(-i\pi/6)$.

[2] Express the following complex numbers in the polar form:

(a) $-1 - i$, (b) $-1 - i\sqrt{3}$, (c) $\frac{1}{\sqrt{3}+i}$,

(d) $1 - i\sqrt{3}$, (e) $1 + \exp(i\alpha)$, (f) $(1 - i)^{17}$,

(g) $\dfrac{\sqrt{3} - i}{1 + i\sqrt{3}}$, (h) $\dfrac{-1 + i}{1 - i\sqrt{3}}$, (i) $\dfrac{\exp(i\pi/3)}{1 - \exp(2i\pi/3)}$.

[3] Express the following numbers in the Cartesian form (such as $a + ib$):

(a) $\sqrt{3}\exp(5\pi i/6)$, (b) $2\sqrt{2}\exp(3\pi i/4)$,

(c) $\sqrt{3}\exp(11\pi i/6)$, (d) $\sqrt{2}\exp(107\pi i/4)$,

(e) $\dfrac{2 + i}{3i + 1}$, (f) $\dfrac{2 + i}{(1 - i)(3 + i)^2}$,

(g) $\left(\dfrac{1 + i}{\sqrt{2}}\right)^{2001}$, (h) $\left(\dfrac{1 - i}{1 + i}\right)^3$.

[4] Let x and y be the real and imaginary parts of a complex number z. Two of the following three statements are equivalent, while the third one is different:

- $y = x - 20$,
- $\arg(z - 20) = \pi/4$,
- $\tan^{-1}[y/(x - 20)] = \pi/4$.

Find the statement which is not equivalent to the other two statements.

[5] Four complex numbers are listed in the first column of the table below. Find $\theta \equiv \arg(z)$ when the range of $\arg(z)$ is selected as indicated by the column heading and complete the table.

z	$0 < \theta < 2\pi$	$\frac{\pi}{2} < \theta < \frac{3\pi}{2}$	$-\pi < \theta < \pi$	$-\frac{3\pi}{2} < \theta < \frac{\pi}{2}$
$(1 - i)^5$				
$(-1 + \sqrt{3}\,i)^5$				
$(-1 - i)^{51}$				
$(1 + i)^{113}$				

Answers

[1] (a) $\exp(i\pi/2) = i$, (b) $\exp(i\pi) = -1$, (c) $\exp(-i\pi/2) = -i$,
(d) $\exp(21\pi i) = -1$, (e) $\frac{\sqrt{3}+i}{2} = \exp(i\pi/6)$, (f) $\frac{1-i\sqrt{3}}{2} = \exp(-i\pi/3)$.

[2] (a) $\sqrt{2}\exp(5\pi i/4)$, (b) $2\exp(4\pi i/3)$, (c) $\frac{1}{2}\exp(-i\pi/6)$,
(d) $2\exp(5i\pi/3)$, (e) $2\cos(\alpha/2)\exp(i\alpha/2)$, (f) $256\sqrt{2}\exp(-i\pi/4)$,
(g) $\exp(-i\pi/2)$, (h) $\frac{1}{\sqrt{2}}\exp(-11\pi i/12)$, (i) $\frac{1}{\sqrt{3}}\exp(i\pi/2)$.

[3] (a) $-\frac{3}{2} + i\frac{\sqrt{3}}{2}$, (b) $-2 + 2i$, (c) $\frac{3}{2} - i\frac{\sqrt{3}}{2}$,
(d) $-1 + i$, (e) $\frac{1}{2} - \frac{i}{2}$, (f) $(13 + 9i)/100$,
(g) $\frac{1}{\sqrt{2}}(1 + i)$, (h) i.

[4] $y = x - 20$ and $\tan^{-1}[y/(x - 20)] = \pi/4$ are equivalent and $\arg(z - 20) = \pi/4$ is different.

[5] The values of $\arg(z)$ for different restrictions are tabulated below:

z	$0 < \theta < 2\pi$	$-\frac{\pi}{2} < \theta < \frac{3\pi}{2}$	$-\pi < \theta < \pi$	$-\frac{3\pi}{2} < \theta < \frac{\pi}{2}$
$(1 - i)^5$	$3\pi/4$	$3\pi/4$	$3\pi/4$	$-5\pi/4$
$(-1 + \sqrt{3}i)^5$	$4\pi/3$	$4\pi/3$	$-2\pi/3$	$-2\pi/3$
$(-1 - i)^{51}$	$7\pi/4$	$-\pi/4$	$-\pi/4$	$-\pi/4$
$(1 + i)^{113}$	$\pi/4$	$\pi/4$	$\pi/4$	$\pi/4$

§§1.2 Exercise

Curves in the Complex Plane

[1] Find complex number z such that

(a) $\arg(z) = \frac{3\pi}{4}, |z| = 2;$

(b) $\arg(z-1-i) = \frac{\pi}{2}, |z-1-i| = 3;$

(c) $\arg(z-1) = \frac{\pi}{2}, |z+2| = 3.$

[2] Draw a diagram to show the set of points

$$\left\{z \middle| \pi \le \arg(z-1-i) \le 3\pi/2 \text{ and } 1 \le |z-1-i| \le \sqrt{2}\right\}.$$

[3] Draw the curves represented by

(a) $\{z|z = 2 + 3\exp(it), 0 \le t \le 2\pi\}$,

(b) $\{z|z = 2 + 3\exp(-it), -\pi/2 \le t \le \pi/2\}$,

(c) $\{z|z = 3 - t\exp(4\pi i/3), t \ge 0\}$,

(d) $\{z|z = 3 + t\exp(-\pi i/3), t \ge 0\}$,

(e) $\{z|z = 2 + 3i\exp(it), -\pi/4 \le t \le \pi/2\}$.

Indicate the direction of increasing t in each case.

[4] Sketch the circles described by the following equations:

(a) $|z+2| = 2|z-2|$,

(b) $|z-4i| = 3|z+2i|$.

[5] Draw the circular arcs given by

$$\left\{z \middle| \text{Arg}\left(\frac{z-i}{z+i}\right) = \alpha\right\}$$

for

(a) $\alpha = -\pi/3, \pi/3, 2\pi/3;$

(b) $\alpha = -\pi/3, \pi/2, \pi/3, \pi/4.$

[6] (a) Show that $\text{Im}\left(\frac{z-\xi_1}{\xi_2-\xi_1}\right) = 0$ represents a straight line passing through the points ξ_1 and ξ_2.

(b) Write the equation for the straight line passing through the point ξ_1 and perpendicular to the line in part (a) of this question.

[7] Show that the set of complex numbers z, satisfying

$$\left|\frac{z-i}{z+i}\right| < 1,$$

coincides with the upper half plane.

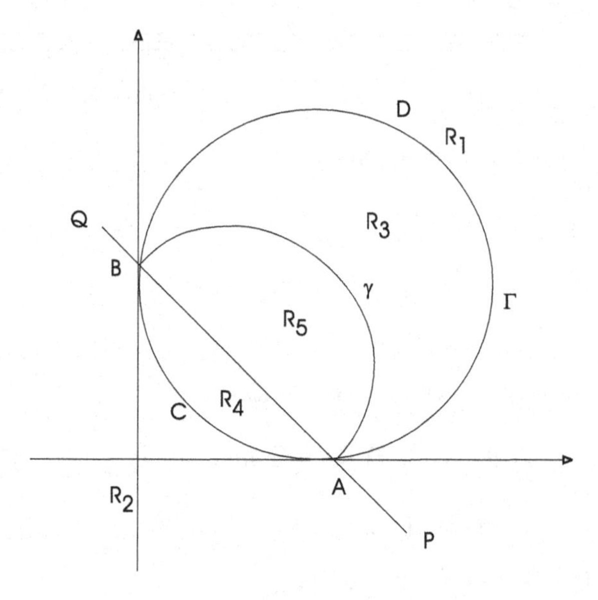

Fig. 1.1

[8] Let α and β be two distinct complex numbers and S be the set of all complex numbers z satisfying

$$\left|\frac{z-\alpha}{z-\beta}\right| < 1.$$

Give a condition on α and β such that S may coincide with

(a) the lower half plane;
(b) the left half plane;
(c) the right half plane.

[9] In Fig. 1.1, the circle Γ touches the axes at the points $z = 1, i$, shown as A and B, and γ is a semicircle with AB as diameter. Specify the subsets, described below, in terms of the function $w(z) = \dfrac{z-1}{z-i}$.

(a) The line segment between A and B;
(b) The union of two infinite line segments obtained by removing AB from the full line PQ;
(c) The two half planes separated by the line AB;
(d) The regions R_1 and R_2, obtained, respectively, by removing the points inside the circle Γ from the two half planes which are separated by the straight line PQ;
(e) The region R_3, bounded by the two circular arcs ADB and the semicircle γ;
(f) The region R_4, bounded by the circular arc ACB and the straight line AB;
(g) The region R_5, bounded by the semicircle γ and the straight line AB.

[10] Let ξ_1 and ξ_2 be two fixed distinct complex numbers and let

$$w(z) = \frac{z - \xi_2}{z - \xi_1}.$$

Sketch the subsets of the complex plane satisfying the constraints implied by the inequalities below:

(a) $2 < |w(z)| < 4$, (b) $\frac{1}{6} < |w(z)| < \frac{2}{3}$, (c) $\frac{2}{3} < |w(z)| < \frac{3}{2}$.

[11] Describe the curves traced by the point z,

$$z = \frac{3\xi - 2}{\xi - 1},$$

when ξ moves so that

(a) $\arg(\xi)$ is kept fixed and $|\xi|$ takes all possible values;
(b) $|\xi|$ is kept fixed and $\arg(\xi)$ takes values in the range $(0, 2\pi)$.

Answers

[1] ⊙ Did you try a graphical method for part (c)?

(a) $\sqrt{2}(-1+i)$, (b) $1+7i$, (c) $1+4i$.

[6] (b) $\mathrm{Re}\left(\dfrac{z-\xi_1}{\xi_1-\xi_2}\right) = 0$.

[8] ⊙ First, find the boundary where the equality holds.

(a) $\alpha = \bar{\beta}$, $\mathrm{Im}\,\alpha < 0$;
(b) $\alpha = -\bar{\beta}$, $\mathrm{Re}\,\alpha < 0$;
(c) $\alpha = -\bar{\beta}$, $\mathrm{Re}\,\alpha > 0$.

[9] (a) $\mathrm{AB} = \{z|\mathrm{Arg}[w(z)] = \pi\}$;
(b) $\mathrm{AP} \cup \mathrm{BQ} = \{z|\mathrm{Arg}[w(z)] = 0\}$;
(c) The half plane above the line PQ is $\{z|\mathrm{Arg}[w(z)] > 0\}$ and that below the line is $\{z|\mathrm{Arg}[w(z)] < 0\}$;.
(d) $R_1 = \{z|0 < \mathrm{Arg}[w(z)] < \pi/4\}$; $R_2 = \{z|-3\pi/4 < \mathrm{Arg}[w(z)] < 0\}$;
(e) $R_3 = \{z|\pi/4 < \mathrm{Arg}[w(z)] < \pi/2\}$;
(a) $R_4 = \{z|-\pi < \mathrm{Arg}[w(z)] < -3\pi/4\}$;
(b) $R_5 = \{z|\pi/2 < \mathrm{Arg}[w(z)] < \pi\}$.

[11] ⊙ Solve for ξ.

§§1.3 Exercise

Complex Numbers and Geometry

[1] Show that four points z_1, z_2, z_3, and z_4 are consecutive vertices of a parallelogram if $z_1 - z_2 + z_3 - z_4 = 0$.

[2] Show that

$$|z_1 + z_2|^2 + |z_1 - z_2|^2 = 2(|z_1|^2 + |z_2|^2)$$

and interpret the relation geometrically.

[3] Geometrically interpret the constraint represented by

$$\operatorname{Arg}\left(\frac{z_2 - z_1}{z_3 - z_1}\right) = \operatorname{Arg}\left(\frac{z_3 - z_2}{z_1 - z_2}\right) = \operatorname{Arg}\left(\frac{z_1 - z_3}{z_2 - z_3}\right),$$

where the three numbers z_1, z_2, and z_3 are all distinct.

[4] Let ABC be a triangle. Using complex numbers prove that

$$\frac{\sin A}{\mathrm{BC}} = \frac{\sin B}{\mathrm{CA}} = \frac{\sin C}{\mathrm{AB}}.$$

Relate the common value of the above expressions to properties of the triangle.

[5] Using properties of complex numbers, show that, for a quadrilateral ABCD, the following properties (of a parallelogram) are pairwise equivalent:

(a) an angle at every vertex is equal to the angle at the opposite vertex;
(b) each side equals the opposite sides;
(c) each side is parallel to the opposite side.

[6] If $|z_1| = |z_2| = |z_3|$, prove that

$$\arg\left(\frac{z_3 - z_2}{z_3 - z_1}\right) = \frac{1}{2}\arg\left(\frac{z_2}{z_1}\right).$$

What is the geometrical interpretation of this relation?

[7] Use complex variable methods to show that two sides of a triangle are equal if and only if the opposite angles are equal.

[8] Prove Pythagoras' theorem and its converse by representing the three vertices of a triangle by the complex numbers z_1, z_2, and z_3 and proving that the angle at the third vertex is $\pi/2$ if and only if the sum of the squares of the first two sides equals the square of the third side.

[9] If four distinct complex numbers z_1, z_2, z_3, and z_4 satisfying

$$|z_1 - z_2| = |z_2 - z_3| = |z_3 - z_4| = |z_4 - z_1|$$

represent consecutive vertices of a quadrilateral, without using any geometrical argument prove that

$$\operatorname{Re}\left(\frac{z_4 - z_2}{z_3 - z_1}\right) = 0.$$

Interpret the above relation geometrically.

[10] Let z_1, z_2, z_3, and z_4 be four complex numbers which satisfy

$$\arg\left(\frac{z_1 - z_2}{z_3 - z_2}\right) = \arg\left(\frac{z_1 - z_4}{z_3 - z_4}\right)$$

and

$$|z_1 - z_2| = |z_3 - z_2| \neq 0, \quad |z_1 - z_4| = |z_3 - z_4| \neq 0.$$

Show that all the four numbers cannot be distinct. One must have either $z_1 = z_3$ or $z_2 = z_4$. Interpret the result geometrically.

Answers

[2] This relation is the parallelogram law. The sum of the squares of all the four sides of a parallelogram equals the sum of the squares of the two diagonals.

[3] The three numbers z_1, z_2, and z_3 form corners of an equilateral triangle because the given condition implies that all the three angles of the triangle, formed at the three points, are equal.

[4] The common value of the three ratios is twice the area of the triangle divided by the product of the length of the three sides of the triangle.

[6] The points z_1, z_2, and z_3 lie on a circle with the center at the origin. The statement to be proven is equivalent to the well-known result in geometry that the angle made by a chord at the center is double the angle subtended in a sector of the circle.

[10] ⊙ Two numbers are equal if and only if they have equal arguments and moduli.

§§1.4 Tutorial

Geometric Representation

[1] In Fig. 1.2 the point P represents a complex number ξ and the point Q is obtained by reflecting P in the origin. The four points A, B, C, and D are at distance d from P and the points E, F, G, and H are at the same distance d from Q. The angles, which are marked, are all equal to α. Six complex numbers, $z_1, \ldots, z_6$, are defined below:

(a) $z_1 = -\xi + d\exp(i\alpha)$;
(b) $z_2 = +\xi + d\exp(i\alpha)$;
(c) $z_3 = -\xi - d\exp(-i\alpha)$;
(d) $\arg(z_4 + \xi) = \pi + \alpha, \ |z_4 + \xi| = d$;
(e) $\arg(\xi - z_5) = -\alpha, \ |\xi - z_5| = d$;
(f) $\arg(z_6 - \xi) = i - \alpha, \ |z_6 - \xi| = d$.

From among the eight points A, B, C, D, E, F, G, and H identify the points corresponding to the numbers $z_1, z_2, \ldots, z_6$. Write the complex numbers corresponding to the remaining two points in terms of ξ, d, and α.

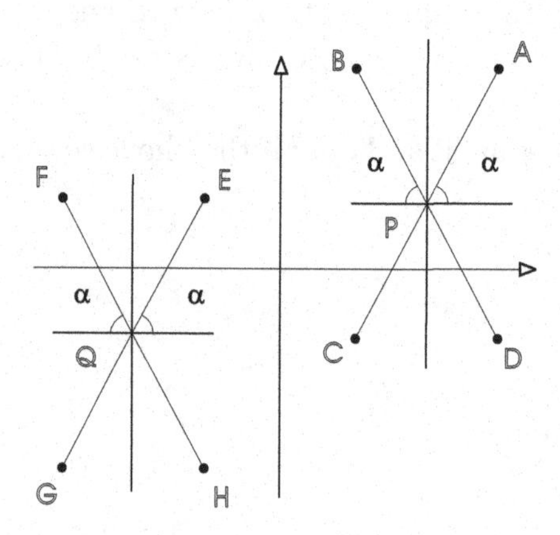

Fig. 1.2

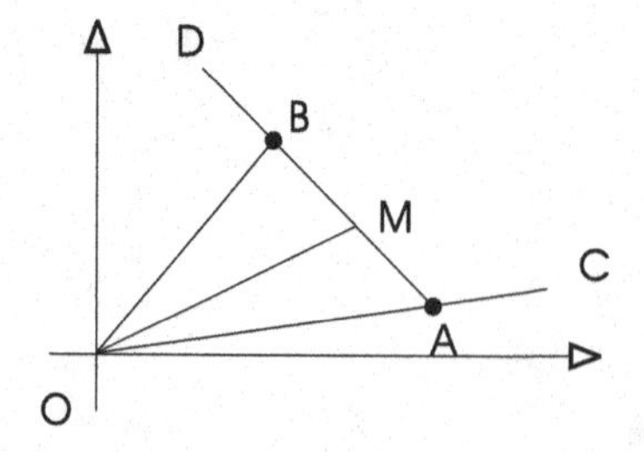

Fig. 1.3

[2] Let two points A and B represent complex numbers z_1 and z_2, respectively. Let M be the middle point of AB, as shown in Fig. 1.3. Match the six angles BAO, OAB, OBA, BMO, CAB, and DBO with the six expressions given below:

(a) $\phi_1 = \operatorname{Arg}(z_2) - \operatorname{Arg}(z_1)$,
(b) $\phi_2 = \operatorname{Arg}[\overline{(z_1 - z_2)z_1}]$,
(c) $\phi_3 = \operatorname{Arg}\left(\dfrac{z_2 - z_1}{z_2}\right)$,
(d) $\phi_4 = \operatorname{Arg}\left(\dfrac{z_1 + z_2}{z_1 - z_2}\right)$,
(e) $\phi_5 = \operatorname{Arg}[\bar{z}_1(z_1 + z_2)]$,
(f) $\phi_6 = \operatorname{Arg}[\bar{z}_1(z_2 - z_1)]$.

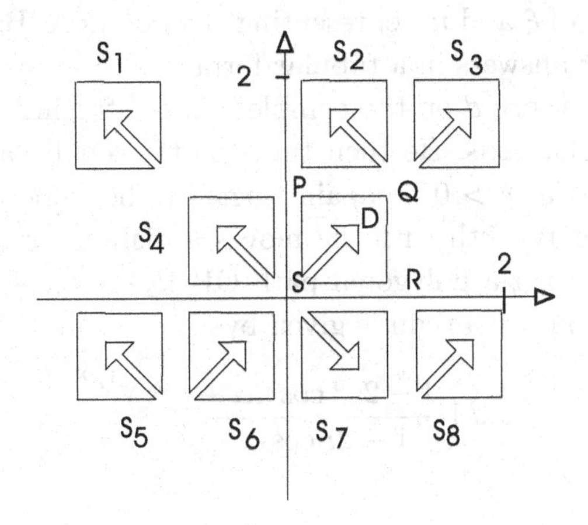

Fig. 1.4

[3] Let D be the set of points inside the square PQRS in Fig. 1.4 but excluding the points lying inside the arrow. The sets $S_1, S_2, \ldots, S_8$ are similar sets of complex numbers, as indicated in the figure. Identify the following sets with the sets from $S_1, S_2, \ldots, S_8$:

(a) $E = \{iz | z \in D\}$,
(b) $F = \{z - 1 - i | z \in D\}$,
(c) $G = \{z + 1 + i | z \in E\}$,
(d) $H = \{-iz | z \in G\}$,
(e) $K = \{z - 2 | z \in G\}$.

Answers

[1] (a) $z_A = z_2$, (b) $z_B = z_5$, (c) $z_C = \xi - de^{i\alpha}$, (d) $z_D = z_6$, (e) $z_E = z_1$, (f) $z_F = z_3$, (g) $z_G = z_4$, (h) $z_H = -\xi + de^{-i\alpha}$.

[2] (a) $\angle AOB = \phi_1$, (b) $\angle OAB = \phi_2$, (c) $\angle OBA = \phi_3$, (e) $\angle BMO = \phi_4$, (f) $\angle AOM = \phi_5$, (g) $\angle CAB = \phi_6$.

[3] (a) $E = S_4$, (b) $F = S_6$, (c) $G = S_2$, (d) $H = S_8$, (e) $K = S_1$.

§§1.5 Quiz

Transformations

[1] Let the midpoint of one of the sides of a 29-sided regular polygon be given by a complex number ξ. Taking the center of the polygon as the origin, find the complex numbers corresponding to the midpoints of the other 28 sides.

[2] In Fig. 1.5 three lines AB, CD, and EF meet at the origin O, which is also the midpoint of all the three lines. The lines AB and EF make equal angles with the real axis. The lines EV, AP, UB, and QF are all parallel to and equal in length to CO. Similarly, RC and DS are parallel to and equal in length to EO. If the numbers ξ and η represent the points E and C, respectively, find the complex numbers, in terms of ξ and η, representing the points A, B, D, F, P, Q, R, S, U, and V. Write your answers in a tabular form.

[3] A man moves a distance d on the complex plane starting from the origin at an angle α with the real axis. He then turns to the left by an angle α, and after moving a distance rd, $r > 0$ he again turns by the same angle α and moves a distance r^2d. After the kth turn, he moves a distance r^kd. Continuing in this fashion he moves along a polygonal path $\mathrm{OP_1P_2 \cdots P}_n$. Show that on reaching P_n, his distance from the origin is given by

$$d\left(\frac{1-2r^n\cos n\alpha+r^{2n}}{1-2r\cos\alpha+r^2}\right)^{1/2}.$$

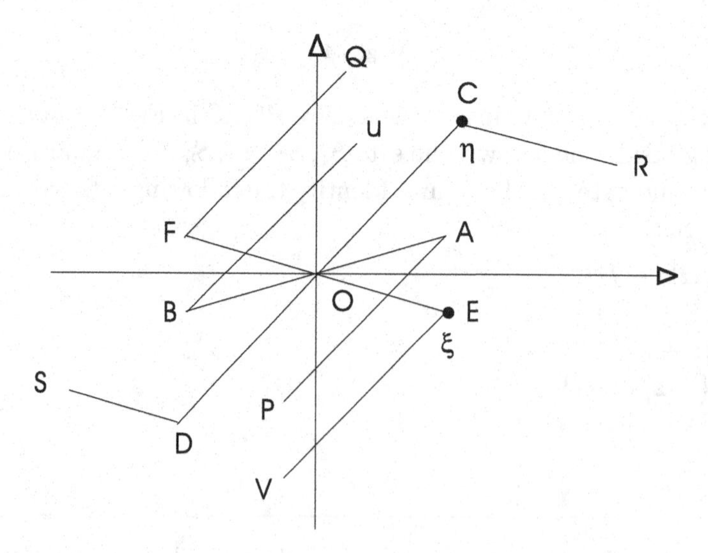

Fig. 1.5

Answers

[1] $\xi \exp\left(\frac{2\pi i}{29}\right)$, $\xi \exp\left(\frac{4\pi i}{29}\right), \ldots, \xi \exp\left(\frac{56\pi i}{29}\right)$.

[2] The points $A, B, C, \ldots$ are given by:

A	B	C	D	E	F	P	Q	R	S	U	V
ξ^*	$-\xi^*$	η	$-\eta$	ξ	$-\xi$	$\xi^* - \eta$	$-\xi + \eta$	$\xi + \eta$	$-\xi - \eta$	$-\xi^* + \eta$	$\xi - \eta$

§§1.6 Exercise

Linear Transformations

[1] The set S of all seventh roots of unity is translated by a complex number ξ. What is the equation satisfied by all the complex numbers in the set T obtained after the translation?

[2] In Fig. 1.6 three sets of seven complex numbers are shown. The set 2 is obtained from the set 1 by applying an anticlockwise rotation by an angle α about the origin; this rotation is represented by the mapping $w = \exp(i\alpha)z$. What is the mapping representing the anticlockwise rotation by an angle α about the point ξ taking the set 1 to the set 3?

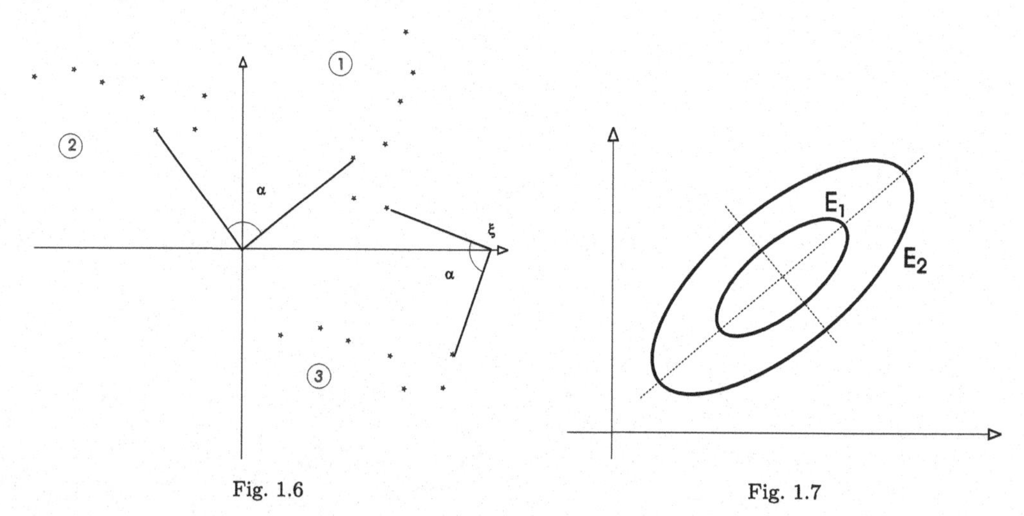

Fig. 1.6 Fig. 1.7

[3] Two ellipses E_1 and E_2, shown in Fig. 1.7, are such that their axes coincide but the ratio of the lengths of the semimajor axes is $1 : p$, and the semiminor axes are in the same ratio. Find the mapping which takes the points on and inside E_2 to the corresponding points on and inside E_1.

[4] For a complex number z introduce "shifted polar variables" ρ and ϕ by

$$\rho = |z - \xi|, \quad \phi = \arg(z - \xi).$$

Geometrically interpret the mapping $z \to z'$, where the shifted polar variables for z' are given by

(a) $\rho' = \rho, \quad \phi' = \phi + \alpha,$ (b) $\rho' = a\rho, \quad \phi' = \phi,$

where a, α are real constants.

[5] Let z_1, z_2, z_3 and z_4 be the consecutive corners of a square ABCD of side L. Draw a figure to show the set of points $\mathcal{V}$ such that

$$\mathcal{V} = \{z||z - z_k| \leq 1.5L, \ \arg(z - z_k) = \pi/3, z_k \in \mathcal{X}\}.$$

[6] Let $\mathcal{B} = \{z||z| \leq 1\}$, and draw a figure to depict the set

$$\mathcal{S} = \{z||z-\xi| \leq 4 \text{ and } \arg(z-\xi) = -\pi/2 \text{ and } \xi \in \mathcal{B}\}.$$

[7] Let

$$\mathcal{B} = \{z|z = r\exp(i\pi/4) \text{ and } 0 < r < 5\}.$$

Geometrically represent the set

$$\mathcal{S} = \{z|z = \xi\exp(i\alpha), \xi \in \mathcal{B} \text{ and } 0 \leq \alpha \leq \pi/4\}.$$

[8] A set of six points, D, in the complex plane is shown in Fig. 1.8. Find the image set $f(D)$ under each of the following mappings:

(a) $f(z) = 1/z$,
(b) $f(z) = \frac{1}{2}(1 + i\sqrt{3})z$,
(c) $f(z) = z + 3$.

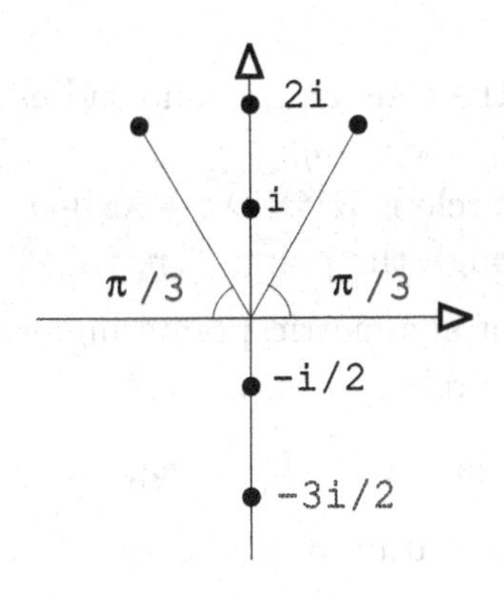

Fig. 1.8

[9] Let $f(z)$ be a function for which the six dots in Fig. 1.8 represent the set of solutions to the equation $f(z) = 0$. Plot the set of solutions to the following equations:

(a) $f(iz) = 0$,
(b) $f(\frac{z}{2} + \frac{3i}{4}) = 0$,
(c) $f(z/4) = 0$.

Answers

[1] $|z-\xi|^7 = 1$.
[2] $z \to w = (z-\xi)\exp(i\alpha) + \xi$.
[3] $z \to w = \frac{1}{p}(z-\xi) + \xi$, where ξ is the common center of the two ellipses.
[4] (a) Anticlockwise rotation about ξ by angle α.
(b) Scaling w.r.t. ξ as the center.

§§1.7 Exercise

Reflections

[1] (a) Sketch the straight line

$$e^{i\alpha}z + e^{-i\alpha}\bar{z} = 2p. \tag{1.1}$$

(b) Find a linear transformation, $w = f(z)$, which takes the given line to the imaginary axis in the w plane.

(c) Use your result in part (b) and the symmetry principle to show that the reflection z^* of a point z in the line (1) is

$$e^{i\alpha}z^* + e^{-i\alpha}\bar{z} = 2p.$$

[2] Find the mirror image of a point z_0 in the straight line

$$ax + by + c = 0.$$

[3] Let z_1 and z_2 be a pair of symmetric points w.r.t. a circle C. Find a relation between z_1 and z_2 when

(a) C is the circle with the center at z_0 and radius R;

(b) C is the circle $|z-\xi| = k|z-\eta|$;

(c) the equation of the circle is $|z|^2 + \bar{\lambda}z + \lambda\bar{z} + a = 0$;

(d) the circle passes through the points ξ, η, and ζ.

[4] If z_1 and z_2 are a pair of symmetric points under reflection in a circle, prove that the equation of the circle can be written as

$$|z - z_1| = k|z - z_2|, \quad k > 0.$$

Give a geometrical interpretation of the constant k.

[5] If the equation of a circle is

$$|z - z_1| = k|z - z_2|, \qquad k > 0,$$

show that z_1 and z_2 are images of each other under reflection in the circle.

[6] Let (z_1, z_2) and (z_3, z_4) be two pairs of symmetric points w.r.t. a circle. Prove that the four points lie on another circle.

Answers

[1] (b) $w = e^{-i\alpha}z - p$,

[2] $\bar{\lambda}z_1 + \lambda\bar{z}_2 + 2c = 0$, $\lambda=(a+ib)/2$.

[3] (a) $(z_1 - z_0)\overline{(z_2 - z_0)} = R^2$,

(b) $(z_1 - \xi)\overline{(z_2 - \xi)} = k^2(z_1 - \eta)\overline{(z_2 - \eta)}$,

(c) $z_1\bar{z}_2 + \bar{\lambda}z_1 + \lambda\bar{z}_2 + a = 0$,

(d) $(z_1, \xi, \eta, \zeta) = \overline{(z_2, \xi, \eta, \zeta)}$.

§§1.8 Mined

Complex numbers

- ⊘ For each question there are sample answers. Read the sample answers carefully and locate the mistakes, if any.
- ⊘ Identify the source of mistakes and give your reasons why you consider a step or a statement incorrect. Be as precise as you can.
- ⊘ Do not write your solution as part of your answers.
- ⊘ State if, in your opinion, a sample answer is correct, partially correct, or completely wrong. Grade each sample answer out of a maximum of 10 points.

[1] Show the set of points satisfying $\arg[z-(1+2i)] = -\pi/3$ in the complex plane.

Sample answer: *It is given that*

$$\begin{aligned} \arg[z-(1+2i)] &= -\frac{\pi}{3}, \\ \Longrightarrow \qquad \tan^{-1}\frac{y-2}{x-1} &= -\frac{\pi}{3}, \\ \frac{y-2}{x-1} &= -\sqrt{3}, \\ y-2 &= -\sqrt{3}(x-1). \end{aligned}$$

This is a line passing through $1+2i$ *and has a slope of* $-\sqrt{3}$. *The required set of points coincides with the line shown in Fig. 1.9.*

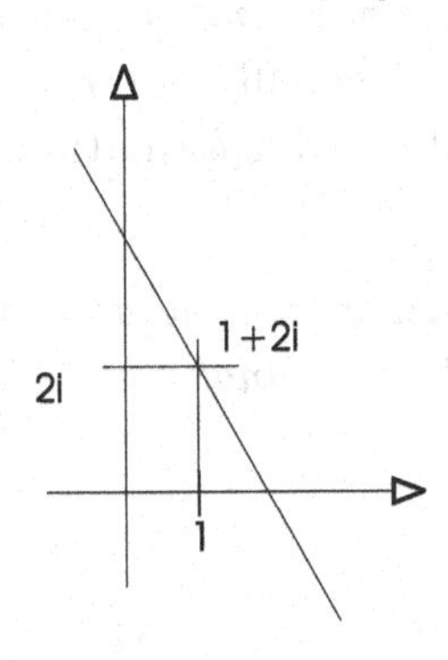

Fig. 1.9

[2] Find the images of the circles $|z| = 1$ and $|z-1/2| = 1$ under the mapping $w = 2\exp(-2\pi i/3)z$.

Sample answer 1. *The given transformation is a combination of a rotation by* $-2\pi/3$ *and a scale transformation by a factor of 2. Under the rotation the circle remains unchanged, while the scale transformation changes the radius of the circle by a factor of 2. The resulting circles are* $|z| = 2$ *and* $|z-1/2| = 2$.

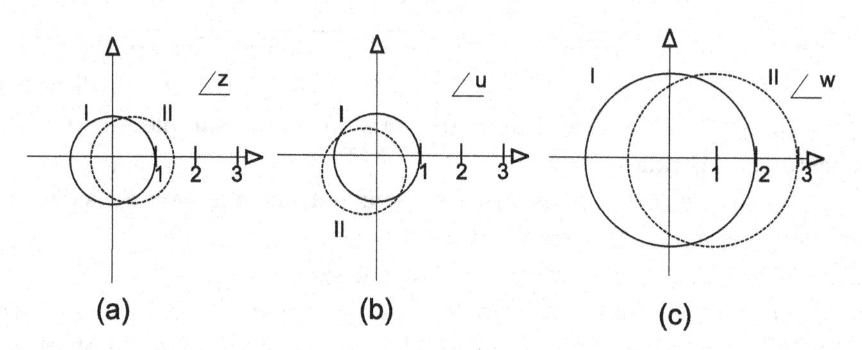

Fig. 1.10

Sample answer 2. *The two circles are* $|z| = 1$ *and* $|z - 1/2| = 1$. *They are plotted in Fig. 1.10(a) and are labeled I and II. Under the given mapping* $w = 2\exp(-2\pi i/3)z$, *the equations of the new circles are*

$$|w| = 2,$$

$$|w - \exp(2\pi i/3)| = 2.$$

The transformation $w = 2\exp(-2\pi i/3)z$ *can be thought of as being a result of two transformations:*

(i) $z \to u = \exp(-2\pi i/3)z$, *which is a clockwise rotation about the origin by* $-2\pi/3$;

(ii) $u \to w = 2u$, *giving scaling by a factor of 2.*

The result of these two transformations is shown in Figs. 1.10(b) and 1.10(c).

[3] Draw a rectangle whose two diagonally opposite corners are at $1 + i$ and $4 + 3i$. Find its image under (a) scaling transformation $z \to w = 0.5z$ and (b) rotation $z \to w = \exp(i\pi/4)z$.

Sample answer. *The given rectangle and its image under the scaling by 0.5 are shown in Fig. 1.11(a). The original rectangle and the image under rotation are shown in Fig. 1.11(b).*

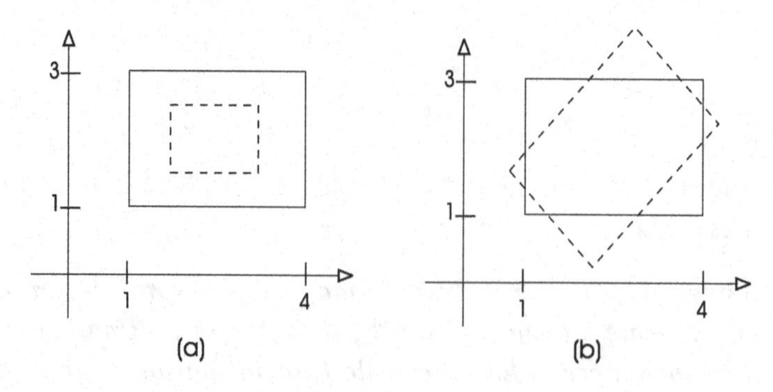

Fig. 1.11

Answers

[1] Only half the ray, directed toward the lower half plane, satisfies the given condition $\arg[z-(1+2i)] = -\pi/3$.

[2] In sample answer 1, only the first circle does not change under the rotation about the origin. Similarly, under the scale transformation by a factor of 2, the radius of the first circle becomes 2 and it is wrong to say that the radius of the second circle becomes double, as it does not have the center at the origin. Attention must be paid to the fact that the rotation and scaling are w.r.t. the origin.

[3] In sample answer 2, the equations of the image are written but not used. In Fig. 1.10(c), image II is not the correct image obtained by scaling by a factor of 2.

[4] The rotation and scaling have been applied w.r.t. the center of the rectangle, whereas the given transformations are w.r.t. the origin.

§§1.9 Mixed Bag

Complex Numbers and Transformations

[1] Provide algebraic proofs of the following inequalities, given in the text in ≀1.10 (p. 32):

(a) $|z| > \text{Re } z$, $|z| > \text{ Im } z$;
(b) $|z_1 + z_2| \leq |z_1| + |z_2|$;
(c) $|z_1 - z_2| \geq ||z_1| - |z_2||$;
(d) $|z_1 + z_2 + \cdots + z_n| \leq |z_1| + |z_2| + \cdots + |z_n|$.

[2] If ξ and η are complex numbers with $|\xi| \leq 1$ and $|\eta| \leq 1$, prove the inequality

$$\left|\frac{\xi - \eta}{1 - \xi\bar{\eta}}\right| \leq 1$$

and show that the equality holds if and only if $|\xi| = 1$ or $|\eta| = 1$.

[3] For nonzero complex numbers $z, z_1, z_2, \ldots, z_n$, find necessary and sufficient conditions so that

$$|z| = |z_1| + |z_2| + |z_3| + \cdots + |z_n|$$

may become equivalent to

$$z = z_1 + z_2 + z_3 + \cdots + z_n.$$

[4] Prove Lagrange's identity:

$$\left|\sum_{k=1}^{n} \xi_k \eta_k\right|^2 = \left|\sum_{k=1}^{n} |\xi_k|^2\right| \left|\sum_{k=1}^{n} |\eta_k|^2\right| - \left|\sum_{1 \leq k \langle j \leq n} |\xi_k^* \eta_k - \xi_k \eta_k^*|\right|.$$

[5] Let $\xi = (\xi_1, \xi_2, \ldots, \xi_n)$ and $\eta = (\eta_1, \eta_2, \ldots, \eta_n)$. Prove the Cauchy–Schwarz inequality

$$|(\xi, \eta)| \leq \|\xi\| \, \|\eta\|,$$

where

$$(\xi, \eta) = \sum_{k=1}^{n} \xi_k^* \eta_k, \quad \|\xi\| = \sqrt{(\xi, \xi)}.$$

[6] Let (ξ, η) be defined as in the previous question. Prove the Hölder inequality

$$|(\xi, \eta)| \leq \left(\sum_{k=1}^{n} |\xi_k|^p\right)^{\frac{1}{p}} \left(\sum_{k=1}^{n} |\eta_k|^q\right)^{\frac{1}{q}},$$

where $\dfrac{1}{p} + \dfrac{1}{q} = 1$.

[7] Prove the Minkowski inequality for complex numbers

$$\left[\sum_{k=1}^{n} |\xi_k + \eta_k|^p\right]^{\frac{1}{p}} \le \left[\sum_{k=1}^{n} |\xi_k|^p\right]^{\frac{1}{p}} + \left[\sum_{k=1}^{n} |\eta_k|^p\right]^{\frac{1}{p}},$$

where $p > 1$.

[8] Let z_A, z_B, and z_C be complex numbers represented by three points A, B, and C respectively, in the complex plane. Write as many conditions on the numbers z_A, z_B, and z_C as you can such that each one implies that the three points A, B, and C are collinear.

[9] Let z_A, z_B, and z_C be the complex numbers representing vertices of an equilateral triangle ABC in the complex plane. Prove that

$$z_A^2 + z_B^2 + z_C^2 = z_A z_B + z_B z_C + z_C z_A.$$

Is the above condition sufficient for the triangle to be an equilateral triangle?

[10] Show that the constraint

$$|z - z_1| = k|z - z_2|$$

represents a circle and find its center z_0 and radius R by demanding that this relation be satisfied by $z = z_0 + R\exp(i\phi)$ for all $\phi \in (0, 2\pi)$.

[11] Show that the radius R and the center X of the circle described by

$$|z + a| = e^{\alpha}|z - a|$$

are given by

$$X = a\coth\alpha, \quad R = a|\text{cosech}\ \alpha|.$$

[12] (a) Show that four distinct, noncollinear points, z_1, z_2, z_3, and z_4, lie on a circle if and only if the "cross ratio"

$$(z_1, z_2, z_3, z_4) \equiv \left(\frac{z_1 - z_2}{z_1 - z_3}\right)\left(\frac{z_4 - z_3}{z_4 - z_2}\right)$$

is real.

(b) Show that η is a reflection of ξ in a circle passing through three points z_1, z_2, and z_3 if

$$\overline{(z_1, z_2, z_3, \eta)} = (z_1, z_2, z_3, \xi).$$

[13] Let A, B, C, and D be four consecutive vertices of a cyclic quadrilateral. Making use of the cyclic property of the quadrilateral in terms of complex numbers, show that

$$\text{AB} \times \text{CD} + \text{AD} \times \text{BC} = \text{AC} \times \text{BD}.$$

[14] (a) Relate the quantity $\text{Im}\,(z_1^* z_2)$ to an area in the complex plane.

(b) Give examples of cases where the area as defined in the previous part is (i) positive and (ii) negative.

(c) Let z_1, z_2, and z_3 be three complex numbers. In addition, let Δ be defined by

$$\Delta = \mathrm{Im}\,(z_1^* z_2 + z_2^* z_3 + z_3^* z_1).$$

Use the result in (a) to show that Δ is proportional to the area of the triangle formed by z_1, z_2, and z_3 in the complex plane.

(d) Generalize the result of part (c) to a polygon with n sides $(n > 3)$ in the complex plane.

[15] A cycloid, an epicycloid, and a hypocycloid are curves traced by a point on a circle when it rolls on a straight line, outside a fixed circle and inside a fixed circle, respectively; see Fig. 1.12 and Fig. 1.13.

(a) Obtain the parametric equation for a cycloid, considering the motion of the point of the moving circle as a combination of translation and rotation.

(b) In the cases of an epicycloid and a hypocycloid, the rolling motion may be regarded as the resultant of two rotations. Use this to find the parametric equations for an epicycloid and a hypocycloid when a circle of radius R/n rolls on a circle of radius R.

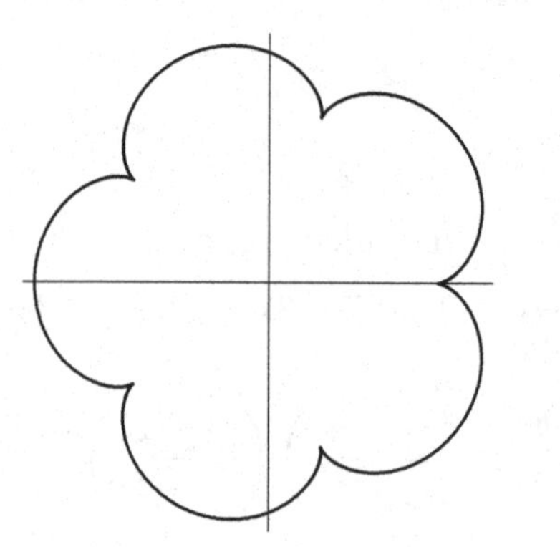

Fig. 1.12 Epicycloid for $n = 5$.

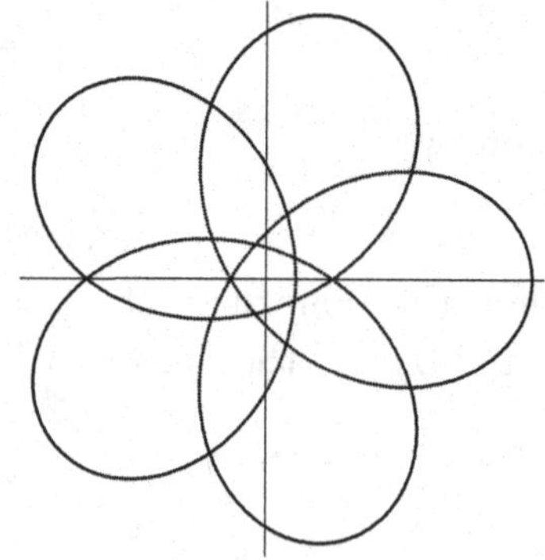

Fig. 1.13 Hypocycloid for $n = 5$.

[16] For fixed ξ and η, show that each circle C_k,

$$C_k = \{z \,|\, |z - \xi| = k|z - \eta|\} \quad k > 0,$$

in the family of circles in ☑1.8 p. 28 is orthogonal to every circle $S_\alpha \bigcup S_{-\alpha}$, described in ✎1.7(d) p. 28, with S_α given by

$$S_\alpha = \left\{ z \,\middle|\, \mathrm{Arg}\left(\frac{z-\xi}{z-\eta}\right) = \alpha \right\}.$$

Answers

[5] $\odot$ Introduce $\lambda = t + is$ and define

$$\chi = (\xi_1 - \lambda\eta_1, \xi_2 - \lambda\eta_2, \ldots)$$

and minimize (χ, χ) w.r.t. t and s.

[6] $\odot$ By finding the minimum of $f(t) = at - p^{-1}t^p$ for $t \geq 0$, first prove that

$$\sum_k |\xi_k \eta_k| \leq \frac{1}{p} \sum |\xi_k|^p + \frac{1}{q} \sum |\eta_k|^q.$$

Next, replace $\xi_k \to s\xi_k$ and $\eta_k \to \eta_k/s$ and minimize the right hand side w.r.t. s.

[7] $\odot$ Use induction and Hölder's inequality.

[9] The given condition is also sufficient for the triangle to be an equilateral triangle.

[14] $\frac{1}{2}\text{Im}\,(z_1^* z_2)$ is the area of the triangle formed by the origin, z_1 and z_2.

[15] (a) The equation for the cycloid is $z = R[\theta + \exp(i\theta)]$.

(b) The equation for the epicycloid is
$z = (R/n)\{(n+1)\exp(i\theta) - \exp[i(n+1)\theta]\}$.
The equation of hypocycloid is $z = (R/n)\{(n-1)\exp(i\theta) + \exp[i(1-n)\theta]\}$.

[16] $\odot$ It is tedious to prove this result directly through calculus, without using knowledge of properties of analytic functions discussed in Ch. 3.

Chapter 2

ELEMENTARY FUNCTIONS AND DIFFERENTIATION

§§2.1 Exercise

De Moivre's Theorem

[1] Prove the following identities for the trigonometric functions of a complex variable:

(a) $\cos^2 z + \sin^2 z = 1$, (b) $\cosh^2 z - \sinh^2 z = 1$,
(c) $\sin(iz) = i \sinh z$, (d) $\cos(iz) = \cosh z$,
(e) $\sinh(iz) = i \sin z$, (f) $\cosh(iz) = \cos z$.

[2] Show that:

(a) $\sin(z_1 + z_2) = \sin z_1 \cos z_2 + \cos z_1 \sin z_2$,
(b) $\cos(z_1 + z_2) = \cos z_1 \cos z_2 - \sin z_1 \sin z_2$,
(c) $\sinh(z_1 + z_2) = \sinh z_1 \cosh z_2 + \cosh z_1 \sinh z_2$,
(d) $\cosh(z_1 + z_2) = \cosh z_1 \cosh z_2 + \sinh z_1 \sinh z_2$.

[3] Prove the following relations:

(a) $\sin(z_1 + z_2) \sin(z_1 - z_2) = \sin^2 z_1 - \sin^2 z_2$,
(b) $\cosh(z_1 + z_2) \cosh(z_1 - z_2) = \cosh^2 z_1 + \cosh^2 z_2$.

[4] Prove the following identities for the real and imaginary parts:

(a) $\sin z = \sin x \cosh y + i \cos x \sinh y$,
(b) $\cosh z = \cosh x \cos y + i \sinh x \sin y$,
(c) $|\cos z|^2 = \cos^2 x + \sinh^2 y$,
(d) $|\sinh z|^2 = \sinh^2 x + \sin^2 y$.

§§2.2 Exercise

Real and Imaginary Parts

[1] Separate the real and imaginary parts of the following functions:

(a) $\cosh^2 z$, (b) $\exp(z^2)$,

(c) $\sin z^2$, (d) $\sin[\exp(z)]$,

(e) $\exp[\exp(z)]$, (f) $\exp(\sin z)$,

(g) $\sinh(\cos z)$, (h) $\cosh(\cosh z)$,

(i) $\cos[\exp(z)]$, (j) $\operatorname{sech} z$,

(k) $\operatorname{cosec} z$, (l) $\coth z^2$.

[2] Which of the following equations are valid for all z? If a relation does not hold for all z, find the set of z values for which it holds.

(a) $\overline{\cos iz} = \cos(i\bar{z})$, (b) $\overline{\sin z} = \sin(\bar{z})$,

(c) $\overline{\tan iz} = \tan(i\bar{z})$, (d) $\overline{\tanh z} = \tanh \bar{z}$.

[3] Prove the following inequalities:

(a) $|\sin z| \geq |\sin x|$,
(b) $|\cos z| \geq |\cos x|$,
(c) $|\sinh y| \leq |\sin z| \leq |\cosh y|$,
(d) $|\sinh y| \leq |\cos z| \leq |\cosh y|$,
(e) $|\sinh x| \leq |\cosh z| \leq |\cosh x|$,
(f) $|\cot z| \leq \dfrac{\cosh y}{(\sin^2 x + \sinh^2 y)^{1/2}}$.

Answers

[1] (a) $\cosh^2 z = \frac{1}{2}(1 + \cosh 2x \cos 2y) + \frac{i}{2} \sinh 2x \sin 2y$,
(b) $\exp(z^2) = \exp(x^2 - y^2)\cos(2xy) + i\exp(x^2 - y^2)\sin(2xy)$,
(c) $\sin z^2 = \sin(x^2 - y^2)\cosh(2xy) + i\cos(x^2 - y^2)\sinh(2xy)$,
(d) $\sin[\exp(z)] = \sin(e^x \cos y)\cosh(e^x \sin y) + i\cos(e^x \cos y)\sinh(e^x \sin y)$,
(e) $\exp(\exp z) = \exp(e^x \cos y)[\cos(e^x \sin y) + i\sin(e^x \sin y)]$,
(f) $\exp(\sin z) = \exp(\sin x \cosh y)[\cos(\cos x \sinh y) + i\sin(\cos x \sinh y)]$,
(g) $\sinh(\cos z) = \sin(\cos x \cosh y)\cosh(\sin x \sinh y)$
$\qquad -i\cos(\cos x \cosh y)\sinh(\sin x \sinh y)$,
(h) $\cosh(\cosh z) = \cosh(\cosh x \cos y)\cos(\sinh x \sin y)$
$\qquad +i\sinh(\cosh x \cos y)\sin(\sinh x \sin y)$,
(i) $\cos[\exp(z)] = \cos(e^x \cos y)\cosh(e^x \sin y) - i\sin(e^x \cos y)\sinh(e^x \sin y)$,

(j) $\operatorname{sech} z = \dfrac{\cosh x \cos y - i \sinh x \sin y}{\sinh^2 x + \cos^2 y}$,

(k) $\operatorname{cosech} z = \dfrac{\sinh x \cos y - i \cosh x \sin y}{\sinh^2 x + \sin^2 y}$,

(l) $\coth z^2 = \dfrac{\sinh 2(x^2 - y^2) - i \sin(4xy)}{\cosh 2(x^2 - y^2) - \cos(4xy)}$.

[2] (a) all values of z,

(b) $z = \pm n\pi$,

(c) $z = \pm n\pi i$,

(d) all values of z.

§§2.3 Questions

Hyperbolic and Trigonometric Functions

⊘ Check if the following statements are true or false and give your comments, if any.

[1] The value of $|\exp(z)|$ is always greater than 1.

[2] The functions $\sin z$ and $\sinh z$ have zeros for real values of z only.

[3] The equation $\cosh z = 1/3$ has no solution in the complex plane because $\cosh z$ is always greater than 1.

[4] The trigonometric functions $\cos z, \sin z$, and $\tan z$ are all periodic functions of z.

[5] The hyperbolic functions $\cosh z$ and $\sinh z$ are not periodic functions.

[6] The equation $\cosh z = 3 + 4i$ has only one solution for z.

[7] There is no value of z in the complex plane such that $\sin z = 3 + 4i$ because the sine function is a real function.

[8] The functions $\tan z$ and $\cot z$ are unbounded.

[9] The function $\sin z$ is bounded, but $\sinh z$ is unbounded in the complex plane.

[10] $|\cosh z| \geq 1$ for all values of z.

[11] $\cos z \leq 1$ for all values of z.

[12] For every complex number λ, we can find a solution to the equation

$$\exp(z) = \lambda.$$

Answers

[1] False, because $|\exp(z)|$ is less than 1 when $y = 0$ and x is negative. When z is complex, $|e^z| = |e^{x+iy}| = e^x$ is less than 1 if $x < 0$.

[2] True for $\sin z$, but false for $\sinh z$.

[3] False; the reason given is applicable only when z is real.

[4] True.

[5] False; these functions are periodic with period $2\pi i$.

[6] False; there are an infinite number of solutions differing by an integral multiple of $2\pi i$.

[7] False; $\sin z$ does become complex when z takes complex values.

[8] True.

[9] $\sin z$ is also unbounded. For example, when $z = iy$ and y tends to infinity, $\sin z = \sinh y$, which tends to infinity.

[10] False.

[11] False, because one cannot compare $\cos z$ with a real (or complex number) when z is a complex number.

[12] A solution can be found for all complex λ except $\lambda = 0$.

§§2.4 Exercise

Solutions to Equations

[1] Find all values of z such that

(a) $e^z = -1 - i$, (b) $e^{(2z-3)} = 2\sqrt{2} - 3$, (c) $e^z + e^{-z} = -2$,

(d) $e^{2z} + e^z + 1 = 0$, (e) $e^{z^4} = (1+i)/\sqrt{2}$, (f) $e^{1/z} = 1 + i$.

[2] Find all complex solutions to the following equations:

(a) $5\sinh z - \cosh z = 5$, (b) $\cos z = \cosh 3$, (c) $\sin z = 2i$,

(d) $\tan z = 2 + i$, (e) $\coth z = 7$, (f) $\sin z = 5/4$.

[3] Assuming α to be real, show that the set of all real roots of $\sin z = \sin\alpha$ is given by

$$z = n\pi + (-1)^n\alpha.$$

Show that there are no other solutions for z in the complex plane, including the case where α is complex.

[4] Find all roots of the following equations:

(a) $\cosh z = \cos\alpha$, (b) $\tan z = \tan\alpha$,

(c) $\sinh z = i\sin\alpha$, (d) $\sinh z = \sinh\alpha$.

[5] Prove the following results:

(a) $\sinh^{-1} z = \log(z \pm \sqrt{z^2+1})$, (b) $\cosh^{-1} z = \log(z \pm \sqrt{z^2-1})$,

(c) $\sin^{-1} z = -i\log(iz \pm \sqrt{1-z^2})$, (d) $\cos^{-1} z = -i\log(z \pm \sqrt{z^2-1})$.

Answers

[1] (a) $\frac{1}{2}\ln 2 + (2n + \frac{5}{4})\pi i$,

(b) $\frac{1}{2}[3 + \ln(3 - 2\sqrt{2}) + (2n+1)\pi i]$,

(c) $(2n+1)\pi i$,

(d) $2(n + \frac{1}{3})\pi i, 2(n + \frac{2}{3})\pi i$,

(e) $\pm\xi, \pm i\xi$, with $\xi = \sqrt[4]{\pi(2n + \frac{1}{4})}\, e^{i\pi/8}$,

(f) $\left[\frac{1}{2}\ln 2 - (2n + \frac{1}{4})\pi i\right] / \left[\frac{1}{4}(\ln 2)^2 + (2n + \frac{1}{4})^2\pi^2\right]$.

[2] (a) $-\ln 2 - (2n+1)\pi i, \ln 3 + 2n\pi i$,

(b) $(2n\pi \pm 3)i$,

(c) $n\pi + (-1)^n \ln(\sqrt{5} - 2)i$,

(d) $(n - \frac{5}{8})\pi + \frac{i}{4}\ln 2$,
(e) $\frac{1}{2}\ln(\frac{4}{3}) + (n + \frac{1}{2})\pi i$,
(f) $(n + \frac{1}{4})\pi \pm \frac{i}{2}\ln 2$.

[3] (a) $2n\pi \pm i\alpha$,
(b) $n\pi + \alpha$,
(c) $n\pi i + (-1)^n \alpha i$,
(d) $n\pi i + (-1)^n \alpha$.

§§2.5 Mined

Solutions to Equations

⊘ For each question there are sample answers. Read the sample answers carefully and locate the mistakes, if any.

⊘ Identify the source of the mistakes and give your reasons why you consider a step or a statement incorrect. Be as precise as you can.

⊘ Do not write your solution as part of your answers.

⊘ State if, in your opinion, a sample answer is correct, partially correct, or completely wrong. Grade each sample answer out of a maximum of 10 points.

[1] Find and plot the zeros of $\sin \pi z^3$ lying inside a circle $|z| = 1.5$.

Sample answer. $\sin \pi z^3 = 0$ *has roots* $z^3 = n$, *where* n *is an integer. Therefore, the zeros of* $\sin z^3$ *are given by* $n^{1/3}, n^{1/3}\exp(2\pi i/3)$, and $n^{1/3}\exp(4\pi i/3)$, *where* n *is an integer. To plot the zeros lying inside the circle* $|z| = 1.5$, *we draw circles of radii* $n^{1/3} = 0, 1, 1.26, 1.4, \ldots$ *The points where these circles intersect the rays* $\arg(z) = 0, 2\pi/3, 4\pi/3$ *are the required zeros inside the circle* $|z| = 1.5$.

Answer

[1] The sample answer is partially correct. It misses the roots corresponding to negative n. On each circle in Fig. 2.1 there are three more zeros.

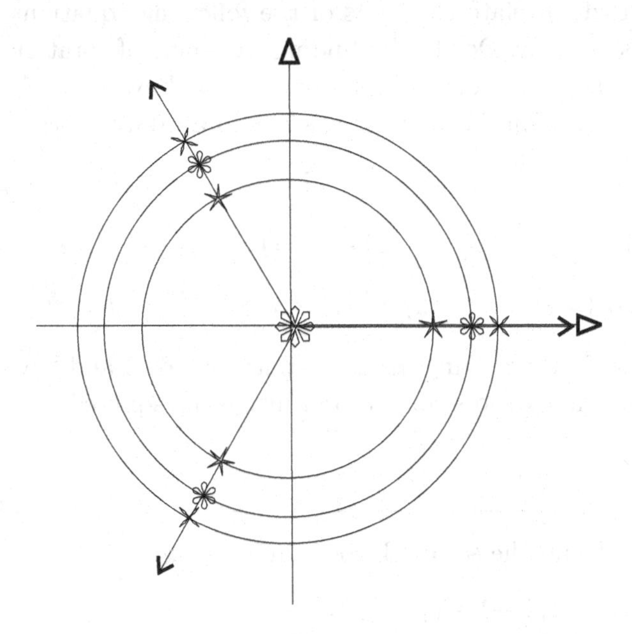

Fig. 2.1

§§2.6 Tutorial

Roots of a Complex Number

⊘ In this tutorial, the N roots of $z^N - 1 = 0$ will be denoted as $r_k, k = 1, \ldots, N$, where

$$r_k = \exp[2\pi i(k-1)/N].$$

[1] Let $\{1, \omega, \omega^2\}$, where $\omega = \exp(2\pi i/3)$, be the cube roots of unity. Show that the cube roots of $8i$ are obtained from the cube roots of unity by scaling by a factor of 2, followed by a rotation by an angle of $\pi/6$.

[2] If ξ is any one solution to the equation $z^N - z_0 = 0$, show that the numbers $\xi r_k, k = 1, 2, \ldots, N$, are solutions of the equation.

[3] Find any one value of each of the following expressions:

(a) $(-1)^{1/4}$, (b) $(-i)^{1/3}$, (c) $(-\sqrt{3}+i)^{1/5}$.

Write your answers for the remaining values of each of the above expressions in terms of the one answer you have already found and the Nth roots of unity for appropriate N.

[4] You are required to relate the roots of the following equations and the corresponding roots of unity. Do this by finding the angle of rotation and the scaling factor which takes the roots of unity, for a suitable value of N, to the roots of the concerned equation. Write your answers in a tabular form.

(a) $z^4 - 16 = 0$, (b) $z^3 + 27 = 0$, (c) $z^5 - 243 = 0$,

(d) $z^4 + 256 = 0$, (e) $z^3 + 8i = 0$, (f) $z^6 + 36i = 0$,

(g) $z^4 = 8(\sqrt{3}+i)$, (h) $z^3 + 1 - i = 0$, (i) $z^5 + 1 + i = 0$.

[5] Find the zeros of $\sin \pi z^2$ lying inside a square of side 5 and having the center at the origin. How many roots do you find inside the square?

Answers

[3] We list one root and the set of all roots in each case:

(a) $\exp(i\pi/4) \times \{1, i, -1, -i\}$.

(b) $i \times \{1, \omega, \omega^2\}$, where $\omega = \exp(2\pi i/3)$.

(c) $\exp(i\pi/6) \times \{1, r_5, r_5^2, r_5^3, r_5^4\}$, where $r_5 = \exp(2\pi i/5)$ ⊙ Writing the answer as $(-1)^{1/6}$ is unacceptable, because the expression does not stand for a unique value.

[4] Write the given equations in the form $z^N = \xi$. Then the answers are as follows:

	N	ξ	Scale Factor $= \xi^{1/N}$	Rotation Angle $= \arg(\xi)/N$
(a)	4	16	2	0
(b)	4	-256	4	$\pi/4$
(c)	4	$8(\sqrt{3}+i)$	2	$\pi/24$
(d)	3	-27	3	$\pi/3$
(e)	3	$-8i$	2	$\pi/2$
(f)	3	$-1+i$	$2^{1/6}$	$\pi/4$
(g)	5	243	3	0
(h)	6	$-36i$	$6^{1/3}$	$\pi/4$
(i)	5	$-1-i$	$2^{1/10}$	$\pi/4$

[5] ⊙ How many zeros did you count? Seven, 13, or 25? Did you find any zeros on the imaginary axis? The zeros are at $z = 0, \pm\sqrt{n}, \pm\sqrt{n}i, n = 1, 2, 3, \ldots$. Do not write your answer as $z = \sqrt{n}$, where n is an integer.

The correct answer is 25.

§§2.7 Quiz

Roots of Unity

In Fig. 2.2 a set of complex numbers, coinciding with the centers of several "stars", is shown. Letting $S_j, j = 1, 2, \ldots, 5$, denote five subsets of complex numbers coinciding with the jth type of star, answer the following questions:

(a) Identify the subset which conicides with a set of Nth roots of unity for some integer N. Find the value of N.
(b) Find the subset which does not coincide with the set of all N^{th} roots for some complex number and for any integer N.
(c) For each of the remaining subsets, find a complex number ξ and an integer N such that the subset coincides with the set of all N^{th} roots of ξ.

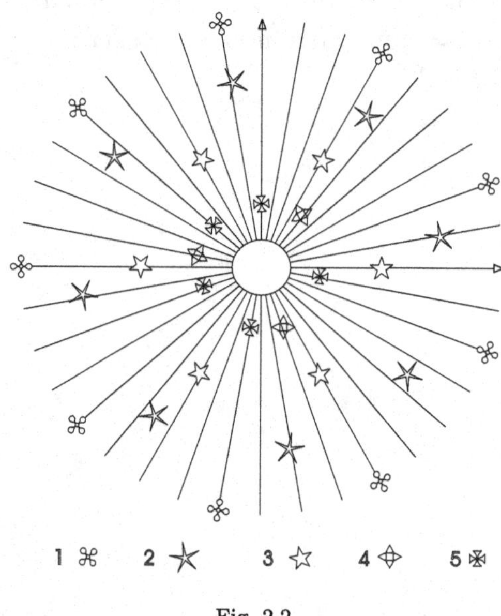

Fig. 2.2

Answers

(a) S_3 is the set of all roots of unity for $N = 6$.
(b) S_5 does not correspond to roots of any complex number.
(c) S_1 is the set of ninth roots of -512;
S_2 is the set of eighth roots of $(6561/256)\exp(4\pi i/9)$;
S_4 is the set of cube roots of $(1/8)\exp(5\pi i/6)$.

§§2.8 Exercise

Continuity and Differentiation

[1] Prove that the function

$$f(z) = \begin{cases} \exp(-z^{-2}), & \text{if } z \neq 0, \\ 0, & \text{if } z = 0, \end{cases}$$

is not continuous at $z = 0$ even though the function $g(x)$, obtained by restricting z to real values,

$$g(x) = \begin{cases} \exp(-1/x^2), & x \neq 0, \\ 0, & \text{if } x = 0, \end{cases}$$

is continuous at $x = 0$.

[2] Show that the following functions are not continuous at (0,0) by demonstrating that the limit $(x, y) \to (0, 0)$, computed in different ways, leads to different answers.

$$\text{(a)}\ f(x,y) = \frac{x^3 - y^3}{x^3 + y^3}, \quad \text{(b)}\ g(x,y) = \frac{x^2 - y^2}{(x+iy)^2}, \quad \text{(c)}\ h(x,y) = \frac{x\,y^3}{x^2 + y^6}.$$

[3] Let three functions $f(x,y)$, $g(x,y)$, and $h(x,y)$ be defined by

$$f(x,y) = x^2 - y^2 + 2ixy,$$
$$g(x,y) = x^3 + y^3,$$
$$h(x,y) = x^4 + y^4 + 2ix^2y^2.$$

(a) Argue that the functions $f, g,$ and h and their first partial derivatives are continuous everywhere.

(b) Which of the above functions satisfy the Cauchy–Riemann equations everywhere and which do not?

(c) Only one of the functions defined above is differentiable with respect to z everywhere. Identify this function.

[4] Find $n \neq 0$, such that the function

$$f(z) = \begin{cases} \exp(-z^{-n}), & \text{if } z \neq 0, \\ 0, & \text{if } z = 0, \end{cases}$$

has partial derivatives which satisfy the Cauchy–Riemann equations at $z = 0$, but the function $f(z)$ itself is not continuous at $z = 0$.

[5] Let a function f be defined by

$$f(x,y) = \begin{cases} \dfrac{xy(x+iy)}{x^2+y^2}, & \text{if } (x,y) \neq (0,0), \\ 0, & \text{if } (x,y) = (0,0). \end{cases}$$

(a) Show that the function is continuous at $z = 0$.
(b) Prove that the partial derivatives of the real and imaginary parts of the function f exist at $z = 0$ and satisfy the Cauchy–Riemann equations at $z = 0$.
(c) Starting from the definition of the derivative as a limit, check if the function f is differentiable w.r.t. z at $z = 0$.
(d) Find out which four partial derivatives of the real and imaginary parts of $f(x, y)$ are continuous and which ones are not.

Answers

[1] Taking $z = iy$ with $y \to 0$ shows that the $\lim f(z)$, as $z \to 0$, does not exist.
[2] (a) The limits along the real and imaginary axes are $+1$ and -1, respectively.
(b) The limits along the real and imaginary axes are equal, but the limit along the straight line $y = mx$ is $\frac{1-m^2}{1+im^2}$, which is different for different values of m.
(c) The limit along the real axis, the imaginary axis as well as along all the straight line $y = mx$ are all equal. The function is discontinuous, because the limit when taken along $x = ay^3$ depends on a and is therefore different for different values of a.
[3] (b) The function $f(z)$ satisfies Cauchy–Riemann equations everywhere. The functions $g(z)$ and $h(z)$ satisfy Cauchy–Riemann equations only at $z = 0$.
(c) Only $f(z)$ is differentiable everywhere.
[4] $4, 8, \ldots$.
[5] The function satisfies Cauchy–Riemann equations at $z = 0$ but is not differentiable at $z = 0$.

§§2.9 Questions

Cauchy–Riemann Equations

⊘ Mark each of the following statements as True or False. In this set the abbreviation CR stands for the Cauchy–Riemann equations.

[1] If a function does not satisfy CR at a point, it cannot be differentiable at that point.

[2] If CR are satisfied for a function at a point, the function must be analytic at that point.

[3] If a function is analytic at a point, it must obey CR at that point.

[4] If a function is analytic at a point, there exists a neighborhood of the point such that the function obeys CR at all points in the neighborhood.

[5] If a function satisfies CR at a point, the function need not be analytic, but it is necessary that its derivative w.r.t. z exists at that point.

[6] If CR are satisfied at a point, the function is differentiable as well as analytic at that point.

[7] For a function to be analytic at a point, it is necessary that the function satisfies the CR at that point.

[8] If a function does not satisfy CR at a point, it cannot be analytic but may be differentiable.

[9] If a function does not satisfy CR at a point, it is neither analytic nor differentiable at that point.

[10] If a function is analytic at a point, it follows that the function satisfies CR at all points in an open neighborhood of that point.

[11] For a function to be analytic at a point, it is necessary that the function satisfies CR in an open neighborhood of that point.

[12] If a function satisfies CR at all points in an open neighborhood of a point z_0, it is analytic at z_0.

[13] CR are only necessary but not sufficient conditions for a function to be differentiable at a point.

[14] CR are neither necessary nor sufficient conditions for a function to be analytic at a point.

[15] The existence of the first order partial derivatives is a sufficient condition for a function to be differentiable at a point in the complex plane.

Answers

[1] True, [2] False, [3] True,

[4] True, [5] False, [6] False,

[7] True, [8] False, [9] True,

[10] True, [11] True, [12] False,

[13] True, [14] False, [15] False.

§§2.10 Quiz

Cauchy–Riemann Equations

⊘ Eight functions are defined below: (a)–(h). For each function, check if the following statements are true or false at the point $z = 0$:

(i) The function is continuous;
(ii) All the four first order partial derivatives appearing in Cauchy–Riemann equations exist at the point $x = 0, y = 0$;
(iii) The four first order partial derivatives are continuous;
(iv) The function obeys Cauchy–Riemann equations;
(v) The function is differentiable;
(vi) The function is analytic.

⊘ Complete the table given at the end, by writing Yes/No in different row and column positions.

(a) $f(x,y) = \sqrt{xy}$,

(b) $f(x,y) = \dfrac{x^3 - iy^3}{x^2 + y^2}$,

(c) $f(z) = \exp(z^*)$,

(d) $f(x,y) = x^2 + 2i(x - y)$,

(e) $f(x,y) = x^4 + y^4$,

(f) $f(x,y) = (x + iy)e^z$,

(g) $f(z) = \dfrac{1}{z+1}$,

(h) $f(z) = \begin{cases} e^{-z^{-4}}, & \text{if } z \neq 0, \\ 0, & \text{if } z = 0. \end{cases}$

Suggested tabular form for the answers:

	(a)	(b)	(c)	(d)	(e)	(f)	(g)	(h)
Continuity								
Existence of partials								
Continuity of all partials								
CR equations obeyed								
Differentiability								
Analyticity								

Answers

	(a)	(b)	(c)	(d)	(e)	(f)	(g)	(h)
Continuity	Yes	Yes	Yes	Yes	Yes	Yes	Yes	No
Existence of partials	No	Yes	Yes	Yes	Yes	Yes	Yes	Yes
Continuity of all partials	No	No	Yes	Yes	Yes	Yes	Yes	No
CR equations obeyed	No	No	No	Yes	Yes	Yes	Yes	Yes
Differentiability	No	No	No	No	Yes	Yes	Yes	No
Analyticity	No	No	No	No	No	Yes	Yes	No

§§2.11 Tutorial

Analytic Functions

⊘ Check, by inspection, if the following functions are analytic everywhere or not. If not, in each case find the set S of points where the function is not analytic.

(a) $x^3 - iy^3 + 3xy(ix - y)$, (b) $y^2 + 2x + i(x^2 + 2y)$, (c) $3ix - 3y + 2i$,
(d) $z + \bar{z}^2$, (e) $x^2 - y^2$, (f) $\dfrac{z-3}{z^2 - 5z + 4}$,
(g) $z^5 + 7z^3 + 8z$, (h) $\sin z$, (i) $\tan z$,
(j) $\coth z$, (k) $\sin(1/z)$, (l) $\cot(1/z)$,
(m) $\exp(1/z^4)$, (n) 2^z, (o) $\exp(-\cos^2 z)$,
(p) $\dfrac{\tan(1/z)}{z^2 + z + 1}$, (q) $\dfrac{z^4 + 2z^2 + 1}{\exp(-\tan z)}$, (r) $\dfrac{\sin z}{z^3 + 1}$,
(s) $\dfrac{z}{\sin z}$, (t) $\dfrac{\exp(z)}{\exp(z) - 1}$, (u) $\dfrac{z^2 - 16}{\sin(\pi z)}$,
(v) $\exp(z^3 + \sin z)$, (w) $\sin z^2$, (x) $\bar{z}^3/z$,
(y) $\operatorname{cosec} z^3$, (z) z.

Answers

(a) Analytic everywhere.
(b) Not analytic anywhere.
(c) Not analytic anywhere.
(d) Analytic everywhere.
(e) Not analytic anywhere.
(f) Singular points are $z = 1, 4$.
(g) Analytic everywhere.
(h) Analytic everywhere.
(i) Singular points are at $z = n\pi/2$, with $n =$ odd integer.
(j) Analytic everywhere except at $z = n\pi i$, where $\sinh z$ vanishes.
(k) Singular only at $z = 0$; analytic elsewhere.
(l) Singular at $z = 0$ and $z = 1/n\pi$.
(m) Singular only at $z = 0$; analytic elsewhere.
(n) $2^z = \exp(z \ln 2)$ is analytic everywhere.
(o) Entire function.
(p) Singular points are $z = 0, \frac{2}{(2n+1)\pi}, \omega, \omega^2$.
(q) Analytic except at $z = (2n + 1)\pi/2$.
(r) Analytic except at $z = -1, -\omega, -\omega^2$.
(s) $\frac{z}{\sin z}$ is singular at $z = n\pi, n = \pm 1, \pm 2, \ldots$.
(t) Singular points are at $z = 2n\pi i$, where the denominator $e^z - 1$ vanishes.

(u) Singular points are at $z = n\pi$, where n is any integer except ± 4.
(v) Entire function.
(w) Entire function.
(x) Not analytic anywhere, because it depends on $\bar{z}$.
(y) Analytic everywhere except at points z, which are solutions to $z^3 = n\pi$.
(z) $f(z) = z$ is an entire function.

§§2.12 Exercise

Harmonic Functions

[1] The following functions are differentiable except possibly at some points. For each of the following functions, by inspection find the points, if any, where the derivative does not exist. Verify that the Cauchy–Riemann equations are satisfied at all other points.

(a) $1/z^2$, (b) $\sin z$,

(c) $\exp(\sin z)$, (d) $\cosh[\exp(z)]$,

(e) $\exp(z^2)$, (f) $\cot z$.

[2] Using results of Q[1] give examples of six functions which are harmonic everywhere.

[3] Find the conjugate harmonic function $v(x, y)$ of each of the following functions:

(a) $u(x, y) = x(-3y^2 + x^2)$,
(b) $u(x, y) = \exp(-y) \sin x$,
(c) $u(x, y) = \sin(x^2 - y^2) \cosh 2xy$,
(d) $u(x, y) = \sin x \sinh y$.

Answers

[1] • For part (a) the derivative does not exist at $z = 0$.
• For (b)–(e) the derivative exists everywhere.
• The derivative of $\cot z$ exists everywhere except at $z = n\pi$.

[2] The real and imaginary parts of the functions in (b)–(e) of Q[1] are harmonic everywhere.

[3] ⊙ Find a function $f(z)$ such that the given function $u(x, y)$ is either a real or an imaginary part of $f(z)$. Such a function $f(z)$ and the conjugate harmonic function $v(x, y)$ are

(a) $f(z) = z^3$, $v(x, y) = -y^3 + 3x^2y$;
(b) $f(z) = -i \exp(iz)$, $v(x, y) = -\exp(-y) \cos x$;
(c) $f(z) = \sin z^2$, $v(x, y) = \cos(x^2 - y^2) \sinh(2xy)$;
(d) $f(z) = i \cos z$, $v(x, y) = \cos x \sinh y$.

§§2.13 Mixed Bag

Differentiation and Analyticity

[1] Using properties of the trigonometric functions alone, without using Euler's formula, prove De Moivre's theorem,

$$(\cos\theta + i\sin\theta)^r = \cos r\theta + i\sin r\theta,$$

when r is an integer.

[2] Use De Moivre's theorem to prove that

(a) $\cos nz = \cos^n z - {}^nC_2 \cos^{n-2} z \sin^2 z + {}^nC_4 \cos^{n-4} z \sin^4 z - \cdots;$
(b) $\sin nz = {}^nC_1 \cos^{n-1} z \sin z - {}^nC_3 \sin^3 z \cos^{n-3} z + \cdots.$

[3] Express $\sin^5 z$ and $\cos^5 z$ in terms of the sine and cosine of multiples of z and prove that

(a) $\cos^5 z = \frac{1}{16}(\cos 5z + 5\cos 3z + 10\cos z);$
(b) $\sin^5 z = \frac{1}{16}(\sin 5z - 5\sin 3z + 10\sin z).$

[4] Prove the following:

(a) $\sinh 7z = 7\sinh z + 56\sinh^3 z + 112\sinh^5 z + 64\sinh^7 z;$
(b) $\cosh 7z = 64\cosh^7 z - 112\cosh^5 z + 56\cosh^3 z - 7\cosh z.$

[5] Find a formula for $\cos^{2n} z$ and $\sin^{2n} z$ in terms of the sine and cosine of multiples of z.

[6] Prove the following summation identities for the trigonometric functions:

(a) $\sum_{k=1}^{n} \sin kz = \sin\frac{(n+1)z}{2} \sin\frac{nz}{2} \operatorname{cosec}\frac{z}{2},$
(b) $\sum_{k=0}^{n} \cos kz = \cos\frac{(n+1)z}{2} \sin\frac{nz}{2} \operatorname{cosec}\frac{z}{2} + 1.$

⊘ Let the set $\{\rho_k | k = 1, 2, \ldots, N\}$ denote the N roots of identity, i.e. solutions to $\rho^N = 1.$

[7] Show that the product of all the roots ρ_k is $(-1)^{N+1}$.

[8] What is the sum of all the Nth roots of unity? In how many different ways can you get your result?

[9] Prove that all the symmetrized sums of the products of the N roots vanish for $2 < n < N$, i.e.

$$\sum \rho_{i_1}\rho_{i_2} = 0,$$

$$\sum \rho_{i_1}\rho_{i_2}\rho_{i_3} = 0,$$

$$\sum_{2<n<N} \rho_{i_1}\rho_{i_2}\cdots\rho_{i_n} = 0.$$

In each of the above expressions, the summation is over all possible distinct terms and each term inside the summation is a product of distinct roots.

[10] Find the real and imaginary parts of all values of

(a) i^i, (b) $(1+i)^{1+i}$,

(c) $((\sqrt{3}+i)/2)^i$, (d) $(-1)^{i\pi}$.

[11] Compute all values of

(a) $\sin^{-1} 2$, (b) $\cosh^{-1} 3$,

(c) $\cot^{-1}(1+i)$, (d) $\sinh^{-1}(i)$,

(e) $\tan^{-1} 2i$, (f) $\tanh^{-1}(1+2i)$.

[12] The function $\sinh z$ is a many-to-one function. Find all and plot any five complex values of z which satisfy $\sinh z = \sinh \alpha$, where α is real.

[13] Prove that the set of all values, of various inverse trigonometric and inverse hyperbolic functions, is given by

(a) $\sin^{-1} z = -i \log \left(iz + \sqrt{1-z^2}\right)$,
(b) $\cos^{-1} z = -i \log \left(z + \sqrt{z^2-1}\right)$,
(c) $\cosh^{-1} z = \log \left(z + \sqrt{z^2-1}\right)$,
(d) $\sinh^{-1} z = \log \left(z + \sqrt{z^2+1}\right)$.

[14] Let u and v be the real and imaginary parts of $w = \sin^{-1} z$. If $z(= x+iy)$ lies in the first quadrant, prove that

$$2\sin u = \sqrt{(x+1)^2+y^2} - \sqrt{(x-1)^2+y^2},$$
$$2\cosh v = \sqrt{(x+1)^2+y^2} + \sqrt{(x-1)^2+y^2}.$$

Do these relations remain correct if z is not in the first quadrant?

[15] (a) The function $f(t) = \exp(-1/t^2)$ tends to 0 as the real variable $t \to 0$. Prove or disprove the corresponding statement,

$$\lim_{z\to 0} \exp(-1/z^2) = 0.$$

for the function $\exp(-1/z^2)$ considered as a function of a complex variable z.

(b) Let $f(z)$ be a function of a complex variable. It is required to compute the limit $f(z)$ as $z \to 0$. Discuss when we can compute the limit and get the correct answer by setting $z = x$ and taking the real variable x to zero.

[16] Give an example of a function for which the Cauchy–Riemann equations are satisfied at all points in the complex plane but the function is not even continuous at some point(s).

[17] If the derivative of a function $g(z)$ exists at a point, show that it can be computed by making use of any one of the following equations:

$$\frac{dg(z)}{dz} = \frac{r}{z}\frac{\partial g}{\partial r}, \quad \frac{dg(z)}{dz} = \frac{1}{iz}\frac{\partial g}{\partial \theta}.$$

[18] If a function is analytic at a point, prove that it must satisfy the Cauchy–Riemann equations

$$\frac{\partial u(r,\theta)}{\partial r} = \frac{1}{r}\frac{\partial v(r,\theta)}{\partial \theta}, \quad \frac{1}{r}\frac{\partial u(r,\theta)}{\partial \theta} = -\frac{\partial v(r,\theta)}{\partial \theta} \quad (r \neq 0)$$

in the polar coordinates.

[19] Let $u(x,y)$ and $v(x,y)$ be the real and imaginary parts of a function $f(z)$ analytic everywhere. Show that the curves $u(x,y) = a$ and $v(x,y) = b$, where a and b are constants, are orthogonal.

[20] Let $f(z)$ and $g(z)$ be two functions analytic everywhere. Let $u(x,y)$ and $v(x,y)$ be the real and imaginary parts of $f(z)$, and $p(x,y)$ and $q(x,y)$ be the real and imaginary parts of $g(z)$. Give a proof in a few lines to show that

$$U(x,y) = u(p(x,y), q(x,y)), \quad V(x,y) = v(p(x,y), q(x,y))$$

satisfy the Cauchy–Riemann equations. Also, verify this statement by an explicit computation.

[21] If $f(z)$ is analytic show that

$$\nabla^2 |f(z)|^2 = 4\left|\frac{df(z)}{dz}\right|^2,$$

where ∇^2 is the Laplacian operator

$$\frac{\partial^2}{\partial x^2} + \frac{\partial^2}{\partial y^2}.$$

Answers

[2] ⊙ Find the real and the imaginary parts of $(\cos z + i \sin z)^n$.

[3] ⊙ Write $\cos z$ and $\sin z$ in terms of exponentials and expand.

[4] ⊙ Write $\cosh 7z = \frac{1}{2}(e^{7z} + e^{-7z})$ and $e^{7z} = \frac{1}{128}(\cosh z + \sinh z)^7$, and expand.

[5] (a) $\cos^{2n} z = \frac{1}{2^{2n}}\left\{\sum_{k=0}^{n-1} 2\binom{2n}{k}\cos 2(n-k)z + \binom{2n}{n}\right\}$,

(b) $\sin^{2n} z = \frac{1}{2^{2n}}\left\{\sum_{k=0}^{n-1} 2(-1)^{n-k}\binom{2n}{k}\cos 2(n-k)z + \binom{2n}{n}\right\}$.

[10] (a) $i^i = \exp(-2\pi n - \frac{\pi}{4})$,

(b) $(1+i)^{(1+i)} = \sqrt{2}\exp(-2\pi n - \frac{\pi}{4})[\cos(\alpha) + i\sin(\alpha)](\alpha = \frac{1}{2}\ln 2 + \frac{\pi}{4})$,

(c) $\left(\frac{\sqrt{3}+i}{2}\right)^i = \exp(-\frac{\pi}{6} - 2\pi n)$,
(d) $(-1)^{\pi i} = \exp(-\pi^2 - 2\pi^2 n)$.

[11] (a) $\sin^{-1} 2 = \frac{\pi}{2} + 2\pi n - i\ln(2 \pm \sqrt{3})$,
(b) $\cosh^{-1} 3 = \ln(3 \pm 2\sqrt{2}) + 2\pi i n$,
(c) $\cot^{-1}(1+i) = \frac{\alpha}{2} + n\pi - \frac{i}{4}\ln 5 (\tan\alpha = 2)$,
(d) $\sinh^{-1}(i) = 2\pi i n + \frac{i\pi}{2}$,
(e) $\tan^{-1}(2i) = (2n+1)\frac{\pi}{2} + \frac{i}{2}\ln 3$,
(f) $\tanh^{-1}(1+2i) = \frac{1}{4}\ln 2 + \frac{3\pi i}{8} + n\pi i$.

[14] ⊙ Plot curves of constants u and v in the z plane.

[15] (b) Whenever the limit exists.

[16] $\exp(-z^{-4})$.

Chapter 3

FUNCTIONS WITH BRANCH POINT SINGULARITY

§§3.1 Questions

Branch Point

⊘ For all the questions in this set, closed paths consisting of circles C_0, C_1, C_2, C_3, and C_4, ellipses E_1 and E_2, and a square S are defined as follows.

⊘ C_0, C_1, C_2, C_3, and C_4 are circles, each of radius 1/3 and having centers at the complex numbers $0, 1, i, -1$, and $-i$, respectively.

⊘ E_1 is an ellipse with the center at $z = 0$ and having the lengths of the semimajor and semiminor axes equal to 3/2 and $\frac{1}{2}$, respectively. It has a major axis along the real axis and a minor axis along the imaginary axis. E_2 is obtained from E_1 by a rotation of $\pi/2$ about the origin.

⊘ S is the square with vertices at $z = \pm 2, \pm 2i$.

[1] Let the angles $\theta_1, \ldots, \theta_4$ be defined by

(a) $\theta_1 = \arg(z)$, (b) $\theta_2 = \arg(z-1)$,

(c) $\theta_3 = \arg(z+1)$, (d) $\theta_4 = (\arg(z-i)+\arg(z+i))/2$.

How do the values of $\theta_1, \ldots, \theta_4$ change when a point z makes a complete loop around the paths (i) C_0, (ii) C_1, (iii) E_1, and (iv) S specified above?

[2] Three functions and three paths are listed below:

(a) $\sqrt{z(z+1)}$, paths C_0, E_1, S;
(b) $\sqrt{z} + \sqrt{z-1}$, paths C_0, E_2, S;
(c) $\sqrt{z}\log(z+1)$, paths C_0, C_2, S.

For each pair of a function and a path given above, answer the following two questions to describe what happens to the value of the function when the point z makes one or more complete loop(s) around the closed path.

Questions. Does the value of the function for a branch return to its original value if a certain minimum number of loops are completed? If Yes, find this minimum number, ν. Enter an asterisk (*) if the function never returns to the initial value for a path.

A sample answer is given below for the function $\sqrt{z(z-1)}$ for all the eight paths (the circles C_0, C_1, C_2, C_3, and C_4, the ellipses E_1 and E_2, and the square S). Write your answers in three similar tables, making one table for each of the cases (a), (b), and (c) listed above.

$$f(z) = \sqrt{z(z-1)}.$$

	C_0	C_1	C_2	C_3	C_4	E_1	E_2	S
ν	2	1	1	2	1	1	2	1

[3] List the branches, and the branch points for

(a) $f(z) = \sqrt{z(z+1)}$, (b) $g(z) = \sqrt{z} + \sqrt{z-1}$, (c) $h(z) = \sqrt{z}\log(z+1)$.

In each case state if the point at infinity is a branch point or not.

[4] (a) For the multivalued function $f(z) = z^{1/4}$, write expressions for the four branches.

(b) Starting from $z_0 = 1$, a point moves in a closed path encircling the origin and comes back to the point z_0 after completing several loops. Find the values for the four branches after the number of completed loops, L, is $0, 1, 2, 3, 4$. Write your answers in a tabular form.

(c) Choose any one branch of $f(z)$ and specify its values after $L = 4m$, $4m+1, 4m+2, 4m+3$ loops are completed where m is an integer. How do these answers compare with the values for the four different branches at the starting point, i.e. with the values in the $L = 0$ column of the table having answers for part (b) of the question.

Answers

[1]

		C_0	C_1	E_1	S
(a)	$\theta_1 = \arg(z)$	$\pm 2\pi$	0	$\pm 2\pi$	$\pm 2\pi$
(b)	$\theta_2 = \arg(z-1)$	0	$\pm 2\pi$	$\pm 2\pi$	$\pm 2\pi$
(c)	$\theta_3 = \arg(z+1)$	0	0	$\pm 2\pi$	$\pm 2\pi$
(d)	$\theta_4 = [\arg(z+i) + \arg(z-i)]/2$	0	0	0	$\pm 2\pi$

[2]

$\sqrt{z(z+1)}$			$\sqrt{z(z-1)}$			$\sqrt{z}\log(z+1)$		
C_0	E_1	S	C_0	E_2	S	C_0	C_2	S
2	1	1	2	2	1	2	1	*

[3] With $\rho_1 = |z+1|, \rho_2 = |z-1|, \phi_1 = \arg(z+1), \phi_2 = \arg(z-1)$, the required branches are

(a) two branches; $\sqrt{z(z+1)} = \pm(r\rho_1)^{1/2}\exp\left[\frac{i}{2}(\theta + \phi_1)\right]$;
$z = 0, -1$; $z = \infty$ is not a branch point.

(b) four branches; $\sqrt{z} + \sqrt{z-1} = \pm r^{1/2}\exp(i\theta/2) \pm \rho_2^{1/2}\exp(i\phi_2/2)$;
$z = 0, 1$, $z = \infty$ are branch points.

(c) infinite branches; $\sqrt{z}\log(z+1) = \pm r^{1/2}e^{i\theta/2}(\log\rho_1 + i\phi_1 + 2i\pi n), n \in \mathbb{Z}$; $z = 0, -1$, $z = \infty$ are branch points.

[4]

Branch	$f(z)$	$L=0$	$L=1$	$L=2$	$L=3$	$L=4$
1st	$\sqrt[4]{r}\exp(i\theta/4)$	1	i	-1	$-i$	1
2nd	$i\sqrt[4]{r}\exp(i\theta/4)$	i	-1	$-i$	1	i
3rd	$-\sqrt[4]{r}\exp(i\theta/4)$	-1	$-i$	1	i	-1
4th	$-i\sqrt[4]{r}\exp(i\theta/4)$	$-i$	1	i	-1	$-i$

§§3.2 Tutorial

Square Root Branch Cut

[1] Let the square root function be expressed as

$$z^{1/2} = r^{1/2} \exp(i\theta/2).$$

Give the location of the branch cut in the complex plane for each of the following choices of ranges for θ values:

(a) $-\pi < \theta < \pi$, (b) $0 < \theta < 2\pi$, (c) $-4\pi/3 < \theta < 2\pi/3$.

[2] The properties of the function $f(z) = \sqrt{z-a}$ are similar to those of the function $\sqrt{z}$. Therefore, introduce the variables ρ and ϕ by

$$z = a + \rho \exp(i\phi).$$

Expressed in terms of ρ and ϕ, the function $f(z)$ becomes

$$f(z) = \lambda \rho^{1/2} \exp(i\phi/2),$$

where the two allowed values of λ are ± 1. Select a value of λ and a range of ϕ such that

(a) $f(z)$ has a branch cut from $z = a$ to $z = \infty$ and the real part of $f(z)$ is negative for z below and close to the cut;
(b) $f(z)$ has a branch cut from $z = -\infty$ to $z = a$ and the imaginary part of $f(z)$ is positive for z just below the cut;
(c) $f(z)$ has a branch cut from $z = -\infty$ to $z = a$ and a negative value on the subset $\{z|\text{Re } z > a\}$ of the real axis.

[3] An svb of the function $f(z) = \sqrt{z+5}$ is defined as

$$f(z) = -\rho^{1/2} \exp(i\phi/2)\,, \quad -\pi < \phi < \pi,$$

with $\rho = |z+5|$ and $\phi = \arg(z+5)$.

(a) Find the branch cut, and
(b) Verify that the function is discontinuous across the cut.

[4] Give the definition of an svb of the function

$$f(z) = \sqrt{z-4}$$

such that the following two conditions hold.

- The branch cut is the subset $\{z|\text{Im } z = 0, \text{ Re } z < 4\}$ on the real axis;
- For $z = x + i\epsilon, \epsilon > 0$, and $x < 0$, $\lim_{\epsilon \to 0} f(z)$ is pure imaginary and the imaginary part of the limit is positive.

[5] For the function $f(z)$, as defined in the previous question, compute the values of the function, and the limits, as indicated below:

(a) $f(z)$ at $z = 29$;

(b) $\lim_{\epsilon\to 0} f(z)$ at $z = -12 + i\epsilon$ $(\epsilon > 0)$;
(c) $\lim_{\epsilon\to 0} f(z)$ at $z = -12 - i\epsilon$ $(\epsilon > 0)$.

Answers

[1] $\sqrt{z} = r^{1/2} \exp(i\theta/2)$.

(a) For $-\pi < \theta < \pi$ the branch cut is along the negative real axis;
(b) For $0 < \theta < 2\pi$ the branch cut is along the positive real axis;
(c) For $-\frac{4\pi}{3} < \theta < \frac{2\pi}{3}$ the branch cut is along the ray $\theta = 2\pi/3$.

[2] $\sqrt{z-a} = \lambda\rho^{1/2} \exp(i\phi/2)$.

(a) $\lambda = 1$, $0 < \phi < 2\pi$;
(b) $\lambda = -1$, $-\pi < \phi < \pi$;
(c) $\lambda = -1$, $-\pi < \phi < \pi$.

⊙ For (a), the branch cut from $z = a$ to ∞ corresponds to the range $0 < \phi < 2\pi$. Just above the cut $\phi \simeq 0$, $\phi > 0$, and $\text{Re}\sqrt{z-a} = \lambda\rho^{1/2}\cos(\phi/2)$ is negative for $\phi \cong 0$ if λ is selected to be -1.

⊙ For (b), the branch cut from $-\infty$ to a corresponds to $-\pi<\phi<\pi$, $\text{Im}\sqrt{z-1} = \lambda\rho^{1/2}\sin\phi/2$. Just below the cut $\phi \cong -\pi + \epsilon$ and $\sin(\phi/2) \cong -\cos(\epsilon/2) < 0$. Hence, $\text{Im}\sqrt{z-1} < 0$ for $\lambda = +1$.

[3] (a) For $-\pi < \phi < \pi$, the branch cut is from -5 to $-\infty$ on the negative real axis.
(b) The value of the discontinuity for $x + 5 < 0$ is $-2i|x+5|^{1/2}$.

[4] With ϕ restricted to $-\pi < \phi < \pi$ and $\lambda = -1$, the imaginary part of $f(z) = \lambda\rho^{1/2}\exp(i\phi/2)$ is negative, on the negative x axis.

[5] $f(z) = -|z-4|^{1/2}\exp(i\phi/2)$.

z	ρ	ϕ	$f(z)$
29	5	0	-5
$-12 + i\epsilon$	4	$\pi - \delta$	$-4i$
$-12 - i\epsilon$	4	$-\pi + \delta$	$4i$

§§3.3 Exercise

Branch Cut for $\sqrt{\frac{z-a}{z+b}}$

⊘ In this set the variables r_1, θ_1, r_2, and θ_2 (see Fig. 3.1) are defined by

$$z - a = r_1 \exp(i\theta_1), \quad z + b = r_2 \exp(i\theta_2),$$

where $a > 0$ and $b > 0$, and g and h will be used to denote

$$g(z) = \sqrt{z-a}, \quad h(z) = \sqrt{z+b}.$$

[1] (a) Draw figures to show the branch cut for $g(z)$ when

(i) $0 < \theta_1 < 2\pi$, (ii) $-\pi < \theta_1 < \pi$.

(b) Find the branch cut for $h(z)$ when θ_2 is restricted to

(i) $0 < \theta_2 < 2\pi$, (ii) $-\pi < \theta_2 < \pi$.

[2] Next, considering $f(z) = \sqrt{\frac{z-a}{z+b}}$ as a quotient of the two functions $g(z)$ and $h(z)$, give a short argument to explain why the function $f(z)$ is continuous at points common to the branch cuts for the two functions $g(z)$ and $h(z)$. Draw figures to show the branch cut for the function $\sqrt{\frac{z-a}{z+b}}$ for the following four cases:

(a) $0 < \theta_1 < 2\pi, \quad 0 < \theta_2 < 2\pi$; (b) $-\pi < \theta_1 < \pi, \quad 0 < \theta_2 < 2\pi$;

(c) $0 < \theta_1 < 2\pi, \quad -\pi < \theta_2 < \pi$; (d) $-\pi < \theta_1 < \pi, \quad -\pi < \theta_2 < \pi$.

[3] For the function $f(z)$ in Q[2], with $a = b = 5$ and the range of angles restricted to

$$0 < \theta_1 < 2\pi, \quad 0 < \theta_2 < 2\pi,$$

compute the discontinuity across the real line,

$$\text{disc}\{f\}(x) \equiv \lim_{\epsilon \to 0}[f(x+i\epsilon) - f(x-i\epsilon)],$$

for the following values of x:

(a) $x = -13$, (b) $x = -3$,

(c) $x = 3$, (d) $x = 13$.

Write your answers in the tabular form given below:

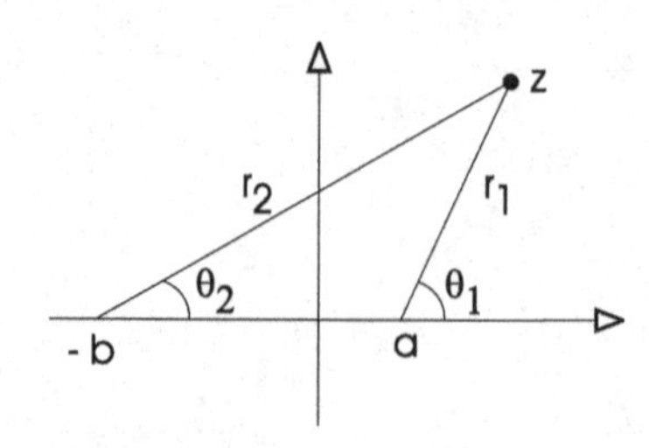

Fig. 3.1

x	z	r_1	r_2	θ_1	θ_2	$f(z)$	$\mathrm{disc}\{f\}(x)$
-13	$-13+i\epsilon$						
	$-13-i\epsilon$						
	$\cdots$						
	$\cdots$						

[4] Taking the restrictions

$$-\pi < \theta_1 < \pi, \quad 0 < \theta_2 < 2\pi$$

on the values of the angles θ_1 and θ_2 and again assuming that $a = b = 5$, repeat the computation of the discontinuity of $f(z)$ across the real line as in Q[3].

Answers

[1] The branch cuts are given by
(a) (i) $\mathrm{Re}\, z > a$, (ii) $\mathrm{Re}\, z < a$,
(b) (i) $\mathrm{Re}\, z > -b$, (ii) $\mathrm{Re}\, z < -b$.

[2] With $x = \mathrm{Re}\, z$, the branch cuts are given by
(a) $-b < x < a$, (b) $x < -b$ and $x > a$,
(c) $x < -b$ and $x > a$, (d) $-b < x < a$.

[3] In this problem $0 < \theta_1 < 2\pi$ and $0 < \theta_2 < 2\pi$, $r_1 + r_2 > (a + b)$.
Values of $f(z)$, listed in the table below, are in $\lim \epsilon \to 0$, where ϵ gives the imaginary part of z, as in the second column.

x	z	r_1	r_2	θ_1	θ_2	$f(z)$	disc f
-13	$-13+i\epsilon$	18	8	π	π	$3/2$	0
	$-13-i\epsilon$	18	8	π	π	$3/2$	
-3	$-3+i\epsilon$	8	2	π	0	$2i$	$4i$
	$-3-i\epsilon$	8	2	π	2π	$-2i$	
3	$3+i\epsilon$	2	8	π	0	$i/2$	i
	$3-i\epsilon$	2	8	π	2π	$-i/2$	
13	$13+i\epsilon$	8	18	0	0	$2/3$	0
	$13-i\epsilon$	8	18	2π	2π	$2/3$	

[4] Values of $f(z)$ are listed below in $\lim \epsilon \to 0$:

x	z	r_1	r_2	θ_1	θ_2	$f(z)$	disc f
-13	$-13+i\epsilon$	18	8	π	π	$3/2$	3
	$-13-i\epsilon$	18	8	$-\pi$	π	$-3/2$	
-3	$-3+i\epsilon$	8	2	π	0	$2i$	0
	$-3-i\epsilon$	8	2	$-\pi$	2π	$2i$	
3	$3+i\epsilon$	2	8	π	0	$i/2$	0
	$3-i\epsilon$	2	8	$-\pi$	2π	$i/2$	
13	$13+i\epsilon$	8	18	0	0	$2/3$	$4/3$
	$13-i\epsilon$	8	18	0	2π	$-2/3$	

⊙ In order to identify the branch cut for $f(z)$, it is necessary to verify that the function is discontinuous all along the branch cut. The computation of the discontinuity helps in checking if the branch cut indeed lies along the part where one had guessed.

§§3.4 Quiz

Discontinuity and Branch Cut

⊘ Let $(z \pm 1)^c$ be defined as

$$(z+1)^c = \rho_1^c \, e^{ic\phi_1}, \qquad (z-1)^c = \rho_2^c \, e^{ic\phi_2},$$

in terms of variables $\rho_1 = |z+1|, \rho_2 = |z-1|, \phi_1 = \arg(z+1)$, and $\phi_2 = \arg(z-1)$. Answer the two questions given below for the functions listed in the second column of the following table and complete the table as per instructions in Q[1] and Q[2].

SN	$f(z)$	$0 < \phi_1 < 2\pi$, $0 < \phi_2 < 2\pi$	$0 < \phi_1 < 2\pi$, $-\pi < \phi_2 < \pi$
1	$(z-1)^{\frac{1}{4}} - (z-1)^{-\frac{1}{4}}$		
2	$(z+1)^{\frac{1}{3}}(z-1)^{\frac{1}{3}}$		
3	$(z+1)^{-\frac{1}{3}}(z-1)^{\frac{1}{3}}$		
4	$(z+1)^{\frac{1}{3}}(z-1)^{\frac{2}{3}}$		
5	$(z+1)^{\frac{3}{4}} + (z-1)^{\frac{1}{4}}$		
Sample entry	$\sqrt{z^2-1}$	(e) $\lvert x\rvert < 1$	(f) $\lvert x\rvert > 1$

[1] Assuming the restrictions on ϕ_1 and ϕ_2 to be

$$0 < \phi_1 < 2\pi, \qquad 0 < \phi_2 < 2\pi,$$

find the branch cut for each of the multiple-valued functions listed in the second column of the table above, select your answer from Figs. 3.2(a)–3.2(g) and write it in the third column in the format suggested by a sample answer in the last row.

[2] Repeat the above question taking the restrictions on ϕ_1 and ϕ_2 to be

$$0 < \phi_1 < 2\pi, \quad -\pi < \phi_2 < \pi.$$

Write your answers in the last column of the given below.

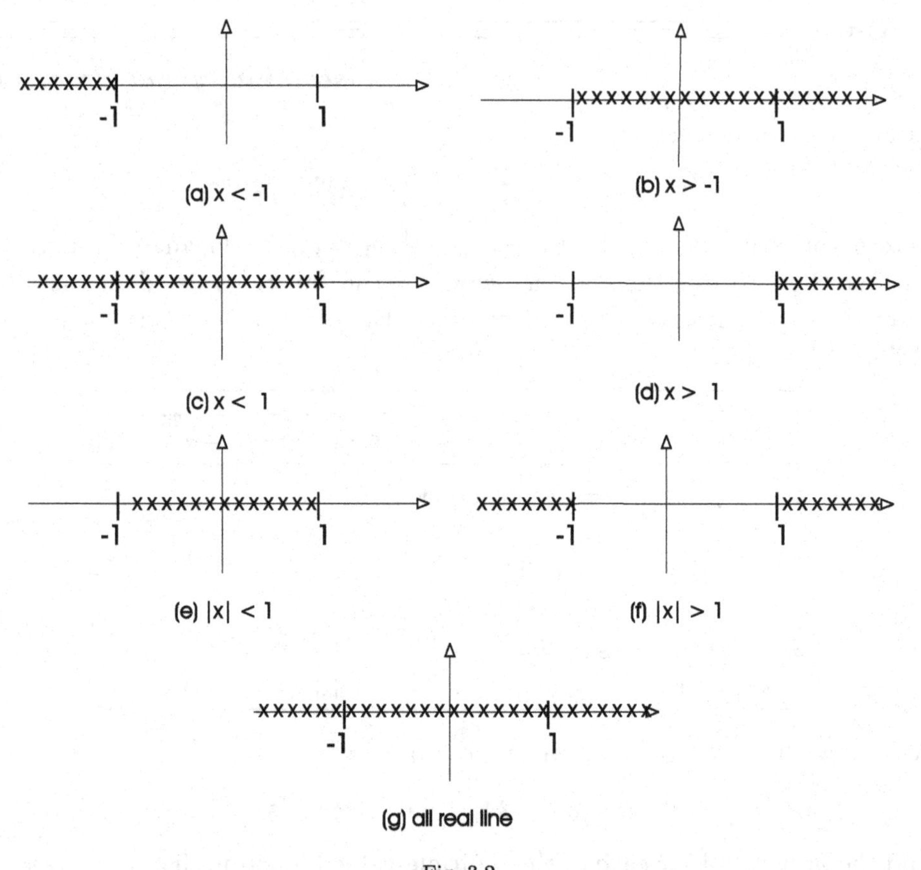

Fig. 3.2

Answers

- ⊙ Find branch cuts for $(z-1)^c$ and $(z+1)^c$ first.
- ⊙ Pay special attention to the values of x where the branch cuts for $(z-1)^c$ and $(z+1)^c$ overlap.

SN	$f(z)$	$0<\phi_1<2\pi$, $0<\phi_2<2\pi$	$0<\phi_1<2\pi$, $-\pi<\phi_2<\pi$
1	$(z-1)^{\frac{1}{4}}-(z-1)^{-\frac{1}{4}}$	(d) $x>1$	(c) $x<1$
2	$(z+1)^{\frac{1}{3}}(z-1)^{\frac{1}{3}}$	(b) $x>-1$	(g) Real axis
3	$(z+1)^{-\frac{1}{3}}(z-1)^{\frac{1}{3}}$	(e) $\lvert x\rvert<1$	(f) $\lvert x\rvert>1$
4	$(z+1)^{\frac{1}{3}}(z-1)^{\frac{2}{3}}$	(e) $\lvert x\rvert<1$	(f) $\lvert x\rvert>1$
5	$(z+1)^{\frac{3}{4}}+(z-1)^{\frac{1}{4}}$	(b) $x>-1$	(g) Real axis

§§3.5 Exercise

Logarithmic Function

⊘ In this set, for Q[1] and Q[2], take the branch cut for the log function along the positive real axis, and the branch selected is specified by

$$\log z = \ln r + i\theta\,, \quad 0 < \theta < 2\pi.$$

[1] (a) Let $\xi = -1 - i$. Compute $\log\xi$, $\log(\xi^2)$, and $\log(\xi^3)$.

(b) Find the differences $\Delta_2 = \log(\xi^2) - 2\log\xi$, $\Delta_3 = \log(\xi^3) - 3\log\xi$.

[2] Let a function $\eta(\xi_1, \xi_2)$ be defined in terms of $\log z$ by

$$\eta(\xi_1, \xi_2) = \log\xi_1 + \log\xi_2 - \log(\xi_1\,\xi_2).$$

Fill in the values of η for each of the 12 pairs of the complex numbers ξ_1 and ξ_2 given in the following table:

SN	ξ_1	ξ_2	$\arg(\xi_1)$	$\arg(\xi_2)$	$\eta(\xi_1, \xi_2)$
1	$1+i$	$1+i\sqrt{3}$			
2	$-1+i$	$-1+i\sqrt{3}$			
3	$-1-i$	$-1-i\sqrt{3}$			
4	$1-i$	$1-i\sqrt{3}$			
5	$1+i$	$-1+i\sqrt{3}$			
6	$-1-i$	$1+i\sqrt{3}$			
7	$-1+i$	$1-i\sqrt{3}$			
8	$1-i$	$1-i\sqrt{3}$			
9	$-1+i$	$-1-i\sqrt{3}$			
10	$-1-i$	$-1+i\sqrt{3}$			
11	$1+i$	$1-i\sqrt{3}$			
12	$1-i$	$1+i\sqrt{3}$			

[3] Let two functions $f(z)$ and $g(z)$ be defined by

$$f(z) = \operatorname{Log}\left(\frac{z-a}{z+a}\right), \quad g(z) = \operatorname{Log}(z-a) - \operatorname{Log}(z+a), \qquad a > 0,$$

where $\operatorname{Log} z$ is the principal value of the logarithm.

(a) What are the branch points of the two functions? Is there a branch point at infinity? Compute the discontinuity across the branch cut.

(b) Compute the discontinuity across the real intervals

(i) $x < -a$, (ii) $-a < x < a$, (iii) $x > 0$.

What can you conclude about the branch cut of the two functions $f(z)$ and $g(z)$?

(c) Compute the discontinuity of $z^n f(z)$ across the branch cut and show that

$$\operatorname{disc}\{z^n f(z)\}(x) = \begin{cases} 0, & \text{if } |x| > a, \\ 2\pi i x^n, & \text{if } |x| > a. \end{cases}$$

[4] Repeat the previous question with

$$f(z) = \mathrm{Log}\,[(z-a)(z+a)], \qquad g(z) = \mathrm{Log}\,(z-a) + \mathrm{Log}\,(z+a).$$

Answers

[1] $\log\xi = \frac{1}{2}\ln 2 + \frac{5\pi i}{4}$, $\log\xi^2 = \ln 2 + \frac{\pi i}{4}$, $\log\xi^3 = \frac{3}{2}\ln 2 + \frac{7\pi i}{4}$,

$$\Delta_2 = \frac{\pi i}{2} - \frac{5\pi i}{2} = -2\pi i, \quad \Delta_3 = \frac{7\pi i}{4} - 3.\frac{5\pi i}{4} = -2\pi i.$$

⊙ The fact that Δ's are not zero shows that, in general,

$$\log\lambda^n \neq n\log\lambda.$$

[2] $\xi_1 = r_1 e^{i\theta_1}$ and $\xi_2 = r_2 e^{i\theta_2}$, and the value of η is given by

$$\begin{aligned}\eta(\xi_1,\xi_2) &= \ln r_1 + \ln r_2 - \ln(r_1 r_2) + i\arg(\xi_1) + i\arg(\xi_2) - i\arg(\xi_1\xi_2)\\ &= i\left[\arg(\xi_1) + \arg(\xi_2) - \arg(\xi_1\xi_2)\right],\end{aligned}$$

where $\arg(z)$ is restricted to the range 0–2π.

SN	ξ_1	ξ_2	$\arg(\xi_1)$	$\arg(\xi_2)$	$\arg(\xi_1\xi_2)$	$\eta(\xi_1,\xi_2)$
1	$1+i$	$1+i\sqrt{3}$	$\pi/4$	$\pi/3$	$7\pi/12$	0
2	$-1+i$	$-1+i\sqrt{3}$	$3\pi/4$	$2\pi/3$	$17\pi/12$	0
3	$-1-i$	$-1-i\sqrt{3}$	$5\pi/4$	$4\pi/3$	$7\pi/12$	$2\pi i$
4	$1-i$	$1-i\sqrt{3}$	$7\pi/4$	$5\pi/3$	$17\pi/12$	$2\pi i$
5	$1+i$	$-1+i\sqrt{3}$	$\pi/4$	$\pi-\pi/3$	$11\pi/12$	0
6	$-1-i$	$1+i\sqrt{3}$	$5\pi/4$	$\pi/3$	$19\pi/12$	0
7	$-1+i$	$1-i\sqrt{3}$	$3\pi/4$	$5\pi/3$	$5\pi/12$	$2\pi i$
8	$-1-i$	$1-i\sqrt{3}$	$7\pi/4$	$5\pi/3$	$17\pi/12$	$2\pi i$
9	$-1+i$	$-1-i\sqrt{3}$	$3\pi/4$	$4\pi/3$	$\pi/12$	$2\pi i$
10	$-1-i$	$-1+i\sqrt{3}$	$5\pi/4$	$2\pi/3$	$23\pi/12$	0
11	$1+i$	$1-i\sqrt{3}$	$\pi/4$	$5\pi/3$	$23\pi/12$	0
12	$1-i$	$1+i\sqrt{3}$	$7\pi/4$	$\pi/3$	$\pi/12$	$2\pi i$

⊙ The nonzero values of $\eta(\xi_1,\xi_2)$ show that, in general, $\log(\xi_1\xi_2) \neq \log\xi_1 + \log\xi_2$.

[3] The branch cut for both the functions is the interval $(-a,a)$, and the discontinuity is $-2\pi i$ for $|x| < a$ and zero otherwise. The values of the two functions coincide in the cut plane.

[4] ⊙ In contrast to the functions in Q[3], the two functions $f(z)$ and $g(z)$ in these questions have very different branch cuts and discontinuities.

§§3.6 Exercise

Discontinuity Across the Branch Cut

[1] Several multivalued functions are listed in the first row of the table given below. Compute the discontinuity

$$\Delta(x) = \lim_{\epsilon\to 0}\{f(x+i\epsilon) - f(x-i\epsilon)\}$$

across the branch cut for two cases corresponding to the branch cut along the positive real axis ($\theta = 0$) and the negative real axis ($\theta = \pi$). Write your answers for the discontinuities in a tabular form:

$f(z)$	$\dfrac{\log z}{1+z^2}$	$\dfrac{z^c}{1+z^2}$	$\dfrac{\log^2 z}{1+z+z^2}$	$\dfrac{\log z}{1+z}$	$\dfrac{z^c}{1+z^3}$	$\dfrac{z^c \log z}{1+z^2}$
$\theta = 0$						
$\theta = \pi$						

[2] For $x > 0$, compute the difference $\lim_{\epsilon\to 0}[g(x+i\epsilon) - g(-x+i\epsilon)]$ for the following functions ($\epsilon > 0$) taking the the principal value for the logarithm function:

(a) $g(z) = \dfrac{\text{Log}\, z}{z+1}$, (b) $g(z) = \dfrac{\text{Log}\, z}{z^2+1}$,

(c) $g(z) = \dfrac{z^c}{z^2+1}$, (d) $g(z) = \dfrac{\text{Log}^2 z}{z^2+z+1}$.

[3] Let $h(z)$ be a single-valued function. Find an expression for the discontinuity of the functions $f(z) = h(z)\log z$ and $g(z) = z^c h(z)$ across the branch cut taking the branch cut for $\log z$ and z^c along

(a) the positive real axis ($\theta = 0$), (b) the negative real axis ($\theta = \pi$).

[4] Find a single-valued function with branch points at $z = \pm a$, with the discontinuity given by

$$\lim_{\epsilon\to 0}\{f(x+i\epsilon) - f(x-i\epsilon)\} = \begin{cases} e^x, & |x| < a, \\ 0, & |x| > a. \end{cases}$$

[5] (a) Find a function with the branch cut along the ray $\theta = 0$ and the discontinuity across the cut given by

$$\lim_{\epsilon\to 0}\{f(x+i\epsilon) - f(x-i\epsilon)\} = \begin{cases} 0, & x < 0, \\ \frac{1}{(\ln x)^2+\pi^2}, & x > 0. \end{cases}$$

(b) Give an example of a function with the branch cut along the ray $\theta = \pi$ and having the discontinuity

$$\lim_{\epsilon\to 0}\{f(x+i\epsilon) - f(x-i\epsilon)\} = \begin{cases} \frac{1}{(\ln |x|)^2+\pi^2}, & x < 0, \\ 0, & x > 0 \end{cases}$$

across the branch cut.

[6] The questions on the discontinuity of a function may be considered as a question about the difference in the limiting values of the function $g(z)$ on two rays at an angle 2π. Sometimes it is useful to consider other, more general, situations. Considering the principal value for z^c, compute the difference

$$f(r) - f(re^{i\alpha}), \qquad r > 0,$$

in the values of the function $f(z) = z^c/(z^5+1)$ at two points, which lie on the rays $\theta = 0, \alpha$ and are at the same distance ($= r$) from the origin. Sketch the rays $\theta = \alpha$ such that the difference is proportional to the value $f(r)$ itself.

Answers

[1] The discontinuities are given in the table below:

SN	$f(z)$	$\theta = 0$	$\theta = \pi$
1	$\dfrac{\log z}{1+z^2}$	$\dfrac{-2\pi i \log x}{1+x^2}$	$\dfrac{2\pi i \log \lvert x\rvert}{1+x^2}$
2	$\dfrac{z^c}{1+z^2}$	$\dfrac{\lvert x\rvert^c(1-e^{2\pi ic})}{1+x^2}$	$2i\left(\dfrac{x^c}{x^2+1}\right)\sin \pi c$
3	$\dfrac{\log z}{1+z}$	$\dfrac{-2\pi i}{1+x}$	$\dfrac{2\pi i \log \lvert x\rvert}{1-x}$
4	$\dfrac{z^c}{1+z^3}$	$\dfrac{(1-e^{2\pi ic})\lvert x\rvert^c}{1+x^3}$	$\dfrac{2i\sin(\pi c)\lvert x\rvert^c}{1-x^3}$
5	$\dfrac{\log^2 z}{1+z+z^2}$	$\dfrac{4\pi(\pi - i\ln x)(1+x^2)}{1+x^2+x^4}$	$\dfrac{-2\pi i \ln \lvert x\rvert}{1+x+x^2}$
6	$\dfrac{z^c \log z}{1+z^2}$	$\dfrac{x^c \log x}{1+x^2}(1-e^{2\pi ic})$ $-\dfrac{2\pi icx^c}{1+x^2}$	$\dfrac{2i\sin(2\pi ic)x^c \log x}{1+x^2}$ $-\dfrac{2i\cos(\pi ic)\lvert x\rvert^c}{1+x^2}$

[2] (a) $\dfrac{-2x\ln x}{1-x^2} - \dfrac{i\pi}{1-x}$, (b) $\dfrac{-i\pi}{x^2+1}$, (c) $(1-e^{i\pi c})\dfrac{x^c}{x^2+1}$,

(d) $-\dfrac{-2x\ln x}{1+x^2+x^4} + \dfrac{\pi^2 - 2\pi i\ln x}{1-x+x^2}$.

[4] $f(z) = \dfrac{1}{2\pi i} e^z \left[\mathrm{Log}\,(z-a) - \mathrm{Log}\,(z+a)\right]$.

[5] (a) $\dfrac{1}{2\pi i}\dfrac{1}{\log z - i\pi}$, (b) $-\dfrac{1}{2\pi i}\dfrac{1}{\mathrm{Log}\, z}$.

§§3.7 Mined

Branch Point Singularity

⊘ For each question there are sample answers. Read the sample answers carefully and locate the mistakes, if any.

⊘ Identify the source of mistakes and give your reasons why you consider a step or a statement incorrect. Be as precise as you can.

⊘ Do not write your solution as part of your answers.

⊘ State if in your opinion a sample answer is correct, partially correct, or completely wrong. Grade each sample answer out of a maximum of 10 points.

[1] Find the sum of values of $\operatorname{Log} z$ at the eighth roots of -2.

Sample answer 1. *One of the eighth roots of* -2 *is* $\eta_0 = \sqrt[8]{2}\exp(\pi i/8)$. *All the other roots are obtained from it by multiplying by powers of* $\exp(2\pi i/8)$. *Thus, the eight roots of* -2 *are*

$$\eta_k = \sqrt[8]{2}\exp\left(\frac{\pi i}{8} + \frac{2k\pi i}{8}\right),$$

where $k = 0, 1, 2, 3, \ldots, 7$. *Therefore, the sum of all the values is*

$$\begin{aligned}\sum_k \operatorname{Log}\eta_k &= \sum_{k=0}^{7}\left[\frac{1}{8}\ln 2 + \frac{1}{8}(2k+1)\pi i\right]\\ &= \ln 2 + 8\pi i.\end{aligned}$$

Sample answer 2. *The product all roots is* -2. *Therefore, we have*

$$\sum_{k=0}^{7}\operatorname{Log}\eta_k = \operatorname{Log}\left(\prod_{k=0}^{7}\eta_k\right) = \operatorname{Log} 2 + i\pi.$$

[2] Under the mapping

$$w = \left(\sqrt{z}\right)_{pv}$$

by the principal value of $\sqrt{z}$, which of the figures (A), (B), (C), and (D) in Fig. 3.3, is the image of the disk $|z| < 1$ in the cut plane?

Sample answer. *Since* $\sqrt{z} = r^{1/2}\exp(i\theta/2)$, *introduce variables* ρ *and* ϕ *for the* w *plane,* $w = \rho\exp(i\phi)$, *as* θ *varies from* 0 *to* 2π, *and* $\phi = \theta/2$ *goes from* 0 *to* π. *Thus, the circle is mapped into the semicircle, as shown in the Fig. 3.3* (A).

[3] Compute the principal value of $z^{\frac{2}{3}}$ for $z = -\sqrt{3} + i, -1 - i$, and $\frac{-\sqrt{3}+i}{-1-i}$.

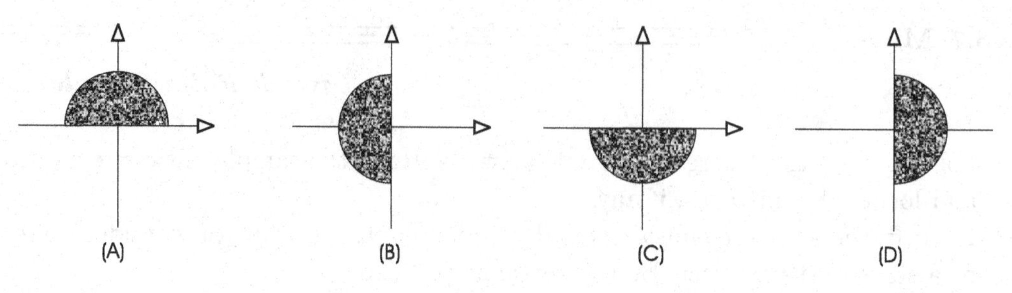

Fig. 3.3

Sample answer. *We first find the polar coordinates for the given numbers and the compute the principal value of the function* $z^{\frac{2}{3}}$ *using*

$$z^{\frac{2}{3}} = r^{\frac{2}{3}} \exp(2i\theta/3).$$

For the principal value, the value of θ *is to be taken in the range* $(-\pi, \pi)$. *The value of* θ *for the last row is obtained by making use of the relation* $\arg(z_1)/\arg(z_2) = \arg(z_1) - \arg(z_2)$. *The values of* $z, r,$ *and* θ *and the function are listed in the table below:*

Table 3.1 Sample answer for Q[3].

z	r	θ	$f(z)$
$-\sqrt{3}+i$	2	$\frac{5\pi}{6}$	$\sqrt[3]{4}\exp(5\pi i/9)$
$-1-i$	$\sqrt{2}$	$-\frac{3\pi}{4}$	$\sqrt[3]{2}\exp(\pi i/2)$
$\frac{-\sqrt{3}+i}{-1-i}$	$\sqrt{2}$	$\frac{19\pi}{12}$	$\sqrt[3]{2}\exp(19\pi i/18)$

Answers

[1] Sample answer 1 is wrong. For $\operatorname{Log} z = \ln z + i\operatorname{Arg}(z)$, the value of the argument should lie in the range $(-\pi, \pi)$, and this has not been ensured. For $k = 4, 5, 6, 7$ the value of $\arg(\eta_k)$ used lies outside the range $(-\pi, \pi)$.
Sample answer 2 is wrong because it uses

$$\sum_{k=0}^{7} \ln \eta_k = \ln\left(\prod_{k=0}^{7} \eta_k\right).$$

Note that $\operatorname{Log}(z_1 z_2)$ is, in general, not equal to $\operatorname{Log} z_1 + \operatorname{Log} z_2$.

[2] The sample answer is wrong because it does not take into account the restriction $-\pi < \theta < \pi$, to be used for the principal value of $\sqrt{z}$. The correct answer is (D).

[3] The values are correctly computed for $-\sqrt{3}+i$ and $-1-i$ but not for the ratio $(-\sqrt{3}+i)/(-1-i)$. The difference between the arg for the first two numbers, $19\pi/2$, does not satisfy the restriction of taking the arg in the range $(-\pi, \pi)$.

§§3.8 Mixed Bag

Multivalued Functions

⊘ In this set, $\operatorname{Log} z$ denotes the principal value of the logarithm of a complex variable z,

$$\operatorname{Log} z = \ln r + i \arg(z),$$

where $\arg(z)$ has the range

$$-\pi < \arg(z) < \pi.$$

[1] (a) Show that the mapping $z \to w = f(z)$, defined by the Joukousky function

$$f(z) = \frac{1}{2}\left(z + \frac{1}{z}\right),$$

is a two-to-one mapping except at ± 1.

(b) Will the mapping $f(z)$ be one-to-one if restricted to the following subset of the complex plane?

(i) left half plane;
(ii) right half plane;
(iii) upper half plane;
(iv) lower half plane;
(v) interior of the unit circle $|z| = 1$;
(vi) exterior of the unit circle $|z| = 1$.

[2] For each of the following functions, find all the branch points, including the branch point at infinity:

(a) $\sqrt{z}/(\sqrt{z} + 2)$,
(b) $(z-1)^{1/3}(z+1)^{2/3}$,
(c) $(z-1)^{1/3}(z+1)^{1/3}$,
(d) $(z-2)^{2/5}(z+3)^{3/5}$,
(e) $\operatorname{Log}(z-a) + \operatorname{Log}(z+a)$,
(f) $\operatorname{Log}\big((z-a)/(z+b)\big)$.

[3] (a) Give a definition of the function

$$f(z) = \sqrt{z^2 + 4}$$

in terms of the angles θ_1, θ_2, r_1, and r_2, as shown in Fig. 3.4, such that the branch cut is located between $z = \pm 2i$, and $f(z)$ is positive for points on the positive real axis.

(b) Is the point at infinity a branch point?

[4] Show that the inverse functions $\sin^{-1} z$, $\cos^{-1} z$, and $\tan^{-1} z$ do not have any branch point other than $z = \pm 1$ in the complex plane. What about the point at ∞?

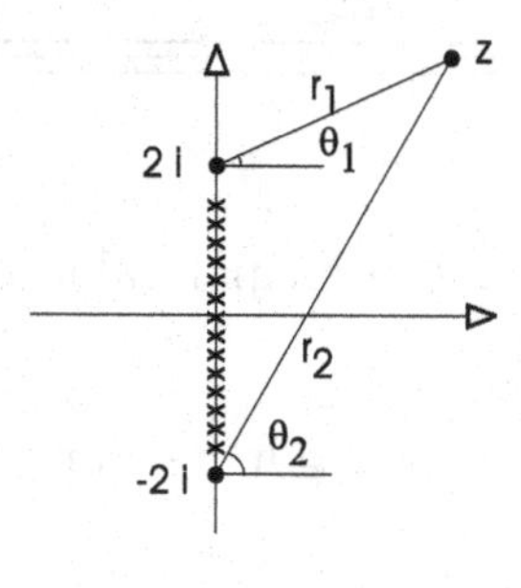

Fig. 3.4

[5] For real x, $|x| < 1$, show that the inverse sine function defined in Eq. (3.32) takes the form

$$\sin^{-1} x = \text{Arg}\left[ix + \sqrt{(1-x^2)}\right].$$

[6] (a) Show that $\log z$ and z^c are analytic at all points in the cut plane.
(b) Compute the derivatives and show that

$$\text{(a)}\ \frac{d}{dz}\log z = \frac{1}{z}, \quad \text{(b)}\ \frac{d}{dz}z^c = cz^{c-1}.$$

[7] (a) Compute the derivative $\frac{d}{dz}\tan^{-1} z$ and show that it does not have any branch point singularity.
(b) For the principal branch of the inverse sine function, which svb of $\sqrt{1-z^2}$ should be selected so that

$$\frac{d}{dz}\sin^{-1} z = \frac{1}{\sqrt{1-z^2}}$$

may hold?

[8] Sketch the curve represented by

(a) Re $\tanh^{-1} z = \text{const}$, (b) Im $\tanh^{-1} z = \text{const}$.

[9] Prove that

$$\text{Log}\,\alpha + \text{Log}\,\beta = \text{Log}\,(\alpha\beta) + \eta(\alpha,\beta),$$

where $\eta(\alpha,\beta)$ is given by

$$2\pi i\theta\{(\text{Im}\,\alpha)\theta(\text{Im}\,\beta)\theta[-\text{Im}\,(\alpha\beta)] - \theta(-\text{Im}\,\alpha)\theta(-\text{Im}\,\beta)\theta[\text{Im}\,(\alpha\beta)]\}$$

and where $\theta(x)$ is the step function

$$\theta(x) = \begin{cases} 0, & \text{if } x < 0, \\ 1, & \text{if } x > 0. \end{cases}$$

[10] If a and b are real and $\epsilon > 0$, show that, in the limit $\epsilon \to 0$, $\text{Log}\,(ab - i\epsilon)$ is equal to

$$\text{Log}\,(b - i\epsilon/a) + \text{Log}\,(a - i\delta).$$

where δ is positive and tends to zero as ϵ tends to zero. How does this result change when ϵ is allowed to be negative?

[11] Let $a_1 < a_2 < a_3 \cdots < a_n$ be real numbers and let $\arg(z)$ be defined as

$$\arg(z) = \theta, \quad 0 < \theta \leq 2\pi.$$

Plot the function $f(x)$ against x, where

$$f(z) = \sum_{k=1}^{n} \arg(z - a_k).$$

[12] Taking the principal value, sketch the curve traced by $w = f(z)$ in the complex plane as x moves along the real axis from $-\infty$ to ∞, where

(a) $f(z) = \sin^{-1} z$,
(b) $f(z) = \cos^{-1} z$,
(c) $f(z) = \sinh^{-1} z$,
(d) $f(z) = \frac{d}{\pi}\left[\sqrt{z^2 - 1} + \text{Log}\,(z + \sqrt{z^2 - 1})\right]$.

[13] Verify that for points on the real axis, the function

$$f(z) = \frac{i\pi}{2} + \sqrt{z(z-1)} - \text{Log}\,(\sqrt{z} + \sqrt{z-1})$$

assumes the values

$$f(x) = \begin{cases} -|x|^{1/2}|x-1|^{1/2} - \ln\left(|x|^{1/2} + |x-1|^{1/2}\right), & x < 0, \\ i\sqrt{x(1-x)} + i\sin^{-1}\sqrt{x}, & 0 < x < 1, \\ \frac{i\pi}{2} + \sqrt{x(x-1)} - \ln(\sqrt{x} + \sqrt{x-1}), & x > 1. \end{cases}$$

Answers

[1] $\odot$ The mapping is not one-to-one if for some ξ in the set $1/\xi$ also belongs to the set. (i) No, not one-to-one; (ii) No; (iii) Yes, one-to-one; (iv) Yes; (v) Yes; (vi) Yes.

[2] The branch point at infinity:

	$f(z)$	A Branch Point at ∞?
(a)	$\sqrt{z}/(\sqrt{z}+2)$	Yes
(b)	$(z-1)^{1/3}(z+1)^{2/3}$	No
(c)	$(z-1)^{1/3}(z+1)^{1/3}$	Yes
(d)	$(z-2)^{2/5}(z+5)^{3/5}$	No
(e)	$\mathrm{Log}\,(z+1)+\mathrm{Log}\,(z-1)$	Yes
(f)	$\mathrm{Log}\,[(z-1)/(z+1)]$	No

[3] (a) $f(z) = 4(r_1 r_2)^{1/2} \exp[i(\theta_1 + \theta_2)/2]$;

$$-\frac{3\pi}{2} < \phi_1 \leq \frac{\pi}{2}, \quad \frac{-3\pi}{2} < \phi_2 \leq \frac{\pi}{2}, \quad r_1 + r_2 > 4;$$

(b) $z = \infty$ is not a branch point.

[11] The plot of the function $f(x)$ looks like a staircase with steps of height 2π at the points a_k.

Chapter 4

INTEGRATION IN THE COMPLEX PLANE

§§4.1 Questions

Range of Parameters in Improper Integrals

Determine the ranges of parameters a, b, c, etc. by inspection, so that the following improper integrals may be well defined:

[1] $\displaystyle\int_{-\infty}^{\infty} \frac{dx}{(x^4+1)^c}$, [2] $\displaystyle\int_0^{\infty} \frac{x^c}{x+2}dx$,

[3] $\displaystyle\int_0^{\infty} \frac{x^c \log x}{(x^2+x)^b}dx$, [4] $\displaystyle\int_0^{\infty} \frac{x^c}{\sinh x}dx$,

[5] $\displaystyle\int_0^{\infty} \frac{\exp(cx)}{\cosh^2 x}dx$, [6] $\displaystyle\int_{-\infty}^{\infty} \frac{\exp(cx)+1}{\exp(-2x)+\exp(7x)}dx$,

[7] $\displaystyle\int_{-\infty}^{\infty} \frac{x^{2n+1}\sin x}{(x^2+1)^a}dx$, [8] $\displaystyle\int_{-\infty}^{\infty} \frac{\cosh ax}{\cosh bx}dx$,

[9] $\displaystyle\int_{-\infty}^{\infty} \frac{\sin^2 x}{x^a(x^2+4)^b}dx$, [10] $\displaystyle\int_0^{\infty} \frac{x^c[\exp(ax)-1]}{(x^3+2)^c}dx$,

[11] $\displaystyle\int_0^{\infty} \frac{x^c}{\exp(x)-1}dx$, [12] $\displaystyle\int_0^{1} x^{a-1}(1-x)^{b-1}dx$,

[13] $\displaystyle\int_0^{\infty} x^{a-1}e^{-x}dx$, [14] $\displaystyle\int_0^{\infty} \frac{dx}{x^a(1+x)^b(1+x)^c}$.

Answers

[1]	$4c > 1$,	[2]	$-1 < c < 0$,	[3]	$b - c < 1$, $2b - c - 1 > 1$,
[4]	$c > 0$,	[5]	$c < 2$,	[6]	$-2 < c < 7$,
[7]	$a > n + 1$,	[8]	$a < b$,	[9]	$a < 3, a + 2b > 1$,
[10]	$a < 0, c > -2$,	[11]	$c > 0$,	[12]	$a > 0, b > 0$,
[13]	$a > 0$,	[14]	$a > 0, a + b + c > 1$.		

§§4.2 Tutorial

Computing Line Integrals in the Complex plane

[1] Integrating the real and imaginary parts of the integrand separately, prove that

$$\int_a^b \exp(\lambda x)dx = \frac{1}{\lambda}[\exp(\lambda b) - \exp(\lambda a)],$$

where λ is a complex number.

[2] Using the definition of a line integral in the complex plane, prove that

$$\int_{z_1}^{z_2} \exp(\lambda z)dz = \frac{1}{\lambda}[\exp(\lambda z_2) - \exp(\lambda z_1)],$$

where the integral is along a path, joining z_1 and z_2, and consisting of two line segments PS and SQ parallel to the two axes (see Fig. 4.1).

[3] Let $\xi = a + ib$ represent the point P in Fig. 4.2. Verify that

$$\int_\Gamma z^2\, dz = \frac{1}{3}(a+ib)^3,$$

where the integral is taken along a contour Γ, starting from the origin and ending at the point P, and Γ consists of

(a) OA along the x axis, followed by AP parallel to the y axis;
(b) OB along the y axis, followed by BP parallel to the x axis;
(c) the straight line joining O and P;
(d) the parabola $y = \frac{b}{a^2}x^2$ from O to P.

[4] Taking Γ to be the ellipse $\dfrac{x^2}{a^2} + \dfrac{y^2}{b^2} = 1$, show that $\displaystyle\oint_\Gamma z^2\, dz = 0.$ ⟨1⟩

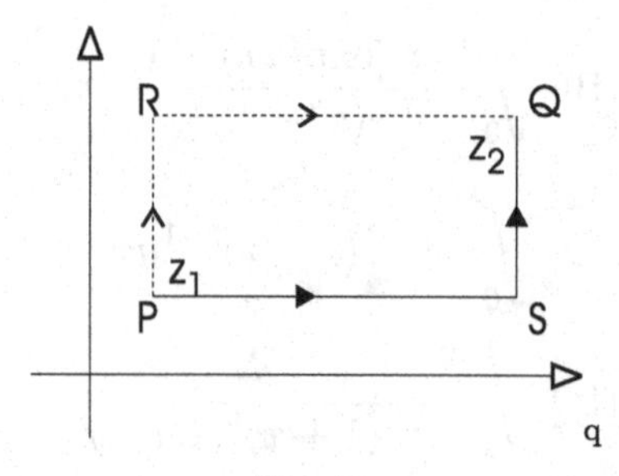

Fig. 4.1

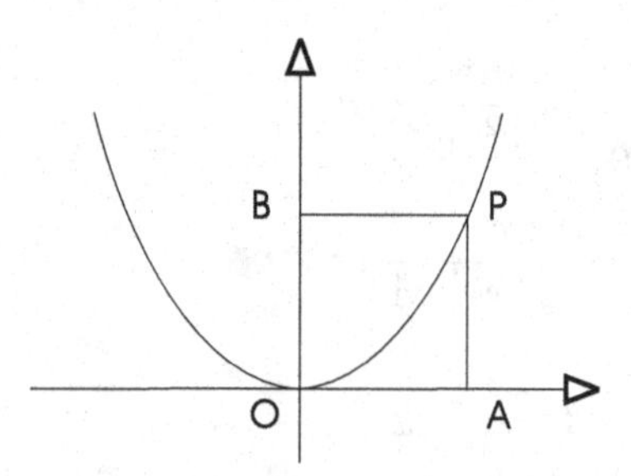

Fig. 4.2

§§4.3 Exercise

Evaluation of Line Integrals

[1] Evaluate the proper integral $\int_0^1 \frac{dt}{t+i}$ over the real parameter t by separating the real and imaginary parts of the integrand.

[2] Compute the line integral

$$\int u(x,y)\,dx - v(x,y)\,dy,$$

where $u(x,y) = x^2 - 2xy$ and $v(x,y) = y^2 + 2xy$, along

(a) the triangle OMNO in Fig. 4.3(a);

(b) the sides AB and BC of the triangle ABC in Fig. 4.3(b).

[3] Integrate $\int_C \bar{z}^2\,dz$, where C is

(a) the circle $|z - i| = 1$;

(b) a rectangle with vertices at $0, 1, 1+2i, 2i$;

(c) the arc of the ellipse $x^2 + \frac{1}{4}y^2 = 1$ in the upper half plane.

[4] By an explicit computation, verify that the value of the integral $\oint_C \bar{z}\,dz$ along the ellipse

$$\frac{(x-x_0)^2}{a^2} + \frac{(y-y_0)^2}{b^2} = 1$$

is $2\pi iab$.

[5] Compute $\int \bar{z}\,dz$ along a line segment joining z_1 to z_2. Use your answer to evaluate $\int \bar{z}\,dz$ around a closed rectangle with corners at $\pm 1 \pm 2i$.

[6] Compute $\int_C \bar{z}^n\,dz$ along the circle C given by $|z - z_0| = R$.

[7] Compute $\int_\gamma \bar{z}^{2n}\,dz$, where

(a) γ is a line segment on the imaginary axis from $-ai$ to ai;

(b) γ is the semicircle $\{z \,|\, |z| = a, \operatorname{Re} z > 0\}$ in the right half plane with end points at $-ai$ to ai.

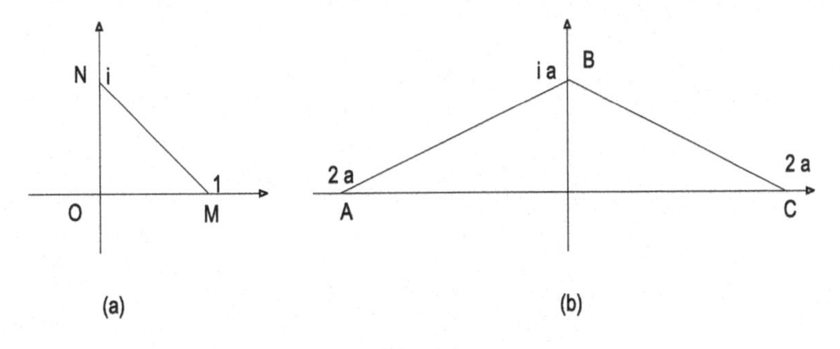

Fig. 4.1

[8] By an explicit computation show that

$$\oint_\Gamma (z - z_0)^{2n}\, dz = 0,$$

where Γ is the ellipse $\dfrac{(x - x_0)^2}{a^2} + \dfrac{(y - y_0)^2}{b^2} = 1$, with the center at z_0.

[9] Let ABCD be a square with successive vertices at $(-1, i)$, $(1, i)$, $(1, -i)$, and $(-1, -i)$. Considering the real and imaginary parts of the *integrand separately*, compute the integrals along the four sides of the square and hence show that

$$\oint_{\text{ABCDA}} \frac{dz}{z} = -2\pi i.$$

[10] Compute the line integral

$$\int_\gamma (z^2 + 1) \exp(-iz^2) dz,$$

where γ is the ray $\theta = -\pi/4$.

[11] Let L be the straight line running parallel to the real axis from $-\infty + 2i$ to $\infty + 2i$. Compute the line integral $\int_L (2z^2 + 3z + 5) \exp(-z^2 + 4iz)\, dz$.

Answers

[1] $\ln 2 - \frac{\pi i}{4}$.

[2] (a) $-2/3$, (b) $-a^3$, $7a^3/3$.

[3] (a) 4π, (b) $-(13 + 16i)/3$, (c) 10.

[5] $16\pi i$.

[6] $2\pi i R^n z_0^{n-1}$.

[7] (a) $2i(-1)^n a^{2n+1}/(2n + 1)$, (b) 0.

[10] $\sqrt{\pi}(4 - i)/2$.

[11] $2\sqrt{\pi}e^{-4}(-1 + 3i)$.

§§4.4 Questions

Deformation of Contours

[1] Ten functions are listed in a table below. Several contours are drawn in Fig. 4.4. PQRS is a square with each side having a length of two units. KFGHL is an arc of a circle with the center at C. For each combination of choices of the contour and function $f(z)$ indicated in the table given below, complete the table by writing Yes or No, as follows:

- Write Yes if the specified contour can be deformed into ABCDA without changing the value of the integral of $f(z)$ around the contour, i.e.

$$\oint_{\Gamma} f(z)dz = \oint_{\text{ABCDA}} f(z)dz.$$

- Write No if the specified contour Γ cannot be deformed into ABCDA without changing the value of the integral of $f(z)$ around the contour, i.e.

$$\oint_{\Gamma} f(z)dz \neq \oint_{\text{ABCDA}} f(z)dz.$$

  Write your answer without working out the integral, taking Γ to be contours given in column headings of table below.

[2] For each function listed in the table, briefly explain your answer given in the table.

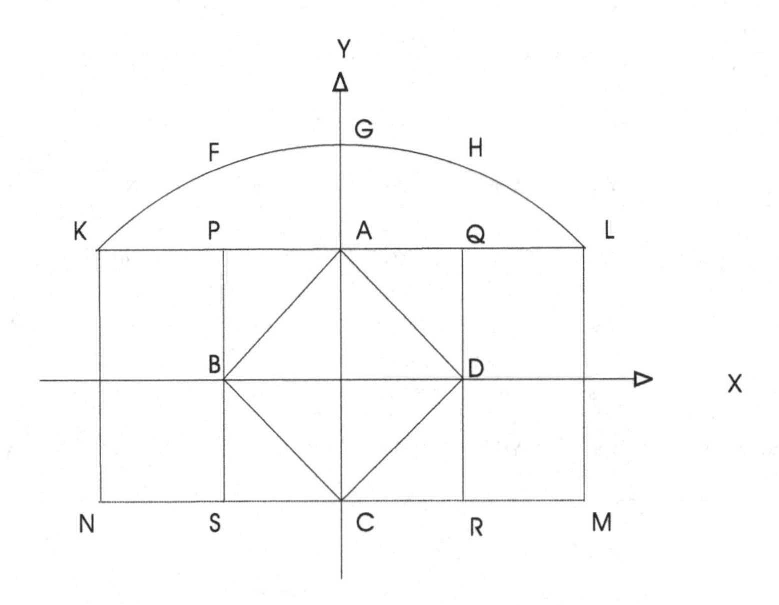

Fig. 4.2

$f(z)\backslash\Gamma$	ANMK	KCLK	AKNCDA	ABCMLA	KNCMLGK	PQRSP
$\bar{z}^2$						
$\lvert z\rvert^2$						
$\sinh z$						
$\dfrac{1}{(z-10)^{10}}$						
$\dfrac{1}{\cos(z/100)}$						
$\dfrac{z^4}{\cosh 100z}$						
$\dfrac{1}{2z-3}$						
$\dfrac{z^3+5z+7}{z^4-9}$						
$\dfrac{5z}{z^2-\sqrt{3}z+1}$						
$\exp(1/z)$						

Answers

<table>
<tr><th>$f(z)\backslash\Gamma$</th><th>ANMK</th><th>KCLK</th><th>AKNCDA</th><th>ABCMLA</th><th>KNCMLGK</th><th>PQRSP</th></tr>
<tr><td>$\bar{z}^2$</td><td colspan="6">No</td></tr>
<tr><td>$\lvert z\rvert^2$</td><td colspan="6">No</td></tr>
<tr><td>$\sinh z$</td><td colspan="6">Yes</td></tr>
<tr><td>$\dfrac{1}{(z-10)^{10}}$</td><td colspan="6">Yes</td></tr>
<tr><td>$\dfrac{1}{\cos(z/100)}$</td><td colspan="6">Yes</td></tr>
<tr><td>$\dfrac{1}{2z-3}$</td><td>Yes</td><td>Yes</td><td>Yes</td><td>No</td><td>No</td><td>Yes</td></tr>
<tr><td>$\dfrac{z^4}{\cosh 100z}$</td><td>Yes</td><td>Yes</td><td>Yes</td><td>Yes</td><td>No</td><td>No</td></tr>
<tr><td>$\dfrac{z^3+5z+7}{z^4-9}$</td><td>Yes</td><td>Yes</td><td>No</td><td>No</td><td>No</td><td>Yes</td></tr>
<tr><td>$\dfrac{5z}{(z^2-\sqrt{3}z+1)}$</td><td>Yes</td><td>Yes</td><td>No</td><td>No</td><td>No</td><td>No</td></tr>
<tr><td>$\exp(1/z)$</td><td>Yes</td><td>Yes</td><td>Yes</td><td>Yes</td><td>No</td><td>Yes</td></tr>
</table>

§§4.5 Exercise

Deformation of Contours

⊘ Use only results in Ch. 4 to solve the following problems. No result from later chapters should be used.

[1] Compute the value of

$$\oint_C \frac{dz}{(z-3)(3z-1)},$$

where C is the unit circle. Use your results to prove that

$$\int_0^{2\pi} \frac{d\theta}{5-3\cos\theta} = \frac{\pi}{2}.$$

[2] For a square S, with corners at $\pm 1 \pm i$, by an explicit computation show

$$\oint_S \frac{1}{z}\exp(pz)\,dz = 2\pi i.$$

Use this result to prove that

$$\int_0^{\pi} \exp(p\cos\theta)\cos(p\sin\theta) = \pi.$$

[3] Find the value of the integral $\oint \frac{dz}{z}$ around the ellipse $\frac{x^2}{a^2}+\frac{y^2}{b^2}=1$ and hence prove that

$$\int_0^{2\pi} \frac{d\phi}{a^2\cos^2\phi + b^2\sin^2\phi} = \frac{\pi}{2ab}.$$

Answer

[1] $\oint_C \frac{dz}{(z-3)(3z-1)} = -\frac{\pi i}{4}.$

§§4.6 Exercise

Cauchy's Theorem

⊘ Use the definition of the integral along a contour in the complex plane as a line integral to compute the integrals in this set.

[1] Compute the following integrals along closed contours, where Γ_1 to Γ_4 are as shown in Fig. 4.5. Verify that the integrals vanish in accordance with Cauchy's fundamental theorem.

(a) $\oint_{\Gamma_1} (z + 3z^2)dz$, (b) $\oint_{\Gamma_2} \exp(z)dz$,

(c) $\oint_{\Gamma_3} (z - 3)dz$, (d) $\oint_{\Gamma_4} z^{10}dz$.

[2] Let Γ_5 and Γ_6 be two contours, OABC and OC, joining the point $z = 0$ to $z = ia$, as shown in Fig. 4.6. Compute the following integrals along the two contours and find out which integrals are not path-independent:

(a) $\int (y + 3ix - 2)dz$, (b) $\int \exp(\pi z^*)dz$,

(c) $\int (z - 3)^2 dz$, (d) $\int \cosh z dz$.

[3] Taking a suitable contour joining the two end points, and consisting of straight line segments only, compute the following integrals:

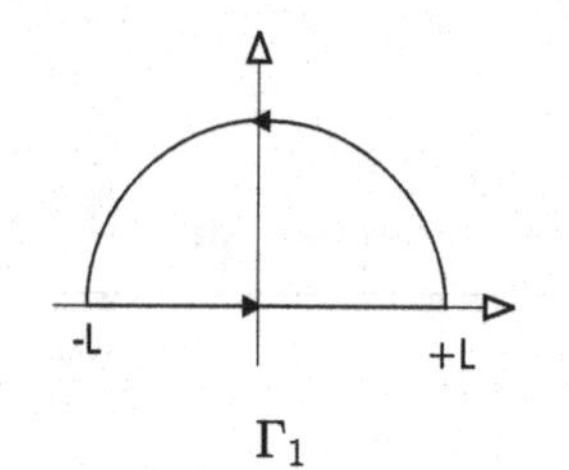

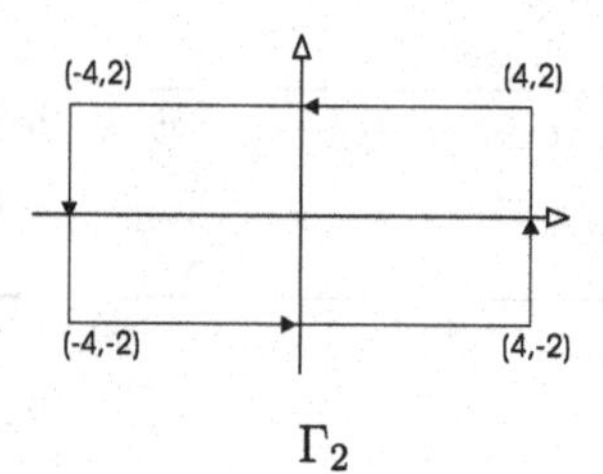

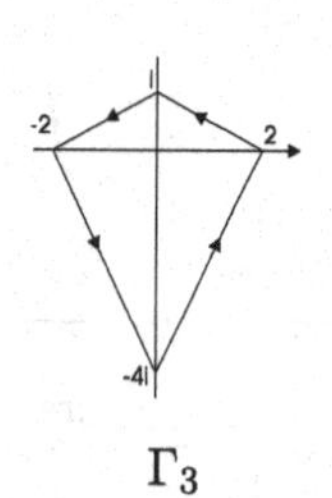

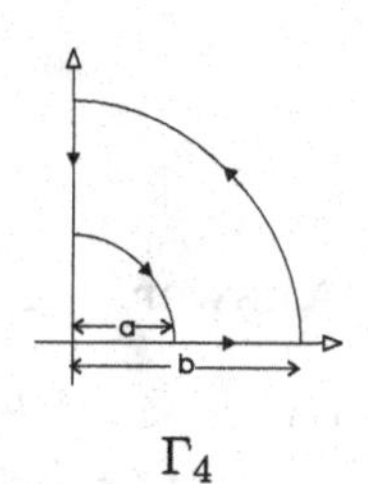

Fig. 4.3

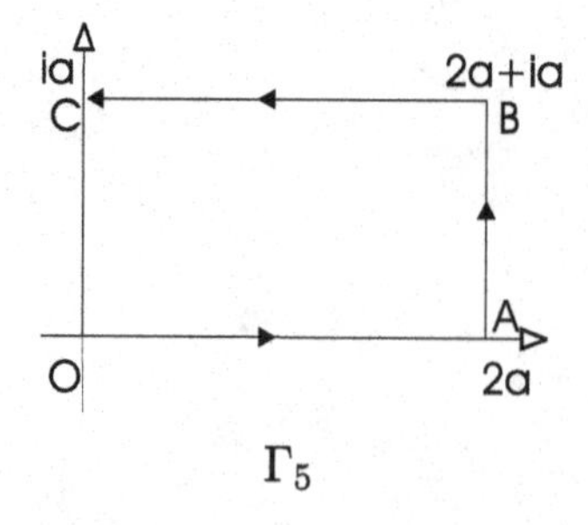

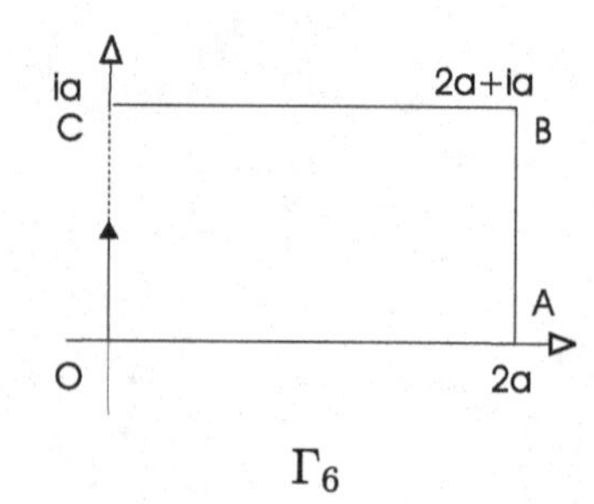

Fig. 4.4

(a) $\int_0^{\xi} dz$, (b) $\int_{\alpha}^{\beta} (z^2 + 3z + 1)dz$,

(c) $\int_0^{\pi+2i} \sin(z/2)dz$, (d) $\int_0^{i/2} \exp(\pi z)dz$.

[4] By integrating along two different suitable paths, prove that

$$\int_{z_1}^{z_2} z^n dz = \frac{z_2^{n+1} - z_1^{n+1}}{n+1}, \quad n > 0.$$

[5] Select a suitable path joining z_1 to z_2 and show that

$$\int_{z_1}^{z_2} \frac{dz}{(z-z_0)^n} = \frac{1}{n-1}\left[\frac{1}{(z_1-z_0)^{n-1}} - \frac{1}{(z_2-z_0)^{n-1}}\right], \quad n > 1.$$

Does the above result hold for all paths joining z_1 and z_2? What restriction, if any, should be imposed on the path selected so that the above result may hold?

Answers

[2] (a) $I_{\Gamma_5} = (-16+i)a^2/2 - (8+2i)a, I_{\Gamma_6} = (a^2/2 - 2a)i$;

(b) $I_{\Gamma_5} = (\frac{i}{\pi} - 1)(e^{2a\pi} - 1) + e^{2a\pi}(1 - e^{-i\pi a}), I_{\Gamma_6} = \frac{i}{\pi}(1 - e^{-i\pi a})$;

(c) $I_{\Gamma_5} = I_{\Gamma_6} = \frac{1}{3}(ia-3)^3 - 9$;

(d) $I_{\Gamma_5} = I_{\Gamma_6} = i \sin z$.

[3] (a) ξ, (b) $\frac{1}{3}(\beta^3 - \alpha^3) + \frac{3}{2}(\beta^2 - \alpha^2) + (\beta - \alpha)$,

(c) $2 + i(e - 1/e)/2$, (d) $1 - i$.

[4] ⊙ Take one of the paths consisting of line segments parallel to the real and imaginary axes. The second path could be made up of a ray(s) from the origin and a circular arc(s) with the center at $z = 0$.

[5] ⊙ A possible choice for the integration path consists of a line segment on a ray from the point z_0 and a circular arc with the center at z_0.

§§4.7 Tutorial

Shift of a Real Integration Variable by a Complex Number

[1] Prove that

$$\int_0^\infty \exp(-px^2)\cos(2bx)\,dx = \frac{1}{2}\sqrt{\frac{\pi}{p}}\exp(-b^2/p), \quad p > 0.$$

For the sake of definiteness, assume that b is positive. Follow the steps given below to obtain the above result.

(a) In the above integral, to be computed, change the integration limits from $(0,\infty)$ to $(-\infty,\infty)$, and write $\cos(2bx)$ in terms of the exponentials and show that the integral is proportional to the real part of an integral I_1, where

$$I_1 = \int_{-\infty}^{\infty} \exp[-p(x+ib/p)^2]\,dx.$$

The integral I_1 can be computed by showing that it is equal to to the Gaussian integral I_2:

$$I_2 = \int_{-\infty}^{\infty} \exp(-px^2)\,dx = \sqrt{\frac{\pi}{p}}, \quad p > 0.$$

(b) To show $I_1 = I_2$, use the equations for the lines AB and DC (see Fig. 4.7),

$$\text{AB} : \{z = x \mid -L \le x \le R\},$$
$$\text{DC} : \{z = x + i(b/p) \mid -L \le x \le R\},$$

and set up the integrals

$$J_{\text{AB}} = \int_{\text{AB}} \exp(-pz^2)\,dz, \quad J_{\text{DC}} = \int_{\text{DC}} \exp(-pz^2)\,dz.$$

Verify that $I_1 = J_{\text{DC}}$ and $I_2 = J_{\text{AB}}$ in the limit $L, R \to \infty$.

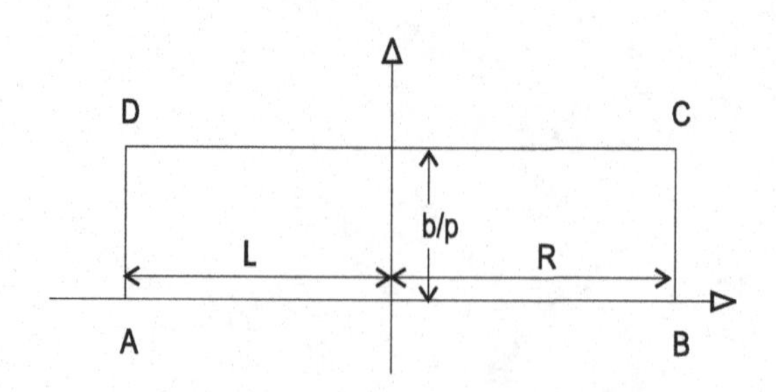

Fig. 4.5

(c) Next, find the value of

$$\oint_{\text{ABCDA}} \exp(-pz^2)\, dz$$

by a simple application of Cauchy's theorem.

(d) Write the integral of $\exp(-pz^2)$ around ABCDA as a sum of four integrals:

$$\oint_{\text{ABCDA}} e^{-pz^2}\, dz = \int_{\text{AB}} e^{-pz^2}\, dz + \int_{\text{BC}} e^{-pz^2}\, dz + \int_{\text{CD}} e^{-pz^2}\, dz + \int_{\text{DA}} e^{-pz^2}\, dz.$$

Prove that the two integrals $\int_{\text{BC}}$ and $\int_{\text{DA}}$, on the right hand side of the equation, vanish when $L, R \to \infty$.

(e) Making use of the results you have obtained so far, prove the result

$$I_1 = I_2$$

and hence show that

$$\int_0^\infty \exp(-px^2 + 2ibx)\, dx = \exp\left(-\frac{b^2}{p}\right) \int_0^\infty \exp(-px^2)\, dx,$$

which gives the desired result.

[12] For $a > b > 0$, prove that

$$\int_0^\infty \exp(-px^2) \sin ax \sin bx\, dx = \frac{1}{4}\sqrt{\frac{\pi}{p}} \left\{\exp\left[-\frac{(a-b)^2}{4p}\right] - \exp\left[\frac{(a+b)^2}{4p}\right]\right\},$$

$$\int_0^\infty \exp(-px^2) \cos ax \cos bx\, dx = \frac{1}{4}\sqrt{\frac{\pi}{p}} \left\{\exp\left[-\frac{(a-b)^2}{4p}\right] + \exp\left[\frac{(a+b)^2}{4p}\right]\right\}.$$

§§4.8 Tutorial

Scaling of a Real Integration Variable by a Complex Number

[1] Use the contour of Fig. 4.8 and the steps outlined below, and prove that

$$\int_0^\infty e^{ipx^2}\, dx = \frac{1}{2}\sqrt{\frac{\pi}{p}}\, e^{i\pi/4}.$$

Hence, derive the values of the following two integrals $\int_0^\infty \cos(px^2)\, dx$ and $\int_0^\infty \sin(px^2)\, dx$.

(a) Use the equation for the arc ACB, (see Fig. 4.8), and show that

$$\int_{\text{ACB}} \exp(ipz^2)\, dz = \frac{iR}{2}\int_0^{\pi/4} \exp[ipR^2\cos 2\theta - pR^2\sin 2\theta]\, d\theta.$$

(b) Using the inequality

$$\sin\theta \geq \frac{2\theta}{\pi},$$

which holds for $\theta \leq \frac{\pi}{2}$, and derive a bound:

$$\left|\int_{\text{ACB}} \exp(ipz^2) dz\right| \leq \int_0^{\pi/2} \exp(-pR^2\theta)\, d\theta.$$

Hence, show that

$$\lim_{R\to\infty}\int_{\text{ACB}} \exp(ipz^2) dz = 0.$$

(c) Argue that the integral

$$\oint_{\text{OABO}} \exp(ipz^2)\, dz$$

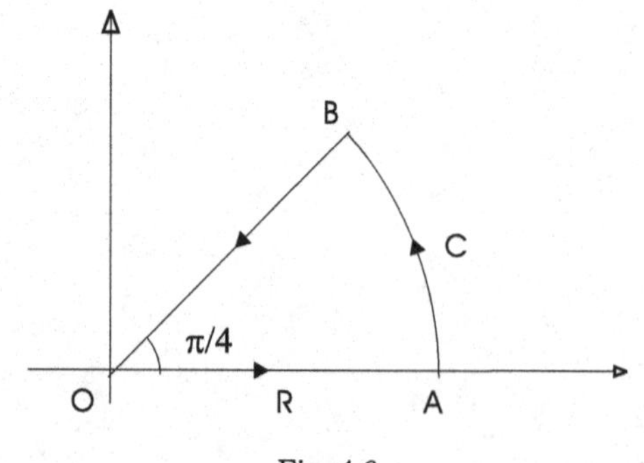

Fig. 4.6

vanishes, where OABO is a closed sector of a circle of radius R and lying between $\theta = 0$ and $\theta = \pi/4$ (see Fig. 4.8).

(d) Split the above integral as a sum of three integrals:

$$\oint_{\text{OABO}} \exp(ipz^2)\, dz = \int_{\text{OA}} \exp(ipz^2)\, dz + \int_{\text{ACB}} \exp(ipz^2)\, dz + \int_{\text{BO}} \exp(ipz^2)\, dz.$$

Taking the limit $R \to \infty$ in the above equation and writing the line integrals along OA and BO as ordinary integrals over a real variables with suitable limits, obtain the desired result. Use this to find values of the integrals in [1](a) and [1](b).

[2] Can one use the above procedure for evaluation of

$$\int_0^\infty e^{-ipx^2}\, dx$$

for the case $p > 0$? Argue that the changes are needed because the integral

$$\int_{\text{ACB}} \exp(-ipz^2)\, dz$$

does not go to zero as $R \to \infty$. Making an appropriate choice of contour, complete the proof of the result

$$\int_0^\infty \exp(-ipx^2)\, dx = \frac{1}{2}\sqrt{\frac{\pi}{p}}\, e^{-i\pi/4}, \quad p > 0,$$

following the steps for Q[1] as closely as possible.

§§4.9 Exercise

Shift and Scaling by a Complex Number

⊘ In this set, a, b, p, and q are real with $p > 0$, and α and β are complex numbers.
⊘ Prove the results in this set of exercises by

- reducing them to Gaussian integrals of the type

$$\int_{-\infty}^{\infty} x^{2m} e^{-px^2}\, dx, \quad \int_{-\infty}^{\infty} x^{2m+1} e^{-px^2}\, dx;$$

- justifying the shift or scaling by a complex number using the Cauchy–Goursat theorem.

[1] $\displaystyle\int_0^\infty e^{-q^2x^2} \sin^2 ax = \frac{1}{2}\frac{\sqrt{\pi}}{q}\left[1 - \exp\left(-\frac{a^2}{q^2}\right)\right].$

[2] $\displaystyle\int_0^\infty e^{-q^2x^2} \begin{bmatrix} \sin p(x+c) \\ \cos p(x+c) \end{bmatrix} = \frac{\sqrt{\pi}}{q}\exp(-p^2/4q^2)\begin{pmatrix} \sin pc \\ \cos pc \end{pmatrix}.$

[3] $\displaystyle\int_0^\infty xe^{-q^2x^2} \sin ax\, dx = \sqrt{\pi}\left(\frac{a}{4q^2}\right)\exp\left(-\frac{a^2}{4q^2}\right).$

[4] $\displaystyle\int_0^\infty x^2e^{-q^2x^2} \cos ax\, dx = \sqrt{\pi}\frac{2q^2 - a^2}{8q^5}\exp\left(-\frac{a^2}{4q^2}\right).$

[5] $\displaystyle\int_0^\infty x^3e^{-q^2x^2} \sin ax\, dx = \sqrt{\pi}\frac{6aq^2 - q^3}{16q^7}\exp\left(-\frac{a^2}{4q^2}\right).$

[6] $\displaystyle\int_0^\infty xe^{-px} \cos(2x^2 + px) = 0.$

[7] $\displaystyle\int_0^\infty xe^{-px} \cos(2x^2 - px) = \frac{p\sqrt{\pi}}{8}\exp\left(-\frac{p^2}{4}\right), \quad p > 0.$

[8] $\displaystyle\int_0^\infty x^2e^{-px} [\sin(2x^2 + px) + \cos(2x^2 + px)] = 0, \quad p > 0.$

[9] $\displaystyle\int_0^\infty xe^{-px} [\sin(2x^2 - px) - \cos(2x^2 - px)] = \frac{\sqrt{\pi}}{16}(2 - p^2)\exp\left(-\frac{p^2}{4}\right), \quad p > 0.$

[10] If $a > 0$, $|\arg\beta| < \frac{\pi}{4}$, show that

(a) $\displaystyle\int_0^\infty xe^{-\beta x} \sin ax^2 \sin\beta x = \frac{\beta}{4}\sqrt{\frac{\pi}{2a^3}}\exp\left(-\frac{\beta^2}{2a}\right);$

(b) $\displaystyle\int_0^\infty xe^{-\beta x} \cos ax^2 \cos\beta x = \frac{\beta}{4}\sqrt{\frac{\pi}{2a^3}}\exp\left(-\frac{\beta^2}{2a}\right).$

Establish the following integration results:

[11] $\displaystyle\int_0^\infty \cos ax^2 \cos 2bx\, dx = \frac{1}{2}\sqrt{\frac{\pi}{2a}}\left[\cos\left(\frac{b^2}{a}\right) + \sin\left(\frac{b^2}{a}\right)\right];$

[12] $\displaystyle\int_0^\infty (\cos ax + \sin ax)\sin(bx)^2\,dx = \frac{1}{2b}\sqrt{\frac{\pi}{2}}\exp\left(-\frac{a^2}{2b}\right);$

[13] $\displaystyle\int_0^\infty \sin ax^2 \sin 2bx \sin 2cx\,dx = \frac{\sqrt{\pi}}{2a}\sin\left(\frac{2bc}{a^2}\right)\cos\left(\frac{b^2+c^2}{a^2}-\frac{\pi}{4}\right);$

[14] $\displaystyle\int_0^\infty x\sin ax^2 \sin 2bx\,dx = \frac{b}{2a}\sqrt{\frac{\pi}{2a}}\left[\cos\left(\frac{b^2}{a}\right)+\sin\left(\frac{b^2}{a}\right)\right];$

[15] $\displaystyle\int_0^\infty \sin ax^2 \cos bx^2\,dx = \begin{cases} \frac{1}{4}\sqrt{\frac{\pi}{2}}\left(\frac{1}{\sqrt{a+b}}+\frac{1}{\sqrt{a-b}}\right), & a > b > 0, \\ \frac{1}{4}\sqrt{\frac{\pi}{2}}\left(\frac{1}{\sqrt{a+b}}-\frac{1}{\sqrt{b-a}}\right), & b > a > 0; \end{cases}$

[16] $\displaystyle\int_0^\infty (\sin^2 ax^2 - \sin^2 bx^2)\,dx = \frac{1}{8}\left(\sqrt{\frac{\pi}{b}}-\sqrt{\frac{\pi}{a}}\right), \quad a > b > 0;$

[17] $\displaystyle\int_0^\infty [\sin(a-x^2)+\cos(a-x^2)]\,dx = \sqrt{\frac{\pi}{2}}\sin a.$

§§4.10 Exercise

Rotation of the Contour

[1] Integrate $\int \exp(-z^2)\,dz$ around a suitable sector contour to prove that

$$\int_0^\infty \exp(-x^2\cos 2\alpha)\cos(x^2\sin 2\alpha) = \frac{\sqrt{\pi}}{2}\cos\alpha,$$

$$\int_0^\infty \exp(-x^2\cos 2\alpha)\sin(x^2\sin 2\alpha) = \frac{\sqrt{\pi}}{2}\sin\alpha.$$

[2] Rotating the contour from the real axis to a ray $\arg z = \alpha$ at a suitable angle α (see Fig. 4.9), prove that

$$\int_0^\infty \exp(-x^p\cos p\lambda)\cos(x^p\sin p\lambda)dx = \frac{1}{p}\cos(\lambda)\Gamma\left(\frac{1}{p}\right),$$

$$\int_0^\infty \exp(-x^p\cos p\lambda)\sin(x^p\sin p\lambda)dx = \frac{1}{p}\sin(\lambda)\Gamma\left(\frac{1}{p}\right).$$

[3] Given the integral

$$\int_0^\infty x^{p-1}\exp(-ax)dx = a^{-p}\Gamma(p), \quad a>0,\ p>0,$$

justify a "complex scaling" of the integration variable $u = ix$ to prove that

(a) $\displaystyle\int_0^\infty x^{p-1}\cos ax\,dx = a^{-p}\Gamma(p)\cos\left(\frac{p\pi}{2}\right), \quad a>0,\ p>0;$

(b) $\displaystyle\int_0^\infty x^{p-1}\sin ax\,dx = a^{-p}\,\Gamma(p)\sin\left(\frac{p\pi}{2}\right), \quad a>0,\ p>0.$

(c) What changes in the steps of the answers to parts (a) and (b) will be needed to derive the following integration formula for $p>0$ and $-\frac{\pi}{2}<\alpha<\frac{\pi}{2}$?

$$\int_0^\infty x^{p-1}\exp(-x\cos\alpha)\cos(x\sin\alpha)dx = \cos(p\alpha)\Gamma(p).$$

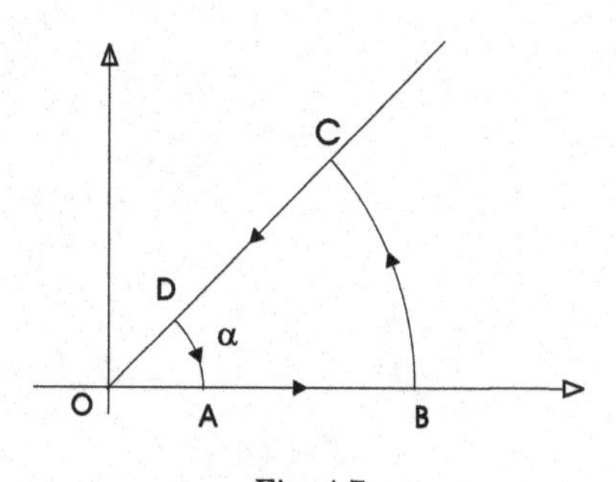

Fig. 4.7

§§4.11 Mixed Bag

Integration in the Complex Plane

[1] For a closed contour Γ show that the integral $\oint_\Gamma \bar{z}\,dz$ has the value $2iA$, where A is the area enclosed by Γ. What area is given by this integral if Γ is not a closed contour?

[2] Study the Milne–Thomson method of finding the conjugate harmonic function in §4.6.3 and answer the following questions:

(a) Use the Milne–Thomson method to find an analytic function whose imaginary part is

$$v(x, y) = \frac{\sin 2x}{\cosh 2y - \cos 2x}.$$

(b) How do you justify setting $y = 0$ in the second step of the Milne–Thomson method?

[3] Considering division of the circle $\gamma = \{z||z - z_0| = R\}$ into n equal parts and starting with the definition of a line integral as a limit of the Riemann sum and taking the limit $n \to \infty$, prove that

$$\oint \frac{dz}{z - z_0} = 2\pi i.$$

[4] (a) Find a bound on the following integral along a closed square with corners at $(\pm 1, \pm i)$:

$$\int \exp(z^*)\,dz.$$

(b) Compute the integral explicitly and verify that its value is indeed less than the bound obtained by you.

[5] Let $p(z)$ and $q(z)$ be polynomials of degree m and n, respectively. Obtain a bound on values of $\frac{p(z)}{q(z)}$ for points on a circle C_R, for large R. Use your result for $m \leq n - 2$ to show that

$$\lim_{R\to\infty} \int_{C_R} \frac{p(z)}{q(z)}\,dz = 0.$$

[6] Let Γ be the arc of the circle $|z| = 3$ lying in the first quadrant. Prove the bound

$$\left| \int_\Gamma \frac{z^2 + z + 1}{z^4 + 1}\,dz \right| < \frac{3\pi}{4}.$$

[7] Obtain a bound on the integral

$$\left| \int_\Gamma \frac{\ln(1-z)}{z} dz \right|,$$

where $\Gamma = \{z||z-1| < \rho,\, \epsilon < \arg(z-1) < 2\pi - \epsilon\}$. Hence, show that

$$\int_\Gamma \frac{\ln(1-z)}{z} dz \longrightarrow 0$$

as $\longrightarrow 0$.

[8] Show that the integral

$$\int_C \frac{a^z}{z^2}\, dz$$

vanishes in the limit where the radius R of the semicircle C, with the center at $z = a$, tends to ∞ and if

(a) $a > 1$ and C is taken to lie to the right of the line $x = a$;
(b) $a < 1$ and C is taken to lie to the left of the line $x = a$.

Hints

[1] Use Green's theorem for a double integral in two dimensions. Or write the integral $\oint \bar{z}\, dz$ as a Riemann sum.

[2] (a) $f(z) = -i\mathrm{cosec}^2 z$.
(b) Regarding $f(z) = u(z - iy, y) + iv(z - iy, y)$ as a formal identity between functions of two variables y and z and setting $y = 0$ gives $f(z)$.

Chapter 5

CAUCHY'S INTEGRAL FORMULA

§§5.1 Exercise

Cauchy's Integral Formula

Compute the following integrals over the specified contours using Cauchy's integral theorem. All the contours are taken to be positively oriented.

[1] $\oint_C \frac{\sin z}{2z-3\pi}\, dz$, where the contour C is the circle $|z| = 6$.

[2] $\oint_C \frac{\sin z}{(2z-3\pi)^3}\, dz$, where the contour C is the circle $|z| = 6$.

[3] $\oint_C \frac{e^{2z}}{(z-\ln\pi)^4}\, dz$, where the contour C is a circle with the center at $z = \ln\pi$ and with radius $1/2$.

[4] $\oint_C \frac{z^2+4}{z^2+6z+5}\, dz$, where the contour C is a rectangle with the corners at $(2, \pm 3)$, and $(\pm 2, 3)$.

[5] $\oint_C \frac{dz}{(z-1)^2(2z+5)}\, dz$, where the contour C is a square with the corners at $(\pm 2, 0)$ and $(0, \pm 2)$.

[6] $\oint_C \frac{\cos^5 z}{(z-\pi/6)^3}\, dz$, where the contour C is the circle $|z-1| = \frac{1}{3}$.

[7] $\oint_C \frac{1}{4z^3+25z}\, dz$, where the contour C is a quadrilateral with the corners at $(-1, 0)$, $(0, \pm 2)$, and $(3, 0)$.

[8] $\oint_C \frac{\exp(\pi z)}{z^2+1}\, dz$, where the contour C is the circle $|z| = 3/4$.

[9] $\oint_C \frac{z\cos\pi z^2}{z^4-16}\, dz$, where the contour C is a rectangle bounded by the lines $x \pm y = -1$ and $x \pm y = 3$.

[10] $\oint_C \frac{dz}{(z^2+25)(z-1)^2}\, dz$, where the contour C is a triangle enclosed by the lines $y = 4$ and $y = \pm 2x - 3$.

Answers

[1] $-\pi i$, [2] $\frac{1}{8}\pi i$, [3] $\frac{8}{3}\pi^3 i$, [4] $\frac{5}{2}\pi i$, [5] $-\frac{4}{49}\pi i$.

[6] 0, [7] $\frac{2}{25}\pi i$, [8] 0, [9] $-\frac{1}{8}\pi i$, [10] $-\frac{1}{169}\pi i$.

§§5.2 Quiz

Circle of Convergence of Taylor Expansion

[1] Several functions are listed in the second column of the table appearing at the end. For each of the functions answer the following questions:

Q1. Does the Taylor series expansion in positive powers of $z - z_0$,

$$f(z) = \sum_{n=0}^{\infty} a_n (z - z_0)^n,$$

exist? The value of z_0 in each case is listed in the third column. Write the answer, Yes/No, in the fourth column.

Q2. If the answer to Q1 is Yes, give the largest value of the radius of convergence of the series in the last column.

SN	$f(z)$	z_0	Ans. 1	Ans. 2
1	$\sin \pi z$	0		
2	$\operatorname{cosech} \pi z$	0		
3	$\dfrac{1}{\lvert z\rvert^2 + 16}$	0		
4	$\dfrac{1}{z-1}$	0		
5	$\dfrac{1}{z^3 + z^2 + z + 1}$	6		
6	$\dfrac{1}{4z - z^3}$	-1		
7	$\exp(z^2)$	$i\pi$		
8	$\dfrac{\sin \pi z}{z+2}$	-2		
9	$\tan(\pi z/2)$	3		
10	$\tan(\pi z/2)$	i		
11	$\dfrac{1}{z^2+1}$	$2+i$		
12	$\dfrac{1}{4z^*(3-z^2)}$	2		

[2] Give a brief explanation for your answers to [1] for each of the functions.

Answers

SN	$f(z)$	z_0	Ans. 1	Ans. 2
1	$\sin \pi z$	0	Yes	∞
2	$\operatorname{cosech} \pi z$	0	No	–
3	$\dfrac{1}{\|z\|^2+16}$	0	No	–
4	$\dfrac{1}{z-1}$	0	Yes	1
5	$\dfrac{1}{z^3+z^2+z+1}$	6	Yes	$\sqrt{37}$
6	$\dfrac{1}{4z-z^3}$	-1	Yes	1
7	$\exp(z^2)$	$i\pi$	YES	∞
8	$\dfrac{\sin \pi z}{z+2}$	-2	Yes	∞
9	$\tan(\pi z/2)$	3	No	–
10	$\tan(\pi z/2)$	i	Yes	$\sqrt{2}$
11	$\dfrac{1}{z^2+1}$	$2+i$	Yes	2
12	$\dfrac{1}{4z^*(3-z^2)}$	2	No	–

§§5.3 Exercise

Using MacLaurin's Theorem

Use MacLaurin's theorem to establish the following series expansions.

[1] $\dfrac{1}{1-z} = \sum_{m=0}^{\infty} z^m = 1 + z + z^2 + z^3 + \cdots,$

[2] $\exp(z) = \sum_{m=0}^{\infty} \dfrac{z^m}{m!} = 1 + z + \dfrac{z^2}{2!} + \dfrac{z^3}{3!} + \cdots,$

[3] $\sin z = \sum_{m=0}^{\infty} (-1)^m \dfrac{z^{2m+1}}{(2m+1)!},$

[4] $\cos z = \sum_{m=0}^{\infty} (-1)^m \dfrac{z^{2m}}{(2m)!},$

[5] $\sinh z = \sum_{m=0}^{\infty} \dfrac{z^{2m+1}}{(2m+1)!},$

[6] $\cosh z = \sum_{m=0}^{\infty} \dfrac{z^{2m}}{(2m)!},$

[7] $\dfrac{1}{(1+z)^n} = 1 - nz + n(n+1)z^2 + \cdots,$

[8] $\cosh z^2 = \sum_{m=0}^{\infty} \dfrac{z^{4m}}{(2m)!} = 1 + \dfrac{z^4}{2!} + \dfrac{z^8}{4!} + \cdots.$

§§5.4 Exercise

Taylor Series Representation

[1] Determine the first four terms in the Taylor series in powers of $z - z_0$, as indicated. Specify the region of the complex plane in which the expansion converges.

(a) $\exp(z)$ in powers of $z - 5$,
(b) $\exp(z) + \exp(2z)$ in powers of $z - i\pi$,
(c) $\cosh z - \cosh 6$ in powers of $z - 6$,
(d) $\sin^2 z$ in powers of $z - \frac{\pi}{2}$,
(e) $\sin z - \cos z$ in powers of $z - \frac{\pi}{4}$.

[2] Use MacLaurin's theorem to derive the following expansions.

(a) $\tan z = z + \dfrac{z^3}{3} + \dfrac{2z^5}{15} + \dfrac{17z^7}{315} + \cdots,$

(b) $\sec^2 z = 1 + z^2 + \dfrac{2z^4}{3} + \dfrac{17z^6}{45} + \cdots,$

(c) $\sec z = 1 + \dfrac{z^2}{2} + \dfrac{5z^4}{24} + \dfrac{61z^6}{720} + \cdots,$

(d) $e^z \cos z = 1 + z - \dfrac{z^3}{3} - \dfrac{z^4}{6} - \dfrac{z^5}{30} + \cdots,$

(e) $\exp(\sin z) = 1 + z + \dfrac{z^2}{2} - \dfrac{z^4}{8} - \dfrac{z^5}{15} + \cdots,$

(f) $\exp(z - z^2) = 1 + z - \dfrac{z^2}{2!} - \dfrac{5z^3}{3!} + \dfrac{z^4}{4!} + \dfrac{41z^5}{5!} + \cdots.$

[3] Using the known series expansions of the sine, cosine and the exponential functions, derive the following expansions.

(a) $\tan z = z + \dfrac{z^3}{3} + \dfrac{2z^5}{15} + \dfrac{17z^7}{315} + \cdots,$

(b) $\sec^2 z = 1 + z^2 + \dfrac{2z^4}{3} + \dfrac{17z^6}{45} + \cdots,$

(c) $z^2 \operatorname{cosec}^2 z = 1 + \dfrac{z^2}{3} + \dfrac{z^4}{15} + \dfrac{2z^6}{189} + \cdots,$

(d) $e^z \sin z = z + z^2 + \dfrac{z^3}{31} - \dfrac{z^5}{30} - \dfrac{z^6}{90} + \cdots,$

(e) $\cos^2 z = 1 - z^2 + \dfrac{z^4}{3} - \dfrac{2z^6}{45} + \dfrac{z^8}{315} + \cdots,$

(f) $\dfrac{1}{1 + \tan z} = 1 - z + z^2 - \dfrac{4z^3}{3} + \dfrac{5z^4}{3} + \cdots,$

(g) $\dfrac{z^2}{\exp(z) - 1} = z - \dfrac{z^2}{2} + \dfrac{z^3}{12} - \dfrac{z^5}{720} + \cdots.$

[4] Expand $\frac{1}{z^3}$ in powers of $z-1$ in a series converging in $|z-1|<1$ by suitably making use of the series

$$\frac{1}{z+a}=\frac{1}{a}\left(1-\frac{z}{a}+\frac{z^2}{a^2}-\frac{z^3}{a^3}+\cdots\right).$$

Write the most general term of the expansion.

[5] Differentiating the series for $(z+a)^{-1}$ term by term $m-1$ times, obtain the expansion of $\frac{1}{(z+a)^m}$ for $|z|<a$ and for $|z|>a$.

Answers

[1] (a) $\exp(z)=e^5\sum_{n=0}^{\infty}\frac{(z-5)^n}{n!}$,

(b) $\exp(z)+\exp(2z)=\sum_{n=0}^{\infty}(2^n-1)\frac{(z-i\pi)^n}{n!}$,

(c) $\cosh z-\cosh 6=\cosh 6\sum_{m=1}^{\infty}\frac{(z-6)^{2m}}{(2m)!}+\sinh 6\sum_{m=0}^{\infty}\frac{(z-6)^{2m+1}}{(2m+1)!}$,

(d) $\sin^2 z=1+\sum_{n=1}^{\infty}(-1)^n 2^{2n-1}\frac{(z-\frac{\pi}{2})^{2n}}{(2n)!}$,

(e) $\sin z-\cos z=\sqrt{2}\sum_{n=0}(-1)^n\frac{(z-\frac{\pi}{4})^{2n+1}}{(2n+1)!}$.

§§5.5 Tutorial

Series Expansion from the Binomial Theorem

[1] It is known that the Taylor series expansion of the function

$$f(z) = \frac{1}{z+a},$$

given by

$$f(z) = \frac{1}{a}(1 - (z/a) + (z/a)^2 - (z/a)^3 + \cdots),$$

converges for $|z| < a$ and diverges for $|z| > a$. We can obtain another expansion, as follows:

$$\begin{aligned} f(z) &= \frac{1}{z(1+\frac{a}{z})} \\ &= \frac{1}{z}\left[1 - \frac{a}{z} + \frac{a^2}{z^2} - \frac{a^3}{z^3} + \cdots\right]. \end{aligned}$$

What is the region of convergence of the above series in the complex plane? Draw a figure to show the region of convergence.

[2] Write the first three nonzero terms in the following expansions:

(i) $1/(z-2)$ in powers of z,
(ii) $1/(z-2)$ in powers of $1/z$,
(iii) $1/(z+3)$ in powers of z,
(iv) $1/(z+3)$ in powers of $1/z$.

Specify the regions of convergence in each case.

[3] Use the answers in Q[2] to get four different kinds of expansions for the function

$$f(z) = \frac{1}{(z-2)(z+3)}.$$

For this function draw figures indicating the regions of convergence for the four cases.

[4] (a) The function

$$\frac{1}{(z-1)(z-2)(z-3)}$$

is to be expanded in powers of z and $1/z$. Find the expansion valid in the

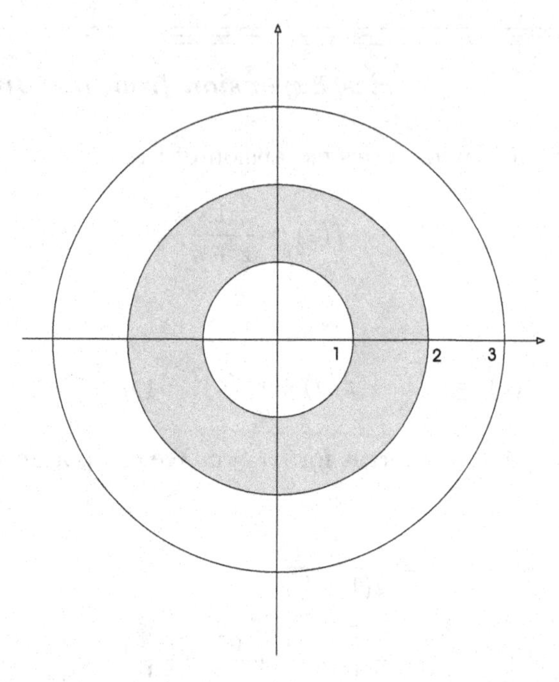

Fig. 5.1

region $1 < |z| < 2$ (see Fig. 5.1).

(b) Find the series in powers of z and $1/z$ and converging in

$$S = \{z||z - \tfrac{5}{2}| < \tfrac{1}{4}\}.$$

Does the series converge in a bigger set? If yes, what is the biggest open set containing S in which the series will converge?

§§5.6 Exercise

Laurent Expansion using the Binomial Theorem

[1] Obtain all the series expansions in powers of z, Taylor as well as Laurent, for the functions given below. Specify the regions of convergence in each case.

(a) $\dfrac{1}{z+3}$, (b) $\dfrac{1}{z(z+4)}$,

(c) $\dfrac{1}{(z+1)(z+2)}$, (d) $\dfrac{1}{z(z^2+1)}$,

(e) $\dfrac{2z}{(z+1)(z+3)}$, (f) $\dfrac{z}{(z+1)^2}$.

[2] Derive Taylor as well as all Laurent series expansions in powers of $z - z_0$ for the functions given below. Specify the regions of convergence in each case.

(a) $\dfrac{1}{z(z+2)}$, $z_0 = -1$; (b) $\dfrac{1}{z^2+5z+6}$, $z_0 = 1$;

(c) $\dfrac{3z}{(z+1)(z^2+2)}$, $z_0 = -1$; (d) $\dfrac{1}{z(3z+2)(z+2)}$, $z_0 = -2/3$.

[3] (a) Obtain all Taylor as well as Laurent series in powers of z for the function

$$f(z) = \frac{1}{(z-2)(z-4)}.$$

(b) Draw figures to sketch subsets of convergence for the expansions of $f(z)$ in powers of z obtained in part (a). State if any one of these expansions can be used to represent the function $f(z)$ in the subset

(i) $\{z||2z-1| < 1\}$,
(ii) $\{z||z+2| < \frac{1}{3}\}$,
(iii) $\{z||z-2| < 5\}$,
(iv) $\{z||z+3| < \frac{1}{3}\}$,
(v) $\{z|5 < \text{Re}\, z < 6, 3 < \text{Im}\, z < 4\}$.

Answers

[1] (a) $0 \le |z| < 3$; $\dfrac{1}{3} - \dfrac{z}{9} + \dfrac{z^2}{27} - \dfrac{z^3}{81} + \cdots$;

$|z| > 3$; $\dfrac{1}{z} - \dfrac{3}{z^2} + \dfrac{9}{z^3} - \dfrac{27}{z^4} + \cdots$;

(b) $0 \lneqq |z| < 4$; $\dfrac{1}{4z} - \dfrac{1}{16} + \dfrac{z}{64} - \dfrac{z^2}{256} + \cdots$;

$|z| > 4$; $\dfrac{1}{z^2} - \dfrac{4}{z^3} + \dfrac{16}{z^4} - \dfrac{64}{z^5} + \cdots$;

(c) $0 \le |z| < 1$; $\quad \frac{1}{2} - \frac{3z}{4} + \frac{7z^2}{8} + \frac{15z^3}{16} + \cdots;$

$1 < |z| < 2$; $\quad \frac{1}{z} - \frac{1}{z^2} + \frac{1}{z^3} - \cdots - \frac{1}{2} + \frac{z}{4} - \frac{z^2}{8} + \frac{z^3}{16} + \cdots;$

$|z| > 2$; $\quad \frac{1}{z^2} - \frac{3}{z^3} + \frac{7}{z^4} - \frac{15}{z^5} + \cdots;$

(d) $0 \lneqq |z| < 1$; $\quad \frac{1}{z} - z + z^3 - z^5 + \cdots;$

$|z| > 1$; $\quad \frac{1}{z^3} - \frac{1}{z^5} + \frac{1}{z^7} - \frac{1}{z^9} + \cdots;$

(e) $|z| \lneqq 1$; $\quad \frac{1}{z} - 2 + 3z - 4z^2 + 5z^3 \cdots;$

$|z| > 1$; $\quad \frac{1}{z^3} - \frac{2}{z^4} + \frac{3}{z^5} - \frac{4}{z^6} + \cdots;$

(f) $0 \lneqq |z| < 1$; $\quad \frac{4}{3} - \frac{10z}{9} + \frac{28z^2}{27} - \frac{82z^3}{81} + \frac{244z^4}{243} - \cdots;$

$1 < |z| < 3$; $\quad \frac{1}{z}\left(1 - \frac{1}{z} + \frac{1}{z^2} - \frac{1}{z^3} + \cdots\right) + \frac{1}{3}\left(1 - \frac{z}{3} + \frac{z^2}{9} - \frac{z^3}{27} + \cdots\right);$

$|z| > 3$; $\quad \frac{2}{z} - \frac{4}{z^2} + \frac{10}{z^3} - \frac{28}{z^4} + \cdots.$

[2] (a) $0 \le |z+1| < 1$; $\quad -[1 + (z+1)^{-2} + (z+1)^{-3} + \cdots];$

$|z+1| > 1$; $\quad \frac{1}{(z+1)^2} + \frac{1}{(z+1)^4} + \frac{1}{(z+1)^6} + \cdots.$

(b) $0 \le |z+1| < 1$; $\quad \frac{1}{2} - \frac{3(z+1)}{4} + \frac{7(z+1)^2}{8} + \frac{15(z+1)^3}{16} + \cdots;$

$1 < |z+1| < 2$; $\quad \frac{1}{z+1} - \frac{1}{(z+1)^2} + \frac{1}{(z+1)^3} - \cdots - \frac{1}{2} + \frac{z+1}{4}$

$$- \frac{(z+1)^2}{8} + \frac{(z+1)^3}{16} + \cdots;$$

$|z+1| > 2$; $\quad \frac{1}{(z+1)^2} - \frac{3}{(z+1)^3} + \frac{7}{(z+1)^4} - \frac{15}{(z+1)^5} + \cdots.$

(c) $|z+1| < \sqrt{3}$; $\quad -\frac{1}{z+1} + \frac{1}{3} - \frac{5(z+1)}{9} + \frac{7(z+1)^2}{27}$

$$-\frac{(z+1)^3}{81} - \frac{23(z+1)^4}{243} + \cdots;$$

$|z+1| > \sqrt{3}$; $\quad \frac{3}{(z+1)^2} + \frac{3}{(z+1)^3} - \frac{3}{(z+1)^4} - \frac{15}{(z+1)^5} + \cdots.$

(d) $0 \lneqq \left|z+\frac{2}{3}\right| < \frac{2}{3}$; $\quad -\frac{3}{8}\left(z+\frac{2}{3}\right)^{-1} - \frac{9}{32} - \frac{81}{128}\left(z+\frac{2}{3}\right)$

$$+\frac{15.27}{16.32}\left(z+\frac{2}{3}\right)^2 - + \cdots;$$

$\frac{2}{3} < \left|z+\frac{2}{3}\right| < \frac{4}{3}$; $\quad \left[-\frac{1}{4}\left(z+\frac{2}{3}\right)^{-1} + \frac{1}{6}\left(z+\frac{2}{3}\right)^{-2} + \frac{1}{9}\left(z+\frac{2}{3}\right)^{-3} + \cdots\right]$

$$+\left[-\frac{1}{6} + \frac{2}{9}\left(z+\frac{2}{3}\right)^2 - \frac{8}{27}\left(z+\frac{2}{3}\right)^3 + \frac{32}{81}\left(z+\frac{2}{3}\right)^4 + \cdots\right];$$

$\left|z+\frac{2}{3}\right| > \frac{4}{3}$; $\quad -\frac{1}{4}\left(z+\frac{2}{3}\right)^{-1} + \frac{1}{3}\left(z+\frac{2}{3}\right)^{-3} - \frac{2}{9}\left(z+\frac{2}{3}\right)^{-4} + \cdots.$

[3] (a) $0 \le |z| < 2$; $\quad \frac{1}{8} + \frac{3z}{32} + \frac{7z^2}{128} + \frac{15z^3}{512} + \frac{31z^4}{2048} + \cdots;$

$2 < |z| < 4$; $\quad -\left(\frac{1}{8} + \frac{1}{z^2} + \frac{2}{z^3} + \frac{4}{z^4} + \cdots\right) - \left(\frac{1}{8} + \frac{z}{32} + \frac{z^2}{128} + \frac{z^4}{512} + \cdots\right);$

$|z| > 4$; $\quad -\frac{1}{z^2} - \frac{6}{z^3} - \frac{28}{z^4} - \frac{120}{z^5} - \cdots.$

(b) (i) Yes,
(ii) No,
(iii) No,
(iv) Yes,
(v) Yes.

§§5.7 Quiz

Subsets for Convergence of Laurent Expansions

Ten pairs of a function and a complex numbers z_0 are specified below. For each pair draw a figure to show nonoverlapping regions of the complex plane, each requiring a different power series expansion in powers of $z - z_0$. Indicate the total number of such regions. If the required number of regions is infinite, draw figures showing three innermost regions, i.e. one containing the point z_0 and the next two regions.

[1] $\dfrac{1}{(z-3)^4} + \dfrac{3}{z-3} + 5z^3, \ z_0 = 0;$

[2] $\dfrac{1}{(z-3)^4} + \dfrac{3}{z-3} + 5z^3, \ z_0 = 3;$

[3] $\dfrac{1}{z^3+1}, \ z_0 = 1;$

[4] $\dfrac{1}{z^3+8}, \ z_0 = \sqrt{3}i;$

[5] $\dfrac{z}{z^{201}+1}, \ z_0 = 0;$

[6] $\dfrac{z+z^2}{z^2-z-6}, \ z_0 = -2;$

[7] $\dfrac{z+z^2}{z^2-z-6}, \ z_0 = 2;$

[8] $f(z) = \operatorname{cosec} \pi z, \ z_0 = \nu;$

[9] $\dfrac{1}{\exp(\pi z)+1}, \ z_0 = \pi i;$

[10] $\operatorname{cosec}(\pi/z), \ z_0 = 1;$

where ν is a nonzero positive integer.

Answers

[1] $\{z||z| < 3\}, \quad \{z||z| > 3\};$

[2] $\{z||z-3| > 0\};$

[3] $\{z||z-1| < 1\}, \quad \{z|1 < |z-1| < 2\}, \quad \{z||z-1| > 2\};$

[4] $\{z||z-\sqrt{3}i| < 1\}, \{z|1 < |z-\sqrt{3}i| < \sqrt{7}\};$

$\{z|\sqrt{7} < |z-\sqrt{3}i| < \sqrt{13}\}, \{z||z-\sqrt{3}i| > \sqrt{13}\};$

[5] $\{z||z| < 1\}, \quad \{z||z| > 1\};$

[6] $\{z|0 < |z+2| < 5\}, \quad \{z||z+2| > 5\};$

[7] $\{z||z-2| < 1\}, \quad \{z|1 < |z-2| < 4\}, \quad \{z||z-2| > 4\};$

[8] $\{z|n < |z-\nu| < n-1\}, \ n = 1, 2, 3, \ldots;$

[9] $\{z||z-\pi i| < 2\}, \quad \{z|2 < |z-\pi i| < 4\}, \cdots;$

[10] $\left\{z|\frac{n}{n+1} < |z-1| < \frac{n-1}{n}\right\}, \ n = 1, 2, 3, \ldots; \{z||z-2| > 2\}.$

§§5.8 Exercise

Laurent Expansion Near a Singular Point

⊘ For the following functions, determine up to the first three nonvanishing terms having positive powers and three terms having negative powers of $z - z_0$ in the Laurent expansion valid in a deleted neighborhood of z_0.

⊘ Use, wherever possible, the known series expansions for the exponential, sine, and cosine and the hyperbolic sine and cosine functions.

[1] $(z^2 + 3z + 5)\exp(1/z), \ z_0 = 0;$

[2] $\dfrac{\cosh z}{z(z^4+1)}, \quad z_0 = 0;$

[3] $z^2 \sin(1/(1-z)), \quad z_0 = 1;$

[4] $\dfrac{z}{\cosh z - 1}, \quad z_0 = 0.$

⊘ For the following functions obtain the Laurent expansion in powers of $z - z_0$ retaining terms up to three nonvanishing positive powers:

[5] $\cot z, \quad z_0 = 0;$

[6] $\text{cosech } z, \quad z_0 = 0;$

[7] $\sec z, \quad z_0 = \pi/2;$

[8] $\dfrac{z+3}{\sin z^2}, \quad z_0 = 0;$

[9] $\dfrac{z}{\sin^3 z}, \quad z_0 = n\pi;$

[10] $\dfrac{z}{\sin \pi z^2}, \quad z_0 = 1.$

⊘ For the following functions obtain the Laurent expansion in powers of $z - z_0$ retaining terms up to $(z - z_0)^3$:

[11] $\dfrac{1}{\exp(z) - 1}, \quad z_0 = 0;$

[12] $\dfrac{1}{[\exp(z) + 1]^2}, \quad z_0 = i\pi;$

[13] Prove that

$$\exp\left[t\left(z + \frac{1}{z}\right)\right] = \sum_{n=-\infty}^{\infty} z^n J_n(t),$$

where $J_n(t)$ are Bessel functions having power series expansion:

$$J_n(x) = \sum_{n=0}^{\infty} \frac{(-1)^k}{(k+n)!k!}\left(\frac{x}{2}\right)^{2n}.$$

[14] Expand $\sin(\pi/z)$ in powers of $z-1$ and show that

$$\sin(\pi/z) = \pi(z-1) - \pi(z-1)^2 + \left(\pi - \frac{\pi^3}{6}\right)(z-1)^3 - \left(\pi - \frac{\pi^3}{2}\right)(z-1)^4 + \left(\pi - \pi^3 + \frac{\pi^5}{120}\right)(z-1)^5 + \cdots.$$

Use the above expansion to obtain expansion of $\operatorname{cosec}(\pi/z)$ in powers of $z-1$ up to terms of order $(z-1)^3$.

Answers

[1] $(z^2+3z+5)e^{1/z} = z^2 + 4z + \frac{17}{2} + \frac{20}{3z} + \frac{73}{24z^2} + \frac{29}{30z^3} + \cdots,$

[2] $\frac{\cosh z}{z(z^4+1)} = \frac{1}{z} + \frac{z}{2!} - \frac{23z^3}{24} - \frac{359z^5}{720} + \cdots,$

[3] $z^2 \sin\left(\frac{1}{1-z}\right) = -2 - (z-1) - \frac{5}{6(z-1)} + \frac{1}{3(z-1)^2} + \frac{19}{120(z-1)^3} - \frac{1}{120(z-1)^4} + \cdots,$

[4] $\frac{z}{\cosh z - 1} = \frac{2}{z} - \frac{z}{6} + \frac{z^3}{120} - \frac{z^5}{3024} + \cdots,$

[5] $\cot z = \frac{1}{z} - \frac{z}{3} - \frac{z^3}{45} - \frac{2z^5}{945} + \cdots,$

[6] $\operatorname{cosech} z = \frac{1}{z} - \frac{z}{6} + \frac{7z^3}{360} - \frac{31z^5}{15120} + \cdots,$

[7] $\sec z = -\frac{1}{z-\frac{\pi}{2}}\left\{1 + \frac{1}{3!}\left(z-\frac{\pi}{2}\right)^2 + \frac{7}{360}\left(z-\frac{\pi}{2}\right)^4 + \frac{31}{15120}\left(z-\frac{\pi}{2}\right)^6 + \cdots\right\},$

[8] $\frac{z+3}{\sin z^2} = \frac{3}{z^2} + \frac{1}{z} + \frac{z^2}{2} + \frac{z^3}{6} + \frac{7z^6}{120} + \frac{7z^7}{360} + \cdots,$

[9] $\frac{z}{\sin^3 z} = \frac{1}{z^2} + \frac{1}{2} + \frac{17z^2}{120} + \frac{457z^4}{15210} + \cdots,$

[10] $\frac{z}{\sin \pi z^2} = \frac{1}{\pi(z-1)}\Big[-\frac{1}{2} - \frac{z-1}{4} + \left(\frac{1}{8} - \frac{\pi^2}{3}\right)(z-1)^2 - \left(\frac{1}{16} + \frac{\pi^2}{2}\right)(z-1)^3 - \left(\frac{1}{32} - \frac{\pi^2}{6} - \frac{7\pi^4}{45}\right)(z-1)^4 \cdots\Big],$

[11] $\frac{1}{\exp(z)-1} = \frac{1}{z} - \frac{1}{2!} + \frac{z}{12} - \frac{z^3}{702} + \cdots,$

[12] $\frac{1}{(e^z+1)^2} = \frac{1}{(z-i\pi)^2} - \frac{1}{z-i\pi} + \frac{5}{12} - \frac{z-i\pi}{12} + \frac{(z-i\pi)^2}{240} + \frac{(z-i\pi)^3}{720} + \cdots.$

§§5.9 Quiz

Regions of Convergence; Taylor and Laurent Series

⊘ The following functions are to be expanded in a Taylor series about the point z_0, as indicated. State if the *Taylor series or Laurent series* expansion exists.

⊘ If for a function the Taylor series expansion exists, give the radius of convergence.

⊘ If for a function the Laurent series expansion(s) exists, draw a figure to indicate the annular regions of convergence of the Laurent expansion(s) in powers of $z - z_0$.

[1] $\sin z^2;\ z_0 = 0;$

[2] $\dfrac{\exp(\pi z) + 1}{z^2 + 625};\ z_0 = 25i;$

[3] $\exp(1/z);\ z_0 = 6 + i8;$

[4] $\sin(\pi z)\sin(\pi/z);\ z = \frac{1}{2};$

[5] $\exp(\tan z);\ z_0 = \frac{9\pi}{2};$

[6] $\cot(1/(z+1));\ z_0 = -1;$

[7] $\sec(\pi/2z);\ z_0 = \frac{2}{3}.$

Answers

[1] The Taylor series converges for all z.

[2] The Laurent series converges for $0 < |z - 25i| < 50$.

[3] The Taylor series; the radius of convergence is 10.

[4] The Taylor series converges for $0 < |z - \frac{1}{2}| < \frac{1}{2}$.

[5] The Laurent series is convergent for $|z - \frac{9\pi}{2}| < \pi$.

[6] Taylor and Laurent series expansions do not exist.

[7] The Taylor series converges for $|z - \frac{2}{3}| < \frac{1}{3}$.

§§5.10 Questions

Region of Convergence for Laurent Expansion

[1] For the function

$$f(z) = \frac{1}{z(z-1)},$$

how many distinct series expansions in positive and negative powers of $z - z_0$, converging in disjoint annular regions, can be written for the following values of z_0?

(a) $z_0 = 1$,　　(b) $z_0 = 0$,

(c) $z_0 = 2$,　　(d) $z_0 = 1 + i$.

In each case depict the nonoverlapping regions of convergence of the series expansion in the complex plane.

[2] Two different series representations of the function

$$f(z) = \frac{1}{z(z+3)}$$

are given below:

$$f(z) = \frac{1}{3z}\sum\left(\frac{z}{3}\right)^m = \frac{1}{3z}\left(1 - \frac{1}{3} + \frac{z^2}{3^2}\cdots\right),$$

$$f(z) = \frac{1}{z^2}\sum\left(-\frac{3}{z}\right)^m = \frac{1}{z^2}\left(1 - \frac{3}{z} + \frac{3^2}{z^2}\cdots\right).$$

Find the regions of the complex plane in which the series expansions converge.

[3] The function

$$f(z) = \frac{1}{(z-1)(z-2)}$$

is represented by the three convergent Laurent expansions given below:

(a) $\sum_0^\infty (1 - 2^{-m-1})z^m$,

(b) $-\sum_0^\infty (z^{-m-1} + 2^{-m-1}z^m)$,

(c) $\sum_0^\infty (2^m - 1)z^{-m-1}$.

How do you reconcile this result with the well-known fact that the Laurent expansion, when it exists, is unique?

[4] (a) For the function

$$\phi(z) = \frac{1}{z^{16}+1},$$

how many distinct *Laurent* expansions of the form

$$\phi(z) = \sum_{n=-\infty}^{\infty} c_n z^n$$

are possible?

(b) What are the regions of validity of the Laurent series expansions for the function $\phi(z)$?

(c) Is there a region in which c_n vanish for all negative n?

[5] The function

$$f(z) = \mathrm{sech}(\pi z/2)$$

is to be represented by a Laurent series in powers of z valid in the set S:

$$S = \{z = x + iy | 3.5 < x < 6.5, 4.5 < y < 8.5\}.$$

(a) Find the number of disjoint subsets of S such that the following two properties hold:

(i) A single power series expansion is valid in each subset;
(ii) Two different subsets have different series expansions?

(b) Draw a figure to show the subsets of S having the two properties given in part (a).

Answers

[1] (a) Two regions: $0 < |z-1| < 1$, $|z-1| > 1$;
(b) Two regions: $0 < |z| < 1$, $|z| > 1$;
(c) Three regions: $|z-2| < 1$, $1 < |z-2| < 2$, $|z-2| > 2$.
(d) Three regions: $|z-(1+i)| < 1$, $1 < |z-(1+i)| < \sqrt{2}$, $|z-(1+i)| > \sqrt{2}$.

[2] (a) $|z| < 3$,
(b) $|z| > 3$.

[3] (a) Converges in $|z| < 1$;
(b) Converges in $1 < |z| < 2$;
(c) Converges in $|z| > 2$.

[4] (a) Two regions: $|z| < 1, \ |z| > 1$;
(b) $|z| > 1$;
(c) $|z| < 1$.

[5] (a) Three subsets given by the intersection of S with
(i) $5 < |z| < 7$, (ii) $7 < |z| < 9$, (iii) $9 < |z| < 11$.

Chapter 6

RESIDUE THEOREM

§§6.1 Questions

Classifying Singular Points

⊘ For each of the functions listed below, find all singular points and classify the singular points as branch points, isolated singular points, and other singular points such as those which are limit points of isolated singular points:

[1] $(3z^5 + 5z^3 + 7)e^{1/z}$; [2] $\dfrac{\exp(z + \frac{1}{z})}{\exp(z) + 1}$, [3] $\dfrac{z^2 + 3z}{(z-2)^2(z^2 - 5z + 6)}$,

[4] $\operatorname{cosec}(\pi/z)$, [5] $\cot(\frac{1}{z}) - \frac{1}{z}$, [6] $\log \sin z$,

[7] $\dfrac{\sin(1/z)}{z^4 - 1}$, [8] $\dfrac{(z^3+1)^{1/3}}{1 + \cos z}$, [9] $\dfrac{\cot z}{1 + z + z^2}$,

[10] $\exp(\cot z)$, [11] $\operatorname{cosec} \pi z^2$, [12] $\operatorname{cosech}(\pi/z)\sin(z)$,

[13] π^z, [14] z^π, [15] $\log(z^2 + 7z + 12)$,

[16] $\dfrac{\cosh z}{\cosh z + \cos \alpha}$, [17] $\dfrac{1}{\sqrt{1 - \sin(z/2)}}$, [18] $\dfrac{\exp(z)}{\sin z + \frac{5}{4}}$,

[19] $\sin(\frac{1}{z} + 1)$, [20] $\sqrt{\sin(\pi/z)}$.

Suggested tabular form for answers:

SN	$f(z)$	Branch Points	Isolated Singular Points	Nonisolated Limits Points of Singular Points

Answers

⊘ Notation used: $m = \pm 1, \pm 2, \ldots$; $n = 0, \pm 1, \pm 2, \ldots$; $N = 1, 2, 3, \ldots$; $\omega = \exp(2\pi i/3)$.

SN	Function	Branch Points	Isolated Singular Points	Other Cases
[1]	$(3z^5 + 5z^3 + 7)\exp(1/z)$		0	
[2]	$\dfrac{\exp(z + \frac{1}{z})}{\exp(z) + 1}$		$0,\ (2n+1)\pi i$	
[3]	$\dfrac{z^2 + 3z}{(z-2)^2(z^2 - 5z + 6)}$		$2, 3$	
[4]	$\operatorname{cosec}(\pi/z)$		$1/m$	0
[5]	$\cot\left(\frac{1}{z}\right) - \frac{1}{z}$		$1/m\pi$	0
[6]	$\log \sin z$	$n\pi$	0	
[7]	$\dfrac{\sin(1/z)}{z^4 - 1}$		$0, \pm 1, \pm i$	
[8]	$\dfrac{\sqrt[3]{(z^3 + 1)}}{1 + \cos z}$	$-1, -\omega, -\omega^2$	$(2n+1)\pi/2$	
[9]	$\dfrac{\cot z}{1 + z + z^2}$		$n\pi,\ \omega,\ \omega^2$	
[10]	$\exp(\cot z)$		$n\pi$	
[11]	$\operatorname{cosec} \pi z^2$		$0, \pm\sqrt{N}, \pm i\sqrt{N}$	
[12]	$\operatorname{cosech}(\pi/z)\sin(z)$		i/m	0
[13]	π^z	none	none	none
[14]	z^π	0		
[15]	$\log(z^2 + 7z + 12)$	$-3,\ -4$		
[16]	$\dfrac{\cosh z}{\cosh z + \cos\alpha}$		$(2n+1)\pi i \pm \alpha i$	
[17]	$[1 - \sin(z/2)]^{-1/2}$	$(4n+1)\pi$		
[18]	$\dfrac{\exp(z)}{\sin z + \frac{5}{4}}$		$2n\pi - \frac{\pi}{2} \pm i\ln 2$	
[19]	$\sin(1/(z+1))$		-1	
[20]	$\sqrt{\sin\left(\frac{\pi}{z}\right)}$	$1/m$		0

§§6.2 Tutorial

Isolated Singular Points

⊘ For each of the functions listed below, find all the isolated singular points and classify them as

- removable singular points;
- essential singular points;
- poles; for each pole find its order.

⊘ Write your answers in the form of a table, as shown below.

[1] $\dfrac{z+1}{(z^2+1)(z^4-1)}$, [2] $\dfrac{z^3+2z^2+\frac{1}{z^3}}{z+1}$, [3] $\dfrac{\exp(1/z)}{(z^4-1)(z^2+1)^2}$,

[4] $z^{-2}\sin^2 z$, [5] $z^{-6}\sin^2 z^2$, [6] $(z-1)^{-2}\sin^2 \pi z$,

[7] $\dfrac{\sin \pi z^2}{(z-1)^2}$, [8] $\dfrac{1}{4\sin z+5}$, [9] $\dfrac{1}{1+\sin z}$,

[10] $z^2\sinh(1/z)$, [11] $z^3\cot^3 z$, [12] $\sec z+\tan z$,

[13] $\dfrac{\exp(z)-1-z}{z^5}$, [14] $\dfrac{\sin \pi z^2}{z^5(z-4)^3}$, [15] $\dfrac{z^2(z^2-4)}{\sin \pi z^2}$,

[16] $\dfrac{z^4+1}{\exp(\pi z)+1}$, [17] $\dfrac{z^2}{[\exp(1/z)-1]^3}$, [18] $\operatorname{cosech}\Big(\dfrac{\pi}{z-1}\Big)$,

[19] $\cot z-\dfrac{1}{z}$, [20] $\dfrac{(z+7)^{2/3}}{z^4+8(1-\sqrt{3}i)}$.

Suggested tabular form with a sample answer:

SN	$f(z)$	Removable Singular Points	Essential Singular Points	Poles: Location	Poles: Order
[79]	$\dfrac{z\cot z\exp[1/(z+1)]}{(z-5)(z^2-7z+10)}$	$z=-1$	$z=0$	(a) $z=n\pi$	1
				(b) $z=2$	1
				(c) $z=5$	2

Answers

⊘ Notation used: $m = \pm 1, \pm 2, \ldots$; $n = 0, \pm 1, \pm 2, \ldots$; $N = 1, 2, 3, \ldots$.

SN	$f(z)$	Removable Singular Points	Essential Singular Points	Poles	
				Location	Order
[1]	$\dfrac{z+1}{(z^2+1)(z^4-1)}$	-1		(a) $\pm i$	2
				(b) 1	1
[2]	$\dfrac{z^3+2z^2+\frac{1}{z^3}}{z+1}$	-1		0	3
[3]	$\dfrac{\exp(1/z)}{(z^4-1)(z^2+1)^2}$		0	(a) $\pm i$	3
				(b) ± 1	1
[4]	$z^{-2}\sin^2 z$	0			
[5]	$z^{-6}\sin^2 z^2$			0	2
[6]	$(z-1)^{-2}\sin^2 \pi z$	1			
[7]	$(z-1)^{-2}\sin \pi z^2$			1	1
[8]	$\dfrac{1}{4\sin z+5}$			$(2n-1/2)\pi \pm i\ln 2$	1
[9]	$\dfrac{1}{\sin z+1}$			$n\pi - (-1)^n\pi/2$	2
[10]	$z^2\sinh(1/z)$		0		
[11]	$z^3\cot^3 z$	0		$m\pi$	3
[12]	$\sec z + \tan z$	$(4n-1)\pi/2$		$(4n+1)\pi/2$	1
[13]	$\dfrac{\exp(z)-1-z}{z^5}$			0	3
[14]	$\dfrac{\sin \pi z^2}{z^5(z-4)^3}$			(a) 0	3
				(b) 4	2
[15]	$\dfrac{z^2(z^2-4)}{\sin \pi z^2}$	$0, \pm 2$		all m except ± 2	1
[16]	$\dfrac{z^4-1}{\exp(\pi z)+1}$	$\pm i$		$\pm(2N+1)i$	1
[17]	$\dfrac{z^2}{[\exp(1/z)-1]^3}$		0	$1/2\pi i m$	3
[18]	$\operatorname{cosech}\left(\frac{\pi}{z-1}\right)$		1	$1+i/n$	1
[19]	$\cot z - \frac{1}{z}$			$m\pi$	1
[20]	$\dfrac{(z+7)^{2/3}}{z^4+8(1-\sqrt{3}i)}$			(a) $\pm 2\exp(i\pi/6)$	1
				(b) $\pm 2i\exp(i\pi/6)$	1

§§6.3 Questions

Selecting Functions with Singularities Specified

⊘ Seventy five functions are listed below. Select functions as specified in requirements I–V. While answering write both the serial number and the function in each case.

I. Five functions that have a branch point.
II. Five functions having nonisolated singular points but no branch point.
III. Ten functions that are analytic everywhere or have only removable singular points.
IV. Ten functions which are meromorphic and may or may not have removable singular points.
V. Ten functions having only isolated singular points, of which at least one is an essential singular point.

[1] $z + z^{*2}$, [2] $z^5 + 7z^3 + 8z - 13$, [3] $z - \frac{3}{z^2} - 5z$,

[4] $x^2 - y^2 + 2ixy$, [5] $x^2 + y^2 - 2ixy$, [6] $\operatorname{cosech}^{10} z$,

[7] $\sin z$, [8] $\tan z$, [9] $\coth z$,

[10] $\sqrt{\sin(1/z)}$, [11] $\cos(1/\sqrt{z})$, [12] $\exp(1/z^4)$,

[13] $\operatorname{cosec}(1/z)$, [14] $\exp(-\cos^2 z)$, [15] $\sin[\exp(z^3)]$,

[16] $\exp(z^3 + 5z + 3)$, [17] 2^z, [18] $z^4 \operatorname{cosec} z^2$,

[19] $\frac{1}{\sqrt{z}} + 1$, [20] $\sin[\sin(1/z)]$, [21] $z^{3/4}$,

[22] $\dfrac{\exp(z)}{\exp(z) - 1}$, [23] $\dfrac{\sin \pi z}{z^2 - 16}$, [24] $\dfrac{\sin^3 z}{z^2}$,

[25] $\dfrac{z^4 + 2z^2 + 1}{\exp(-\tan z)}$, [26] $\dfrac{\exp[1/(z-1)]}{\exp(z) - 1}$, [27] $\dfrac{(z+1)(z-2)(z+4)}{\exp(2\pi i z) - 1}$,

[28] $\dfrac{1}{\sin z^3}$, [29] $\dfrac{\exp(5z)}{(z-1)^2}$, [30] $\exp\left(\dfrac{1}{z^3 + 5z + 6}\right)$,

[31] $\dfrac{z^2 - 4z + 4}{z^2 - 2z}$, [32] $\dfrac{\cos(\pi z/2)}{z+1}$, [33] $\dfrac{1 - \exp(2z)}{z^4 - 8z^3}$,

[34] $\dfrac{\tan z}{z^2 + z + 1}$, [35] $\dfrac{z}{\cos z + 5}$, [36] $\dfrac{z^2 + \pi^2}{\exp(z)}$,

[37] $\exp(z) - \operatorname{cosec}^2 z$, [38] $z^2 \operatorname{cosec} z$, [39] z,

[40] $z^2 \operatorname{cosech}(1/z)$, [41] $z^3 \operatorname{cosec} \sqrt{z}$, [42] $z^3 \operatorname{cosec}^2 z$,

[43] $\dfrac{z^2 \sin z}{\exp(z)-1}$, [44] $\dfrac{\exp(z^2)}{\exp(z)+1}$, [45] $\sin\left(\dfrac{1}{z+1}\right)^2$,

[46] $e^z \cos(\frac{1}{z-2})$, [47] $\operatorname{cosech}(\pi z/4)$, [48] $\tanh z^2$,

[49] $\sin z^4$, [50] $\sin(z)\exp(1/z)$, [51] $\cos(\sqrt{z^2+3z+1})$,

[52] $\dfrac{z+1}{z^2-2z}$, [53] $\dfrac{1}{(z+1)^5}$, [54] $\dfrac{1}{(z^2-25)(z+1)}$,

[55] $\dfrac{1}{z^3+1}$, [56] $\dfrac{\exp(\cot z)}{\tan z+1}$, [57] $\dfrac{\cos \pi z}{(z-3)^7}$,

[58] $\dfrac{\exp(5z)-\frac{1}{e^5}}{(z+1)^2}$, [59] $\dfrac{\exp(z)}{\exp(z)+1}$, [60] $\dfrac{e^{z^2}}{z\cosh z^2}$,

[61] $\log(z+z^2)\sin z$, [62] $\sqrt{z}\operatorname{cosec}\sqrt{z}$, [63] $z\log(\cos z^2)$,

[64] $\dfrac{\exp(z)-1}{\sin^2 z}$, [65] $\dfrac{z^2}{\sin^2 z+2}$, [66] $\dfrac{\exp(i\pi/z)+i}{z+2}$,

[67] $\cos(1/z^2)$, [68] $\exp(\operatorname{cosec}^3 z)$, [69] $\operatorname{cosech}^{5/2}(\pi z/4)$,

[70] $\dfrac{\cosh z}{\sinh z+\sqrt{3}}$, [71] $\dfrac{z^{4/3}}{(z-2)^{10}}$, [72] $\dfrac{1-\exp(2z^2)}{z^4}$,

[73] $\dfrac{\sin \pi z}{(z-13)^3}$, [74] $\dfrac{\cos \pi z}{(z-7)^4}$, [75] $\coth\left(\dfrac{1}{z+4}\right)$.

Answers

⊙ To save space, only the serial numbers of the functions are given for this set.

I. Functions that have a branch point:
[10], [19], [21], [41], [61], [63], [69], [71]

II. Functions having a nonisolated singular point but not a branch point:
[1], [5], [10], [11], [13], [40], [75]

III. Functions that are analytic everywhere:
[2], [4], [7], [11], [14], [15], [16], [17], [36], [39], [49], [51]

Functions having only removable singular points:
[23], [24], [32], [38], [66]

IV. Functions which are meromorphic:
[3], [6], [8], [9], [18], [22], [27], [26], [28], [29], [31], [33], [34], [35], [37], [38], [40], [42], [43], [44], [47], [48], [52], [53], [54], [55], [57], [58], [59], [60], [62], [64], [65], [70], [72], [73], [74]

V. Functions having only isolated singular points, of which at least one is an essential singular point:
[11], [12], [16], [20], [25], [30], [45], [46], [50], [56], [66], [67], [68]

§§6.4 Tutorial

Residues at Simple Poles

⊘ Find the residues at the simple poles indicated:

[1] $\dfrac{5}{z-3}+\dfrac{16}{(z-3)^2(z+5)}$, at $z=3$;

[2] $\dfrac{5}{z-3}+\dfrac{16}{(z-3)^2(z+5)}$, at $z=-5$;

[3] $\dfrac{z+3}{(z-1)(z^2-1)}$, at $z=-1$;

[4] $\dfrac{\cos \pi z}{3\,z-1}$, at $z=\dfrac{1}{3}$;

[5] $\dfrac{\sin(\frac{\pi}{z})}{(9-4z^2)(z+1)^2}$, at $z=\dfrac{3}{2}$;

[6] $\dfrac{\cos(z)}{\sqrt{2}\,\sin(z)-1}$, at $z=\dfrac{\pi}{4}$;

[7] $\dfrac{1}{\exp(z)-\exp(2z)}$, at $z=0$;

[8] $\dfrac{\sin(2z^2)}{z^3}$, at $z=0$;

[9] $\dfrac{z^3}{\sin(2z)^4}$, at $z=0$;

[10] $\dfrac{z^2}{\sin(\pi z^3)}$, at $z=2$;

[11] $\dfrac{z}{\sin(\pi z)(e^{2\,z}-1)}$, at $z=0$;

[12] $\dfrac{z^3}{z^{17}+1}$, at $z=-1$;

[13] $\dfrac{\sinh(z)}{\cosh(z)-\cosh(4)}$, at $z=4$;

[14] $\sec\left(\dfrac{\pi}{z}\right)$, at $z=2$.

⊘ Find all poles and the residues at the poles of the functions given below:

[15] $\operatorname{cosec} \pi z$,

[16] $\tanh \pi z$,

[17] $\dfrac{\exp(z)+1}{\sinh z}$,

[18] $\dfrac{1}{z(z^5+1)}$,

[19] $\dfrac{\cos z}{\exp(z)+1}$,

[20] $\dfrac{\sin z}{5\cos z-4}$.

Answers

[1] $\dfrac{19}{4}$, [2] $\dfrac{1}{4}$, [3] $\dfrac{1}{2}$, [4] $\dfrac{1}{6}$,

[5] $\dfrac{-1}{50\sqrt{3}}$, [6] $\dfrac{1}{\sqrt{2}}$, [7] -1, [8] 2,

[9] $\frac{1}{16}$, [10] $\frac{1}{3\pi}$, [11] $\frac{1}{2\pi}$, [12] $-\frac{1}{17}$,

[13] 1, [14] $\frac{4}{\pi}$, [15] $\left\{m\pi, \frac{(-1)^m}{\pi}\right\}$, [16] $\left\{in, \frac{1}{\pi}\right\}$.

[17] $\{2n\pi, 2\}$,

[18] $\{0,1\}; \{e^{(2k\pi i/5)},\ k = 0, \ldots, 4, -\frac{1}{5}\}$,

[19] $\{(2n+1)\pi i; -\text{sech}(2n+1)\pi\}$,

[20] $\{2n\pi \pm \tan^{-1}(3/4); -1/5\}$.

§§6.5 Tutorial

Residues at Multiple Poles

[1] Find the residues at the poles by using the formula for the residue at a pole:

(a) $\dfrac{(z^2+1)^{10}}{z^5}$, at $z=0$; (b) $\dfrac{1}{\sinh \pi z^3}$, at $z=3\,i$;

(c) $\dfrac{\exp(kz)}{[\exp(z)+1]^2}$, at $z=i\,\pi$; (d) $\dfrac{1}{z\,\sin(z)(e^z-1)}$, at $z=0$.

[2] Find the residues at the points indicated by expanding in a Laurent series:

(a) $\dfrac{z^{12}-12z^6+8}{(z-10)^{10}}$, at $z=10$; (b) $\dfrac{\exp(z)-\exp(5z)}{(z-1)^4}$, at $z=1$;

(c) $\dfrac{\sin(\pi z)}{(2z-3)^3}$, at $z=\dfrac{3}{2}$; (d) $\dfrac{\cos(\pi z)^2}{(3z-1)^3}$, at $z=\dfrac{1}{3}$.

[3] Calculate the residue at $z=0$ using any method that you find best suited to the question:

(a) $\dfrac{1}{z^{10}(z-a)(z+b)}$, at $z=0$; (b) $\dfrac{\sin(z)^3}{z^6}$, at $z=0$;

(c) $\dfrac{z^2}{\sin(z)^5}$, at $z=0$; (d) $\dfrac{1}{z\,[\exp(z)-1]}$, at $z=0$.

Answers

[1] (a) 45, (b) $\dfrac{1}{27\pi}$, (c) $e^{ik\pi}(-1+k)$, (d) $\dfrac{1}{4}$;

[2] (a) 220,000, (b) $\dfrac{e-125\,e^5}{6}$, (c) $\dfrac{\pi^2}{16}$, (d) $\dfrac{\pi^2}{54}$;

[3] (a) $\dfrac{b^{-10}-a^{-10}}{a+b}$, (b) $-\dfrac{1}{2}$, (c) $\dfrac{5}{6}$, (d) $-\dfrac{1}{2}$.

§§6.6 Exercise

Computation of Integrals

[1] Compute the integral

$$\oint_C \frac{(3z^2+2)dz}{(z-4)(z^2+9)}$$

for the positively oriented circle C given by

(a) $|z-2|=3$; (b) $|z|=5$;

(c) $|z+1|=4$; (d) $|3z-2|=1$.

[2] Calculate the integral of f around the unit circle in the *clockwise sense* with the origin as the center, where $f(z)$ is

(a) $z^{-2}\exp(z)$; (b) $z^{-1}\operatorname{cosec} z$;

(c) $(z^2+5z+7)\exp(1/z)$; (d) $\dfrac{z}{\cosh z - \cos(\pi/20)}$.

[3] If C is the circle $|z|=2$ described in the positive sense, compute the following integrals:

(a) $\oint_C \tan z dz$; (b) $\oint_C \dfrac{dz}{\sinh 2z}$;

(c) $\oint_C \dfrac{\cosh \pi z dz}{z(z^2+1)}$; (d) $\oint_C \sec(\pi z) dz$;

(e) $\oint_C \dfrac{dz}{z^3(z-1)}$; (f) $\oint_C \dfrac{\exp(-z)}{z} dz$;

(g) $\oint_C z^4 \exp(1/z) dz$; (h) $\oint_C \dfrac{z^2 \cosh \pi z}{(1+z^2)\sinh(\pi z/4)} dz$.

[4] (a) Use Cauchy residue theorems and compute the integral

$$\oint_C z^{-2}\exp(i\alpha z/2)dz,$$

where C is an anticlockwise circle of radius 7 with the center at the origin.

(b) Next, compute the same integral after making a change of variable to $w = 1/z$ and rewriting the above integral as an integral over the new variable w.

[5] Let C be the positively oriented circle $|z|=1/10$ of radius $1/10$ and with center $z=0$. Note that

$$\oint_C z^{-2}(1+5z)^{-1}\sin(\pi/2z)dz = 2\pi i R,$$

where R is the residue of the integrand at $z=0$. Evaluate R by computing the above integral making the change of variable to $w=1/z$.

Answers

[1] (a) $4\pi i$, (b) $6\pi i$, (c) $2\pi i$, (d) 0.

[2] (a) $-2\pi i$, (b) 0, (c) $-58\pi i/3$, (d) $\dfrac{\pi^2}{10}\operatorname{cosec}\left(\dfrac{\pi}{20}\right)$.

[3] (a) $4\pi i$, (b) $-\pi i$, (c) $4\pi i$, (d) 0,
(e) 0, (f) $2\pi i$, (g) $\pi i/3$, (h) $\sqrt{8}\pi i$.

[4] $\pi\alpha$, [5] 5.

§§6.7 Questions

Residue Theorem

⊘ Five statements (A–E) and 20 integrals are given below. C_R denotes the circle of radius R and with the center at $z = 0$. For each of the integrals, identify the statement which is applicable to the 20 integrals listed.

(A) The integral is zero, because the function is analytic everywhere.
(B) The integrand in not analytic everywhere, but the integral is zero not because the function is analytic but because all the singular points lie outside the contour of integration.
(C) There are one or more singular points inside the contour but the integral vanishes because it is obvious that the residue at *each* singular point is separately zero.
(D) There are one or more singular points inside the contour with nonzero residues. The integral is found to be zero, because the sum of all the residues adds up to zero.
(E) The integral is nonzero. In this case compute the value of the integral.

[1] $\oint_{C_7} \sin z\, dz$, [2] $\oint_{C_3} \cos(1/z) dz$, [3] $\oint_{C_{79}} (z^3 + 3z + 7) dz$,

[4] $\oint_{C_7} \frac{dz}{(z+2\pi)^9}$, [5] $\oint_{C_{5/2}} \frac{dz}{2-3z}$, [6] $\oint_{C_3} \frac{dz}{(z+\pi)(z^3+216)}$,

[7] $\oint_{C_1} \tan z\, dz$, [8] $\oint_{C_{50}} \sinh z^3 dz$, [9] $\oint_{C_{37}} \exp(\sin^2 z) dz$,

[10] $\oint_{C_{2/3}} \frac{e^z dz}{e^z+1}$, [11] $\oint_{C_3} \frac{dz}{(z-2)(z+1)}$, [12] $\oint_{C_{1/2}} \frac{dz}{z^4+1}$,

[13] $\oint_{C_7} \frac{dz}{z^2+9}$, [14] $\oint_{C_{1/20}} \frac{dz}{\sinh 15z}$, [15] $\oint_{C_6} \frac{z\, dz}{(z+1)^3}$,

[16] $\oint_{C_{1/2}} \frac{dz}{\cos^2 z - 1}$, [17] $\oint_{C_{5/2}} \frac{dz}{z^3+1}$, [18] $\oint_{C_{14}} \frac{\exp(z)}{\exp(z)-1} dz$,

[19] $\oint_{C_{30}} \operatorname{sech}(30\pi z) dz$, [20] $\oint_{C_5} e^{1/z}\left(z^4 + 5z^2 + \frac{1}{24}\right) dz$.

Answers

[1] A, [2] C, [3] A, [4] C, [5] E,
[6] B, [7] E, [8] A, [9] A, [10] B,
[11] D, [12] B, [13] D, [14] E, [15] C,
[16] C, [17] D, [18] E, [19] D, [20] E.

Answers for nonzero integrals:

[5] $-2\pi/3i$, [7] $2\pi i$, [14] $2\pi/15i$, [18] $10\pi i$, [20] $53\pi i/30$.

§§6.8 Tutorial

Integrals of Trigonometric Functions

[1] In this tutorial we shall learn how to evaluate the integrals of trigonometric functions over a finite range. The first integral to be calculated is

$$I = \int_0^{2\pi} \frac{d\theta}{13 + 12\cos\theta}.$$

Follow the steps given below to compute the above integral.

(a) Changing the variables to $z = \exp(i\theta)$, write the given integral I in terms of an integral,

$$J = \oint \frac{dz}{6z^2 + 13z + 6},$$

over the positively oriented unit circle $|z| = 1$ in the complex plane.

(b) Find the two poles of the integrand of J and the values of the corresponding residues. Do both the poles lie inside the unit circle?

(c) Use Cauchy's residue theorem to compute the integral J.

(d) Using the above results, show that

$$\int_0^{2\pi} \frac{d\theta}{13 + 12\cos\theta} = \frac{2\pi}{\sqrt{5}}.$$

[2] Recompute the integral in problem [1] using the transformation $z = \exp(-i\theta)$ in the first step. Verify that the final answer can be correctly obtained by this method also.

[3] Compute the integral

$$\int_0^{\pi} \frac{d\theta}{(k + \cos\theta)^2}, \quad k > 1.$$

What changes will be needed in the evaluation if $k < -1$?

[4] Using the method outlined in Q[1], evaluate the integral

$$\int_0^{\pi} \cos^4\theta \, d\theta.$$

§§6.9 Exercise

Integrals of the Type $\int_0^{2\pi} f(\cos\theta, \sin\theta)d\theta$

Evalute the following integrals involving trigonometric functions using the method of contour integration:

[1] $\displaystyle\int_0^{2\pi} \frac{d\theta}{5+4\cos\theta}$, [2] $\displaystyle\int_0^{2\pi} \frac{d\theta}{2\cos\theta+3}$,

[3] $\displaystyle\int_0^{\pi} \frac{d\theta}{3-\sin\theta}$, [4] $\displaystyle\int_{-\pi}^{\pi} \frac{d\theta}{1+\sin^2\theta}$,

[5] $\displaystyle\int_0^{2\pi} \sin^4\theta\, d\theta$, [6] $\displaystyle\int_0^{\pi/2} \sin^4\theta \cos^4\theta\, d\theta$.

Establish the following integration formulae:

[7] $$\int_0^{2\pi} \frac{d\theta}{(a+b\cos\theta)^2} = \frac{2\pi a}{(a^2-b^2)^{3/2}}, \quad a>0,\ |a/b|>1;$$

[8] $$\int_0^{2\pi} \frac{d\theta}{(a+b\cos^2\theta)^2} = \frac{\pi(2a+b)}{a^{3/2}(a+b)^{5/2}}, \quad a>0,\ b>0,\ |a/b|>1;$$

[9] $$\int_0^{\pi/2} \sin^{2m}\theta' d\theta = \frac{\pi}{2}\frac{(2m-1)!!}{(2m)!!};$$

[10] $$\int_0^{2\pi} \frac{d\theta}{1-2p\cos\theta+p^2} = \begin{cases} 2\pi/(1-p^2), & 0<p<1; \\ 2\pi/(p^2-1), & p>1; \end{cases}$$

[11] $$\int_0^{2\pi} \frac{\cos^2 2\theta}{1-2p\cos\theta+p^2} d\theta = \pi\frac{1+p^4}{1-p^4}, \quad 0<p<1;$$

[12] $$\int_0^{2\pi} \frac{d\theta}{a^2\sin^2\theta+b^2\cos^2\theta} = \frac{2\pi}{ab};$$

[13] $$\int_0^{\pi} \frac{\cos n\theta}{1+a\cos\theta} d\theta = \frac{\pi}{\sqrt{1-a^2}}\left(\frac{\sqrt{1-a^2}-1}{a}\right)^n, \quad a^2<1;$$

[14] $$\int_0^{\pi} \frac{\cos n\theta\cos\theta}{1-2a\cos\theta+a^2} d\theta = \begin{cases} \frac{\pi}{2}\frac{1+a^2}{1-a^2}a^{n-1}, & a<1, \\ \frac{\pi}{2}\frac{a^2+1}{a^2-1}\frac{1}{a^{n+1}}, & a>1. \end{cases}$$

Answers

[1] $3\pi/4$, [2] $2\pi/\sqrt{5}$, [3] $\pi/\sqrt{2}$,
[4] $\sqrt{2}\pi$, [5] $3\pi/4$, [6] $3\pi/256$.

§§6.10 Exercise

Integrals Using the Residue at Infinity

⊘ Using the principal value for the square root function, $\sqrt{z-\xi}$, with $\xi = a, b$, and taking a contour enclosing the interval (a, b), establish the results given below:

[1] $\oint \sqrt{(z-a)(z-b)}\, dz = -\frac{\pi i}{4}(a-b)^2,$

[2] $\oint \sqrt{\frac{z-a}{z-b}}\, dz = -\pi i(a-b),$

[3] $\oint \frac{z}{\sqrt{(z-a)(z-b)}} dz = \pi i(a+b),$

[4] $\oint \frac{z^2}{\sqrt{(z-a)(z-b)}} dz = \frac{\pi i}{4}(3a^2 + 2ab + 3b^2).$

⊘ Apply the residue theorem to appropriately chosen contour integrals and compute the following integrals:

[5] $\int_0^1 \frac{x^{1-p}\,(1-x)^p}{(1+x)^3}\, dx,$ [6] $\int_0^1 \frac{x^{1-p}\,(1-x)^p}{1+x^2}\, dx,$

[7] $\int_0^1 \frac{x^{1-p}(1-x)^p}{(1+x^2)^2}\, dx,$ [8] $\int_0^1 \frac{1}{(1+x)^3[x^2\,(1-x)]^{\frac{1}{3}}}\, dx,$

[9] $\int_0^1 \frac{x^{2n}}{(1+x^2)\sqrt{1-x^2}}\, dx,$ [10] $\int_0^1 \frac{x^{2n}}{[x(1-x^2)]^{\frac{1}{3}}} dx,$

[11] $\int_0^1 \frac{1}{(1-x^n)^{\frac{1}{n}}}\, dx.$

Answers

[5] $-[2^{-3+p}\,(-1+p)\,p\,\pi\,\csc(p\,\pi)],$ [6] $\pi\left[-1+2^{\frac{p}{2}}\,\cos\left(\frac{p\,\pi}{4}\right)\right]\csc(p\,\pi),$

[7] $2^{\frac{-3+p}{2}}\, p\pi\csc(p\pi)\,\sin\left(\frac{\pi - p\pi}{4}\right),$ [8] $\frac{13\,\pi}{9.2^{\frac{1}{3}}\,\sqrt{3}},$

[9] $\frac{\pi}{3}\frac{1.4\ldots(3n-2)}{3.6\ldots 3n},$ [10] $\frac{\Gamma\left(\frac{2}{3}\right)\Gamma\left(\frac{1}{3}+n\right)}{2\,\Gamma(1+n)},$

[11] $\frac{\pi\csc\left(\frac{\pi}{n}\right)}{n}.$

§§6.11 Quiz

Finding Residues

⊘ Compute the residues for the functions listed below at the specified point in each case. If you can complete all the problems in 20 minutes, you deserve a medal!

[1] $\sin(1/z^2)$, at $z = 0$;

[2] $z^3 \exp[-1/(z+1)^2]$, at $z = -1$;

[3] $\dfrac{\sin(\pi^2\alpha/z)\exp[(3z) - 1]}{z + i\pi}$, at $z = -i\pi$;

[4] $\dfrac{\exp(i\pi z)}{(z+1)^{77}}$, at $z = -1$;

[5] $\dfrac{\exp(z^{13}) - 1}{z^{53}}$, at $z = 0$;

[6] $\dfrac{z^{10} + 7z^8 + 5}{z^9(1 - z^2)}$, at $z = 0$;

[7] $\dfrac{z^{17}}{\sin^{23}(z^{19})}$, at $z = 0$;

[8] $\dfrac{1}{\sinh(z)^{100}}$, at $z = 0$;

[9] $\dfrac{\exp(1/z)}{z^2(z-1)}$, at $z = 0$;

[10] $\dfrac{\sin z}{3\cos z + 2}$, at any one pole.

Answers

[1] $-\dfrac{5}{2}$, [2] 0, [3] $2i\sinh(\pi\alpha)$, [4] $\dfrac{\pi^{76}}{76!}$, [5] $\dfrac{1}{24}$,

[6] 12, [7] 0, [8] 0, [9] $-e$, [10] $-1/3$.

§§6.12 Mined

How to Compute Residues

- ⊘ For each question there are sample answers. Read the sample answers carefully and locate the mistakes, if any.
- ⊘ Identify the source of mistakes and give your reasons why you consider a step or a statement incorrect. Be as precise as you can.
- ⊘ Do not write your solution as part of your answers.
- ⊘ State if in your opinion a sample answer is correct, partially correct, or completely wrong. Grade each sample answer out of a maximum of 10 points.

[1] Compute the residue of

$$f(z) = \frac{1}{z^2(z-3)},$$

at $z = 0$.

Sample answer. *I shall expand the given function in powers of z and find the coefficient of $1/z$.*

$$f(z) = \frac{1}{z^2(z-3)} = \frac{1}{z^3}\frac{1}{(1-\frac{3}{z})}$$
$$= \frac{1}{z^3} \times \left(1 - \frac{3}{z} + \frac{9}{z^2} + \cdots\right).$$

As the residue is defined to be the coefficient of $1/z$, the answer for the residue is zero.

[2] Two functions f and g have expansions

$$f(z) = \frac{1}{z} + \frac{1}{3!z^3} + \frac{1}{5!z^5} + \cdots + \frac{1}{(2n+1)!z^{2n+1}} + \cdots,$$
$$g(z) = \frac{3}{2z} - \frac{9}{4z^2} + \cdots + (-1)^n\frac{3^n}{2^n z^n} + \cdots.$$

From these expansions what can be said about the nature of singularities at $z = 0$? Give reasons in support of your answers.

Sample answer 1. *Since the given Laurent expansion for both the functions have an infinite number of terms in negative powers of z, both the functions have an essential singular point at $z = 0$.*

Sample answer 2. *We first determine the region of convergence of the series in each case. The series for $f(z)$ converges for all $z \neq 0$. Therefore, there is no other singular point and the point $z = 0$ is an isolated essential singular point. In fact, the function $f(z)$ can be easily seen to be $\cosh(1/z)$. The second series converges for $|z| > 3/2$. In this case no conclusion about the nature of* *singularity at $z = 0$ can be drawn because the region of convergence does not include a deleted neighborhood of $z = 0$.*

[3] Find the poles of the function $\dfrac{1}{\sin z + 3}$.

Sample answer. *The value of* $\sin z$ *lies in the range of* -1 *to* 1*; therefore,* $\sin z + 3$ *does not vanish for any value of* z*. Hence, the given function does not have a singular point.*

[4] Find the location and the nature of all the singular points of the function

$$\frac{\sin z - 2i}{\cos z + \sqrt{5}}.$$

Sample answer. *This function has poles where the denominator vanishes. These are given by*

$$\begin{aligned}\cos z + \sqrt{5} &= 0,\\ \exp(iz) + \exp(-iz) + 2\sqrt{5} &= 0,\\ \exp(2iz) + 2\sqrt{5}\exp(iz) + 1 &= 0,\\ \exp(iz) &= -\sqrt{5} \pm 2.\end{aligned}$$

Hence,

$$iz = \ln(\sqrt{5} \pm 2) + (2n+1)\pi i,$$

where n *is any integer. The singular points are all poles and are located at the points*

$$z = -i\ln(\sqrt{5} - 2) + (2n+1)\pi, -i\ln(\sqrt{5} + 2) + (2n+1)\pi.$$

Answers

[1] The sample answer is wrong, because the Laurent series derived is not the correct Laurent series to be used for computation of the residue at $z = 0$. The series converges for $|z| > 3$, which does not include any deleted neighborhood of the origin.

[2] Sample answer 1 is wrong; sample answer 2 is 100% correct.

[3] The sample answer is wrong, because for complex z the values of $\sin z$ are not restricted to the range $(-1, 1)$.

[4] The sample answer is partially correct. The mistake is that at the points $z = -i\ln(\sqrt{5} + 2) + (2n+1)\pi$ the numerator also vanishes and these points are removable singular points and not poles.

§§6.13 Mixed Bag

Residues and Integration in the Complex Plane

[1] Give examples of functions which have a 10th order pole at $z = -2$ and at least one other singular point such that:

(a) the residue at $z = -2$ is 0;

(b) the residue at $z = -2$ is 6.

[2] Give two examples of meromorphic functions which have an infinite number of poles and where the residue at each pole vanishes.

[3] Find the residues at the points indicated:

(a) $\dfrac{z^{10} + 3z^6 + 7}{(z-5)^{11}}$, at $z = 5$;

(b) $\dfrac{1}{(z^3+1)(z^6-1)}$, at $z = -1$;

(c) $\dfrac{1}{z}\left(z + \dfrac{1}{z}\right)^N$, at $z = 0$ for $N > 1$;

(d) $\dfrac{z^{10} + 5z^5 + 1}{z^N(z-a)(z+b)}$, at $z = 0$ for $N > 1$.

[4] The expansions

$$\frac{1}{z} - 1 + z - z^2 + \cdots, \qquad \text{for } |z| < 1,$$

$$\frac{1}{z^2} - \frac{1}{z^3} + \frac{1}{z^4} - \cdots, \qquad \text{for } |z| > 1,$$

for the function $f(z) = \frac{1}{z(z+1)}$ can be easily derived using the binomial expansion. Derive the above expansions as Laurent series using the results (5.42) and (5.43), for the expansion coefficients.

[5] Compute the integral

$$I = \oint_C z^3 \exp(1/z) dz$$

around the unit circle $C = \{z||z| = 1\}$ in the following three ways:

(a) directly using the residue theorem;

(b) by first making a transformation $w = 1/z$;

(c) by computing the residue at ∞.

You must get the same answer by the three methods!

[6] Without an explicit computation prove that

$$\oint_C \frac{dz}{z^n + 1} = 0,$$

where $n > 1$ and C is any closed contour lying entirely in the region $\{z||z| > 1\}$.

[7] Show that the residue of $\frac{\text{Log}\, z}{(z^2+1)\cosh(\pi z/2)}$ at $z = \pm i$ is $\frac{\pm 4i+\pi}{4\pi}$.

[8] Compute the residue at infinity for the following functions:

(a) $\dfrac{a_0 z^N + a_1 z^{N-1} + \cdots + a_N}{b_0 z^N + b_1 z^{N-1} + \cdots + b_N}$, (b) $\dfrac{a_0 z^N + a_1 z^{N-2} + \cdots + a_N}{b_0 z^{N+1} + b_1 z^N + \cdots + b_{N+1}}$.

[9] Show that the sum of all residues of a rational function, including the residue at infinity, vanishes.

[10] Compute the residue at infinity:

(a) e^z, (b) $e^z \text{Log}\left(\dfrac{z-a}{z-b}\right)$.

[11] Prove that

$$\exp\left[\frac{c}{2}\left(z - \frac{1}{z}\right)\right] = \sum_{n=-\infty}^{\infty} a_n z^n,$$

where

$$a_n = \frac{1}{2\pi}\int_0^{2\pi} \cos(n\theta - c\sin\theta)d\theta.$$

[12] Show that

$$\exp\left(\frac{z+c^2}{2z^2}\right) = \sum_{n=-\infty}^{\infty} a_n z^n,$$

where

$$a_n = \int_0^{2\pi} \exp[c(\cos\theta + \cos^2\theta)]\cos[c\sin\theta(1-\cos\theta) - n\theta]d\theta.$$

[13] Let $f(z) = \chi(z)/\phi(z)$, where the two functions $\chi(z)$ and $\phi(z)$ are the analytic point z_0.

(a) If both χ and ϕ have a zero of order m at z_0, show that

$$\lim_{z\to z_0} f(z) = \frac{\chi^{(m)}(z_0)}{\phi^{(m)}(z_0)}.$$

This analog of l'Hopital's rule is very useful for computing the residues.

(b) Let both χ and ϕ vanish at the point z_0 and $f(z)$ have a second order pole at z_0, then

$$\text{Res}\{f(z)\}_{z=z_0} = 2\frac{\chi'}{\phi''} - \frac{2}{3}\frac{\chi\phi'''}{\phi''^2}.$$

(c) If ϕ has a zero of order m at a point z_0 and χ has a zero of order $m-1$ at the same point, so that the function f has a first order pole at z_0, show that

$$\text{Res}\{f(z)\}_{z=z_0} = \left.\frac{m\chi^{(m-1)}}{\phi^{(m)}}\right|_{z=z_0}.$$

[14] Prove that the function $\frac{1}{z}\log(1+z)$ has a pole for most definitions of the logarithm. Find a definition for which $z=0$ is not a singular point.

Answers

[1] (a) $\frac{1}{(z+2)^{10}} + \frac{1}{(z-1)^7}$,

(b) $\frac{5}{(z+2)^{10}} + \frac{6}{z+2} + \frac{1}{z}$.

[2] $\text{cosec}^2 z, \dfrac{1}{\sin z + 1}$.

[3] (a) 1;

(b) $-7/36$;

(c) ${}^N C_{N/2}$ if N is even and 0 if N is odd;

(d) $\left(\frac{1}{a-b}\right)\left(\frac{1}{b^N} - \frac{1}{a^N}\right)$, if $N \le 5$;

$\left(\frac{1}{a-b}\right)\left[\frac{1}{b^N} - \frac{1}{a^N} + \left(\frac{5}{b^{N-5}}\right) - \left(\frac{5}{a^{N-5}}\right)\right]$, if $5 < N \le 10$;

$\left(\frac{1}{a-b}\right)\left[\frac{1}{b^N} - \frac{1}{a^N} + \left(\frac{5}{b^{N-5}}\right) - \left(\frac{5}{a^{N-5}}\right) + \frac{1}{b^{N-10}} - \frac{1}{a^{N-10}}\right]$, if $N > 10$.

[8] (a) $(b_0 a_1 - b_1 a_0)/b_0^2$,

(b) a_0/b_0.

[10] (a) 0,

(b) $e^a - e^b$.

Chapter 7

CONTOUR INTEGRATION

§§7.1 Tutorial

$\int_0^\infty Q(x)\,dx$

[1] Follow the steps outlined below and show that

$$\int_0^\infty \frac{dx}{x^2+1} = \frac{\pi}{2}. \tag{7.1}$$

(a) Transform the given integral into a line integral along an open contour in the complex plane. Using the property that the integrand is even, relate the given integral in (7.1) to

$$\lim_{R\to\infty} \int_{\text{AOB}} \frac{dz}{z^2+1}, \tag{7.2}$$

where AOB is the part of the real axis from $-R$ to R (see Fig. 7.1).
(b) Close the contour by adding a semicircular arc BCA of radius R. This addition does not alter the value of the integral in the limit $R \to \infty$ and leads to the closed contour AOBCA, Γ of Fig. 7.1. Prove this statement by using Darboux's theorem and showing that

$$\lim_{R\to\infty} \int_{\text{BCA}} \frac{dz}{z^2+1} = 0. \tag{7.3}$$

(c) The closed contour integral is a sum of integrals along the parts AOB and BCA:

$$\oint_\Gamma \frac{dz}{z^2+1} = \int_{\text{BCA}} \frac{dz}{z^2+1} + \int_{\text{AOB}} \frac{dz}{z^2+1}. \tag{7.4}$$

Use the results obtained above to arrive at

$$\int_0^\infty \frac{dx}{x^2+1} = \lim_{R\to\infty} \frac{1}{2} \oint_\Gamma \frac{dz}{z^2+1}. \tag{7.5}$$

(d) Apply the residue theorem and compute the value of the integral. The integrand on the right hand side of (5) has poles at $z = \pm i$ and only the pole at $z = i$ is enclosed inside the contour. Hence, you would get the required answer:

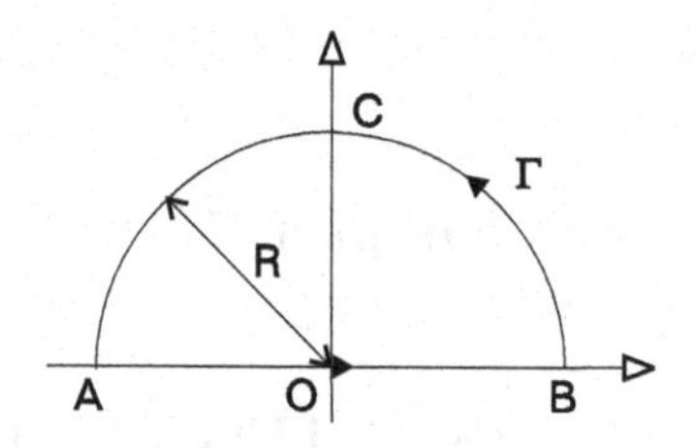

Fig. 7.1

$$\int_0^\infty \frac{dx}{x^2+1} = \frac{\pi}{2}. \tag{7.6}$$

[2] Prove that

$$\int_0^\infty \frac{\cos ax}{x^2+1}\,dx = \frac{\pi}{2}\,e^{-|a|}. \tag{7.7}$$

(a) Rewrite the integral and verify that it can be written as

$$\int_0^\infty \frac{\cos ax}{x^2+1}\,dx = \frac{1}{2}\text{Re}\int_{\text{AOB}} \frac{\exp(iaz)dz}{z^2+1}. \tag{7.8}$$

Next, note that the behavior of $\exp(iaz) = \exp(iax - ay)$ for large x, y depends on the sign of a. First, assume that $a > 0$ and follow the same steps as given for Q[1] and show that the value of the integral on the right hand side is

$$\int_{-\infty}^\infty \frac{\exp(iax)dx}{x^2+1} = \oint_\Gamma \frac{\exp(iaz)dz}{z^2+1} = \pi e^{-a}. \tag{7.9}$$

(b) Now, consider $a < 0$ and explain why the integral

$$\int_{-\infty}^\infty \frac{\exp(iax)}{x^2+1}\,dx, \tag{7.10}$$

appearing above, cannot be computed using the contour of Fig. 7.1, and instead the contour in Fig. 7.2 has to be used.

(c) Evaluating the integral (7.10) explicitly, complete the proof of Eq. (7.7) for $a < 0$.

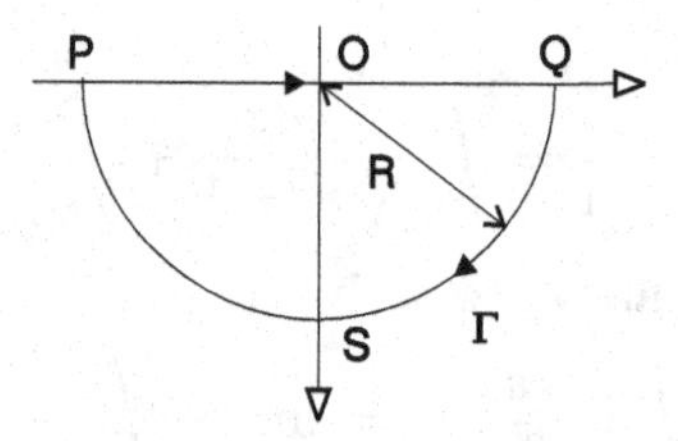

Fig. 7.2

§§7.2 Tutorial

Improper Integrals of Rational Functions

[1] Prove that

$$\int_0^\infty \frac{1}{(x^2+x+1)(x+1)}\,dx = \frac{\pi}{3\sqrt{3}}. \tag{7.11}$$

Follow the steps given below to complete the integration.
(a) Initial setup: Let $f(z)$ be the single-valued branch, defined by

$$f(z) = \frac{\log z}{(z^2+z+1)(z+1)},$$

taking

$$\log z = \ln r + i\theta, \quad 0 < \theta < 2\pi. \tag{7.12}$$

Compute the discontinuity of $f(z)$ and show that

$$\int_0^\infty \frac{-1}{(x^2+x+1)(x+1)}\,dx = \frac{-1}{2\pi i}\left[\int_{\text{AB}} f(z)\,dz - \int_{\text{ED}} f(z)\,dz\right], \tag{7.13}$$

where AB and ED are the line segments, parallel to the real axis, in the two circles' contour of Fig. 7.3.
(b) Use Darboux's theorem and show that the integral of $f(z)$ along the two circular arcs BCD and EFA have the limits

$$\lim_{\rho\to 0}\int_{\text{EFA}} f(z)\,dz = 0, \tag{7.14}$$

$$\lim_{R\to\infty}\int_{\text{BCD}} f(z)\,dz = 0. \tag{7.15}$$

(c) Setup the integral along the two circles' contour and write it as a sum of line integrals along the individual straight line segments, and circular arc and

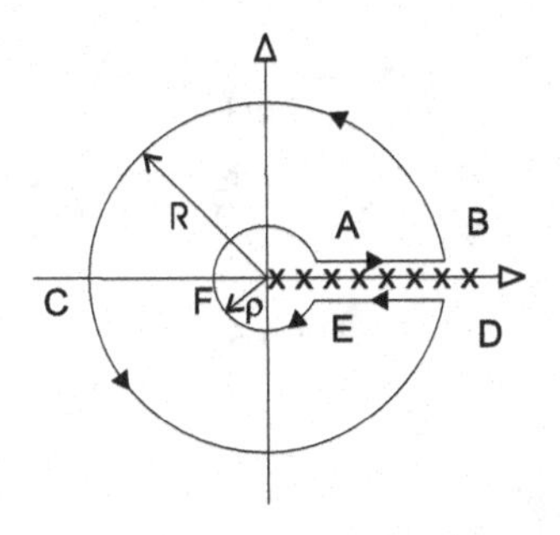

Fig. 7.3

show that

$$\int_0^\infty \frac{1}{(x^2+x+1)(x+1)}\,dx = \frac{1}{2\pi i}\lim \oint f(z)dz. \tag{7.16}$$

(d) Use the residue theorem to compute the integral in the above equation and hence show that

$$\int_0^\infty \frac{1}{(x^2+x+1)(x+1)}\,dx = \frac{2\pi}{3\sqrt{3}}. \tag{7.17}$$

[2] Follow the steps outlined below and compute the integral

$$I = \int_0^\infty \frac{dx}{x^3+1}. \tag{7.18}$$

(a) Transforming the integral by replacing x with complex variable z, the integrand in (7.18) becomes $(z^3+1)^{-1}$, which has the same value along the rays OA and OB, corresponding to arg $z = 0, 2\pi/3$ (also along the ray arg $z = 4\pi/3$). Use this to verify that

$$\int_{\mathrm{OA}} \frac{dz}{z^3+1} = I, \quad \int_{\mathrm{OB}} \frac{dz}{z^3+1} = \exp\left(\frac{2\pi i}{3}\right) I \tag{7.19}$$

hold in the limit $R \to \infty$.

(b) Use Darboux's theorem and show that

$$\lim_{R\to\infty} \int_{\mathrm{ACB}} \frac{dz}{z^3+1} = 0. \tag{7.20}$$

(c) To close the contour combine the integrals in Eqs. (7.19) and (7.20) with suitable signs to get an integral along the closed contour OACBO (Fig. 7.4):

$$\int_{\mathrm{OA}} \frac{dz}{z^3+1} + \int_{\mathrm{ACB}} \frac{dz}{z^3+1} + \int_{\mathrm{BO}} \frac{dz}{z^3+1} = \oint_{\mathrm{OABCO}} \frac{dz}{z^3+1}. \tag{7.21}$$

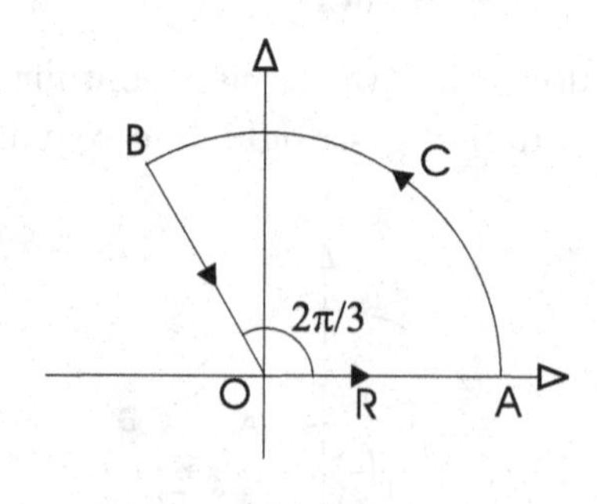

Fig. 7.4

Collecting the results in Eqs. (7.19)–(7.21) and taking the limit, prove the relation

$$I = \frac{1}{1 - \exp(2\pi i/3)} \lim_{R\to\infty} \oint_{\mathrm{OABCO}} \frac{dz}{z^3 + 1}. \tag{7.22}$$

(d) Apply the residue theorem to complete the integral on the right hand side. There is only one — simple — pole at $\exp(2\pi i/3)$ enclosed in the contour OACBO. Compute the residue at the pole and show that

$$\int_0^\infty \frac{dx}{x^3 + 1} = \frac{\pi}{3\sqrt{3}}. \tag{7.23}$$

§§7.3 Exercise

Integrals of Type $\int Q(x)dx$

⊘ Compute the following integrals using the method of contour integration:

[1] $\displaystyle\int_0^\infty \frac{1}{(x^2+p^2)^2}\,dx;$

[2] $\displaystyle\int_0^\infty \frac{x}{x^4+1}\,dx;$

[3] $\displaystyle\int_{-\infty}^\infty \frac{1}{x^4+x^2+1}\,dx;$

[4] $\displaystyle\int_{-\infty}^\infty \frac{1}{(x^2+x+1)^2}\,dx;$

[5] $\displaystyle\int_{-\infty}^\infty \frac{x^2+2}{x^4+10\,x^2+9}\,dx;$

[6] $\displaystyle\int_0^\infty \frac{x-2}{x^4+10\,x^2+9}\,dx;$

[7] $\displaystyle\int_0^\infty \frac{x^2}{x^6+1}\,dx;$

[8] $\displaystyle\int_0^\infty \frac{1}{(x^2+p^2)(x^2+q^2)}\,dx;$

[9] $\displaystyle\int_0^\infty \frac{x^2+1}{(x^4+1)^2}\,dx;$

[10] $\displaystyle\int_0^\infty \frac{x^2}{x^{2N+1}+1}\,dx;$

[11] $\displaystyle\int_0^\infty \frac{x^4}{x^{2N}+1}\,dx, \quad N>2;$

[12] $\displaystyle\int_0^\infty \frac{1}{x^N+1}\,dx, \quad N>2;$

[13] $\displaystyle\int_0^\infty \frac{x^{2M}}{x^{2N}+1}\,dx, \quad N>M>0;$

[14] $\displaystyle\int_0^\infty \frac{1}{(p+q\,x^2)^N}\,dx, \quad p,q>0.$

Answers

[1] $\dfrac{\pi}{4p^3}$, [2] $\dfrac{\pi}{4}$, [3] $\dfrac{\pi}{\sqrt{3}}$,

[4] $\dfrac{4\,\pi}{3\,\sqrt{3}}$, [5] $\dfrac{5\,\pi}{12}$, [6] $\dfrac{1}{24}(\ln 27-2\,\pi)$,

[7] $\dfrac{\pi}{6}$, [8] $\dfrac{\pi}{2\,pq\,(p+q)}$, [9] $\dfrac{\pi}{2\,\sqrt{2}}$,

[10] $\dfrac{\pi}{2N+1}\,\text{cosec}\left(\dfrac{3\,\pi}{2\,N+1}\right)$, [11] $\dfrac{\pi}{2N}\,\text{cosec}\left(\dfrac{5\pi}{2N}\right)$, [12] $\dfrac{\pi}{N}\,\text{cosec}\left(\dfrac{\pi}{N}\right)$,

[13] $\dfrac{\pi}{2N}\,\text{cosec}\left(\dfrac{2M+1}{2N}\right)\pi$, [14] $\sqrt{\dfrac{\pi p}{q}}\,\dfrac{\Gamma(N-\frac{1}{2})}{2\,p^n\,\Gamma(N)}$.

§§7.4 Exercise

Integrals of $\binom{\sin x}{\cos x}$ *with Rational Functions*

⊘ For the problems in this set (assume that $p > 0$ and $\alpha > 0$), compute the following integrals using the method of contour integration:

[1] $\displaystyle\int_0^\infty \frac{x \sin px\, dx}{x^2+1}$, [2] $\displaystyle\int_0^\infty \frac{\cos px\, dx}{(x^2+1)^2}$,

[3] $\displaystyle\int_0^\infty \frac{\cos px\, dx}{(x^2+1)(x^2+4)}$, [4] $\displaystyle\int_0^\infty \frac{x^3 \sin x\, dx}{(x^2+4)(x^2+9)}$,

[5] $\displaystyle\int_0^\infty \frac{x^2 \cos px\, dx}{x^4+1}$, [6] $\displaystyle\int_0^\infty \frac{x \sin px\, dx}{x^4+1}$,

[7] $\displaystyle\int_0^\infty \frac{x \cos \alpha x\, dx}{x^2-2x+10}$, [8] $\displaystyle\int_0^\infty \frac{x \sin \alpha x\, dx}{x^2+4x+20}$.

⊘ Prove the following results using the method of contour integration ($a > 0$, $b > 0$):

[9] $$\int_0^\infty \frac{\cos ax\, dx}{(x^2+b^2)^2+c^2} = \frac{\pi}{2c}\frac{\exp(-aA)}{\sqrt{b^4+c^2}}\left[B\cos(aB) + A\sin(aB)\right];$$

[10] $$\int_0^\infty \frac{x \sin ax\, dx}{(x^2+b^2)^2+c^2} = \frac{\pi}{2c} e^{-aA}\sin(aB),$$

where A and B in the above two questions are defined by

$$A^2 = \frac{1}{2}\left(\sqrt{b^4+c^2}+b^2\right), \quad B^2 = \frac{1}{2}\left(\sqrt{b^4+c^2}-b^2\right).$$

[11] $$\int_0^\infty \frac{\sin^2 ax \cos^2 bx}{\beta^2+x^2}dx = \frac{\pi}{16\beta}\left[2+2e^{-2b\beta}-2e^{-2a\beta}-e^{-2(a+b)\beta}-e^{-2|a-b|\beta}\right];$$

[12] $$\int_0^\infty \frac{x \sin 2ax \cos^2 bx}{\beta^2+x^2}dx = \frac{\pi}{8}\left[2e^{-a\beta}+e^{-2(a+b)\beta}-e^{-2|a-b|\beta}\right].$$

Answers

[1] $\dfrac{\pi}{2}e^{-p}$,

[2] $\dfrac{\pi}{4}(1+p)e^{-p}$,

[3] $\dfrac{\pi}{6}(2e^{-p}-e^{-2p})$,

[4] $\dfrac{\pi}{10e^3}(9-4e)$,

[5] $\dfrac{\pi}{\sqrt{2}}e^{-\frac{p}{\sqrt{2}}}\left[\cos\left(\dfrac{p}{\sqrt{2}}\right)-\sin\left(\dfrac{p}{\sqrt{2}}\right)\right]$,

[6] $\dfrac{\pi}{2}e^{-\frac{p}{\sqrt{2}}}\sin\left(\dfrac{p}{\sqrt{2}}\right)$,

[7] $\dfrac{\pi}{3}\cos\alpha e^{-3\alpha}$,

[8] $\dfrac{\pi}{2}(2\cos 2\alpha+\sin 2\alpha)e^{-4\alpha}$.

§§7.5 Tutorial

Integration Around the Branch Cut

[1] The improper integral $I_1 = \displaystyle\int_0^\infty \frac{x^{-c}}{(x+1)}\,dx, 0 < c < 1$, is to be calculated by making use of the two circles' contour in Fig. 7.5 as suggested in steps given below.

(a) Initial setup: Let $f(z) = \dfrac{z^{-c}}{z+1}$, with z^{-c} defined by taking the svb

$$z^{-c} = r^{-c}\exp(-ic\theta), \quad \text{for} \quad 0 < \theta < 2\pi.$$

Show that the discontinuity across the branch cut is given by

$$\lim_{\epsilon\to 0}\{f(x+i\epsilon) - f(x-i\epsilon)\} = (2ie^{-ic\pi}\sin c\pi)\frac{x^c}{x+1}. \tag{7.24}$$

(b) Transform the integral I_1, which is related to the integral on the right hand side of Eq. (7.24) from ρ to R. Thus, show that

$$(2ie^{-ic\pi}\sin c\pi)I_1 = \int_{\mathrm{AB}} f(z)dz - \int_{\mathrm{CD}} f(z)dz, \tag{7.25}$$

where AB and CD are rays from ρ to R, as indicated in Fig. 7.5.
(c) To use Darboux's theorem, by C_ρ and C_R denote the two arcs CSA and BPQRD, of radii ρ and R, respectively, as shown in Fig. 7.5. Prove that

$$\lim_{\rho\to 0}\int_{C_\rho}\frac{z^{-c}}{z+1}dz = 0, \quad \lim_{R\to\infty}\int_{C_R}\frac{z^{-c}}{z+1}dz = 0. \tag{7.26}$$

(d) Choose closed contour Γ_1 of Fig. 7.5 and write

$$\oint_{\Gamma_1}\frac{z^{-c}}{z+1}dz = \int_{\mathrm{AB}}\frac{z^{-c}}{z+1}dz + \int_{C_R}\frac{z^{-c}}{z+1}dz + \int_{\mathrm{DC}}\frac{z^{-c}}{z+1}dz + \int_{C_\rho}\frac{z^{-c}}{z+1}dz, \tag{7.27}$$

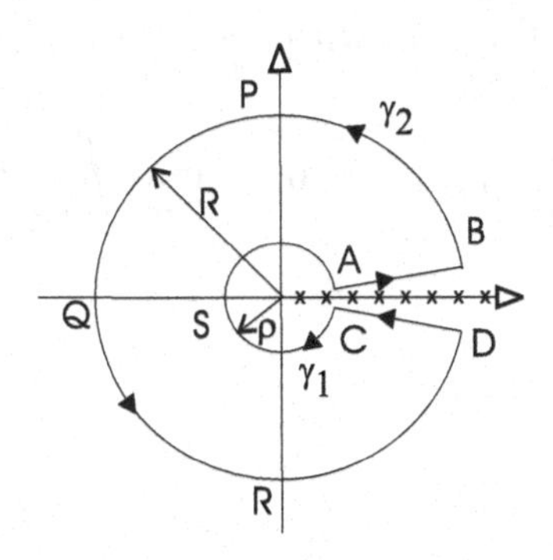

Fig. 7.5

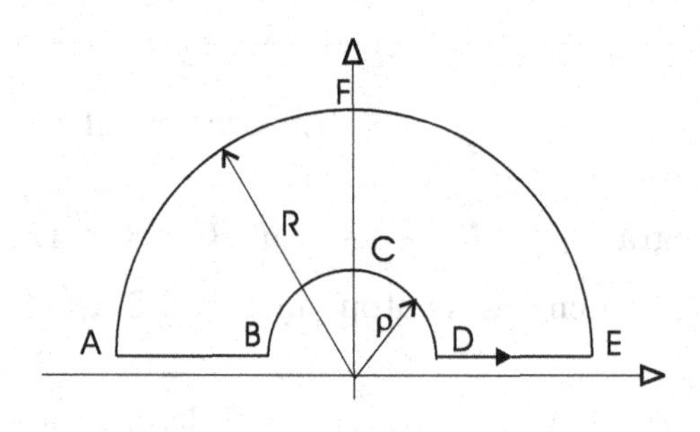

Fig. 7.6

and make use of the results obtained above to show that

$$I_1 = \frac{-ie^{-ic\pi}}{2\sin c\pi} \oint_{\Gamma_1} f(z)dz. \tag{7.28}$$

(e) Use the residue theorem and show that the integral on the right hand side has the value

$$\oint_{\Gamma} \frac{z^{-c}}{z+1}\, dz = 2\pi e^{ic\pi}, \tag{7.29}$$

and hence obtain

$$\int_0^{\infty} \frac{x^{-c}}{x+1}\, dx = \frac{\pi}{\sin c\pi}. \tag{7.30}$$

[2] Set up $\oint \frac{\log z}{(z^2+1)^2}\, dz$ along the contour ABCDEFA ($\equiv \Gamma_2$) of Fig. 7.6 and hence find the value of the integral

$$I_2 = \int_0^{\infty} \frac{\log x}{(x^2+1)^2} dx.$$

(a) First, note that the integral over Γ_2, of any function $f(z)$, can be written as

$$\oint_{\Gamma_2} f(z)dz = \int_{\text{AB}} f(z)dz + \int_{\text{BCD}} f(z)dz + \int_{\text{DE}} f(z)dz + \int_{\text{EFA}} f(z)dz. \tag{7.31}$$

Next, apply Darboux's theorem with the choice $f(z) = \dfrac{\log z}{(z^2+1)^2}$ to show that

$$\lim_{\rho\to 0} \int_{\text{BCD}} f(z)dz = 0, \quad \lim_{R\to\infty} \int_{\text{EFA}} f(z)dz = 0. \tag{7.32}$$

(b) Writing the remaining integrals in Eq. (7.31) as integrals over a real parameter gives

$$\int_{\rm BA} f(z)\,dz = \int_\rho^R f(re^{i\pi})\,e^{i\pi}dr \Rightarrow \int_{\rm AB} f(z)\,dz$$

$$= -\int_\rho^R f(re^{i\pi})\,e^{i\pi}dr, \tag{7.33}$$

$$\int_{\rm DE} f(z) = \int_\rho^R f(r)\,dr. \tag{7.34}$$

Hence, show that, in the limit $\rho \to 0, R \to \infty$, Eq. (7.31) gives

$$\oint_{\Gamma_2} f(z)\,dz = 2\int_0^\infty \frac{\log x}{(x^2+1)^2}\,dx + i\pi \int_0^\infty \frac{dx}{(x^2+1)^2}. \tag{7.35}$$

(c) Use the residue theorem and compute the integral on the left hand side of the above equation to obtain

$$2\int_0^\infty \frac{\log x\,dx}{(x^2+1)^2} + i\pi \int_0^\infty \frac{dx}{(x^2+1)^2} = \left(-\frac{\pi}{2} + i\frac{\pi^2}{4}\right).$$

(d) The final answer is obtained by taking the real part of the above equation,

$$\int_0^\infty \frac{\log x\,dx}{(x^2+1)^2} = -\frac{\pi}{4},$$

and, as a byproduct, equating the imaginary part gives

$$\int_0^\infty \frac{dx}{(x^2+1)^2} = \frac{\pi}{4}.$$

§§7.6 Exercise

Integrals of the Type $\int x^a Q(x)dx$

⊘ Evaluate the following integrals:

[1] $\displaystyle\int_0^\infty \frac{x^{\frac{1}{2}}}{(x+4)(x+25)}\,dx$;

[2] $\displaystyle\int_0^\infty \frac{x^{p-1}}{x^2+2x+2}\,dx, \quad 0<p<2$;

[3] $\displaystyle\int_0^\infty \frac{x^{p-1}}{(x+1)^2}\,dx, \quad 0<p<2$;

[4] $\displaystyle\int_0^\infty \frac{x^p}{x^2+a^2}\,dx, \quad -1<p<1$;

[5] $\displaystyle\int_0^\infty \frac{x^p}{(x^2+a^2)^2}\,dx, \quad -1<p<3$;

[6] $\displaystyle\int_0^\infty \frac{x^p}{x^4+1}\,dx, \quad -1<p<3$;

[7] $\displaystyle\int_0^\infty \frac{x^p}{x^6+1}\,dx, \quad -1<p<5$;

[8] $\displaystyle\int_0^\infty \frac{x^{\frac{1}{2}}}{(x^2+1)^2(x^2+4)}\,dx$;

[9] $\displaystyle\int_0^\infty \frac{x^{p-1}}{1+x^q}\,dx, \quad 0<p<q$;

[10] $\displaystyle\int_0^\infty \frac{x^{p-1}}{(x+1)^3}\,dx, \quad 0<p<3$;

[11] $\displaystyle\int_1^\infty \frac{(x-1)^{p-1}}{x^2}\,dx, \quad 0<p<2$;

[12] $\displaystyle\int_1^\infty \frac{(x-1)^{p-1}}{x^3}\,dx, \quad 0<p<3$.

[13] Show that

$$\int_0^\infty \frac{x^{p-1}dx}{x^3+b^3} = \frac{\pi b^{p-3}}{3\sin \pi p}[1+2\cos(2\pi p/3)],$$

by integrating $z^{p-1}/(z^3+b^3)$ around the branch cut taken along $\theta=0$, the positive real axis. What is the range of p for which the above result is valid? Explain why.

[14] For $-1<p<1$ show that

$$\int_0^\infty \frac{x^p dx}{x^2+2x\cos\alpha+1} = \pi \sin p\alpha \operatorname{cosec} p\pi \operatorname{cosec}\alpha.$$

Answers

[1] $\dfrac{\pi}{7}$,

[2] $2^{\frac{p-1}{2}}\,\pi \operatorname{cosec} p\pi \sin\left[\dfrac{(1-p)\pi}{4}\right]$,

[3] $(1-p)\,\pi \operatorname{cosec} p\pi$,

[4] $\dfrac{1}{2}\, a^{p-1}\,\pi \sec\left(\dfrac{p\pi}{2}\right)$,

[5] $a^{p-3}(p-1)\left(\frac{\pi}{4}\right)\sec\left(\frac{p\pi}{2}\right)$,

[6] $\frac{\pi}{4}\ \text{cosec}\left[\frac{(p+1)\pi}{4}\right]$,

[7] $\frac{\pi}{6}\ \text{cosec}\left[\frac{(p+1)\pi}{6}\right]$,

[8] $\frac{\pi}{72}(4-\sqrt{2})$,

[9] $\frac{\pi}{q}\ \text{cosec}\left(\frac{p\pi}{q}\right)$,

[10] $\frac{1}{2}(p-2)(p-1)\pi\ \text{cosec}\ p\pi$,

[11] $(1-p)\,\pi\ \text{cosec}\ p\pi$,

[12] $\frac{1}{2}p(1-p)\pi\ \text{cosec}\ p\pi$.

§§7.7 Exercise

Integrals of the Type $\int \log x Q(x) dx$

⊘ Evaluate the following integrals:

[1] $\displaystyle\int_0^\infty \frac{\log x}{x^2+1}\, dx,$ [2] $\displaystyle\int_0^\infty \frac{\log x}{x^3+1}\, dx,$

[3] $\displaystyle\int_0^\infty \frac{\log^2 x}{x^2+1}\, dx,$ [4] $\displaystyle\int_0^\infty \frac{\log^4 x}{x^2+1}\, dx,$

[5] $\displaystyle\int_0^\infty \frac{\log^6 x}{x^2+1}\, dx,$ [6] $\displaystyle\int_0^\infty \frac{\log^8 x}{x^2+1}\, dx,$

[7] $\displaystyle\int_0^\infty \frac{\log(p\,x)}{x^2+q^2}\, dx,$ [8] $\displaystyle\int_0^\infty x^2 \frac{\log x}{(\,x^2+1)^2}\, dx,$

[9] $\displaystyle\int_0^\infty \frac{\log x}{(x^2+1)\,(1+q^2\,x^2)}\, dx,$ [10] $\displaystyle\int_0^\infty \frac{\log^2 x}{(x^2+1)^2}\, dx.$

[11] Prove the result $\displaystyle\int_0^\infty \frac{\log x\; dx}{x^2+2ax\cos\alpha+a^2} = \frac{\alpha\log a}{a\sin\alpha}.$

[12] Show that $\displaystyle\int_0^\infty \frac{x^2\log x}{(a^2+b^2x^2)(1+x^2)}\, dx = \frac{a\pi q\log(b/a)}{2b(b^2-a^2)}.$

Answers

[1] 0, [2] $\dfrac{-2\pi^2}{27}$, [3] $\dfrac{\pi^3}{8}$, [4] $\dfrac{5\pi^5}{32}$,

[5] $\dfrac{61\pi^7}{128}$, [6] $\dfrac{1385\pi^9}{512}$, [7] $\dfrac{\pi\,\log(pq)}{2\,q}$, [8] $\dfrac{\pi}{4}$,

[9] $-\dfrac{\pi q\ln q}{2(1-q^2)}$, [10] $\dfrac{\pi^3}{16}$.

§§7.8 Tutorial

Hyperbolic Functions

[1] Our first example in this tutorial is evaluation of the integral

$$I = \int_{-\infty}^{\infty} \frac{\exp(kx)}{1+\exp(x)} dx, \quad 0 < k < 1,$$

using the closed rectangular contour ABCDA in Fig. 7.7 and following the steps suggested below.

(a) Break up the integral around the rectangle ABCDA as a sum of four line integrals along open pieces making up the rectangle:

$$\oint \phi(z)dz = \int_{\mathrm{AB}} \phi(z)dz + \int_{\mathrm{BC}} \phi(z)dz + \int_{\mathrm{CD}} \phi(z)dz + \int_{\mathrm{DA}} \phi(z)dz,$$

where

$$\phi(z) = \frac{\exp(kz)}{1+\exp(z)}.$$

(b) Process the line integrals along different parts of the contour. Show that both of the integrals $\int_{\mathrm{AB}} \phi(z)dz$ and $\int_{\mathrm{CD}} \phi(z)dz$ are related to the integral I to be evaluated and that the sum of these two integrals is

$$\int_{\mathrm{AB}} \phi(z)\, dz + \int_{\mathrm{CD}} \phi(z)\, dz = \{1 - \exp(2ik\pi)\} \int_{-\infty}^{\infty} \frac{\exp(kx)}{1+\exp(x)} dx.$$

(c) Use Darboux's theorem and show that the two integrals $\int_{\mathrm{BC}} \phi(z)dz$ and $\int_{\mathrm{DA}} \phi(z)dz$ become zero in the limit $L \to \infty$ when k is in the range $0 < k < 1$.

(d) Assemble the above results and show that

$$I = \frac{1}{1-\exp(2ik\pi)} \oint \frac{\exp(kz)}{1+\exp(z)} dz.$$

(e) Using the residue theorem compute the integral $\oint \phi(z)dz$ around the rectangle and hence obtain the desired result

$$I = \frac{\pi}{\sin k\pi}.$$

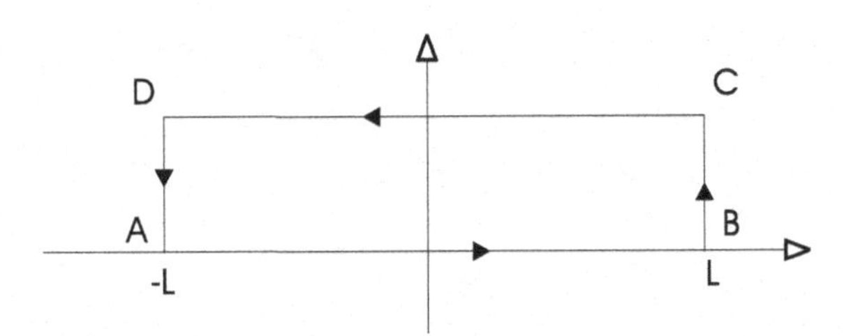

Fig. 7.7

[2] Integrating $\phi(z) = \dfrac{1}{z \cosh \pi z}$ around a rectangular contour, prove that

$$\int_0^\infty \frac{dx}{(x^2+1)\cosh \pi x} = \frac{1}{2}(4-\pi). \tag{7.36}$$

Select a contour such that the integral of $\phi(z)$ along the two horizontal sides combines to become proportional to the integrand in Eq. (7.36). One such choice is the rectangle EFGH, with consecutive vertices at $-L-i, L-i, L+i$, and $-L+i$, as in Fig. 7.8, where a limit $L \to \infty$ is taken at the end. Verify this statement by showing that

$$\int_{\mathrm{EF}} \phi(z)dz + \int_{\mathrm{GH}} \phi(z)dz = -4i \int_0^\infty \frac{dx}{(x^2+1)\cosh \pi x}.$$

Now, proceed as in the previous question and complete the remaining steps.

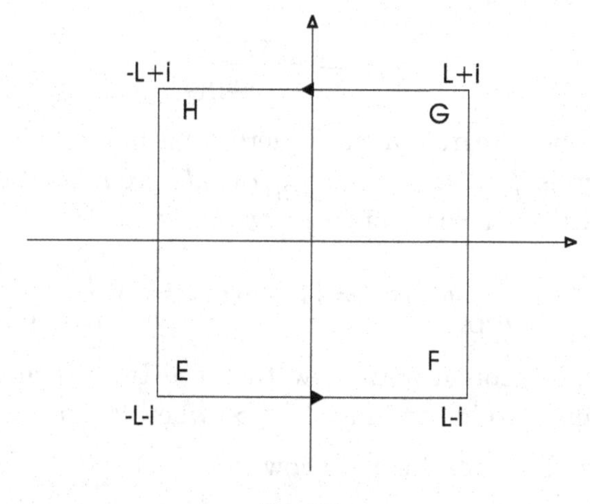

Fig. 7.8

§§7.9 Exercise

Integrals Involving Hyperbolic Functions

⊘ Compute the following integrals using an indented contour wherever necessary:

[1] $\displaystyle\int_{-\infty}^{\infty} x\,\frac{\exp(-x)}{1+\exp(-4x)}\,dx;$

[2] $\displaystyle\int_{-\infty}^{\infty} x\,\frac{\exp(-x)}{1+\exp(-6x)}\,dx;$

[3] $\displaystyle\int_{-\infty}^{\infty} x\,\frac{\exp(px)}{1+\exp(x)}\,dx, \quad 0<p<1;$

[4] $\displaystyle\int_{-\infty}^{\infty} x^2\,\frac{\exp(px)}{1+\exp(x)}\,dx, \quad 0<p<1;$

[5] $\displaystyle\int_{0}^{\infty} x\,\frac{[1-\exp(-x)]\exp(-x)}{1+\exp(-3x)}\,dx;$

[6] $\displaystyle\int_{0}^{\infty} \frac{x\{\exp(-qx)+\exp[(q-p)x]\}}{1-\exp(-px)}\,dx, \quad 0<q<p.$

⊘ Successively prove the results:

[7] $\displaystyle\int_{0}^{\infty} \frac{x^2}{\cosh \pi x}\,dx = \frac{1}{8},$ [8] $\displaystyle\int_{0}^{\infty} \frac{x^4}{\cosh \pi x}\,dx = \frac{5}{32},$

[9] $\displaystyle\int_{0}^{\infty} \frac{x^6}{\cosh \pi x}\,dx = \frac{61}{128},$ [10] $\displaystyle\int_{0}^{\infty} \frac{x^8}{\cosh \pi x}\,dx = \frac{1385}{512}.$

⊘ Integrating around a suitable rectangular contour, evaluate the following integrals ($b<a$):

[11] $\displaystyle\int_{0}^{\infty} \frac{x\sinh ax}{\cosh bx}\,dx,$ [12] $\displaystyle\int_{0}^{\infty} \frac{\cosh ax}{\cosh bx}\,dx,$

[13] $\displaystyle\int_{0}^{\infty} \frac{x^2\cosh ax}{\cosh bx}\,dx,$ [14] $\displaystyle\int_{0}^{\infty} \frac{dx}{\cosh ax+\cos t}\,dx.$

Answers

[1] $\dfrac{\pi^2}{8\sqrt{2}},$ [2] $\dfrac{\pi^2}{6\sqrt{3}},$

[3] $\pi^2 \cot p\pi \operatorname{cosec} p\pi,$ [4] $\dfrac{\pi^3}{2}(3+\cos 2p\pi)\operatorname{cosec}^3(p\pi),$

[5] $\dfrac{2\pi^2}{27}$, [6] $\dfrac{\pi^2}{p^2}\operatorname{cosec}^2\!\left(\dfrac{\pi q}{p}\right)$,

[11] $\dfrac{\pi^2}{4b^2}\sin\left(\dfrac{a\pi}{2b}\right)\sec^2\left(\dfrac{a\pi}{2b}\right)$, [12] $\dfrac{\pi}{2b}\sec\left(\dfrac{a\pi}{2b}\right)$,

[13] $\dfrac{\pi^3}{8b^3}\left[2\sec^3\left(\dfrac{a\pi}{2b}\right)-\sec\left(\dfrac{a\pi}{2b}\right)\right]$, [14] $\dfrac{t}{a}\operatorname{cosec} t,\quad 0<t<\pi.$

§§7.10 Tutorial

Principal Value Integrals

[1] Prove that

$$\int_0^\infty \frac{\sin^2 x}{x^2}\, dx = \frac{\pi}{2}.$$

(a) Note that the given integral is well defined and equals half of the limiting value of

$$\int_{-R}^{-\rho} \frac{\sin^2 z}{z^2}\, dz + \int_{\rho}^{R} \frac{\sin^2 z}{z^2}\, dz,$$

as $\rho \to 0$ and $R \to \infty$.
(b) Set up the integral in the complex plane. Since we can write

$$\frac{\sin^2 x}{x^2} = \frac{1-\cos 2x}{2x^2} = \frac{1}{2}\mathrm{Re}\left[\frac{1-\exp(2ix)}{x^2}\right],$$

we first set up the sum of integrals of $\frac{1-\exp(2ix)}{x^2}$ along AB and DE, see Fig. 7.9, in the complex plane. This leads us to the principal value integral

$$J_{\mathrm{PV}} = \lim \left[\int_{\mathrm{AB}} \frac{1-\exp(2iz)}{z^2}\, dz + \int_{\mathrm{DE}} \frac{1-\exp(2iz)}{z^2}\, dz\right],$$

and we make a transition to the closed contour ABCDEFA, as in Fig. 7.9.
(c) Break up the integral over the closed contour ABCDEFA,

$$J_\Gamma = \oint_\Gamma \frac{1-\exp(2iz)}{z^2}\, dz,$$

into integrals over parts of the contour. The contour consists of four pieces:
(i) the line AB from $z = -R$ to $z = -\rho$,
(ii) the semicircle γ_o of radius ρ and with the center at the origin,
(iii) the line from $z = \rho$ to $z = R$, and
(iv) the semicircle γ_R of radius R and with the center at the origin.

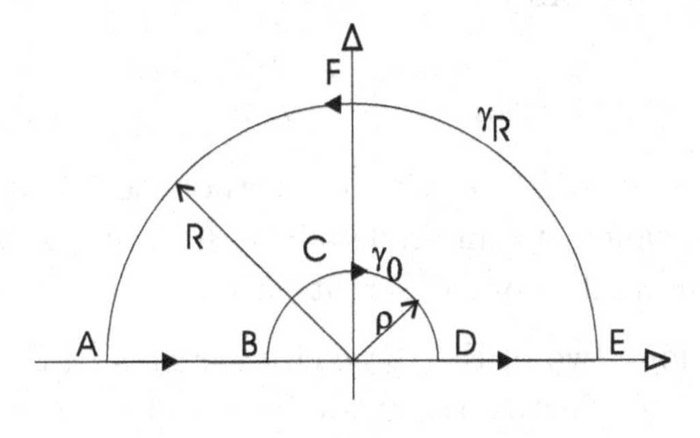

Fig. 7.9

Hence,

$$J_\Gamma = J_{\rm AB} + J_{\rm BCD} + J_{\rm DE} + J_{\rm EFA}.$$

(d) Process the integrals over parts. Complete the following steps:

- Use Darboux's theorem and prove that

$$\lim_{R\to\infty} J_{\rm EFA} = 0; \tag{7.37}$$

- Use the result of §6.5.1 to obtain

$$\lim_{\rho\to 0} J_{\rm BCD} = -\pi i \,\text{Res}\left\{\frac{1-\exp(2iz)}{z^2}\right\}_{z=0} = -2\pi. \tag{7.38}$$

(e) Add the above results obtained and show that

$$\text{PV}\int_{-\infty}^{+\infty} \frac{1-e^{2ix}}{x^2}\,dx + \int_{\rm BCD} \frac{1-e^{2iz}}{z^2}\,dx = \oint_\Gamma \frac{1-e^{2iz}}{z^2}\,dz. \tag{7.39}$$

(f) *Use the residue theorem* to evaluate the integral J_Γ along the contour Γ and hence complete the proof of the result

$$\int_0^\infty \frac{\sin^2 x}{x^2}\,dx = \frac{\pi}{2}.$$

[2] Integrating $\phi(z) = \dfrac{\log z}{z^2-1}$ around a suitably indented contour and following the steps outlined, prove that PV $\displaystyle\int_0^\infty \frac{\log x}{x^2-1} = \frac{\pi^2}{4}$.

(a) Choose a contour. To compute the integrals use the indented contour shown in Fig. 7.10 and write the integral along the closed contour as

$$\oint \phi(z)\,dz = \int_{\rm AB} \phi(z)\,dz + \int_{\rm BCD} \phi(z)\,dz + \int_{\rm DE} \phi(z)\,dz$$
$$+ \int_{\rm EFG} \phi(z)\,dz + \int_{\rm GP} \phi(z)\,dz + \int_{\rm PQA} \phi(z)\,dz.$$

Here EFG, PQA, and BCD are circular arcs of radii R, ρ, and δ, respectively. Select a suitable branch cut and a definition for the logarithm function.

(b) Process the integrals along different parts:

- The integral along two of the three circular arcs vanishes in the limit $R \to \infty, \rho \to 0, \delta \to 0$. Identify these two arcs and prove your statement using Darboux's theorem.

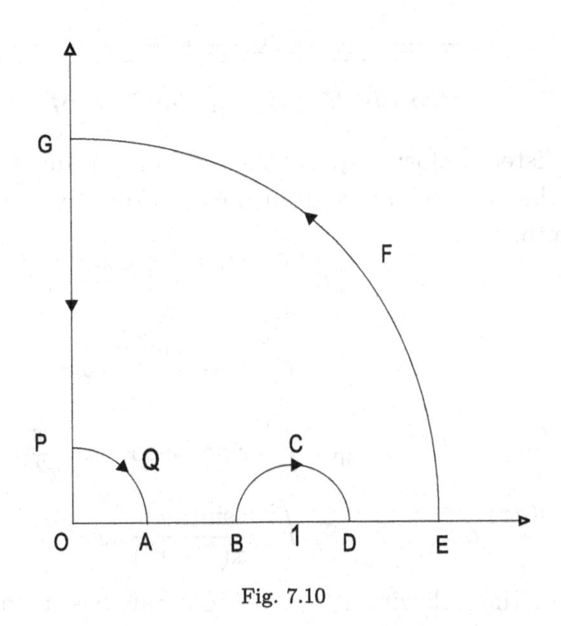

Fig. 7.10

- The integral along the third circular arc is also zero. Prove this by using §6.5.1 to write

$$\lim_{\delta\to 0}\int_{\mathrm{BCD}} \phi(z)\,dz = -\left(\frac{\pi i}{4}\right) \mathrm{Res}\{\phi(z)\}_{z=0}$$

and computing the residue of $\phi(z)$.
- Set up the line integrals over the three line segments and show that the integral in (a) is

$$\lim \int_{\mathrm{AB}} \phi(z)\,dz + \lim \int_{\mathrm{DE}} \phi(z)\,dz = \mathrm{PV}\int_0^\infty \phi(x)\,dx,$$

$$\lim \int_{\mathrm{GP}} \phi(z)\,dz = -\int_{\mathrm{PG}} \phi(z)\,dz = i\int_0^\infty \frac{\log x}{x^2+1}\,dx - \frac{\pi}{2}\int_0^\infty \frac{1}{x^2+1}\,dx.$$

Thus, arrive at

$$\oint \phi(z)\,dz = \mathrm{PV}\int_0^\infty \phi(x)\,dx + i\int_0^\infty \frac{\log x}{x^2+1}\,dx - \frac{\pi}{2}\int_0^\infty \frac{1}{x^2+1}\,dx.$$

(c) Cauchy's theorem implies that the integral on the left hand side, along the closed contour, is zero (why?). Where are the singular points of the integrand? Where did you select the branch cut for the logarithm?

(d) Take the real and imaginary parts, after substituting the value of $\displaystyle\int_0^\infty \frac{1}{x^2+1}\,dx$, to get the required result.

§§7.11 Exercise

Integrals Requiring the Use of Indented Contours

⊘ The integrals listed below require the use of an indented contour when computed by the method of contour integration. Evaluate them using an appropriate contour:

[1] $\displaystyle\int_0^\infty \frac{1-\cos ax}{x^2}\,dx, \quad (a>0),$ [2] $\displaystyle\int_0^\infty \frac{\sin x - x\cos x}{x^2}\,dx,$

[3] $\displaystyle\int_0^\infty \frac{\sin^3 x}{x^3}\,dx,$ [4] $\displaystyle\int_0^\infty \frac{x^3-\sin^3 x}{x^5}\,dx,$

[5] $\displaystyle\int_0^\infty \sin px \, \sin qx\,\frac{dx}{x^2},$ [6] $\displaystyle\int_0^\infty \sin^2 px\, \cos 2qx\,\frac{dx}{x^2},$

[7] $\displaystyle\int_0^\infty \frac{\cos(px)-\cos(qx)}{x^2}\,dx,$ [8] $\displaystyle\int_0^\infty \frac{\sin(\pi x)}{x(1-x^2)}\,dx.$

⊘ The integrand of the following improper integrals has a singular point in the range of integration and the integrals are not defined. However, the Cauchy principal value exists. Find the Cauchy principal value for each integral:

[9] $\displaystyle\int_0^\infty \frac{1}{1-x^4}\,dx,$ [10] $\displaystyle\int_0^\infty \frac{x^{p-1}}{1-x^3}a\,dx.$

⊘ Use a suitably indented rectangular contour to complete the following integrals:

[11] $\displaystyle\int_0^\infty \frac{x}{\sinh x}\,dx,$ [12] $\displaystyle\int_0^\infty \frac{\sin ax}{\sinh \pi x}\,dx,$

[13] $\displaystyle\int_0^\infty \frac{x^2(e^x-e^{-x}+2)}{(e^x-1)^2}\,dx,$ [14] $\displaystyle\int_o^\infty \frac{\cos \pi x}{1-4x^2}\,dx.$

[15] Integrating $\dfrac{z^{p-1}}{1+z}$ around a suitable contour and using the value for the integral

$$\int_0^\infty \frac{x^{p-1}}{1+x}\,dx = \pi \operatorname{cosec} p\pi,$$

find the principal value of $\displaystyle\int_0^\infty \frac{x^{p-1}}{1-x}\,dx.$

Answers

[1] $\frac{a\pi}{2}(a > 0)$, [2] 1, [3] $\frac{3\pi}{8}$,

[4] $\frac{13\pi}{32}$, [5] $\frac{\pi}{4}(|p+q| - |p-q|)$, [6] $\frac{\pi}{4}(|p-q| - 2|q| + |p+q|)$,

[7] $\frac{\pi}{2}(|q| - |p|)$, [8] π, [9] $\frac{\pi}{4}$,

[10] $\frac{\pi}{3}\cot\left(\frac{p\pi}{3}\right)$ [11] $\frac{\pi^2}{4}$, [12] $\frac{1}{2}\tanh\left(\frac{a}{2}\right)$,

[13] $\frac{2}{3}(\pi^2 - 3)$, [14] $\frac{\pi}{4}$, [15] $\pi \cot p\pi$.

§§7.12 Exercise

Series Summation and Expansion

⊘ Obtain the following Mittag–Leffler expansion formulae:

[1] $\pi \operatorname{cosec} \pi z = \dfrac{1}{z} + \sum_{m \neq 0} (-1)^m \left(\dfrac{1}{z-m} + \dfrac{1}{m} \right),$

[2] $\pi \cot \pi z = \dfrac{1}{z} + \sum_{m \neq 0} \left(\dfrac{1}{z-m} + \dfrac{1}{m} \right),$

[3] $\pi \coth \pi z = \dfrac{1}{z} + \sum_{m=1}^{\infty} \dfrac{2z}{z^2+m^2},$

[4] $\operatorname{cosec}^2 z = \sum_{-\infty}^{\infty} \dfrac{1}{(z-m\pi)^2}.$

⊘ Prove the following series summation results:

[5] $\dfrac{1}{1^2} + \dfrac{1}{2^2} + \dfrac{1}{3^2} + \cdots + \dfrac{(-1)^{m+1}}{m^2} + \cdots = \dfrac{\pi^2}{6},$

[6] $\dfrac{1}{1^2} - \dfrac{1}{2^2} + \dfrac{1}{3^2} - \cdots + \dfrac{(-1)^{m+1}}{m^2} + \cdots = \dfrac{\pi^2}{12},$

[7] $\dfrac{1}{1^2} + \dfrac{1}{3^2} + \dfrac{1}{5^2} + \cdots + \dfrac{1}{(2m-1)^2} + \cdots = \dfrac{\pi^2}{8},$

[8] $\dfrac{1}{1^2} - \dfrac{1}{3^2} + \dfrac{1}{5^2} - \cdots + \dfrac{(-1)^{m+1}}{(2m-1)^2} + \cdots = \dfrac{\pi^2}{32},$

[9] $\dfrac{1}{1^4} + \dfrac{1}{2^4} + \dfrac{1}{3^2} + \cdots + \dfrac{1}{m^4} + \cdots = \dfrac{\pi^4}{90},$

[10] $\dfrac{1}{1^4} + \dfrac{1}{3^4} + \dfrac{1}{5^4} + \cdots + \dfrac{1}{(2m-1)^2} + \cdots = \dfrac{\pi^4}{96},$

[11] $\dfrac{1}{1^5} - \dfrac{1}{3^5} + \dfrac{1}{5^5} - \cdots + \dfrac{(-1)^{m+1}}{(2m-1)^5} + \cdots = \dfrac{5\pi^5}{1536}.$

§§7.13 Exercise

What You See Is Not What You Get

⊘ While attempting to get a result on an improper integral I by the method of contour integration, you would start with a function $\phi(z)$ of a complex variable integrated around a closed contour Γ. You would then compute the integral $\oint_\Gamma \phi(z)\,dz$ and take a suitable limit. Find the missing entries of the following table and specify the limit to be taken, if any.

SN	What You See	Γ	Result You Get
1	$\oint_\Gamma \frac{dz}{(z-a)(z-\frac{1}{a})}$	Contour?	$\int_0^{2\pi} \frac{d\theta}{1+a^2-2a\cos\theta} = \frac{2\pi}{1-a^2}$
2	Input?	Fig. 7.14	$\int_0^\infty \frac{x\sin mx}{x^4+a^4}\,dx = \frac{\pi}{2a^2}e^{-ma/\sqrt{2}}\sin\left(\frac{am}{\sqrt{2}}\right)$
3	$\oint_\Gamma \frac{\operatorname{sech} z}{z-i\pi}\,dz$	Contour?	$\int_0^\infty \frac{\operatorname{sech} x}{x^2+\pi^2}\,dx = \frac{2}{\pi} - \frac{1}{2}$
4	$\oint_\Gamma \frac{\exp(iz^2)}{\sin\sqrt{\pi}z}\,dz$	Fig. 7.13 $a = \frac{\sqrt{\pi}}{2}$, $b = R$	What do you get?
5	$\oint_\Gamma \frac{1}{z\log z}\,dz$	Contour?	$\int_0^\infty \frac{dx}{x[(\ln x)^2+\pi^2]} = 1$
6	Input?	Fig. 7.15	$\int_0^\infty \frac{\cos\log x}{x^2+1} = \frac{\pi}{\cosh(\pi/2)}$
7	Input?	Fig. 7.16	$\int_0^\infty \frac{\cot^{-1} x}{x^2+1}\,dx = \frac{\pi^2}{2}$
8	Input?	Fig. 7.11	$\int_0^\infty x^{p-1}e^{-x\cos\alpha}\cos(x\sin\alpha)\,dx = \cos(p\alpha)\Gamma(p)$
9	$\oint_\Gamma \frac{z\,dz}{a-\exp(-iz)}$	Fig. 7.12 $a = \pi$, $b = n$	What do you get?
10	$\oint_\Gamma \frac{\exp(-z^2)}{z}\,dz$	Fig. 7.13 $a = R$, $b = p$	What do you get?
11	$\oint_\Gamma \frac{1-e^{iaz}\,dz}{z(\log z - i\frac{\pi}{2})}$	Fig. 7.16	What do you get?

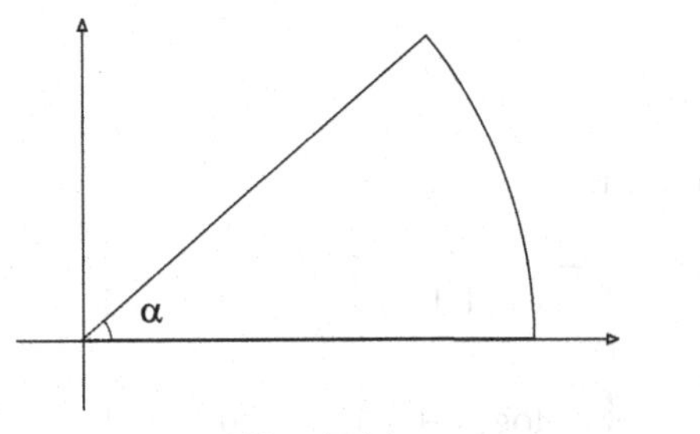

Fig. 7.11

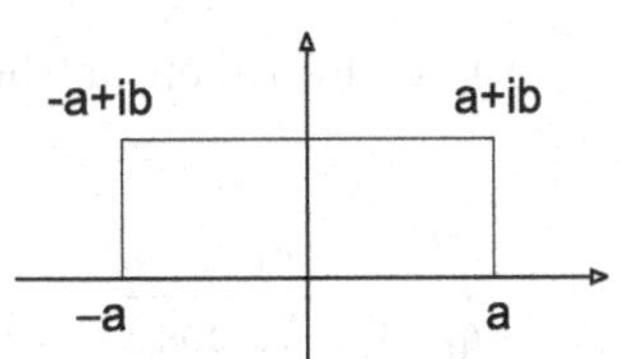

Fig. 7.12

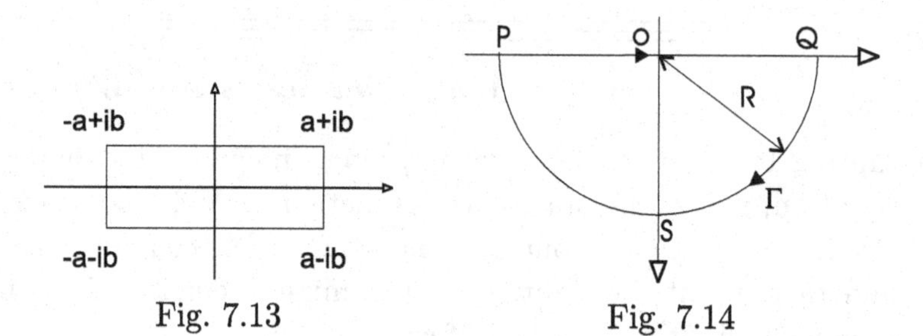

Fig. 7.13 Fig. 7.14

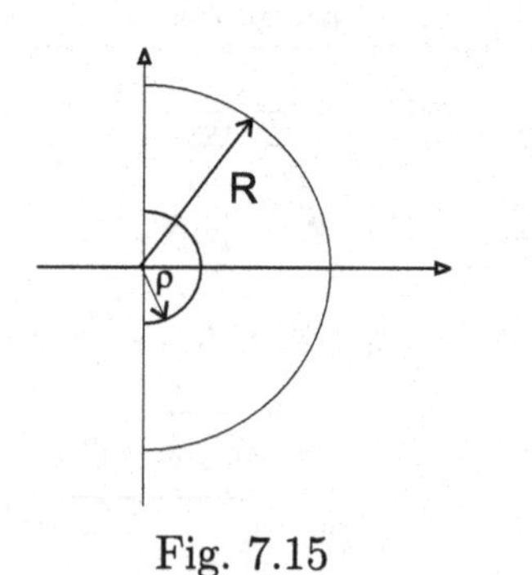

Fig. 7.15

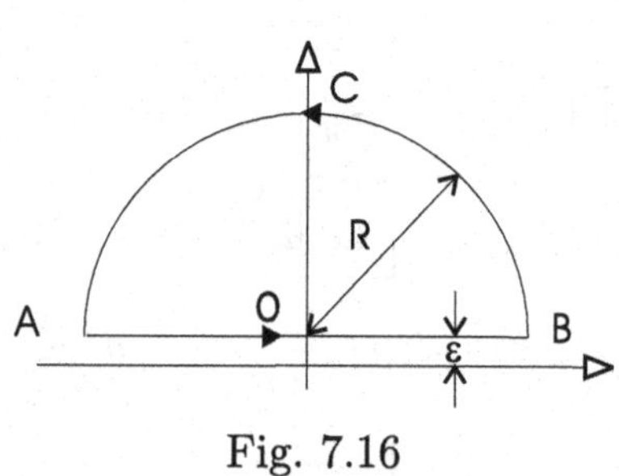

Fig. 7.16

Answers

⊘ The limit $R \to \infty$, $\rho \to 0$ is understood.

[1] Unit circle, $\{z||z| = 1\}$;

[2] $\oint \frac{ze^{-imz}}{z^4 + a^4}\, dz$;

[3] Figure 7.12, with $a = R, b = 2\pi i$;

[4] $\int_0^\infty \cos t^2\, dt = \int_0^\infty \sin t^2\, dt = \sqrt{\frac{\pi}{8}}$;

[5] Figure 7.16;

[6] $\oint \frac{z^i\, dz}{z^2 - 1}$;

[7] $\oint \frac{\text{Log}\,(z + i)}{z^2 + 1}\, dz$;

[8] $\oint z^p e^{-z}\, dz$;

[9] limit $n \to \infty$ is to be taken at the end;

$$\int_0^\pi \frac{x \sin x\, dx}{1 - 2a\cos x + a^2} = \begin{cases} \frac{\pi}{a} \log\left(1 + \frac{1}{a}\right), & \text{if } a > 1, \\ \frac{\pi}{a} \log(1 + a), & \text{if } a < 1; \end{cases}$$

[10] $$\int_0^\infty e^{-x^2} \frac{p\cos 2px + x\sin 2px}{x^2+p^2}\,dx = \frac{\pi}{2}e^{-p^2};$$

[11] $$\int_0^\infty \frac{\pi(1-\cos ax) - 2\log x \sin ax}{x[(\ln x)^2 + \frac{\pi^2}{4}]}\,dx = 2\pi(1-e^{-a}).$$

§§7.14 Exercise

Integrals from Statistical Mechanics

⊘ The method of contour integration is a powerful technique for many improper integrals which cannot be computed directly. However, examples can be given where this method does not turn out to be the simplest one.

[1] The value of the integral

$$\int_0^\infty \frac{e^x}{(e^x+1)^2}\,dx$$

is easily seen to be 1 by a straightforward integration. The examples in §§7.8 and §§7.9 suggest integrating $\dfrac{e^z}{(e^z+1)^2}$ around a rectangular contour with corners at $\pm R, \pm R + 2\pi i$ and taking $R \to \infty$. Try this to see what you get! How do you complete the evaluation of the integral by contour integration?

[2] Another integral which is easy to evaluate directly is

$$\int_0^\infty \frac{x\,dx}{e^x+1}.$$

The range of integration cannot be changed to the full real line. How do you proceed to make use of the methods learnt for hyperbolic functions? Use contour integration to show that the value of the integral is $\pi^2/6$.

[3] By the method of contour integration, derive a reduction formula for

$$I_m = \int_{-\infty}^\infty \frac{x^m e^x\,dx}{(e^x+1)^2},$$

which will relate I_m to I_k $(0 \le k < m)$. Use the reduction method to show that

$$\text{(a) } I_2 = \frac{\pi^2}{3}, \qquad \text{(b) } I_4 = \frac{7\pi^4}{15}.$$

[4] In order to compute the integrals I_m, one may compute the generating function

$$J(t) = \int_{-\infty}^\infty \frac{e^{itx}e^x\,dx}{(e^x+1)^2}.$$

An expansion of $J(t)$ in powers of t,

$$J(t) = \sum_m \frac{(it)^m}{m!} I_m,$$

gives the value of I_m for different m. Use contour integration to show that

$$J(t) = \frac{\pi t}{\sinh \pi t}.$$

[5] (a) Use an appropriately indented contour to show that

$$\int_0^\infty \frac{x^3 e^x}{(e^x - 1)^2}\, dx = \frac{\pi^4}{15}.$$

(b) Next, evaluate the above integral by expanding the denominator in powers of e^{-x} and integrating the series term by term and summing the resultant series.

⊗ The integrals of this type given here appear in statistical mechanics computations.

Hints

[1] ⊙ Integrate $\frac{ze^z}{(e^z+1)^2}$ around a rectangular contour.

[2] ⊙ Integrate by parts once and then use a rectangular contour.

§§7.15 Exercise

Alternate Routes to Improper Integrals

[1] Compare the answers for the integrals $\int (\log x)^n/(1+x^2)\,dx$ in problems [3]–[6] of §§7.7 with the answers for $\int_0^\infty x^n/\cosh \pi x\,dx$ and in problems [7]–[10] of §§7.9, and establish a connection between the two sets of problems, the results, and their respective contours.

[2] The two integral I and J,

$$I = \int_0^\infty \frac{x^{p-1}}{x^2+1}\,dx, \quad J = \int_0^\infty \frac{e^{px}}{e^{2x}+1}\,dx,$$

can be evaluated independently by using two circles' contour and a rectangular contour, respectively. Establish a correspondence between the two integrals and the contours used for their evaluation.

[3] The integrals

$$I = \int_0^\infty \frac{dt}{x[(\log x)^2 + \pi^2]}, \quad J = \int_{-\infty}^\infty \frac{dt}{t^2+1}$$

are related by a simple transformation. Use this information and find a contour for the integral I and complete its computation by the method of contour integration.

[4] An alternative method for evaluation of improper integrals of the form $\int Q(x)\cos kx\,dx$, where $Q(x)$ is a general rational function, has been outlined on p. (173), after ☑7.2. This method consists in writing $\cos kx$ in terms of exponentials $\exp(\pm ikx)$ to get

$$\int_{-\infty}^\infty Q(x)\cos kx\,dx = \frac{1}{2}\int_{-\infty}^\infty Q(x)\exp(ikx)\,dx + \frac{1}{2}\int_{-\infty}^\infty Q(x)\exp(-ikx)\,dx.$$

The evaluation of the improper integral, $\int Q(x)\cos kx\,dx$, can be completed by closing the contour in the upper half plane for one of the two terms on the right hand side and in the lower half plane for the other term. Adopt this approach to show that

$$\text{(a)} \int_{-\infty}^\infty \frac{\cos px}{x^2+q^2} = \frac{\pi}{q}\exp(-pq), \qquad \text{(b)} \int_{-\infty}^\infty \frac{\sin^3 x}{x^3}dx = \frac{3\pi}{4}.$$

Also evaluate the second integral in the above, using the approach discussed in ☑7.6 on p. (172). Which of the two approaches works better?

§§7.16 Open-Ended

Killing Two Birds with One Stone

⊘ Very often, two improper real integrals can be computed by doing a single contour integration. This set collects a few examples of this type.

[1] Compute the two integrals given below using the result (4.154) for the gamma function:

(a) $\displaystyle\int_0^\infty x^{p-1}\cos ax\,dx, \quad a>0,\ 0<p<1;$

(b) $\displaystyle\int_0^\infty x^{p-1}\sin ax\,dx, \quad a>0,\ -1<p<1.$

[2] Prove that

(a) $\displaystyle\int_0^\infty \frac{\cos(ax^2)-\sin(ax^2)}{x^4+b^4}\,dx = \frac{\pi\exp(-ab^2)}{2\sqrt{2}b^3},$

(b) $\displaystyle\int_0^\infty \frac{x^2[\cos(ax^2)+\sin(ax^2)]}{x^4+b^4}\,dx = \frac{\pi\exp(-ab^2)}{2\sqrt{2}b}.$

[3] Use a single contour integration to relate the following two integrals and derive their values:

(a) $\displaystyle\int_{-\infty}^\infty \frac{x\,dx}{a^2e^x+b^2e^{-x}} = \frac{\pi}{2ab}\ln\left(\frac{b}{a}\right),$

(b) $\displaystyle\int_{-\infty}^\infty \frac{x\,dx}{a^2e^x-b^2e^{-x}} = \frac{\pi^2}{4ab}.$

[4] Selecting a pair of integrals from the following list, integrate a single chosen function around a closed contour to compute the values of relate the two integrals ($0<p<q$):

(a) $\displaystyle\int_0^\infty \frac{x^{p-1}}{1+x^q}\,dx = \frac{\pi}{q}\operatorname{cosec}\left(\frac{p\pi}{q}\right),$

(b) $\displaystyle\int_0^\infty \frac{x^{p-1}}{1-x^q}\,dx = \frac{\pi}{q}\cot\left(\frac{p\pi}{q}\right),$

(c) $\displaystyle\int_0^\infty \frac{x^{p-1}\ln x}{1+x^q}\,dx = -\frac{\pi^2}{q^2}\cot\left(\frac{p\pi}{q}\right)\operatorname{cosec}\left(\frac{p\pi}{q}\right),$

(d) $\displaystyle\int_0^\infty \frac{x^{p-1}\ln x}{1-x^q} = -\frac{\pi^2}{q^2}\operatorname{cosec}^2\left(\frac{p\pi}{q}\right).$

Do this for as many pairs as you can.

§§7.17 Open-Ended

Food for Your Thought

⊘ The problems in this set go a little beyond the problems in solved examples, tutorials, and exercises. Establish the following identities:

[1] $\displaystyle\int_0^\infty \frac{x\tan(ax)}{x^2+b^2}\,dx = \frac{\pi}{e^{2ab+1}},\quad a>0, b>0;$

[2] $\displaystyle\int_0^\infty \frac{x}{(x^2+b^2)\sin(ax)}\,dx = \frac{\pi}{2\sinh(ab)},\qquad b>0;$

[3] $\displaystyle\int_0^\infty \frac{\sin(ax)}{\sin(bx)}\frac{dx}{x^2+c^2} = \left(\frac{\pi}{2c}\right)\frac{\sinh(ac)}{\sinh(bc)},\qquad 0<a<b,\ c>0;$

[4] $\displaystyle\int_0^\infty \frac{1}{1-2a\cos(bx)+a^2}\frac{dx}{x^2+c^2} = \left(\frac{\pi}{2c}\right)\frac{1}{1-a^2}\frac{1+ae^{-bc}}{1-ae^{bc}},\qquad a^2<1;$

[5] The contour integration method enables us find several integrals by relating them to the Gaussian integral $\int_{-\infty}^{\infty}\exp(-x^2)\,dx$. For example, all of the problems §§4.7–§§4.9 are completed in this fashion. How about finding a contour integral method to obtain the value $\sqrt{\pi}$ of the Gaussian integral itself? To begin, integrate $\frac{\exp(iz^2/\pi)}{\sin z}$ around a rectangular contour and prove that

$$\int_0^\infty \sin x^2 dx = \int_0^\infty \cos x^2 dx = \sqrt{\frac{\pi}{8}}.$$

Next, integrate the same function around a parallelogram to obtain

$$\int_{-\infty}^{\infty} \exp(-x^2)dx = \sqrt{\pi}.$$

[6] (a) Set up the integral of $f(z) = 1/(z^2+1)\cosh(\pi z/2)$ around a square S_N with corners at $\pm N$ and $\mp N + 2iN$, where N is an integer, and show that the integrals along three of the four sides of the square tend to zero as $N \to \infty$. Can you now complete the proof of the following integral?

$$I = \int_0^\infty \frac{dx}{(x^2+1)\cosh(\pi x/2)} = \ln 2.$$

(b) The integral I in the previous problem can also be written as an integral of discontinuity of

$$f(z) = \frac{\operatorname{Log} z}{(z^2+1)\cosh(\pi z/2)}$$

across the branch cut of $\operatorname{Log} z$. Using the method of ☑7.4 on p. (175), investigate if setting up the integral of $f(z)$ along two circles' contour leads to the value of the integral I.

[7] Use the method of contour integration to prove that

(a) $$\int_0^1 \frac{x^{p-1} + x^{q-p-1}}{1 + x^q}\,dx = \frac{\pi}{q}\,\mathrm{cosec}\left(\frac{p\pi}{q}\right),$$

(b) $$\int_0^1 \frac{x^p - x^{-p}}{1 + x^2}\,x\,dx = \frac{1}{p} - \frac{\pi}{2}\,\mathrm{cosec}\left(\frac{p\pi}{2}\right).$$

[8] Suggest a scheme for computing the proper integral $\int_a^b f(x)dx$ by writing it as a contour integral of a multivalued function having its branch cut running from a to b. What restrictions, if any, would you require for the function $f(x)$? Illustrate your scheme by showing that

$$\int_{x_1}^{x_2} x^n\,dx = \frac{x_2^{n+1} - x_1^{n+1}}{n+1}.$$

§§7.18 Mixed Bag

Improper Integrals

⊘ Using a semicircle with the line joining $c - iR$ and $c + iR$ as a diameter (see Fig. 7.17), prove the results in Q[1]–Q[4].

[1] $\displaystyle\int_{c-\infty}^{c+i\infty} \frac{a^z}{z} dz = \begin{cases} 1, & \text{if } a > 1, \\ 0, & \text{if } 0 < a < 1. \end{cases}$

[2] $\displaystyle\int_{c-i\infty}^{c+i\infty} \frac{a^z}{z^2} dz = \begin{cases} \log a, & \text{if } a > 1, \\ 0, & \text{if } 0 < a < 1. \end{cases}$

[3] $\displaystyle\int_{c-i\infty}^{c+i\infty} \frac{1}{a^z \sin \pi z} dz = \frac{1}{\pi(a+1)}.$

[4] $\displaystyle\int_{c-i\infty}^{c+i\infty} \frac{\exp(zs)}{(z^2+\nu^2)^2} dz = \frac{1}{2\nu^3}\left(\sin \nu - \frac{t}{2\nu^2} \cos \nu t\right).$

[5] Prove that

$$\int_{-\infty}^{\infty} \frac{dx}{x^{4n} + 2x^{2n}\cos\alpha + 1} = \frac{\pi}{2n} \frac{\sin[(2n-1)\alpha/n]}{\sin\alpha \sin(\pi/2n)}.$$

[6] For a and b real prove that

$$\int_0^{\infty} \frac{\cos ax \cosh bx}{\cosh \pi x} dx = \frac{\cosh(a/2)\cos(b/2)}{\cosh a + \cos b}.$$

[7] Prove the following results using the integration in the complex plane ($a > 0$):

(a) $\displaystyle\int_0^{\infty} \frac{x \sin^{2m} x \sin(2m+2)x}{x^2+a^2} dx = \frac{(-1)^m \pi}{2^{2m+1}} e^{-2a}(1-e^{-2a})^{2m},$

(b) $\displaystyle\int_0^{\infty} \frac{x \sin^{2m} x \sin(4mx)}{x^2+a^2} dx = \frac{(-1)^m \pi}{2} e^{-4ma} \sinh^{2m} a,$

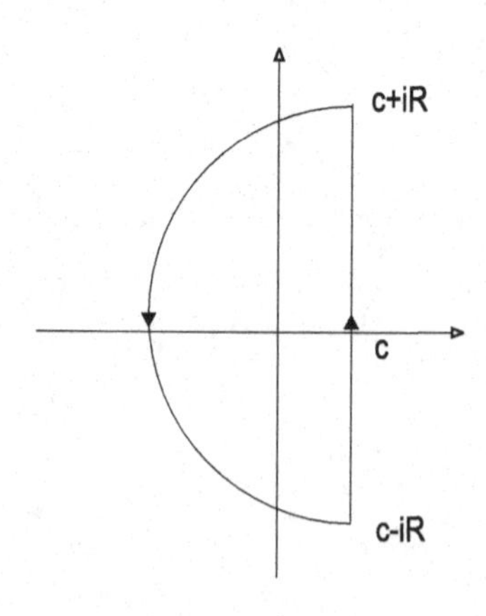

Fig. 7.17

(c) $\displaystyle\int_0^\infty \frac{\sin^{2m} x \cos 2mx}{x^2+a^2}\,dx = \frac{(-1)^m \pi}{2^{2m+1}a}(1-e^{-2a})^{2m}.$

[8] The following four integrals appear to be very similar, but application of contour integration to their evaluation reveals that they are very different. While some of them can be evaluated easily by minor modifications in the approach to hyperbolic functions in ☑7.9, others require a very different approach — figure out why.

(a) $\displaystyle\int_0^\infty \frac{dx}{(1+x^2)\cosh x}$, (b) $\displaystyle\int_0^\infty \frac{dx}{(1+x^2)\cosh(\pi x/2)}$,

(c) $\displaystyle\int_0^\infty \frac{x\,dx}{(1+x^2)\sinh x}$, (d) $\displaystyle\int_0^\infty \frac{x\,dx}{(1+x^2)\sinh(\pi x/4)}$.

[9] For $q > p > -1$, use an indented semicircular contour with the diameter along the imaginary axis to prove that

$$\int_0^{\frac{\pi}{2}} \cos^p\theta \cos q\theta \; d\theta = \frac{\pi\Gamma(p+1)}{2^{p+1}\Gamma(\frac{p+q+2}{2})\Gamma(\frac{p-q+2}{2})}.$$

[10] If $S_k = \displaystyle\sum_{m=1}^{\infty} \frac{1}{(4m^2-1)^k}$, show that

(a) $S_1 = \dfrac{1}{2}$, (b) $S_2 = \dfrac{\pi^2-8}{12}$,

(c) $S_3 = \dfrac{32-3\pi^2}{64}$, (d) $S_4 = \dfrac{\pi^4+30\pi^2-384}{768}$.

Chapter 8

ASYMPTOTIC EXPANSIONS

§§8.1 Exercise

Integration by Parts

[1] (a) Integrating by parts, show that the function $\Gamma(a,x)$, defined by

$$\Gamma(a,x) = \int_x^\infty e^{-t} t^{a-1}\, dt,$$

satisfies

$$\Gamma(a,x) = \sum_{m=1}^{n} \frac{\Gamma(a)}{\Gamma(a-m+1)} e^{-x} x^{a-m} + \frac{\Gamma(a)}{\Gamma(a-n)} \Gamma(a-n,x).$$

(b) Show that the difference

$$\left| \Gamma(a,x) - \sum_{m=1}^{n} \frac{\Gamma(a)}{\Gamma(a-m+1)} e^{-x} x^{a-m} \right|,$$

for a fixed n, tends to zero as $x \to \infty$.

(c) Use your results to obtain asymptotic expansion of the incomplete gamma function

$$\gamma(a,x) = \int_0^x e^{-t} t^{a-1}\, dt.$$

[2] (a) Show that the exponential integral function $Ei(x) = \int_x^\infty \frac{e^{-t}}{t}\, dt$, after n integration by parts, becomes

$$Ei(x) = e^{-x} \left\{ \frac{1}{x} - \frac{1}{x^2} + \frac{2!}{x^3} + \cdots + \frac{(-1)^{(r-1)}(r-1)!}{x^r} \right\} + (-1)^r r! \int_x^\infty \frac{e^{-t}}{t^{r+1}}\, dt.$$

(b) Use the above result to show that the series

$$Ei(x) = e^{-x}\left\{\frac{1}{x} - \frac{1}{x^2} + \frac{2!}{x^3} + \cdots + \frac{(-1)^r r!}{x^{r+1}} + \cdots\right\}$$

is asymptotic to $Ei(x)$.

[3] Show that the cosine integral function $Ci(x) = \int_x^\infty \frac{\cos t}{t}\,dt$ has the asymptotic expansion

$$Ci(x) \sim \sin x\left(-\frac{1}{x} + \frac{2!}{x^3} - \frac{4!}{x^5} - \cdots\right) + \cos x\left(\frac{1}{x^2} - \frac{3!}{x^4} + \frac{5!}{x^6} + \cdots\right).$$

[4] (a) Use integration by parts to derive the asymptotic expansion

$$\int_x^\infty t^{-a}\,e^{it}\,dt \sim i\frac{e^{ix}}{x^a}\sum_{m=0}^{\infty}\frac{\Gamma(a+m)}{\Gamma(a)(ix)^m}.$$

(b) What happens if we attempt to use integration by parts to obtain the asymptotic expansion of $\int_0^x \frac{\cos t}{\sqrt{t}}\,dt$?

Answers

[1](c) $\gamma(a,x) = \Gamma(a) - \Gamma(a,x)$:

$$\Gamma(a,x) \sim \sum_{m}^{\infty}\frac{\Gamma(a)}{\Gamma(a-m+1)}e^{-x}x^{a-m}.$$

[4](b) The method will fail for $\int_0^x \frac{\cos t}{\sqrt{t}}\,dt$. To get its asymptotic expansion, use

$$\int_0^x \frac{\cos t}{\sqrt{t}}\,dt = \int_0^\infty \frac{\cos t}{\sqrt{t}}\,dt - \int_x^\infty \frac{\cos t}{\sqrt{t}}\,dt$$

and results in part (a).

§§8.2 Exercise

Dominant Term

[1] Use Laplace's method and the integral representation of the gamma function

$$\Gamma(x+1) = \int_0^\infty e^{-t} t^x \, dt$$

to obtain the Stirling formula

$$n! \sim \sqrt{2\pi n}\, n^n e^{-n}.$$

[2] Use Laplace's method and show that the dominant term of $\int_0^\infty \frac{dt}{(1+t^2)^n}$, as $n \to \infty$, is $\sqrt{\pi/4n}$.

⊘ Use integration by parts, or Laplace's method, and obtain the dominant term of the asymptotic expansion of the integrals given below as $x \to \infty$:

[3] $\int_0^\infty \frac{e^{-xt}}{1+t^2}\, dt \sim \frac{1}{x}$,

[4] $\int_0^\infty e^{-x\cosh t}\, dt \sim \sqrt{\frac{\pi}{2x}}\, e^{-x}$,

[5] $\int_{-\frac{\pi}{2}}^{\frac{\pi}{2}} e^{x\cos\theta}\, d\theta \sim \sqrt{\frac{2\pi}{x}}\, e^x$,

[6] $\int_0^\pi e^{-x\sin^2\theta} d\theta \sim \sqrt{\frac{\pi}{x}}$,

[7] $\int_0^{\frac{\pi}{2}} (\sin t)^n dt \sim \sqrt{\frac{\pi}{2n}}$,

[8] $\int_0^{\frac{\pi}{2}} t^2 e^{x\cos t}\, dt \sim \sqrt{\frac{\pi}{2x^3}}\, e^x$,

[9] $\int_x^\infty e^{-t^4}\, dt \sim \frac{e^{-x^4}}{4x^3}$,

[10] $\int_0^1 t^{-\frac{1}{2}} e^{-ixt}\, dt \sim \frac{-i}{\sqrt{x}} e^{-ix}$.

⊘ Use the method of stationary phase and obtain the dominant term of the asymptotic expansion of the integrals given below for $x \to \infty$:

[11] $\int_0^1 \frac{e^{ixt}}{1+t^2}\, dt$,

[12] $\int_0^{\frac{\pi}{2}} e^{ix\sin\theta}\, d\theta$,

[13] $\int_{-\frac{\pi}{2}}^{\frac{\pi}{2}} e^{ix\cos\theta}\, d\theta$,

[14] $\int_0^\pi \cos(xt^2 + t)\, dt$.

Answers

[11] $\frac{1}{2ix}(e^{ix} - 2)$,

[12] $\sqrt{\pi/2x}e^{i\pi/4}e^{ix}$,

[13] $\sqrt{2\pi/x}e^{-i\pi/4}e^{ix}$,

[14] $\sqrt{\pi/8x}$.

§§8.3 Exercise

Laplace's Method

[1] Use Laplace's method to obtain the asymptotic expansion of

$$\operatorname{Erfc}(x) = \int_x^{\infty} e^{-t^2}\, dt.$$

[2] Find the asymptotic expansion of

$$I(x) = \int_0^{x} e^{t^2}\, dt.$$

[3] Obtain the asymptotic expansion of

$$I(x) = \int_0^{\infty} \frac{e^{-xt}}{1+t^2}\, dt,$$

for large x, using Watson's lemma.

[4] Show that the asymptotic expansion of $I_\nu(x)$ as $x \to \infty$, where

$$I_\nu(x) = e^{-x^2/4} \int_0^{\infty} t^{-\nu-1} e^{-t^2/2} e^{-xt}\, dt,$$

is given by

$$I_\nu(x) = e^{-x^2/4} \sum_{m=0}^{\infty} (-1)^m \frac{\Gamma(2m-\nu)}{2^m m! x^{2n-\nu}}.$$

[5] Prove that

$$\int_0^{\infty} \frac{e^{-zt^2}}{1+t^2}\, dt \sim \frac{1}{2}\sqrt{\frac{\pi}{z}} \left\{ 1 - \frac{1}{2z} + \frac{1.3}{(2z)^2} - \frac{1.3.5}{(2z)^5} + \cdots \right\}.$$

[6] Show that for large values of x

$$\int_0^{\infty} \frac{dt}{(1+t^2)^z} \sim \frac{\sqrt{\pi}}{2} \frac{1}{\sqrt{z}} \left\{ 1 + \frac{3}{8z} + \frac{25}{128z^2} + \cdots \right\}.$$

[7] Find the asymptotic expansion of

$$I(x) = \int_0^{\pi/2} \exp\left(-x \sin^2 t\right) dt,$$

for large x.

Answers

[1] $\operatorname{Erfc}(x) = \dfrac{e^{-x^2}}{2} \displaystyle\sum_{m=0}^{\infty} \frac{(-1)^m \Gamma(m+\frac{1}{2})}{\sqrt{\pi} x^{2m+1}}$,

[2] $I(x) \sim \frac{e^{x^2}}{2x}\left\{1 + \frac{1}{2x^2} + \frac{1}{2}\frac{3}{2}\frac{1}{x^4} + \cdots + \frac{1.3.5...(2n-1)}{(2x^2)^n} + \cdots\right\}$,

[3] $I(x) \sim \frac{1}{x} - \frac{2!}{x^3} + \frac{4!}{x^5} - \frac{6!}{x^7} + \cdots$,

[7] $\frac{1}{2}\sum_{m=0}^{\infty}\frac{[\Gamma(m+\frac{1}{2})]^2}{m!\sqrt{\pi}x^{m+\frac{1}{2}}}$.

§§8.4 Exercise

Steepest Paths

⊘ Complete the steps [A]–[D] suggested below leading to the steepest paths which could be used for deriving the asymptotic expansion of the integral in each of the problems [1]–[3] in the limits as specified.

[A] Find the directions of steepest ascent and descent at the end points of the interval, whenever the integration limit is finite.

[B] Find saddle points, if any, and the directions of steepest descent and ascent.

[C] Draw a figure showing short line segments along the directions of steepest descent and ascent you have found.

[D] From the figures in Fig. 8.1, select a contour which correctly joins the line segments and which could be used to derive the asymptotic expansion.

[1] $\displaystyle\int_{-1}^{1} e^{ixt}\phi(t)\,dt$ as $x \to \infty$.

[2] $\displaystyle\int_{-\frac{\pi}{2}}^{\frac{\pi}{2}} e^{in\cos\tau}\phi(\tau)\,d\tau$ as $n \to \infty$.

[3] $\displaystyle\int_{-\pi}^{\pi} e^{ix(\tau-\sin\tau)}\phi(\tau)\,d\tau$ as $x \to \infty$.

⊘ The two Hankel functions, $H_\nu^{(1)}(x)$ and $H_\nu^{(2)}(x)$, which appear in the problems below, are given by

$$H_\nu^{(1)}(x) = \int_{C_1} \exp\left(-ix\sin\tau + i\nu\tau\right)d\tau, \quad H_\nu^{(2)}(x) = \int_{C_2} \exp\left(-ix\sin\tau + i\nu\tau\right)d\tau,$$

where C_1 and C_2 are the contours shown in Fig. 8.2.

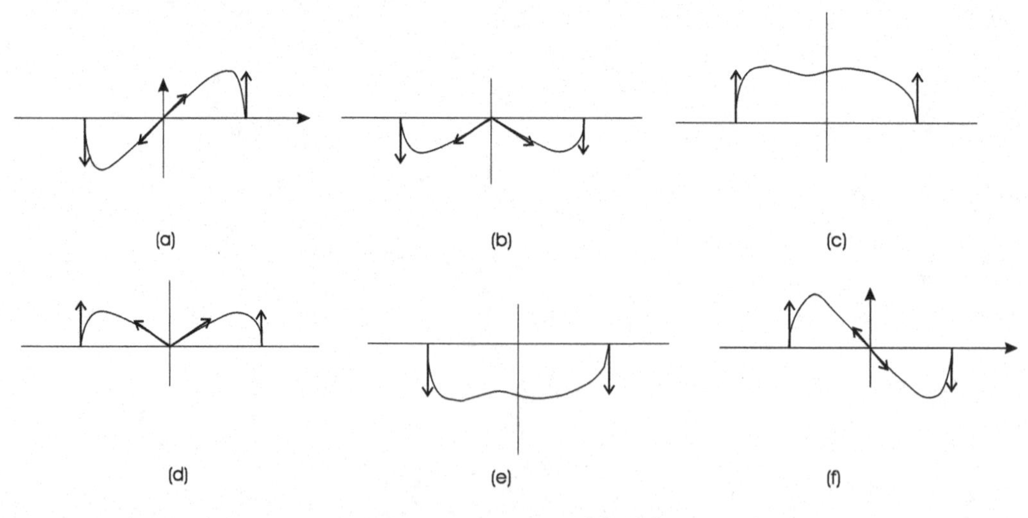

Fig. 8.1 Figures for Q[1]–Q[3].

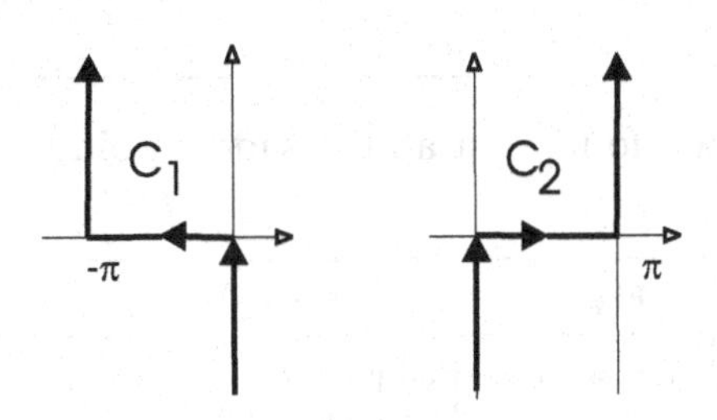

Fig. 8.2

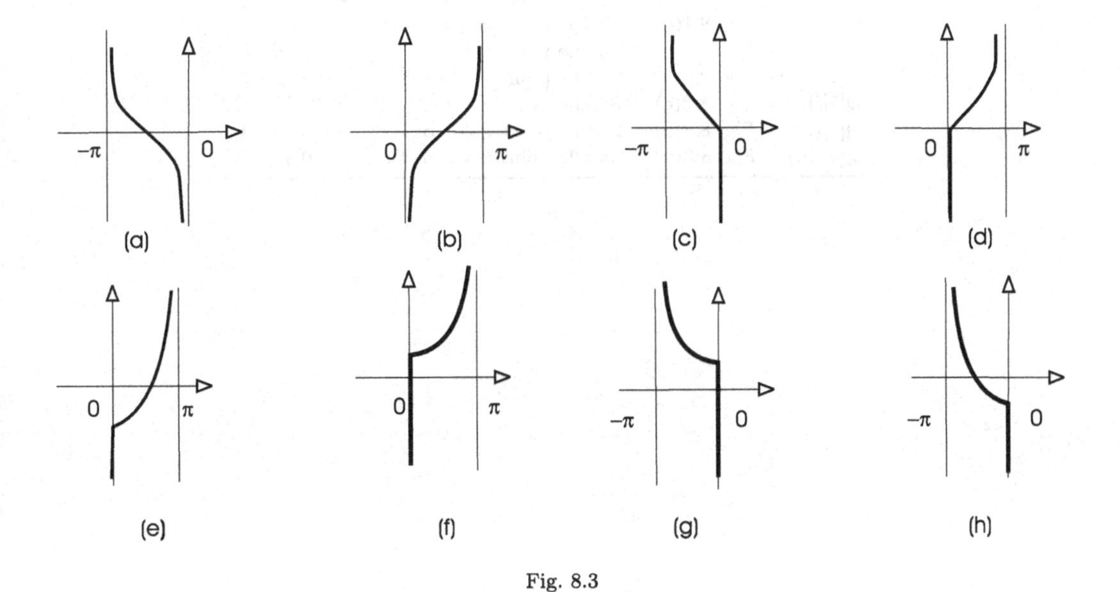

Fig. 8.3

[4] Using the integral representation given above, answer the following questions for the Hankel function $H_\nu^{(1)}(x)$ as $x \to \infty$ and ν remains fixed:

- Find saddle points, if any, and the directions of steepest descent and ascent.
- From the contours in Fig. 8.3 given at the end, select a contour which correctly joins the line segments through the saddle point(s) and which could be used to derive the asymptotic expansion.
- Indicate the location of the saddle point(s) and find the values of the angles that the directions of steepest descent make with the real axis.

[5] Consider the Hankel function integral of the $H_\nu^{(1)}(x)$ and repeat Q[4] for $x \to \infty, \nu \to \infty$ and taking $p = \nu/x$ fixed for the following three cases:

(i) $p = 1$,
(ii) $p < 1$ (write $p = \cos\alpha$),
(iii) $p > 1$ (write $p = \cosh\alpha$).

[6] Using your results for the previous question, or otherwise, discuss the differences you find for the Hankel function $H_\nu^{(2)}(x)$ for the three cases (i)–(iii).

Answers

In the answers have we do not list all the saddle points; only the relevant saddle point is listed.

	Figure	Point(s)	Angles
Q[1]	Fig. 8.1(c)	End point $t = -1$	$\pi/2$
		End point $t = 1$	$\pi/2$
Q[2]	Fig. 8.1(f)	End point $t = -\pi/2$	$\pi/2$
		End point $t = \pi/2$	$3\pi/2$
		Saddle point $t = 0$	$-\pi/4, 3\pi/4$
Q[3]	Fig. 8.1(d)	End points $t = \pm\pi$	$\pi/2$
		Saddle point $t = 0$	$\pi/6, 5\pi/6$
Q[4]	Fig. 8.3(a)	Saddle point $\tau = -\pi/2$	$-\pi/4, 3\pi/4$
Q[5](i)	Fig. 8.3(c)	Saddle point $\tau = 0$	$\pi/6, 5\pi/6, 3\pi/2$
Q[5](ii)	Fig. 8.3(a)	Saddle point $\tau = -\alpha$	$-\pi/4, 3\pi/4$
Q[5](iii)	Fig. 8.3(h)	Saddle point $\tau = -i\alpha$	$0, \pi$

§§8.5 Tutorial

Method of Steepest Descent

[1] Starting from the integral representation of the Bessel function $J_0(x)$,

$$J_0(x) = \frac{1}{\pi}\int_{-1}^{1} \frac{e^{ix\tau}}{\sqrt{1-\tau^2}}\,d\tau,$$

use the method of steepest descent to find the asymptotic expansion as $x \to \infty$.
(a) Find the directions of steepest descent at the two end points. Verify that these are correctly given by the line segments AM and BN running upward parallel to the imaginary axis; see Fig. 8.4. Note that there are no saddle points. Use Cauchy's theorem to write the integral for $J_0(x)$ as an integral along ABNM:

$$\frac{1}{\pi}\int_{-1}^{1} \frac{e^{ix\tau}}{\sqrt{1-\tau^2}}\,d\tau = \left\{\frac{1}{\pi}\int_{\text{AM}} + \frac{1}{\pi}\int_{\text{MN}} - \frac{1}{\pi}\int_{\text{BN}}\right\} \frac{e^{ix\tau}}{\sqrt{1-\tau^2}}\,d\tau.$$

(b) Show that the integral $\int_{\text{MN}} \cdots\, d\tau$ does not contribute to the asymptotic expansion. Therefore,

$$\frac{1}{\pi}\int_{-1}^{1} \frac{e^{ix\tau}}{\sqrt{1-\tau^2}}\,d\tau \sim \frac{1}{\pi}\int_{\text{AM}} \frac{e^{ix\tau}}{\sqrt{1-\tau^2}}\,d\tau - \frac{1}{\pi}\int_{\text{BN}} \frac{e^{ix\tau}}{\sqrt{1-\tau^2}}\,d\tau.$$

(c) The integral along AM becomes

$$\int_{\text{AM}} \frac{e^{ix\tau}}{\sqrt{1-\tau^2}}\,d\tau = \int_{-1}^{-1+i\delta} \frac{e^{ix\tau}}{\sqrt{1-\tau^2}}\,d\tau$$

and similarly for the second integral along BN.
(d) Extend the range of integration in the two integrals to an infinite range by letting $\delta \to \infty$. This gives

$$\int_{-1}^{1} \frac{e^{ix\tau}}{\sqrt{1-\tau^2}}\,d\tau \sim \lim_{\delta\to\infty}\left\{\int_{-1}^{-1+i\delta} \frac{e^{ix\tau}}{\sqrt{1-\tau^2}}\,d\tau - \frac{1}{\pi}\int_{1}^{1+i\delta} \frac{e^{ix\tau}}{\sqrt{1-\tau^2}}\,d\tau\right\}.$$

Also, verify that in this limit the integral $\int_{\text{BN}} \cdots\, d\tau \to 0$. Change the integration variables to $\tau \pm 1$ in the two integrals so as to make the limits of integration from 0 to ∞.

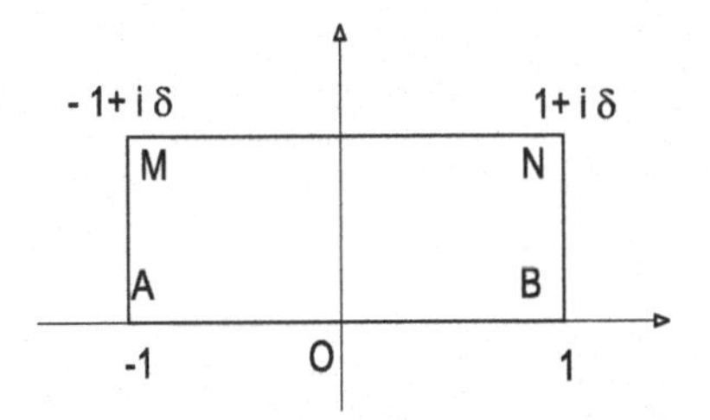

Fig. 8.4

(e) Expand the square root in the denominator and, keeping the first term, show that

$$J_0(x) \sim \frac{2}{\sqrt{\pi x}} \cos\left(x - \frac{\pi}{4}\right).$$

(f) Use Watson's lemma to integrate term by term and show that the full asymptotic expansion is given by

$$J_0(x) \sim \frac{1}{\sqrt{2\pi x}} \left[f(x)\cos\left(x - \frac{\pi}{4}\right) + g(x)\sin\left(x - \frac{\pi}{4}\right)\right],$$

where

$$f(x) = \frac{1}{\pi} \sum_{m=0}^{\infty} (-1)^m \frac{\left[\Gamma\left(2m + \frac{1}{2}\right)\right]^2}{(2m)!(2x)^{2m}},$$

$$g(x) = \frac{1}{\pi} \sum_{m=0}^{\infty} (-1)^m \frac{\left[\Gamma\left(2m + \frac{3}{2}\right)\right]^2}{(2m+1)!(2x)^{2m+1}}.$$

§§8.6 Tutorial

Saddle Point Method

[1] The Legendre polynomial $P_n(x)$ has the integral representation

$$P_n(x) = \frac{1}{2^{n+1}\pi i}\int_C \frac{(\tau^2-1)^n}{(\tau-x)^{n+1}}\,d\tau, \tag{8.1}$$

where C is a contour enclosing the point x. For $x < 1$ ($x \equiv \cos\alpha$) and large n show that

$$P_n(\cos\alpha) \sim \sqrt{\frac{2}{\pi n \sin\alpha}}\,\sin\bigl[(n+1/2)\alpha + \pi/4\bigr]. \tag{8.2}$$

⊘ To prove the above result, follow the steps given below and complete the problem.

(a) Compare the integral in Eq. (8.1) with the form $\displaystyle\int e^{xh(\tau)}\phi(\tau)\,dt$ by identifying $h(\tau)$ and $\phi(\tau)$ with

$$h(\tau) = \log(\tau^2-1) - \log(\tau - x), \quad \phi(\tau) = \frac{1}{\tau - x}. \tag{8.3}$$

(b) Set $x = \cos\alpha$ and find saddle points by solving $\frac{dh}{d\tau} = 0$. How many saddle points are there? Show that they are all equally important.
Select the unit circle as a contour passing through the two saddle points. Writing $x = \cos\alpha$ show that $P_n(\cos\alpha)$ can be transformed into

$$P_n(\cos\alpha) = \frac{1}{2^n\pi}\,\mathrm{Im}\int_\Gamma e^{nh(\tau)}\phi(\tau)\,d\tau, \tag{8.4}$$

where $\Gamma = \{\tau \big| |\tau| = 1,\ 0 \le \arg(\tau) \le \pi\}$ is the unit semicircle in the upper half plane. This form has the advantage that we need to compute the contribution of only one of the saddle points at $\tau_0 \equiv e^{i\alpha}$.
(c) Write $\tau - \tau_0 = \rho e^{i\phi}$ and show that, in a small neighborhood of the saddle point $e^{i\alpha}$,

$$\Delta h(\tau) = h(\tau) - h(\tau_0) \approx -\left(\frac{i}{2}e^{i(2\phi-\alpha)}\mathrm{cosec}\ \alpha\right)\rho^2, \tag{8.5}$$

and get directions of steepest descent.
(d) In a small neighborhood of the saddle point $e^{i\alpha}$, find the values of ϕ which correspond to:

(a) $\mathrm{Re}\,\Delta h = 0$,
(b) $\mathrm{Im}\,\Delta h = 0$,
(c) $\mathrm{Im}\,\Delta h = 0$ and $\mathrm{Re}\,\Delta h > 0$,
(d) $\mathrm{Im}\,\Delta h = 0$ and $\mathrm{Re}\,\Delta h < 0$.

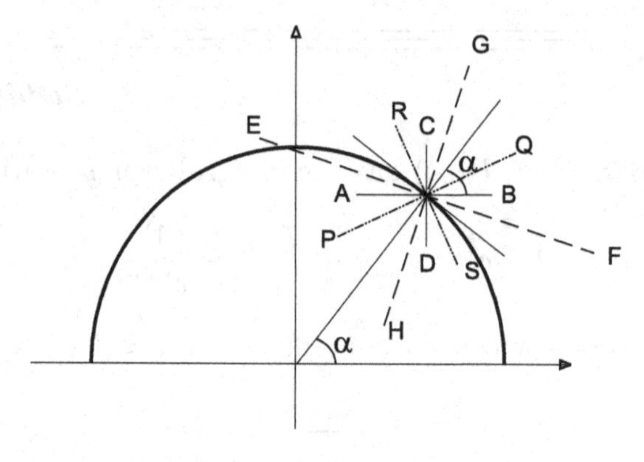

Fig. 8.5

(e) Identify steepest path directions: Several directions are shown in Fig. 8.5, where the semicircle is $|\tau| = 1$. Identify them with the directions found above. Which of these are the directions of steepest descent and steepest ascent?
(f) Verify that the directions of steepest descent through the saddle point $e^{i\alpha}$ relevant to the present case are given by $\phi = \frac{\alpha}{2} + \frac{\pi}{4} \pm \frac{\pi}{2}$ and correspond to the dashed line EF in the figure through $e^{i\alpha}$.
(g) To integrate over a line segment given by

$$\left\{\tau \middle| \tau = e^{i\alpha} + \rho e^{i\phi}, \quad -\delta \leq \rho \leq \delta\right\},$$

select a value of ϕ which corresponds to the correct steepest descent path through the saddle point $e^{i\alpha}$.
(h) Finally, extend the range of integration over ρ to an infinite range and derive the desired result for the dominant term.

§§8.7 Exercise

Steepest Descent and Saddle Point Method

[1] Use the method of steepest descent to show that the asymptotic expansion of

$$I(x) = \int_0^1 \sqrt{t} e^{ixt}\, dt$$

is given by

$$\sqrt{\frac{\pi}{2x^3}}\, e^{\frac{3i\pi}{4}} - \frac{i}{x} e^{ix} - \frac{ie^{ix}}{2x\sqrt{\pi}} \sum_{m=0}^{\infty} \Gamma(m + \tfrac{1}{2})(ix)^{-m-1}.$$

[2] Using the method of steepest descent, show that

$$\int_0^1 \frac{1}{\sqrt{t}} \exp\left[ix(t + t^2)\right] dt \sim \sqrt{\frac{\pi}{x}} e^{i\frac{\pi}{4}} \left(1 - \frac{3i}{4x} - \frac{105}{32x^2} + \cdots\right)$$
$$- \frac{i}{3x} e^{2ix} \left(1 - \frac{7i}{18x} - \frac{37}{108x^2} + \cdots\right),$$

where x is real and positive and the path of integration is along the real axis.

[3] (a) Find the saddle points of

$$I(x) = \int_{-\frac{\pi}{2}}^{\frac{\pi}{2}} \exp\left(ix \cos \tau\right) d\tau.$$

(b) In which of the two figures in Fig. 8.6, do the thick lines correctly show the directions of steepest descents at the three points $-\frac{\pi}{2}, 0$, and $\frac{\pi}{2}$?

(c) Compute the contribution of the two end points and the saddle point at $\tau = 0$, and prove that the dominant term comes from the neighborhood of the saddle point and is given by

$$I(x) \sim \sqrt{2\pi/x} \exp\left(ix - \frac{i\pi}{4}\right).$$

[4] The Bessel function $J_n(x)$, for integer n, has the representation

$$J_n(x) = \frac{1}{2\pi} \int_{-\pi}^{\pi} \exp\left(in\tau - ix \sin \tau\right) d\tau.$$

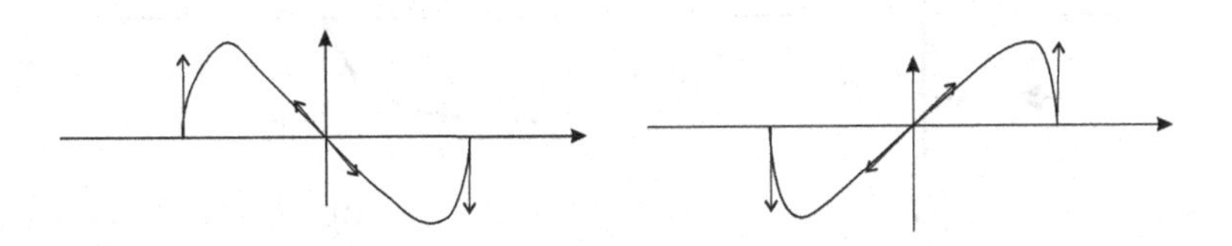

Fig. 8.6

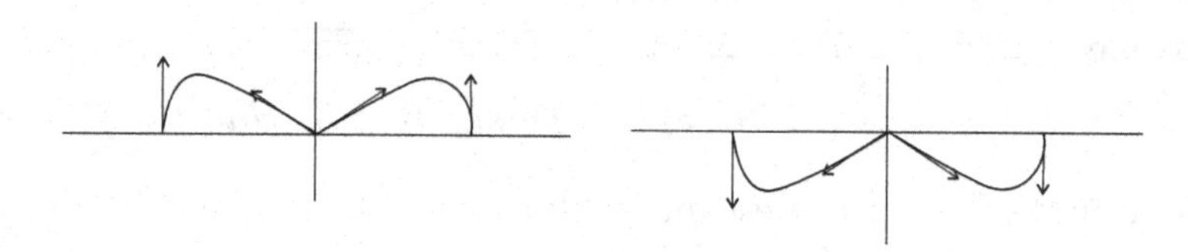

Fig. 8.7

(a) For $x = n$, in which of the two figures in Fig. 8.7 do the thick lines correctly represent the steepest, descent path through the end points and the saddle point at $\tau = 0$?

(b) Prove that, for $x = n \to \infty$,

$$J_n(n) \sim \frac{\Gamma(\frac{1}{3})}{\pi 2^{\frac{2}{3}} 3^{\frac{1}{6}} n^{\frac{1}{3}}}.$$

[5] The gamma function has the integral representation

$$\frac{1}{\Gamma(x)} = \frac{1}{2\pi i} \int_C e^{\tau} \tau^{-x} \, d\tau,$$

where C is the branch cut of $\tau^{-x} = \exp[-x \log \tau)]$ taken along the negative real axis. Perform the transformation $\tau = x\xi$ and use that saddle point method to show that

$$\Gamma(x) \sim \sqrt{(2\pi x)}\, x^{x-1}\, e^{-x}.$$

[6] Show that the representation

$$J_0(x) = \operatorname{Re} \frac{1}{\pi i} \int_{-\frac{i\pi}{2}}^{\frac{i\pi}{2}} \exp(ix \cosh \tau) \, d\tau$$

can be transformed into the representation

$$J_0(x) = \operatorname{Re} \frac{1}{\pi i} \int_C \exp(ix \cosh \tau) \, d\tau,$$

where C is the contour [shown in Fig. 8.8(a)], which starts from $-\infty - i\frac{\pi}{2}$ and ends at $\infty + i\frac{\pi}{2}$. Find the saddle point(s) and use the comtour in Fig. 8.8(b) to

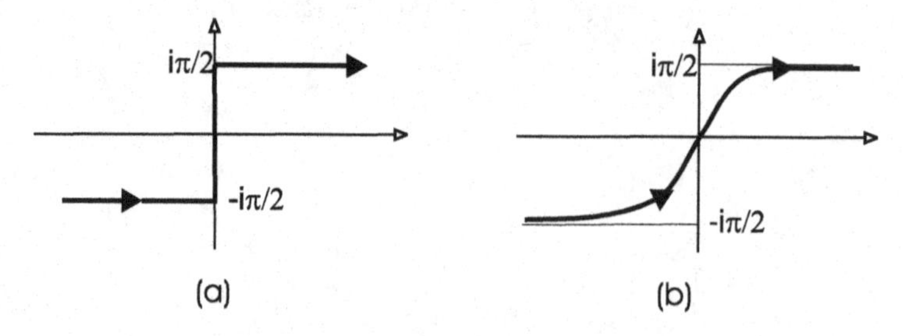

Fig. 8.8

derive the asymptotic expansion

$$J_0(x) \sim \text{Re}\,\frac{e^{-i\frac{\pi}{4}}e^{ix}}{\sqrt{x\pi^3}}\sum_{m=0}^{\infty}\frac{\left[\Gamma(n+\frac{1}{2})\right]^2}{n!}\left(\frac{1}{2ix}\right)^n,$$

as $x \to \infty$.

[7] Show that the asymptotic expansion of the integral

$$I(x) = \int_0^1 e^{ix\tau^3}\,d\tau,$$

as $x \to \infty$, is given by

$$I(x) \sim -x^{-1/3}\,e^{\frac{1}{6}\pi i}\Gamma(\tfrac{4}{3}) + \frac{1}{3}\,e^{ix}\sum_{0}^{\infty}\frac{\Gamma(n+\frac{2}{3})}{\Gamma(\frac{2}{3})}\,(ix)^{-n-1}.$$

Chapter 9

CONFORMAL MAPPING

§§9.1 Tutorial

Inversion Map

⊘ The complex plane is divided into several disjoint subsets, as described below in Q[1]–Q[5]. Draw figures to show the subsets and their images under the inversion map $w = 1/z$.

[1] The four quadrants bounded by the real and imaginary axes.

[2] The eight sectorial subsets bounded by the x axis and the y axis, and the four rays $\theta = \frac{\pi}{4}, \frac{3\pi}{4}, \frac{4\pi}{3}, \frac{5\pi}{3}$.

[3] Four subsets bounded by the circles $|z| = \frac{1}{2}, 1, \frac{3}{2}, 2$.

[4] Two half planes and four strips obtained by drawing the parallel lines $y = -1, -\frac{1}{2}, 0, \frac{1}{2}, \frac{3}{2}$.

[5] The unit circle $|z| = 1$ divides each quadrant into two subsets, giving a total of eight subsets of the complex plane.

[6] In Fig. 9.1 two circles $|z| = 1$ and $|z + i| = 1$ and a line $y = \frac{1}{2}$ are drawn in the complex plane. Draw a figure to show the images of the circles and the line, indicating the images of the 12 points A, B, ... , L under the mapping $w = \frac{1}{z}$.

[7] Sketch the images of the circles in Fig. 9.2 under the map $w = i/z$ and indicate the orientations of the image circles in the w plane.

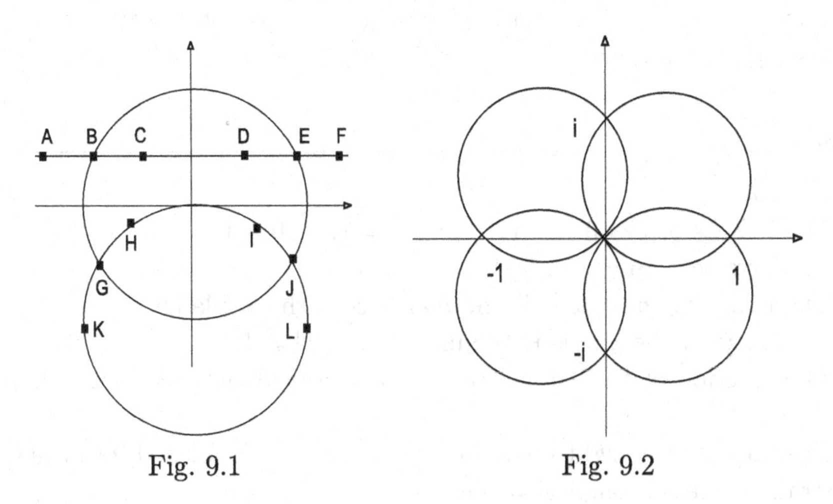

Fig. 9.1 Fig. 9.2

§§9.2 Exercise

Map $\frac{1+z}{1-z}$

⊘ Answer all the questions in this set for the mapping $w = \frac{1+z}{1-z}$.

[1] Answer the following questions:

(a) Find the fixed points of the mapping.
(b) What are the images of $z = 0, 1, -1, \infty$?
(c) Find the inverse map.

[2] Sketch the images of

(a) the x axis and the y axis; show the correspondence between different parts of the axes and their images;
(b) the four quadrants;
(c) the interior and the exterior of the unit circle $|z| = 1$.

[3] Geometrically represent the images of a few members of the family $\mathcal{F}$ of straight lines

(a) passing through the point $z = 1$;
(b) intersecting at $z = -1$;
(c) parallel to the x axis;
(d) parallel to the y axis.

[4] Describe the image of a family of circles

(a) passing through $z = 1$ and not through $z = -1$;
(b) passing through $z = -1$ and not through $z = 1$;
(c) passing through both of the points $z = -1$ and $z = 1$.

[5] Sketch the set of points in the z plane which are inverse images of rays

(a) drawn through $w = 0$, and having $\arg w = 2\pi/3, \pi/3, -\pi/3, -2\pi/3$;
(b) drawn through $w = 1$, and having $\arg(w - 1) = -\pi/4, \pi/6, \pi/3$;
(c) passing through $w = -1$, and having $\arg(w + 1) = -\pi/3, -5\pi/4, \pi/4$.

Answers

[1] (a) $z = \pm i$.
(b) The images of $z = 0, 1, -1, \infty$ are $w = 1, \infty, 0, -1$.
(c) The inverse map is $z = \frac{w-1}{w+1}$.

[2] (a) The image of the x axis is the real axis in the w plane; the image of the y axis is the unit circle $|w| = 1$.
(b) The interior of the circle $|z| < 1$ is mapped onto the right half plane, $\mathrm{Re}\, w > 0$.

[3] (a) The family of straight lines through $z = 1$ is transformed into the family of straight lines through $w = -1$.

(b) The image is a family circles intersecting at $w = 0$ and $w = -1$.
(c) The image is a family of circles touching the real axis at $w = -1$.
(d) The image is a family of circles touching at $w = -1$ and having the centers on the real axis in the w plane.

[4] (a) Straight lines not passing through $w = 0$;
(b) circles intersecting at $w = 0$;
(c) pencil of rays emerging from $w = 0$.

§§9.3 Exercise

Bilinear Transformation I

[1] Give the form of bilinear transformation which maps three points z_1, z_2, and z_3 on w_1, w_2, and w_3 in the limits when

(a) one of the points z_k goes to infinity;
(b) one of the points w_k goes to infinity,

[2] (a) What is the most general form of a bilinear transformation having only one point as a fixed point?
(b) Find the most general bilinear transformation having two points ξ and η as fixed points.
(c) Is there bilinear transformation with more than two fixed points?

[3] Find the condition(s) under which the transformation $w = \frac{\alpha z+\beta}{\gamma z+\delta}$ has only one fixed point. You may assume that $\alpha\delta - \beta\gamma = 1$.

[4] Prove that a bilinear transformation which has both 0 and ∞ as fixed points must reduce to a linear transformation of the form $w = \lambda z$.

[5] Show that a bilinear transformation $w = \frac{\alpha z+\beta}{\gamma z+\delta}$ which maps the unit circle into a straight line must satisfy $|\delta| = |\gamma|$. Interpret this condition geometrically.

[6] Give a condition on α, β, γ, and δ so that the image of the circle $|z-3| = 2|z-5|$, under the map $w = \frac{\alpha z+\beta}{\gamma z+\delta}$, may be a straight line.

[7] Show that the image of the circle $|z-1| = 5/2$, under the mapping $w = \frac{8z+17}{4z+16}$, is the circle $|w-1| = 1/2$.

Answers

[1] The map is $(w, w_1, w_2, w_3) = (z, z_1, z_2, z_3)$ and as z_1, z_2, or z_3 tends to ∞ the cross ratio assumes the three limiting forms

$$\lim_{z_1\to\infty}(z, z_1, z_2, z_3) = \frac{z_3 - z_2}{z - z_2}, \qquad \lim_{z_2\to\infty}(z, z_1, z_2, z_3) = \frac{z - z_1}{z_3 - z_1},$$

$$\lim_{z_3\to\infty}(z, z_1, z_2, z_3) = \frac{z - z_1}{z - z_2}.$$

[2] (a) $\frac{1}{w-z_0} = \frac{\lambda}{z-z_0} + \mu$, (b) $\frac{w-\xi}{w-\eta} = \lambda\frac{z-\xi}{z-\eta}$.

[3] $\delta + \alpha = \pm 2$.

[5] When $|\delta| = |\gamma|$, the point $z = -\delta/\gamma$ lies on the unit circle, which is sent to ∞ in the w plane by the map $z \to w(z)$.

[6] The circle must pass through the point $z = -\delta/\gamma \Rightarrow |\delta + 3\gamma| = 2|\delta + 5\gamma|$.

§§9.4 Questions

Bilinear Transformation II

⊘ Answer all the questions in this set for the bilinear mapping

$$w = \lambda\left(\frac{z-\xi}{z-\eta}\right), \qquad \xi \neq \eta.$$

[1] For fixed values of ξ, η, and λ, when is

(a) the image of a straight line a straight line?
(b) the image of a straight line not a straight line?
(c) the image of a circle a straight line?
(d) a circle transformed into a circle again?

[2] Describe the images of the following families:

(a) Family of circles passing through the two points $z = \xi$ and $z = \eta$;
(b) Family of lines parallel to the straight line passing through ξ and η;
(c) Family of a pencil of rays passing through η;
(d) Family of circles touching at $z = \eta$.

[3] Does a family $\mathcal{F}$, with the image $\mathcal{F}'$ under the map $w(z)$, exist satisfying the requirement:

(a) $\mathcal{F}$ and $\mathcal{F}'$ are families of parallel straight lines?
(b) $\mathcal{F}$ and $\mathcal{F}'$ are families of concentric circles?
(c) $\mathcal{F}$ is a family of parallel straight lines and $\mathcal{F}'$ is a family of straight lines passing through a given point?
(d) $\mathcal{F}$ is a family of parallel straight lines and $\mathcal{F}'$ is a family of concentric circles?
(e) $\mathcal{F}$ is a family of lines intersecting at a point and $\mathcal{F}'$ is also a family of lines intersecting at a point?

[4] Answer the following questions assuming that the image of a circle C is again a circle C' in the w plane:

(a) Does the center of C get mapped into the center of C' for every circle C?
(b) Do the images of every diameter of C always correspond to a diameter of C'?
(c) Is there always a diameter of circles C whose end points are mapped into the end points of a diameter of C'? If yes, how do you locate this diameter of C?

Answers

⊙ Note that the points $z = \xi, \eta, \infty$ are mapped onto $w = 0, \infty, \lambda$.

[1] The image of a generalized circle γ is again a generalized circle γ'. The image will be a straight line if and only if the image passes through the point at ∞. This is possible if and only if γ passes through η.

[2] (a) Rays passing through the origin in the w plane;
(b) Family of circles touching at $w = \lambda$;
(c) Family of rays through $w = \lambda$;
(d) Family of parallel straight lines.

[3] (a) No.
(b) Yes, concentric circles with the center at $z = \eta$ are mapped onto a family of concentric circles with the center at $w = \lambda$.
(c) Yes.
(d) No.
(e) Yes, the lines intersecting at η are mapped onto lines meeting at $w = \lambda$.

[4] (a) No.
(b) No.
(c) Yes, the diameter of C passing through η is mapped onto a straight line and by conformal property its image must be the diameter of C'.

§§9.5 Exercise

Symmetry Principle

[1] Show that a bilinear transformation maps two points P and Q symmetric w.r.t. a generalized circle C into points P′ and Q′ which are symmetric w.r.t. the image of C.

[2] (a) Show that the most general bilinear transformation that maps the unit circle $|z| = 1$ into the unit circle $|w| = 1$ is of the form

$$w = e^{i\alpha}\left(\frac{z-\xi}{\bar{\xi}z-1}\right), \qquad |\xi| \neq 1.$$

(b) What is the image of the interior of the unit circle when

(i) $|\xi| < 1$? (ii) $|\xi| > 1$?

(c) What happens when $|\xi| = 1$?

[3] (a) What should be the condition satisfied by $|\xi|$, so that a mapping of the form

$$w = e^{i\alpha}\left(\frac{z-\xi}{z-\bar{\xi}}\right)$$

transforms the upper half plane onto the unit disk $|w| < 1$?

(b) If the lower half plane is to be mapped onto the unit disk, what condition must be placed on $|\xi|$?

[4] Find a map such that the image of the y axis is the circle $|w - w_1| = k|w - w_2|$. Which part of the complex z plane is mapped onto the interior of the circle? Is your map the most general one?

[5] Construct a bilinear transformation which maps the family of circles

$$|z - z_1| = k|z - z_2|,$$

with z_1 and z_2 fixed, into a family of concentric circles with the center at w_0. Is your map unique? Is it the most general map?

Answers

[2] (b) When $|\xi| < 1$, the interior of the circle $|z| < 1$ is mapped onto the interior of the circle $|w| < 1$.

[3] For $\operatorname{Im}\xi > 0$, the upper half plane is mapped into the interior of the unit circle $|w| = 1$.

[4] Take two points symmetric w.r.t. the y axis, ia and $-ia$, and find a map which sends these points to w_1 and w_2.

[5] ⊙ Map z_1 to w_0 and z_2 to ∞. So $w - w_0 = \lambda\left(\frac{z-z_1}{z-z_2}\right)$, where λ is arbitrary.

§§9.6 Exercise

Elementary Functions

[1] Using the properties of the mapping by $\sin z$, obtain the image of a domain D described below under a map $w = f(z)$ as specified in each case:

(a) $D = \left\{ z \middle| 0 < x < L, y > 0, f(z) = \cos\left(\frac{\pi z}{L}\right) \right\}$;

(b) $D = \left\{ z \middle| x > 0, 0 < y < L, f(z) = \cosh\left(\frac{\pi z}{L}\right) \right\}$.

In each case draw a figure showing the images of different parts of D.

[2] Writing $\tan^2(z/2)$ in terms of $\cos z$, verify that the function $f(z) = \tan^2(z/2)$ maps the half strip $\{z | 0 < x < \pi/2, y > 0\}$ onto the interior of a semicircle in the upper half plane, as indicated in Fig. 9.3.

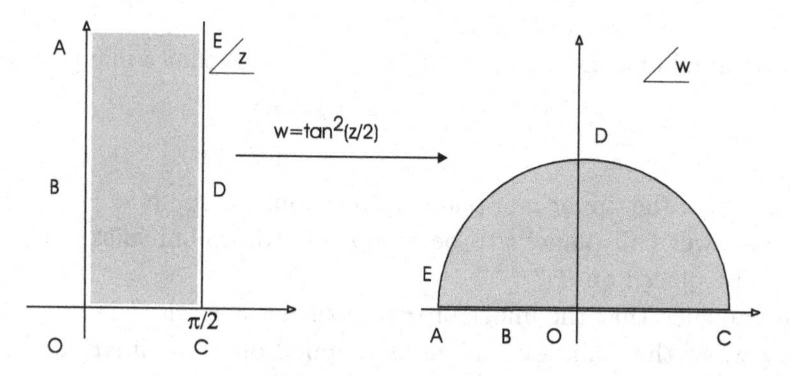

Fig. 9.3

[3] Find a mapping $w = f(z)$ which transforms the half strip $\{z | 0 < x < \pi/2, y > 0\}$ into a domain D', which is

(a) the interior of the unit circle lying in the first quadrant;

(b) the infinite strip $\{w | 0 < \operatorname{Im} w < \pi\}$.

[4] Sketch the image of the semi-infinite strip $\{z | x > 0, -\pi < y < \pi\}$ under the map $w = \coth(z/2)$ using the properties of the exponential map.

[5] Writing $f(z) = \coth(\pi/z)$ as a result of several successive maps, show that the image of the upper half plane, with a disk $|z - i| \leq 1$ removed, is the upper half plane with parts of the boundary mapped, as shown in Fig. 9.4.

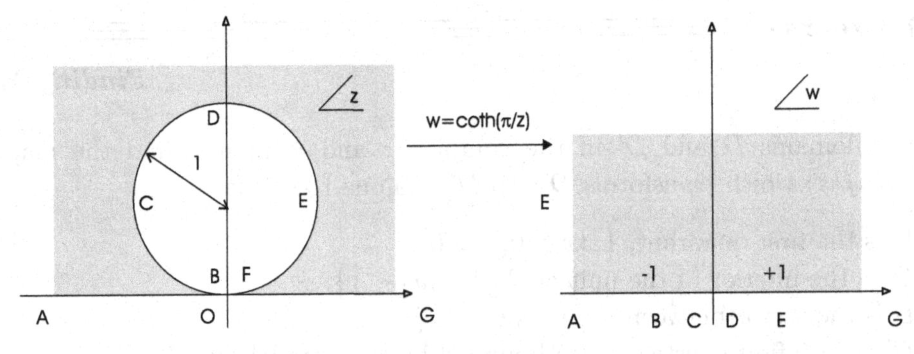

Fig. 9.4

Answers

[3] (a) $w = \tan(z/2)$, (b) $w = -i\sin^2 z$.

[4] Right half plane with a slit from 0 to -1.

§§9.7 Exercise

Finding Maps

⊘ For domains D and D' in the complex z and w planes, find the mapping $w = f(z)$ which transforms D into D' in a one-to-one fashion.

[1] D is the first quadrant, $\{z|x > 0, y > 0\}$;
D' is the interior of the unit circle $\{w||w| < 1\}$.

[2] D is the sector between the rays $\theta = \pm\frac{\pi}{3}$;
D' is an infinite vertical strip bounded by the parallel lines $\operatorname{Re} w = \pm\frac{\pi}{3}$.

[3] D is the domain exterior to the two circles, each of radius 1, and touching at $z = 0$;
D' is the upper half plane.

[4] D is the interior of the semicircle, $\{z||z| < R, y > 0\}$, in the upper half plane;
D' is the upper half plane.

[5] D is the domain bounded by two circles touching at $z = i$, shown in Fig. 9.5;
D' is the semi-infinite strip $\{w| -\frac{\pi}{2} < \operatorname{Re} w < \frac{\pi}{2}, \operatorname{Im} w > 0\}$.

[6] D is the lens-like set bounded by circular arcs shown in Fig. 9.6;
D' is the infinite strip $\{w| -\alpha < \operatorname{Im} w < \alpha\}$.

[7] D is the domain bounded between the branches of the hyperbola $\frac{x^2}{\sin^2\alpha} - \frac{y^2}{\cos^2\alpha} = 1$;
D' is the infinite strip $\{w| -\alpha < \operatorname{Im} w < \alpha\}$.

[8] D is the interior of the ellipse $\frac{x^2}{a^2} + \frac{y^2}{b^2} = 1$;
D' is the upper half w plane.

[9] D is the semi-infinite strip $\{z| -a < \operatorname{Re} z < a, \operatorname{Im} z > 0\}$;
D' is the unit disk $\{w||w| < 1\}$.

[10] D is the domain inside an ellipse with foci at $\pm c$ and outside the slit $-c < x < c$;
D' is an annular region between two circles.

[11] D is the z plane with a slit along $(-\infty, -1)$;
D' is the interior of the unit circle $\{w||w| < 1\}$.

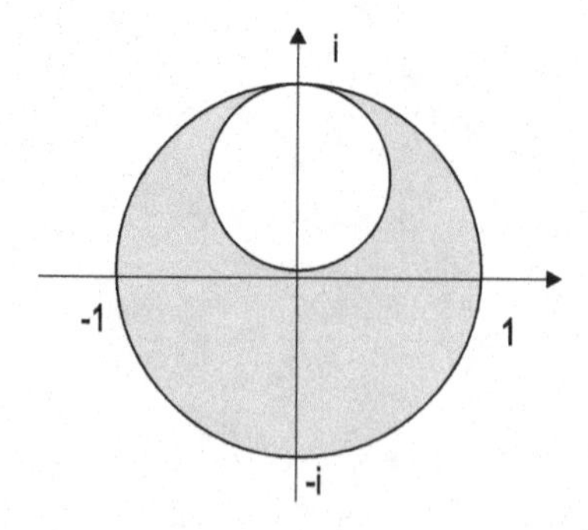

Fig. 9.5

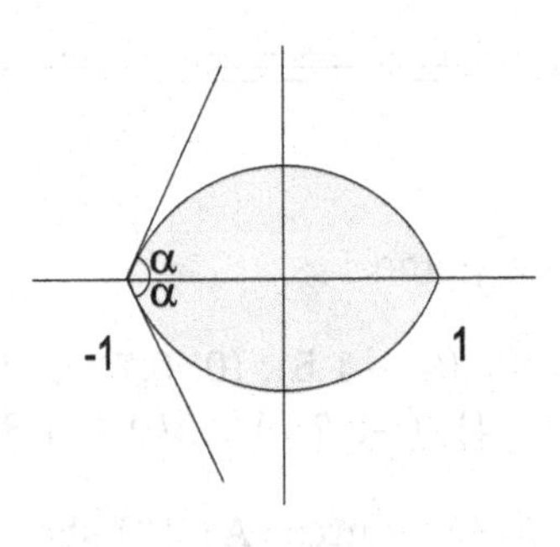

Fig. 9.6

[12] D is the z plane with two slits along $(-\infty, -1)$ and $(1, \infty)$;
D' is the right half plane.

Answers

⊘ The required map can be found as a composite map of several others, as given below.

[1] $z \to t = z^2 \to w = \frac{i-t}{i+t}$.
[2] $z \to t = \operatorname{Log} z \to w = it$.
[3] $z \to t = \frac{1}{z} \to u = t + \frac{1}{2} \to v = iu \to w = \exp(\pi v)$.
[4] $z \to t = \frac{R+z}{R-z} \to w = t^2$.
[5] $z \to t = \frac{1}{z-i} \to u = t - \frac{i}{2} \to v = \exp(2\pi u) \to w = \sin^{-1} v$.
[6] $z \to t = \frac{1+z}{1-z} \to w = \operatorname{Log} t$.
[7] $z \to t = z + \sqrt{z^2 - 1} \to s = -it \to w = \operatorname{Log} s$, where $z = \frac{1}{2}(t + \frac{1}{t})$.
[8] $z \to t \to w = i\frac{R+t}{R-t}$, where t is given by $z = \frac{1}{2}(t + \frac{c^2}{t})$ and $c^2 = a^2 - b^2$, $R = a + b$.
[9] $z \to t = \sin(\pi z/2a) \to w = \frac{i-t}{i+t}$.
[10] Use inverse of the Joukowski map $z = \frac{1}{2}(w + c^2/w)$.
[11] $z \to t = z + 1 \to u = \sqrt{t} \to w = \frac{u-1}{u+1}$.
[12] $z \to t = \frac{1+z}{1-z} \to w = \sqrt{t}$.

§§9.8 Quiz

Matching Aerofoils

⊘ A set of points in the z plane,

$$(0,0),\ (0,\ -0.8),\ (0,\ -1.5),\ (0,\ 0.7),\ (-0.85,\ 0),\ (1.5,\ 0),$$
$$(-0.4,\ -0.3),\ (-0.6,\ -0.4),\ (-0.7,\ 0.2),\ (0.4,\ 1.8),\ (0.5,\ -0.9),\ (0.7,\ -0.7),$$

is shown in the figure below. Figures (A)–(J) show the images, in the w plane, of 10 circles in the z plane under the map $w = \frac{1}{2}\left(z - \frac{1}{z}\right)$. It is given that each circle has the center at one of the above points and passes through the point $z = i$. For each image, find the center of the corresponding circle. Explain your answers briefly.

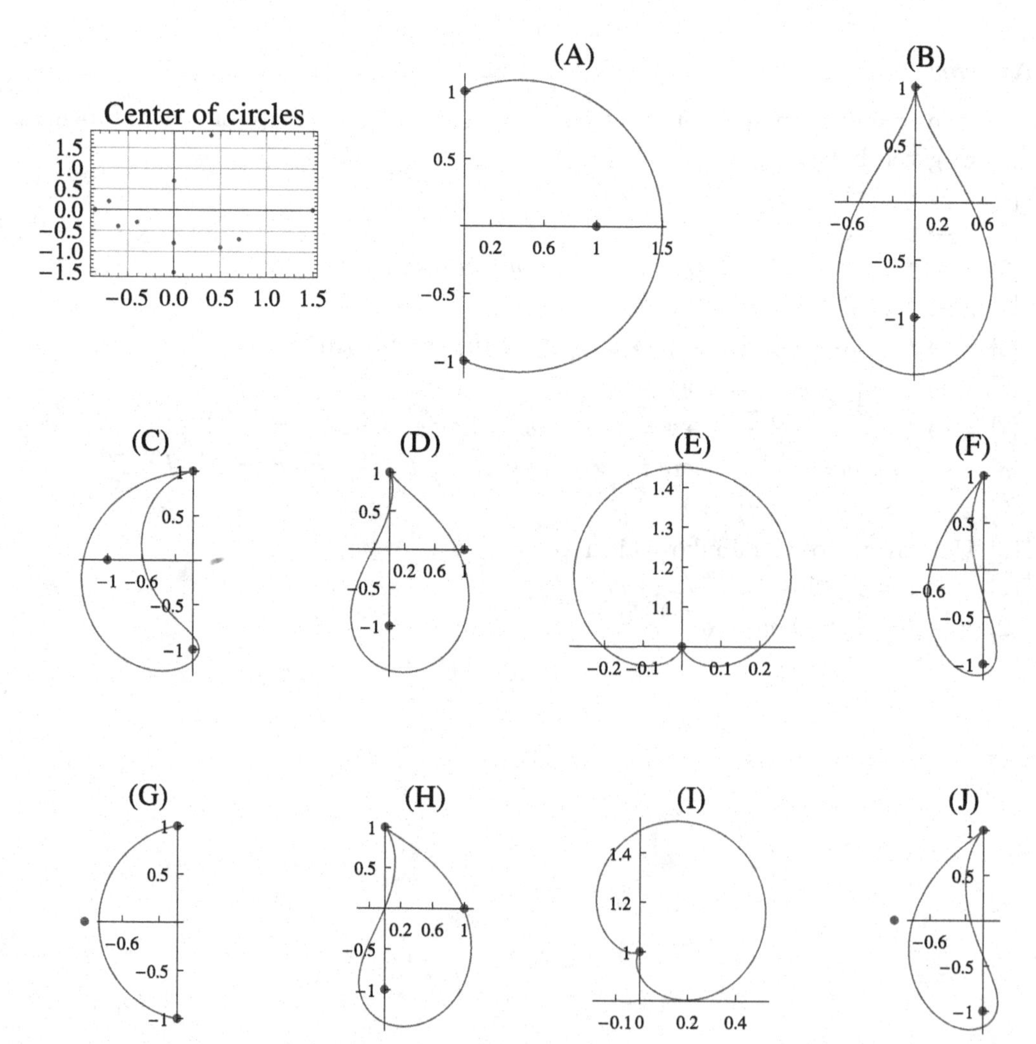

Answers

[A] (1.5, 0), [B] (0, −0.8), [C] (−0.7, 0.2), [D] (0.5, −0.9),
[E] (0.0, 0.7), [F] (−0.4, −0.3), [G] (−0.85, 0), [H] (0.7, −0.7),
[I] (0.4, 1.8), [J] (−0.6, −0.4).

§§9.9 Tutorial

Schwarz–Christoffel Transformation

[1] Find the Schwarz–Christoffel transformation which maps the upper half plane onto the interior of triangle ABC with interior angles $\alpha\pi, \beta\pi$, and $\gamma\pi$ and the three sides having lengths a, b, and c, as shown in the figure. Taking the three vertices A, B, and C as the images of the points $z = 0, 1, \infty$, respectively, write the Schwarz–Christoffel mapping function in the form

$$w(z) = \lambda \int_{z_0}^{z} \xi^{-c_1}(\xi - 1)^{-c_2}\, d\xi + \mu, \tag{9.1}$$

and give the values of the exponents c_1 and c_2. Set $\lambda = 1$ and proceed as follows to determine the lengths of the sides of the triangle.

(a) Fix μ so that $w(1) = 0$, i.e. the vertex B is at $w = 0$.

(b) Locate the vertex A by computing $w(0) - w(1)$ and expressing it in polar form. Also compute the length of the side AB.

(c) Next, locate the vertex C by computing $w(\infty) - w(1)$, and hence compute the length of the side BC.

(d) Check your result on the vertex C by showing that C is the image of $-\infty$ too.

(e) Using the results found above, express the lengths of the three sides of the triangle in terms of the gamma function.

(f) Using properties of the gamma function, prove the result

$$\frac{a}{\sin A} = \frac{b}{\sin B} = \frac{c}{\sin C} = \frac{1}{\pi}\Gamma(\alpha)\Gamma(\beta)\Gamma(\gamma).$$

[2] What is the image of the real line under the mapping

$$w(z) = \int_{-1}^{z} (s+1)^{-1/6} s^{5/12} (s-1)^{-1/4}\, ds?$$

First, try giving your answer without doing any computation. Next, follow the steps given below to verify correctness of your guess.

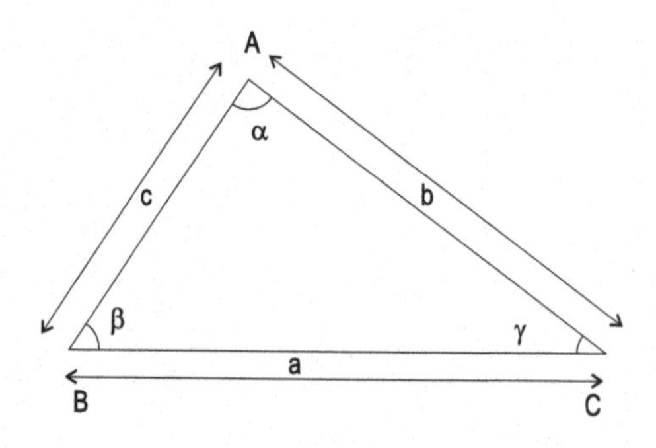

Fig. 9.7 Figure for Q[1].

(a) Compute $w(0)-w(-1)$ and $w(1)-w(0)$, and hence locate the vertices which are the images of the points -1, 0, and 1.

(b) Compute $\Delta_1(x) \equiv w(x) - w(-1)$ for $x < -1$ and show that $\Delta_1(x)$ is real and tends to $-\infty$ as $x \to -\infty$.

(c) Compute $\Delta_2(x) \equiv w(x) - w(1)$ for $x > 1$ and show that $\Delta_2(x)$ is real and tends to ∞ as $x \to \infty$.

(d) Draw the image of the real line. Is it a closed polygon?

§§9.10 Quiz

Schwarz–Christoffel Mapping

⊘ Ten functions, $f_1, \ldots, f_{10}$, are listed here. For each function, state if the Schwarz–Christoffel transformation given by the function maps the upper half plane onto

- the interior of a triangle;
- the interior of a closed polygon having more than three vertices;
- an unbounded subset of the upper half plane.

[1] $f_1(z) = \int_{-1}^{z} (z-1)^{-1/2} z^{-5/6} (z+1)^{-2/3}\, dz,$

[2] $f_2(z) = \int_{-1}^{z} (z+1)^{-1/2} z^{-2/3}\, dz,$

[3] $f_3(z) = \int_{-1}^{z} (z-1)^{-5/6} z^{1/2} (z+1)^{-1/3}\, dz,$

[4] $f_4(z) = \int_{-1}^{z} (z-1)^{-1/2} z^{-1/4} (z+1)^{-1/3}\, dz,$

[5] $f_5(z) = \int_{-1}^{z} (z-1)^{-1/2} z^{-2/3} (z+1)^{-3/4} (z+2)^{-1/4}\, dz,$

[6] $f_6(z) = \int_{-1}^{z} (z-1)^{-1/2} z^{-3/4} (z+1)^{-2/3}\, dz,$

[7] $f_7(z) = \int_{-1}^{z} (z-1)^{-1/6} z^{1/2} (z+1)^{-1/3}\, dz,$

[8] $f_8(z) = \int_{-1}^{z} (z-1)^{-5/6} (z+1)^{-2/3}\, dz,$

[9] $f_9(z) = \int_{-1}^{z} (z-1)^{-1/2} (z+1)^{-1/2}\, dz,$

[10] $f_{10}(z) = i \int_{-1}^{z} (z+1)^{-1/6} z^{1/3} (z-1)^{-1/6}\, dz.$

Answers

f_1: Triangle with three vertices
f_2: Triangle with three vertices; the third vertex is the image of ∞
f_3: Unbounded region, integral divergent for large real z
f_4: Quadrilateral, with one vertex corresponding to the point at ∞
f_5: Pentagon, with one vertex corresponding to the point at ∞
f_6: Quadrilateral, with one vertex corresponding to the point at ∞
f_7: Unbounded region
f_8: Triangle
f_9: Unbounded region
f_{10}: Unbounded region

§§9.11 Exercise

Schwarz–Christoffel Mapping

⊘ For each of the following problems, verify that the Schwarz–Christoffel transformation, as specified there, gives the function $w(z)$ for mapping of the upper half z plane onto the domain shown on the right with the "vertices" as the images of points x_k on the real line.

⊘ In the following tables, x_0 is an unknown vertex to be fixed.

[1] $w(z) = \dfrac{d}{\pi}\Big[\sqrt{z^2-1} + \mathrm{Log}\,\big(z+\sqrt{z^2-1}\big)\Big]$

Vertices w_k	P	A	B	Q
Points x_k	$-\infty$	-1	1	∞

[2] $w(z) = \dfrac{d}{\pi}(1+z+\mathrm{Log}\,z)$

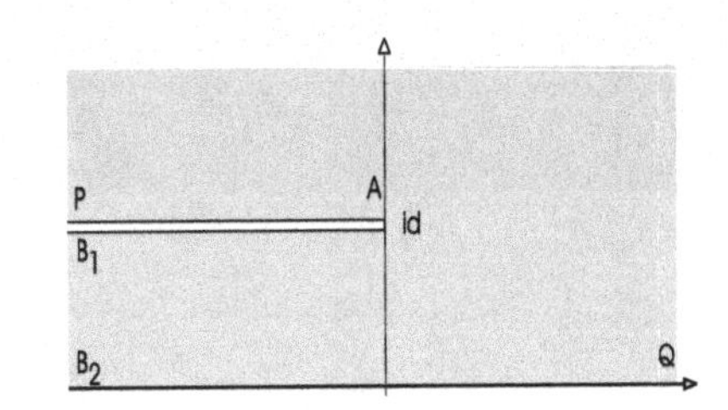

P	A	(B₁, B₂)	Q
$-\infty$	-1	0	∞

[3] $w(z) = \dfrac{d}{\pi}\left[2\sqrt{z+1} + \mathrm{Log}\left(\dfrac{\sqrt{z+1}-1}{\sqrt{z+1}+1}\right)\right]$

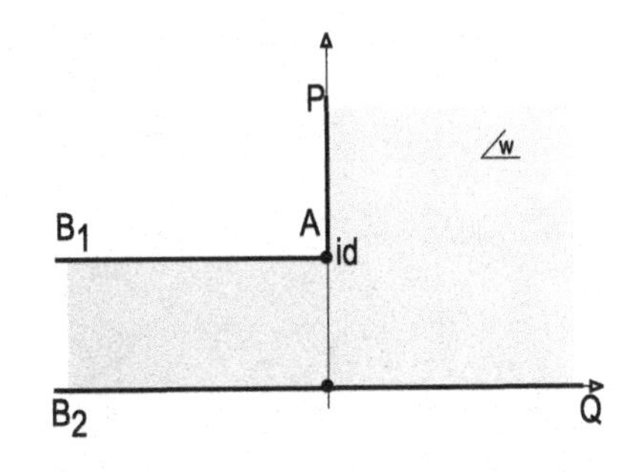

P	A	(B₁, B₂)	Q
$-\infty$	-1	0	∞

[4] $z^2 = 1 + \exp(-2w\pi/d)$

P	(A₁, A₂)	B	(C₁, C₂)	Q
$-\infty$	-1	x_0	1	∞

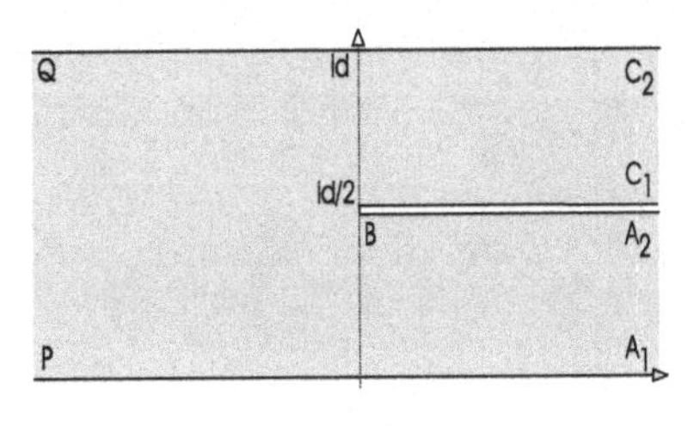

[5] $w(z) = \frac{b}{\pi}\text{Log}\left(\frac{t-1}{t+1}\right) - \frac{a}{\pi}\text{Log}\left(\frac{at-b}{at+b}\right), \quad t^2 = \frac{az-b}{a(z-1)}.$

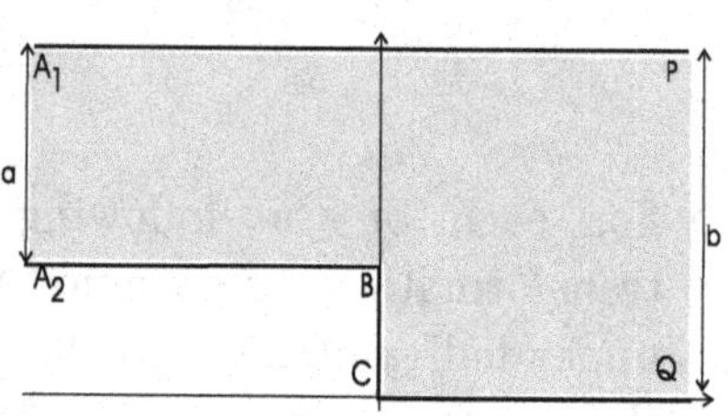

P	(A_1, A_2)	B	C	Q
$-\infty$	0	1	x_0	∞

[6] $w(z) = \frac{2d}{\pi}\left[\sqrt{z(z-1)} - \text{Log}\left(\sqrt{z}+\sqrt{z-1}\right) + \frac{i\pi}{2}\right]$

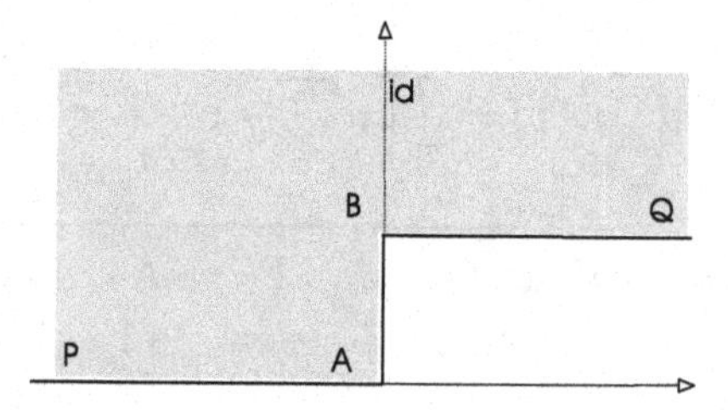

P	A	B	Q
$-\infty$	0	1	∞

§§9.12 Mixed Bag

Conformal Mappings

[1] A bilinear transformation

$$w = \frac{\alpha z + \beta}{\gamma z + \delta}$$

has only one fixed point $z = \xi$. Show that the function $w(z)$ can be written in the form

$$\frac{1}{w - \xi} = \frac{1}{z - \xi} + \lambda.$$

[2] What is the image of the square with vertices at $(\pm 1, \pm i)$ under

(a) the reflection in the unit circle?

(b) the mapping $w = \dfrac{z+1}{z-1}$?

[3] What relationship exists between reflections in the circle and bilinear maps? Given a bilinear transformation, can it be related to a reflection in some circle? Conversely, is a reflection in a circle expressible in terms of a bilinear transformation? Explain by relating a reflection in the circle $|z - \lambda| = k|z - \mu|$ to a bilinear transformation.

[4] Let C be a circle that passes through the points $z = 1$ and $z = -1$, and γ_1 and γ_2 be the two arcs of C which lie in the upper and lower half planes. Show that γ_1 and γ_2 are mapped into each other by inversion. Use this property to show that the image of the circle C under the Joukowski map $\frac{a}{2}(z + \frac{1}{z})$ is a circular arc traced twice.

[5] Select a suitable svb of the $\sqrt{z^2 - 1}$ to be used in the map

$$g(z) = \frac{1}{2}(z + \sqrt{z^2 - a^2}),$$

inverse to the Joukowski map, and prove the properties of the inverse map stated in §9.6.

[6] Show that a bilinear transformation carrying four distinct points z_1, z_2, z_3, and z_4 into four specified images w_1, w_2, w_3, and w_4 exists if and only if

$$\begin{vmatrix} w_1 z_1 & z_1 & w_1 & 1 \\ w_2 z_2 & z_2 & w_2 & 1 \\ w_3 z_3 & z_3 & w_3 & 1 \\ w_4 z_4 & z_4 & w_4 & 1 \end{vmatrix} = 0.$$

[7] Why is the condition $\alpha\delta - \beta\gamma \neq 0$ imposed on the map

$$z \to w = \frac{\alpha z + \beta}{\gamma z + \delta}?$$

What happens when $\alpha\delta - \beta\gamma = 0$?

[8] If a bilinear transformation

$$z \to w = \frac{\alpha z + \beta}{\gamma z + \delta}$$

maps the points z_1, z_2, and z_3 into the points w_1, w_2, and w_3, prove that

$$\alpha\delta - \beta\gamma = (w_1 - w_2)(w_1 - w_3)(w_2 - w_3)(z_1 - z_2)(z_1 - z_3)(z_2 - z_3).$$

[9] Let ξ and η be a pair of symmetric points w.r.t. inversion in a circle C. Prove that every circle and the straight line passing through the two points ξ and η intersect C orthogonally.

[10] Let ξ and η be a pair of symmetric points w.r.t. inversion in a circle C, and ξ' and η' and C' be their images under a bilinear map. Show that the points ξ' and η' are symmetric w.r.t. inversion in C'.

[11] Under the exponential map $f(z) = \exp(z)$, the family of lines parallel to the real axis is transformed into rays through the origin. Explain why it does not contradict the fact that the exponential map is conformal everywhere. If the angle is preserved everywhere, how can parallel lines be transformed into intersecting lines?

[12] Use the correspondence given in the table below to find mapping of the real line onto the shaded domain in Fig. 9.8. Determine the unknown vertex x_0 and show that it is given by $x_0 = 1 + b^2/a^2$. See also Q[2], on p. 496.

Mapping of vertices for the domain in Fig. 9.8:

Vertices of the domain	P	O	(A, B)	C	Q
Points x_k on the real line	$-\infty$	0	1	x_0	∞

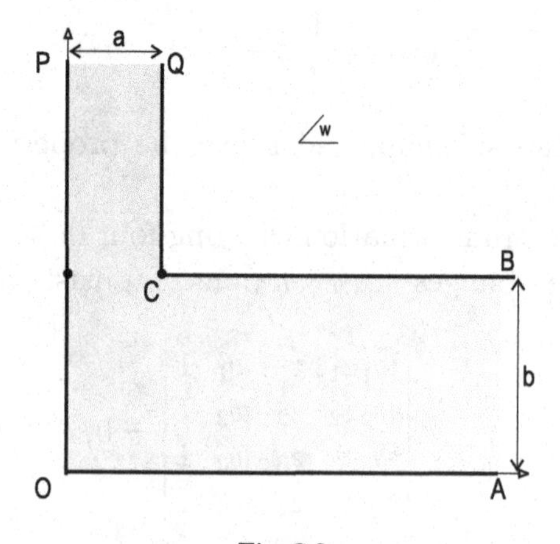

Fig. 9.8

[13] Show that the map $z \to w = f(z)$, which transforms the interior of the unit circle $\{z||z| = 1\}$ onto the interior of a square with a diagonal of length L, is given by

$$f(z) = \int_0^z \frac{d\xi}{(1-\xi^4)^{1/2}},$$

where $L = \left[\Gamma(\frac{1}{4})\right]^2/2\sqrt{2\pi}$.

Chapter 10

PHYSICAL APPLICATIONS OF CONFORMAL MAPPINGS

§§10.1 Tutorial

Temperature Distribution

⊘ Complete the solution of boundary value problems following the steps outlined in each case.

[1] Find the temperature distribution between two long parallel conductors when parts of the boundary are held at temperatures as specified in Fig. 10.1.

$$T(x,y) = \begin{cases} T_1, & \text{if } y = \pi, \\ T_2, & \text{if } x < 0 \text{ and } y = 0, \\ T_3, & \text{if } x > 0 \text{ and } y = 0. \end{cases}$$

(a) Find a map $w = f(z)$ which maps the given problem of interest in D onto a model problem in the upper half w plane.
(b) Write boundary conditions in the w plane and give the complex potential Ω as a function of w.
(c) Express the complex potential as a function of z. Show that for $T_1 = 0$, $T_2 = -T_0/2$, and $T_3 = T_0/2$ the complex potential assumes the form

$$\Omega(z) = \frac{T_0}{2} + i\frac{T_0}{\pi}\text{Log}\,\sinh(z/2).$$

(d) Show that the isotherms are given by

$$\tan(y/2) = k\tanh(x/2),$$

where k is a constant. Sketch a few isothermal lines.

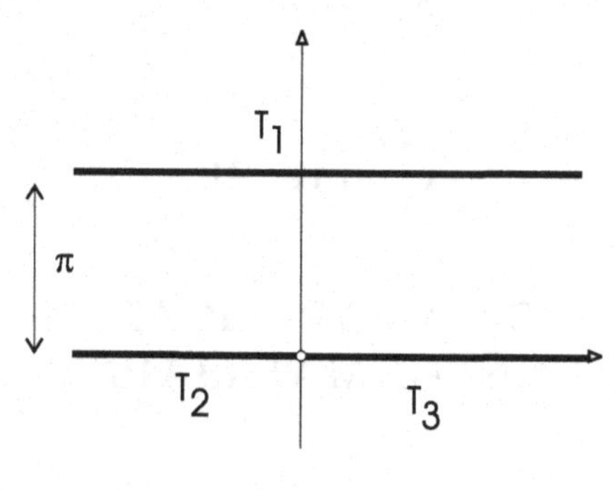

Fig. 10.1

[2] Solve the heat conduction problem in domain D in the upper half plane outside the semicircle, of radius R and with the center at $z = 0$, and extending to infinity with the boundary held at temperatures as specified in Fig. 10.2.

(a) Verify that the map

$$t = \frac{z - R}{z + R}$$

transforms the given problem into a boundary value problem in the first quadrant of the t plane. Write the boundary conditions for the t plane problem.

(b) Next, note that the map $w = t^2$ relates the w plane problem to one of the model problems in §10.1.1. Use this to show that the complex potential is given by

$$\Omega(z) = T_1 - 2i\left(\frac{T_2 - T_1}{\pi}\right) \text{Log}\left(\frac{z - R}{z + R}\right).$$

(c) Taking the real part, find the required temperature distribution in the upper half z plane outside the semicircle.

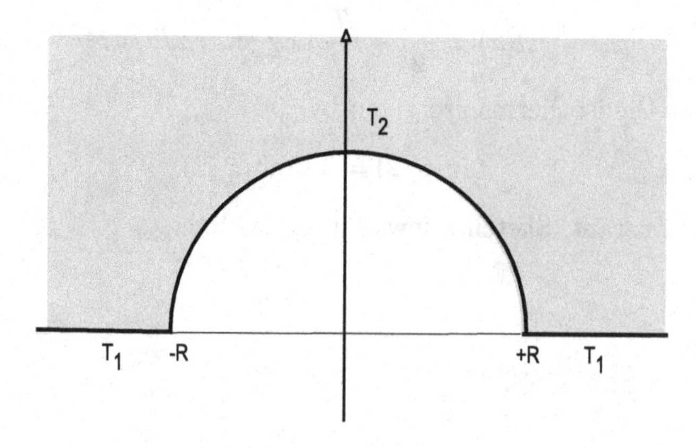

Fig. 10.2

§§10.2 Exercise

Steady State Temperature

⊘ Several problems of heat conduction are stated below, in Q[1]–Q[8]. Find the complex temperature Ω with specified boundary conditions by mapping the problems into one of the model problems of §10.1.1. The insulated part of the boundary is marked with xxxxxxxx.

⊘ Complete the problems by taking the real parts of the complex temperature found.

[1] Find the temperature distribution in the rectangular strip D (Fig. 10.3),

$$D = \left\{ z \middle| 0 < x < \frac{\pi}{2}, y > 0 \right\},$$

subject to the boundary conditions

$$T(x,0) = T_1, \qquad 0 \le x \le \frac{\pi}{2};$$

$$T(0,y) = T_2, \qquad y > 0; \tag{10.1}$$

$$T(\pi/2,y) = T_1, \qquad y > 0. \tag{10.2}$$

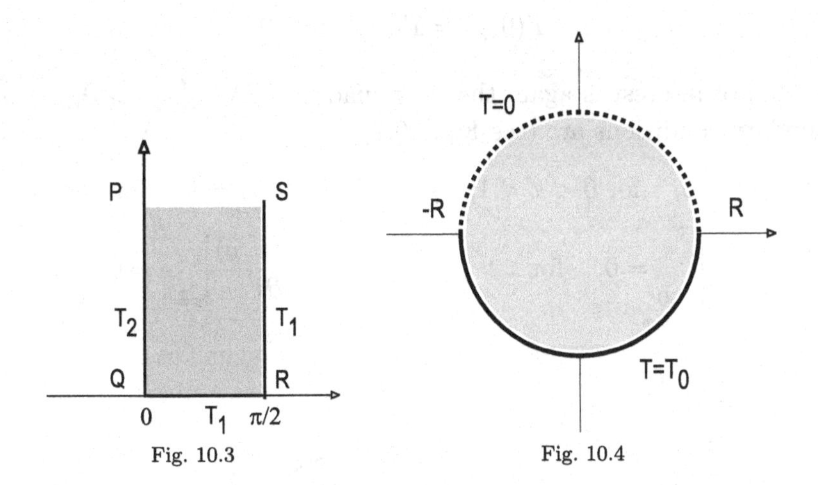

Fig. 10.3

Fig. 10.4

[2] A cylindrical solid, with the circle $\{z \big| |z| = R\}$ as the boundary, has the upper and lower halves kept at the temperatures $T = 0$ and $T = T_0$; see Fig. 10.4. Find the temperature distribution inside the cylinder.

[3] The temperature distribution is required in the upper half plane subject to the boundary conditions (Fig. 10.5)

$$T(x,0) = \begin{cases} T_1, & \text{for } x < 0, \\ T_2, & \text{for } 0 < x < 1, \\ T_1, & \text{for } x > 1. \end{cases} \tag{10.3}$$

Solve it by use of the mapping $w = \frac{z}{1-z}$. Note that this itself is a model problem; see Eq. (10.9). Compare your answer with the answer given there.

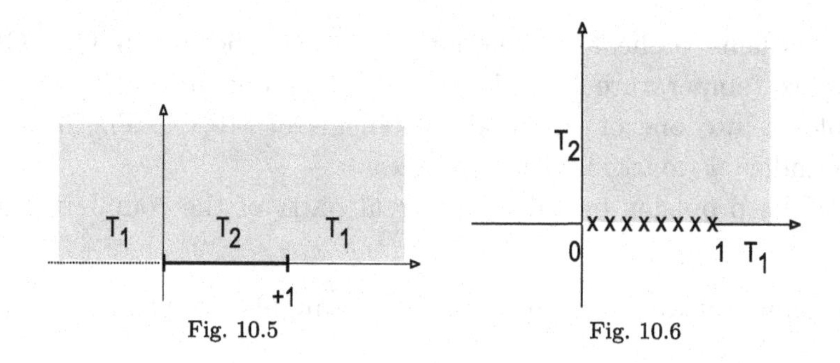

Fig. 10.5 Fig. 10.6

[4] The temperature distribution is needed in the first quadrant with the following boundary conditions (see Fig. 10.6):

$$T(x,0) = T_1, \quad \text{for } x > 1; \tag{10.4}$$

$$\left.\frac{\partial T(x,y)}{\partial y}\right|_{y=0} = 0, \quad \text{for } 0 < x < 1; \tag{10.5}$$

$$T(0,y) = T_2, \quad y > 0. \tag{10.6}$$

[5] The domain of interest is again the first quadrant, $D = \{z|x > 0, y > 0\}$, and the boundary conditions are (see Fig. 10.7)

$$T(x,0) = 0, \quad \text{for } 0 < x < 1; \qquad T(0,y) = 1, \quad \text{for } 0 < y < 1;$$

$$\left.\frac{\partial T(x,y)}{\partial y}\right|_{y=0} = 0, \quad \text{for } x > 1; \qquad \left.\frac{\partial T(x,y)}{\partial x}\right|_{x=0} = 0, \quad \text{for } y > 1.$$

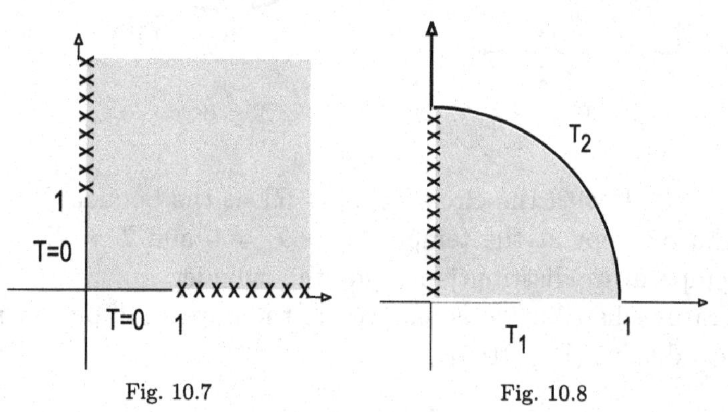

Fig. 10.7 Fig. 10.8

[6] A cylindrical solid is cut into four equal parts by perpendicular planes passing through the axis of the cylinder; see Fig. 10.8. Find the complex potential

and the temperature distribution inside one of the parts with the boundary conditions as specified in the figure.

[7] Use the Joukowski map to solve the heat conduction problem in the domain D in the upper half plane outside the semicircle, of radius R and with the center at $z = 0$, and extending to infinity with the boundary held at temperatures as specified in Fig. 10.9.

[8] Repeat the above question for the domain D in the first quadrant with the temperatures T_1 and T_2 on the boundary, as sketched in Fig. 10.10.

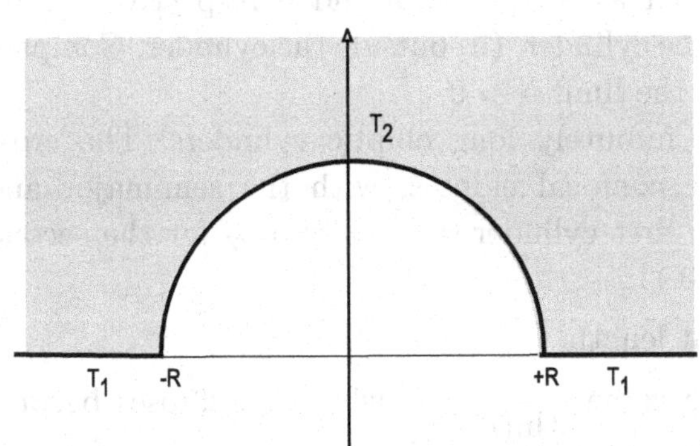

Fig. 10.9

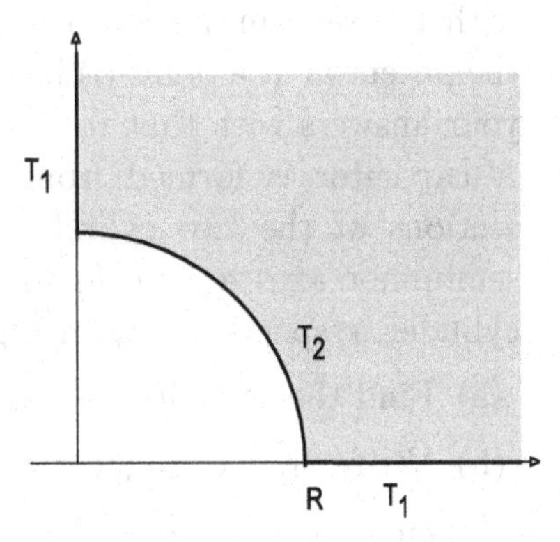

Fig. 10.10

Answers

[1] $\phi = \dfrac{T_2 - T_1}{\pi} \tan^{-1}\big(\cot x \tanh y\big) + T_1,$

[2] $T = \dfrac{T_0}{\pi} \tan^{-1}\left(\dfrac{R^2 - r^2}{2rR\sin\theta}\right),$

[3] $T = T_1 + \dfrac{T_2 - T_1}{\pi} \arg(z-1) + \left(\dfrac{T_1 - T_2}{\pi}\right) \arg z,$

[4] $T = T_2 + \dfrac{2}{\pi}(T_1 - T_2) \sin^{-1}\Big(|z+1| - |z-1|\Big),$

[5] $T = \dfrac{1}{2} + \dfrac{1}{\pi} \sin^{-1}\big(|z^{-2} - 1| - |z^{-2} + 1|\big),$

[6] $\Omega(z) = -i\left(\sin^{-1} w - \dfrac{\pi}{2}\right), \quad w = \dfrac{1 + 6z^2 + z^4}{(1 - z^2)^2},$

[7] $T = T_1 + 2\left(\dfrac{T_2 - T_1}{\pi}\right) \arg\left(\dfrac{z - R}{z + R}\right),$

[8] $T = T_2 + 2\left(\dfrac{T_2 - T_1}{\pi}\right) \arg\left(\dfrac{z^2 - R}{z^2 + R}\right).$

§§10.3 Exercise

Electrostatics

[1] A charged conducting infinite strip has a cross section of the line segment $\{z | -c < x < c, y = 0\}$. Show that the potential at any point is given by

$$\phi(z) = -2q \operatorname{Re}\operatorname{Log}(z + \sqrt{z^2 - c^2}),$$

where q is the charge per unit height of the strip.

[2] Repeat Q[1] for an infinite conducting cylinder having an elliptic cross section, with the semi-major and semi-minor axes equal to a and b, respectively. Find the potential at a point (i) inside the cylinder, (ii) outside the cylinder. Compare your answers with that to Q[1] in the limit $b \to 0$.

[3] A capacitor is formed from two infinitely long elliptic cylinders. The cross sections of the two cylinders are confocal ellipses, with the semimajor and semiminor axes a_1 and b_1 for the first cylinder, and a_2 and b_2 for the second cylinder, respectively, as in Fig. 10.11.

(a) Find the capacitance per unit length.

(b) Verify that C reduces to the value $\dfrac{1}{2\ln(r_2/r_1)}$ when the ellipses become circles of radii r_1 and r_2.

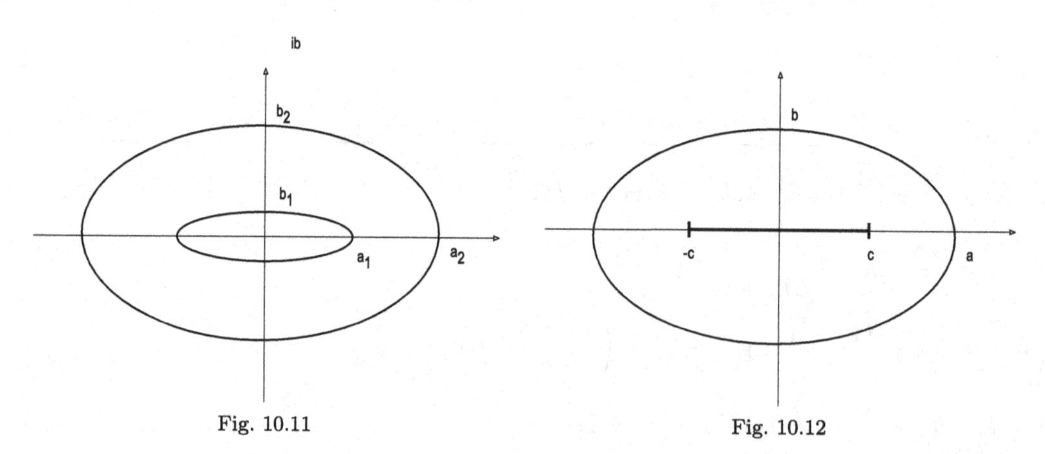

Fig. 10.11

Fig. 10.12

[4] (a) Find the complex potential in domain D inside an elliptic cylinder and outside a flat plate of length $2c$ (as shown in Fig. 10.12), where $c = \sqrt{a^2 - b^2}$. The potential of the plate is $-V_0$ and that of the elliptic cylinder is V_0.

(b) Show that the capacitance per unit length is given by $1/C = \frac{1}{2}\ln\left(\frac{a+b}{c}\right)$.

[5] A hyperbolic condenser is formed of two infinite plates having two branches of the hyperbola

$$\frac{x^2}{c^2\cos^2\alpha} - \frac{y^2}{c^2\sin^2\alpha} = 1.$$

The two plates of the capacitor are raised to potentials $-V_0$ and V_0. Show that the complex electric field at any point is given by

$$E = -\frac{V_0}{\alpha}\frac{1}{\sqrt{\bar{z}^2 - c^2}}.$$

Find the points where the electric field is maximum.

Answer

[2] $\Omega(z) = -2q\text{Log}\left(z + \sqrt{z^2 - R^2}\right), \phi(z) = -2q\text{Log}\left|z + \sqrt{z^2 - R^2}\right|, R^2 = a^2 - b^2.$

[3] (a) $1/C = 2\text{Log}\left(\dfrac{a_2 + b_2}{a_1 + b_1}\right).$

[4] (a) $\Omega(t) = 2V_0\left[\dfrac{\text{Log}\,(z/c)}{\text{Log}\,(R/c)}\right] - V_0.$

[5] The electric field is maximum at the vertices $z = \pm c$.

§§10.4 Quiz

Nine Problems and ...

⊘ Nine boundary value problems of finding the electrostatic potential in two dimensions, in nine given domains, are to be solved by the method of conformal mapping.

⊘ The domains of interest are bounded by conductors along curves or slits, as described below. All the conductors extend to infinity on both sides in the third dimension perpendicular to the x–y plane.

⊘ Assume general values for the potentials of the conductor(s).

⊘ *A single clue is sufficient for the solution to all the nine problems. Identify this clue.*

⊘ Find the map(s) for relating each of the nine problems to one of the model problems of §10.1.1. Specify the map(s) to be used and briefly describe the model problems that need to be solved in each of the nine cases.

⊘ Details of the solution are not needed.

[1] Between the two branches, in the lower half plane, of two different confocal hyperbola having the focus at $-i$.

[2] The whole plane except for slits along the real axis and the infinite interval $(i, i\infty)$ on the imaginary axis.

[3] The whole plane except for two semi-infinite slits — one from $-i\infty$ to $-i$ and the other from i to $i\infty$.

[4] Inside an elliptic cylinder with the foci at $\pm i$.

[5] The whole plane except for a slit of finite width from $-i$ to i.

[6] Between a hyperbola with the focus at $-i$ and the real axis.

[7] Between two confocal ellipses with the foci at $\pm i$.

[8] The domain of interest is inside an ellipse with the foci at $\pm i$ except for a slit between the foci.

[9] Between two branches of a single hyperbola with the foci at $\pm i$.

§§10.5 Exercise

Flow of Fluids

[1] Find the complex potential for steady flow past a cylinder of radius R if the fluid flow at infinity is uniform and if the velocity at infinity is v_0 making an angle α with the x axis, as shown in Fig. 10.13.

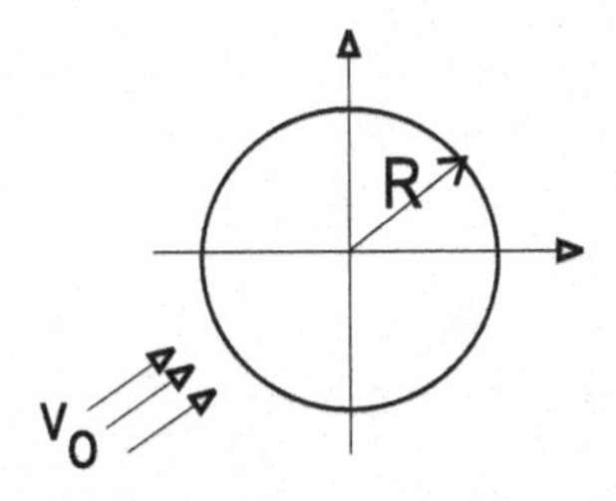

Fig. 10.13

[2] (a) Use the results of Q[1] to obtain the complex potential for the flow of a fluid past an elliptic cylinder if the velocity at infinity is v_0. Show that the complex potential is

$$\Omega = v_0 \left[te^{-i\alpha} + \frac{(a+b)^2}{4t} e^{i\alpha} \right],$$

where

$$t = \frac{1}{2}\left(z + \sqrt{z^2 + b^2 - a^2}\right).$$

(b) Taking the limit $b \to 0$ in part (a), show that the flow against a plate between $-L$ and L is given by

$$\Omega = v_0\left(z \cos\alpha - i\sqrt{z^2 - L^2} \sin\alpha\right).$$

Find the velocity at the edges of the plate and locate the stagnation points, if any.

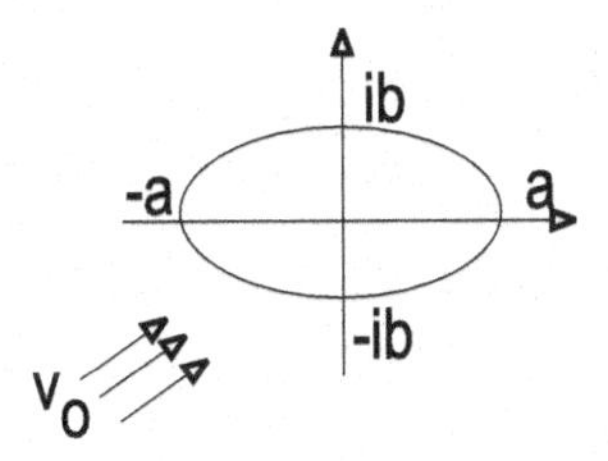

Fig. 10.14

[3] Solve the electrostatic problems analogous to those in Q[1]–Q[2] taking the electric field at infinity to be $(E_0 \cos\alpha, E_0 \sin\alpha)$ and assuming the boundaries to be perfect conductors held at zero potential.

Answer

[1] $\Omega(z) = v_0\left(ze^{-i\alpha} + \dfrac{R^2 e^{-i\alpha}}{z}\right).$

§§10.6 Exercise

Method of Images

[1] A line charge q is located at a point ξ in the first quadrant. The positive parts of the real and imaginary axes are held at zero potential. Compute the electric field in the first quadrant.

[2] (a) A line charge q is placed at $z = 1$ between two infinite plane conductors forming a wedge-shaped region corresponding to the sector $\{z| - \alpha < \arg z < \alpha\}$]. The two conductors are kept at zero potential. Find the complex potential everywhere inside the sector using an appropriate conformal mapping.

(b) Using your answer in part (a), find the system of images and their locations which will solve the problem for $\alpha = \pi/6$.

[3] A line charge q is located at z_0 between two parallel grounded conducting infinite planes at $y = 0$ and $y = d$. Find a conformal map which transforms the strip $\{z|0 < y < d\}$ into a half plane and thus obtain the electrostatic potential between the plates.

[4] A line charge q is placed at $z = a$ outside an infinite cylindrical conductor of radious R held at zero potential with its axis parallel to the line charge. Use a mapping to transform this problem to that of line change in the presence of a grounded conducting plane and hence find the complex potential. Verify that the electrostatic potential satisfies the boundary condition $\phi = 0$ on the boundary of the conductor.

[5] Find the steady state temperature distribution for a point source of strength k located at $z = i$ in the upper half plane. Assume that the positive x axis is held at zero temperature and the negative x axis is insulated.

Answers

[1] $\overline{\mathcal{E}(z)} = \dfrac{2q}{z-\xi} + \dfrac{2q}{z+\xi} - \dfrac{2q}{z-\bar{\xi}} - \dfrac{2q}{z+\bar{\xi}}.$

[2] (a) $\phi(z) = -2q\text{Re Log}\left(\dfrac{z^{\pi/2\alpha} - 1}{z^{\pi/2\alpha} + 1}\right);$

(b) The image charges are charge q at $e^{2\pi i/3}$ and $e^{4\pi i/3}$ each; $-q$ each at the three points where $z^3 = -1$.

[3] $2q\text{Re Log}\left[\dfrac{\exp(\pi z/d) - \exp(\pi \bar{z}_0/d)}{\exp(\pi z/d) - \exp(\pi z_0/d)}\right].$

[4] $\Omega(z) = -2q\big[\text{Log}\,(z-a) - \text{Log}\,(z-\frac{R^2}{a}) - \text{Log}\,(R/a)\big]$;
$\phi(z) = -2q\big[\text{Log}\,|z-a| - \text{Log}\,|z-\frac{R^2}{a}| - \text{Log}\,(R/a)\big]$.

[5] $T = k\text{Log}\left[\dfrac{(\sqrt{z}-\xi)(\sqrt{z}+\bar{\xi})}{(\sqrt{z}+\xi)(\sqrt{z}-\bar{\xi})}\right]$, $\xi = 1 + i/\sqrt{2}$.

§§10.7 Open-Ended

Using Ideas from Gauss' Law

⊘ This set is about *understanding* the answers for the integrals $\oint \frac{dz}{z-z_0}$ utilizing only the definition as a Riemann sum and is inspired by Gauss' law in electrostatics.

⊘ Use only simple considerations involving a Riemann sum to argue for the correctness of the following results. Considering subdivisions of closed contour C into n parts by means of points $z_k, k = 1, \ldots, n$, the Riemann sum representing the integral can be cast in the form

$$\oint_C \frac{dz}{z - z_0} = \sum_k \left(\frac{z_{k+1} - z_k}{z_{k+\frac{1}{2}} - z_0} \right).$$

Here $z_{k+\frac{1}{2}}$ stands for a point on C between z_k and z_{k+1}, and C is a simple closed contour.

⊘ Do not make use of results in integration such as Green's theorem in two dimensions or of Cauchy's fundamental theorem. *Use ideas from electrostatics freely.*

[1] First, prove that the value of the integral is $2\pi i$ when C is taken as a circle γ having the center at z_0. Suitably select the points z_k to divide the circle into n parts. Taking the limit $n \to \infty$ explicitly, compute the value of the integral.

[2] Next, consider the case where C is an arbitrary simple closed contour enclosing the point z_0. Draw a circle γ centered at z_0 and lying completely inside C. Find a subdivision of C and a corresponding subdivision of γ in such a way that the two Riemann sums, for C and γ, are easily seen to be equal.

[3] Next, consider arbitrary closed contour C such that the point z_0 does not lie in its interior. By choosing a suitable subdivision of the contour, argue that the integral vanishes because the terms in the Riemann sum cancel pairwise when the number of subdivisions tends to infinity.

[4] When n is a positive integer different from 1, contour integral $\oint \frac{dz}{(z-z_0)^n}$ vanishes. Is there a way of getting this result using the answers to the Q[2] and Q[3]?

Hints

[1] Divide the circle into n equal parts.

[2] Show that the contributions of parts of C, and r in a sector, bounded by two rays drawn from z_0 with an infinitesimal opening angle, are equal.

⊙ See Eq. (10.36) in §10.4, about the electric flux in two dimensions. See also the section on Gauss' law in [Feynman *et al.* (1965)].

§§10.8 Tutorial

Boundary Value Problems

⊘ Complete the solution of boundary value problems following the steps outlined in each case.

[1] A conducting plate is kept along the real axis in the electric field which becomes uniform at infinity and makes an angle α with the real axis; see Fig. 10.16. If the end points of the plate are at the points ± 2, find the complex potential, the points where the electric field becomes maximum, and the magnitude of the electric field at the origin.
(a) Use the Joukowski map $z \to t$ to relate the given problem to that of finding the electric field outside the unit sphere in the complex t plane.
Apply a rotation $t \to s$ so that the electric field becomes parallel to the real axis in the s plane and follow the method of ☑10.11.

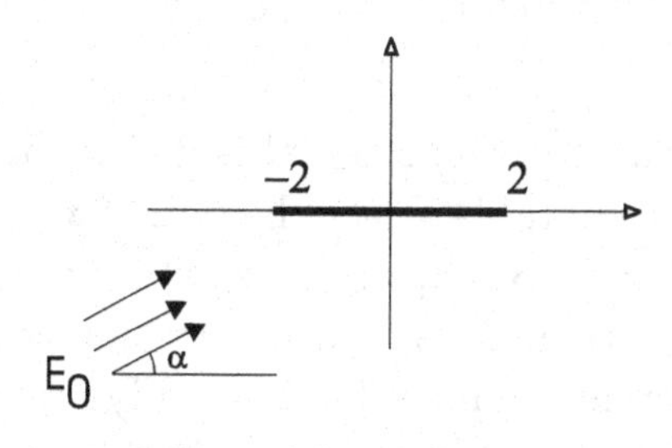

Fig. 10.16

(b) Write the complex potential as a function of s and obtain an expression for the electric field. Verify that the electric field satisfies the correct boundary condition at infinity.
Find the points where the electric field becomes maximum, assuming that $\alpha \neq 0, \pi$.
(c) Show that the magnitude of the electric field at the origin is $E_0 \sin\alpha$.

[2] Solve the problem of ideal fluid flow in the upper half plane with a triangular trough in the path of the fluid. (Fig. 10.17). Assume that AB = BC and $\angle$ABC = $\pi/2$, and describe how the speed of the fluid varies along the path of the flow. Find the stagnation points and sketch a few stream lines.
(a) Write the Schwarz–Christoffel transformation as $z = f(t)$, with the points -1, 0, and 1 in the t plane having the images A, B, and C, respectively.
(b) Find the complex potential as a function of t and show that the complex velocity as a function of t is given by

$$\bar{\mathcal{V}} = \frac{v_0 t^{1/2}}{(t^2-1)^{1/4}}.$$

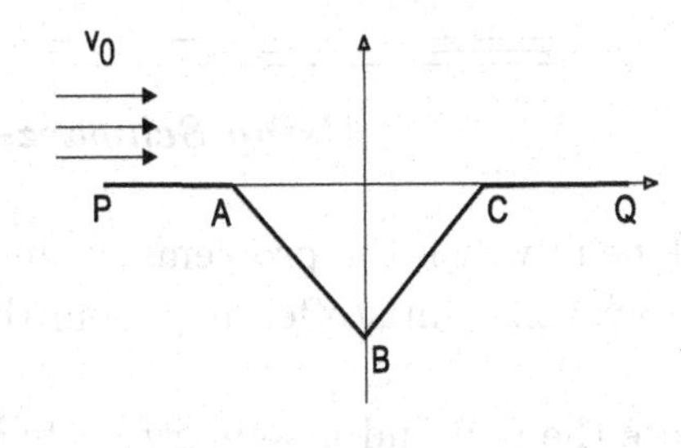

Fig. 10.17

(c) Verify that the stagnation point(s) and the point(s) where the speed becomes maximum coincide with the corners. Sketch a few stream lines.

§§10.9 Exercise

Using Schwarz–Christoffel Mapping

⊘ Denote the complex plane in which the problems in this set are to be solved by the w plane and use a Schwarz–Christoffel map from the upper half z plane to the given domain.

⊘ As usual, u and v denote the real and imaginary parts of w.

[1] Use the Schwarz–Christoffel transformation to map the problem of flow in a river, onto the problem of flow in the upper half plane if the river bed has

(a) a jump discontinuity as shown in Fig. 10.18;
(b) a discontinuity in the slope, as in Fig. 10.19;
(c) a spike at $x = 0$; see Fig. 10.20.

Find the complex velocity if it is given that the velocity of the fluid tends to v_0 as $x \to -\infty$, and describe how the velocity varies near the discontinuity.

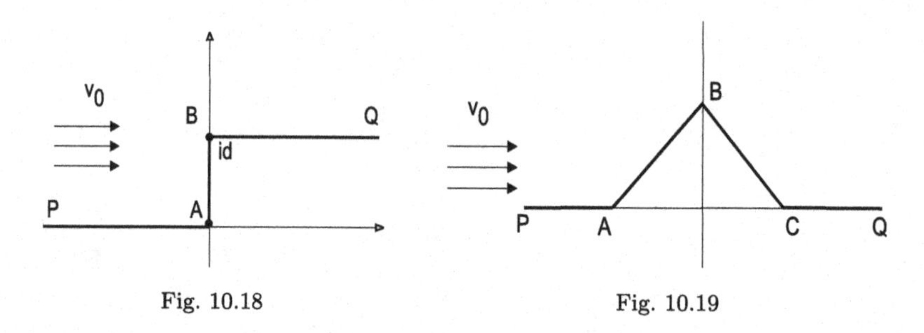

Fig. 10.18 Fig. 10.19

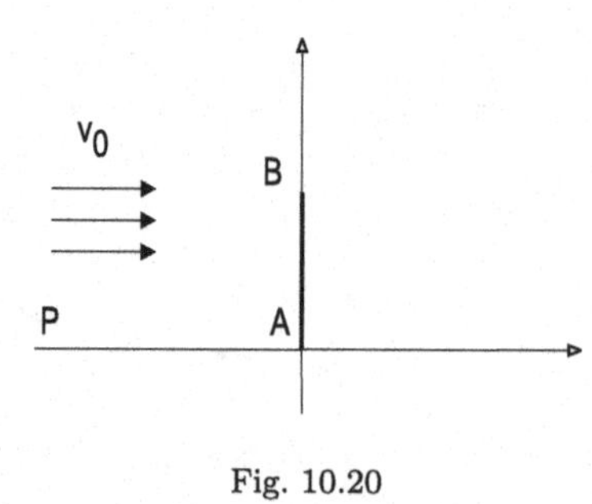

Fig. 10.20

[2] (a) Show that the Schwarz–Christoffel transformation, defined by

$$\frac{df}{dz} = A\,\frac{\sqrt{z+1}}{z},$$

maps the real line in the z plane onto a w plane domain D, shown in Fig. 10.21.

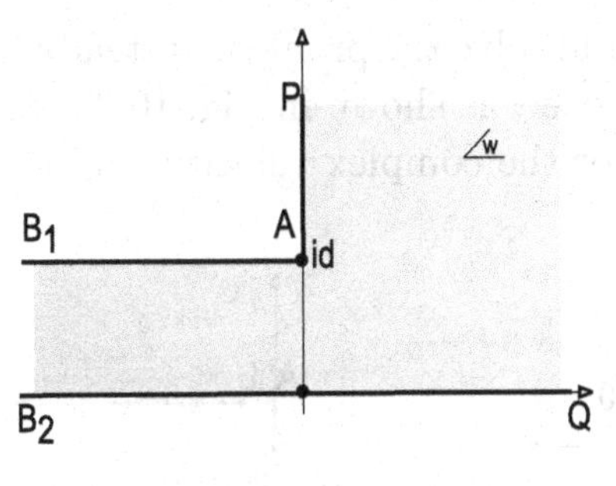

Fig. 10.21

(b) Solve the problem of fluid flow in the domain D, given that the velocity is uniform near the points B_1 and B_2 and far away from A.

[3] Find the electric complex potential and the electric field inside the domain D (Fig. 10.21), if the potential on PAB_1 is $-V_0$ and on B_2Q is V_0 and if the field near B_1 and B_2 is uniform and parallel to the real axis. Sketch a plot of equipotential curves.

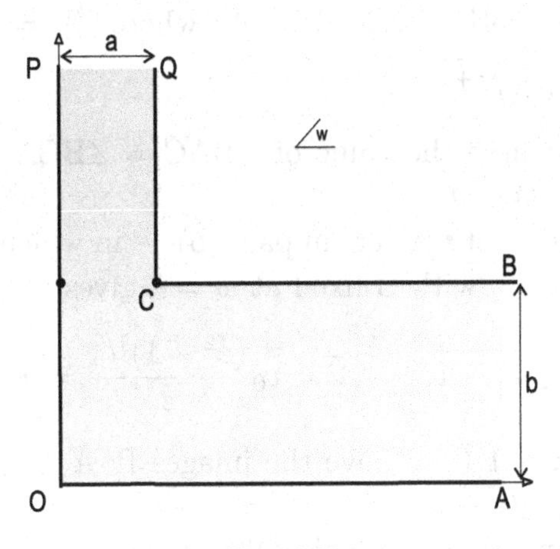

Fig. 10.22

[4] Use Schwarz–Christoffel mapping to solve the problem of ideal fluid flow in a channel with a 90° bend (Fig. 10.22). Compute the complex potential and complex velocity, and use it to show that the fluid speed v_1 near P and Q and v_2 near A and B, far away from the bend, satisfy $v_1 a = v_2 b$.

[5] A condenser consists of two parallel semi-infinite thin plates separated by a distance $2d$, with lines $\{w|u < 0, v = id\}$ and $\{w|u < 0, v = -id\}$ forming the cross section of the plates. Find the complex potential and show that

(i) the electric field tends to zero as $u \to \infty$;

(ii) as $u \to -\infty$, the electric field tends to the value expected for an infinite parallel plate condenser.

[6] Use results of Q[1](a) and solve the problem of steady flow of fluid against a semi-infinite rectangular barrier, as shown in Fig. 10.23. Find the complex potential, obtain an expression for the complex velocity, and find the stagnation points.

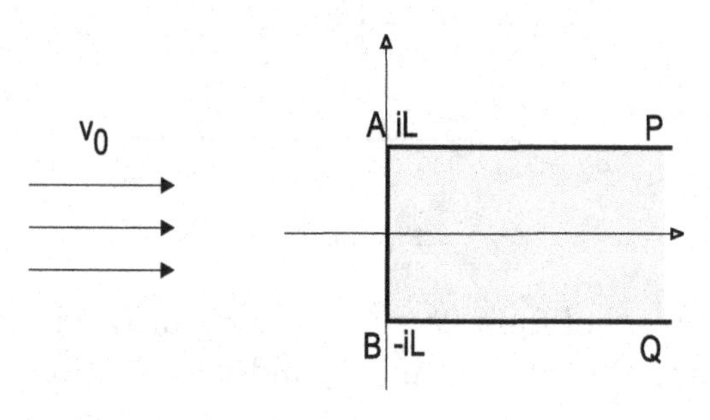

Fig. 10.23

Answers

[1] (a) The complex potential is $\Omega(w) = Az$ where $\frac{dw}{dz} = \frac{d}{\pi}\sqrt{\frac{z-1}{z+1}}$. The complex velocity is $\bar{\mathcal{V}} = v_0\sqrt{\frac{z+1}{z-1}}$.

(b) $\bar{\mathcal{V}} = v_0\frac{(z^2-1)^k}{z^{2k}}$, where the value of $\angle$BAC $= \angle$BCA is taken to be $k\pi$ and the point B is taken at $w = i$.

(c) Taking the limit — of answers in part (b) — in which the angles $\angle$BAC and $\angle$BCA become $\pi/2$, with B fixed at $w = i$, gives

$$w = \sqrt{z^2-1}, \qquad \bar{\mathcal{V}} = v_0\frac{(z^2-1)^{1/2}}{z} = v_0\frac{w}{(w^2+1)^{1/2}}.$$

[2] The points $z = -\infty, -1, 0, \infty$ have the images P, A, (B_1, B_2), Q, respectively, and $\Omega = Az, \bar{\mathcal{V}} = v_0/\sqrt{z+1}$.

[3] ⊘ Use the same transformation as in Q[2]. This gives $\Omega(w) = iV_0\big(1 - \frac{2}{\pi}\text{Log}\, z\big)$ and $\bar{\mathcal{E}} = \frac{2iV_0}{d}\frac{1}{\sqrt{z+1}}$.

[4] ⊘ Use the Schwarz–Christoffel map of Q[12] on §§9.12 The complex velocity is

$$\bar{\mathcal{V}} = iv_0\sqrt{\frac{z}{z-x_0}}.$$

[5] ⊘ Use results of ☑10.15, p. (270)

[6] ⊘ The symmetry of the problem under reflection in the real axis reduces the given problem to that of Q[1](a). The solution is the same as in Q[1](a), with d replaced by L. Here $w = 0$ is a stagnation point.

§§10.10 Exercise

Utilizing Conservation of Flux of Fluids

[1] Consider a channel having a cross section with a discontinuity as shown in Fig. 10.24. Taking the Schwarz–Christoffel transformation $z \to w = f(z)$, and mapping the upper half plane onto the domain between the channel, to be of the form

$$f(z) = A\int_0^z \frac{1}{z}\sqrt{\frac{z-1}{z-x_0}}\,dz, \qquad 0 < x_0 < 1,$$

where A is a constant, show that the unknown vertex x_0 can be determined using flux considerations and is given by

$$x_0 = \left(\frac{b}{a}\right)^2.$$

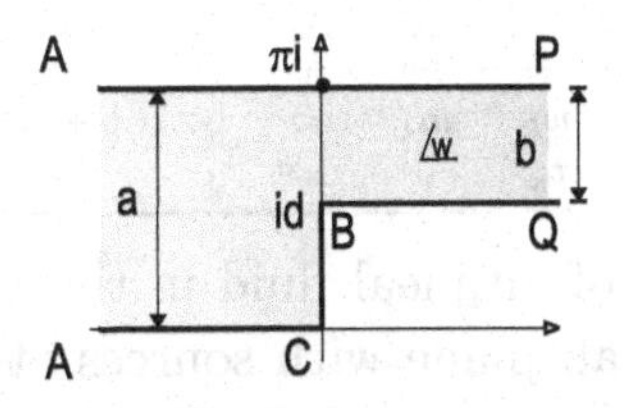

Fig. 10.24

[2] It is given that the real line can be mapped onto a channel with a 90° bend (Fig. 10.25) by means of the Schwarz–Christoffel transformation $z \to w = f(z)$, where

$$\frac{df}{dz} = \left(\frac{A}{z-1}\right)\sqrt{\frac{x_0 - z}{z}}, \qquad x_0 > 1,$$

and A is a constant. Prove that $x_0 = 1 + \left(\frac{b}{a}\right)^2$ using flux arguments for an ideal flow of liquid in the channel.

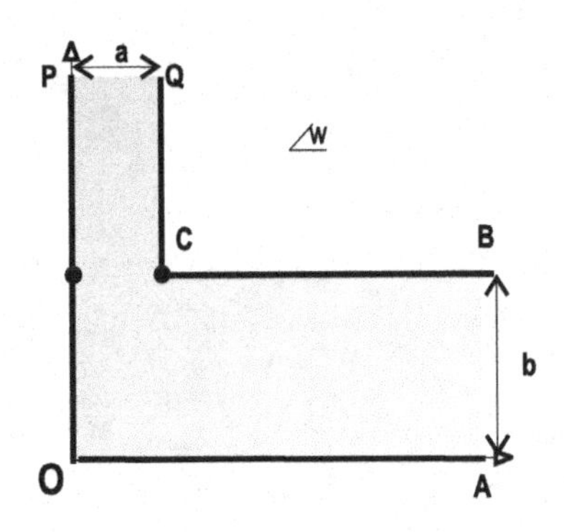

Fig. 10.25

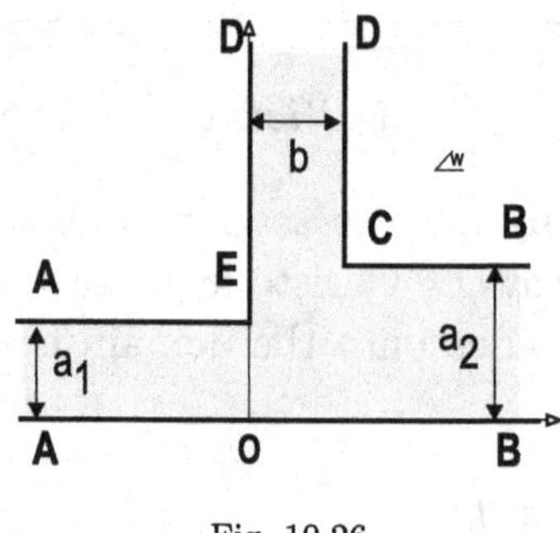

Fig. 10.26

[3] (a) Using the correspondence of vertices of domain of Fig. 10.26 as shown in table below, find the exponents in the Schwartz Christoffel mapping

$$w(z) = \lambda \int z^{-c_1}(z-1)^{-c_2}(z-c)^{-c_3}(z+d)^{-c_4}\, dz,$$

	E	A	B	C	D
w_k	ia_1	∞	∞	$b+ia_2$	∞
x_k	0	1	c	∞	$-d$

(b) Noting that the flow of an ideal fluid in the given domain is equivalent to a flow in the upper half plane with sources at $1, c$, and $-d$, and using flux arguments, show that

$$\frac{\lambda\pi}{(c-1)(1+d)} = a_1, \quad \frac{\lambda\pi\sqrt{c}}{(c-1)(c+d)} = a_2, \quad \frac{\lambda\pi\sqrt{d}}{(c+d)(d+1)} = b.$$

§§10.11 Mixed Bag

Boundary Value Problems

[1] Let $\phi(x, y)$ be a harmonic function such that its normal derivative is a nonzero constant along a curve C. If $w = f(z)$ is a conformal mapping and C' is the image of C under w, then show that the directional derivative of ϕ along the normal to C' need not be constant.

[2] Let C be an oriented curve and the direction of the normal at any point be chosen such that an anticlockwise rotation of the normal gives that tangent to the curve. In addition, let $V = (V_x, V_y)$ be a vector field in two dimensions. Show that the line integrals of the tangential and the normal components of V, $\int_C V_t \, ds$ and $\int_C V_n \, ds$, are given by

$$\int_C V_t \, ds = \text{Re} \int_C \bar{\mathcal{V}} \, dz,$$

$$\int_C V_n \, ds = \text{Im} \int_C \bar{\mathcal{V}} \, dz,$$

where $\mathcal{V} = V_x + iV_y$ is the complex vector field.

[3] Write the Schwarz–Christoffel transformation for mapping the interior of the unit disk onto the exterior of a square of side a and with the center at the origin. Use your result to find the electric field outside a square in the presence of a point charge q located at

(a) a point z_0 outside the square;
(b) the origin inside the square.

[4] Show that the complex force, $\bar{F}$, and the torque, τ, on an obstacle in a fluid flow described by complex potential Ω are given by

$$\bar{F} = \frac{i\rho}{2} \oint_\Gamma \left(\frac{d\Omega}{dz} \right)^2 dz,$$

$$\tau = -\frac{\rho}{2} \text{Re} \left[\oint_\Gamma z \left(\frac{d\Omega}{dz} \right)^2 dz \right],$$

where Γ is the boundary and ρ the density of the obstacle.

[5] For an obstacle with the circular boundary $|z| = R$ in a fluid flow described by the complex potential

$$\Omega(z) = v_0 \left(z + \frac{R^2}{z} \right) + \frac{iK}{2\pi} \text{Log}\, z,$$

show that the complex force on the body of the obstacle is given by $K v_0 \rho$, where ρ is the density of the fluid.

BIBLIOGRAPHY

Ablowitz, M. J., and Fokas, A. S., *Complex Variables* (Cambridge University Press, 2003).

Ahlfors, L. V., *Complex Analysis* (McGraw-Hill, Kogakusha, Tokyo, 1966).

Bender, C. M., and Orsaz, S. A., *Advanced Mathematical Methods for Scientists and Engineers* (Springer-Verlag, New York, 1999).

Carrier, G. F., Krook, M., and Pearson, C. E., *Functions of a Complex Variable* (McGraw-Hill, New York, 1966).

Churchill, R. V., *Complex Variables and Applications* (McGraw-Hill, Kogakusha, Tokyo, 1964).

Feynman, R. P., Leighton, R. B., and Sands, M., *The Feynman Lectures on Physics*, Vol. II, (Addison–Wesley, 1965).

Fraleigh, J. B., *A First Course in Abstract Algebra* (Addison–Wesley, 1973).

Hardy, G. H., *A Course in Pure Mathematics* (English Language Book Society, London, 1971).

Henricki, P., *Applied and Computational Complex Analysis*, Vol. I (John Wiley and Sons, New York, 1974).

Lepage, W. R., *Complex Variables and Laplace Transform for Engineers*, (Dover, 1980).

MacRobert, T. M., *Functions of a Complex Variable* (Macmillan, London, 1954).

Needham, T., *Visual Complex Analysis* (Clarendon, Oxford, 1997).

Nehari, Z., *Conformal Mapping* (McGraw-Hill, New York, 1952).

Polya, G., *Complex Variables* (John Wiley and Sons, (New York, 1974).

Spiegel, M. R., *Theory and Problems of Complex Variables* (McGraw-Hill, Singapore, 1964).

Titchmarsh, E. C., *Theory of Functions* (Oxford University Press, London, 1939).

Vilkovyskii, L. I., Lunts, G. L., and Aramanovich, I. G., *A Collection of Problems on Complex Analysis* (Pergamon, Oxford, 1965).

Whittaker, E. T., and Watson, G. N., *Modern Analysis* (Cambridge University Press, London, 1950).

INDEX